Pattern Formation

PATTERN FORMATION   2/35   Diane Malakinski

PATTERN FORMATION
A Primer in Developmental Biology

George M. Malacinski, Ph.D., Editor
Indiana University

Susan V. Bryant, Ph.D., Consulting Editor
University of California, Irving

Macmillan Publishing Company
A Division of Macmillan, Inc.
NEW YORK

Collier Macmillan Publishers
LONDON

Copyright © 1984 by
Macmillan Publishing Company
A Division of Macmillan, Inc.

Macmillan Publishing Company
A Division of Macmillan, Inc.
866 Third Avenue, New York, N.Y. 10022

Collier Macmillan Canada, Inc.

Printed in the United States of America

printing number

1 2 3 4 5 6 7 8 9 10

Library of Congress Cataloging in Publication Data
Main entry under title:

Pattern formation.

    Bibliography: p.
    Includes index.
    1. Developmental biology.   2. Morphogenesis.   3. Cell
differentiation.   I. Malacinski, George M.   II. Bryant,
Susan V.
QH491.P38   1984   574.3   83–25553
ISBN 0-02-949480-X

Dedication

This book is respectfully dedicated to the "old timers"—the early embryologists and experimental morphologists. Their careful and detailed observations provided a foundation for modern studies in biological *pattern formation*.

Contents

Preface

Primers in Developmental Biology was conceived as a series of textbooks in contemporary biological research areas with the following aims:

A. To *identify* rapidly emerging and potentially important disciplines and subdisciplines of developmental biology.

B. To *organize* those disciplines for students and scholars who might be seeking both an introduction to current research problems and a broad overview of a discipline.

C. To provide *insight* into the thought processes of several of the important contributors to the discipline.

Rather than solicit a small number of thoroughly referenced, highly pedantic chapters that might represent only a limited cross section of a research area, an alternative approach was formulated. It is best described by describing the procedure employed to develop Volume I, *Pattern Formation*. First, the field of pattern formation was surveyed by examining the few textbooks and collections of review articles currently existing. Then, computer searches were carried out employing both key words and author names to collect information. Finally, Sue V. Bryant (consulting editor) and I compiled subject lists and made selections, with the aid of colleagues, of appropriate authors. Once an invitation list was compiled, it was further reviewed by several colleagues before final invitations were extended. Both S. V. B. and I were delighted with the initial response, for almost everybody invited agreed to participate.

The following "guidelines" were formulated:

1. Chapters would be written at a level that would be the most meaningful to the graduate student, postdoctoral fellow, or new *Pattern Formation* investigator.

2. Authors would be urged to speculate, and to provide their personal views or interpretations.

3. Literature references would be limited in scope, to encourage authors to generalize.

4. Chapters would be cross-referenced, wherever possible, so that a careful reading of the entire book would give the reader both a comprehensive and coherent view of contemporary studies in *Pattern Formation*.

To generate an "insider's" view of the discipline, the editors (especially myself) talked informally (and candidly) with several of the contributors. A summary of several of those discussions is included in the following pages. Likewise, all authors were requested to provide answers to two general questions about their chapters. Those answers are included at the end of each chapter.

Using this approach, an advanced level textbook for *Pattern Formation* was formulated. It was designed, like a textbook should, to be general in scope. Yet it is high in detailed information content, so it should also serve as a reference book.

I am grateful to the contributors for their willingness to participate in this project, by not only contributing a chapter, but also by abiding by the guidelines. Their efforts permitted the editors to generate a unique spirit—that Volume I of *Primers in Developmental Biology*, "Pattern Formation," should provide a learning experience for not only the readers, but for the individual contributors, as well as the editors.

The contributors also deserve special praise for being prompt with their manuscripts and for their willingness to offer their frank opinions of "pattern formation" as a discipline.

Kathy Swan provided many of the illustrations and Diane Malacinski participated in much of the manuscript editing and provided the fine art print. Sarah Greene, our editor at Macmillan, provided expert professional publishing advice.

George M. Malacinski

Introductory Comments

One of the major distinguishing characteristics of living organisms is the high degree of orderliness of their constituent parts. The particular arrangement of organelles, cells, tissues, or organs within each species is faithfully repeated in each individual. Furthermore, each individual goes through a process of development during which the complex, but orderly final form emerges from a more simple, unicellular starting point. Hence it seems inescapable that any meaningful understanding of developmental biology should involve an understanding of how biological organization emerges. That is, it should involve the study of *pattern formation*. It may therefore take one by surprise, given that developmental biology has a solid record of experimentation reaching back more than 100 years and given the central importance of pattern formation to this discipline, to learn that the present volume represents one of less than a handful of books that are entirely devoted to the topic of pattern formation.

The publication of this book reflects the renewed interest in pattern formation that has come about in the last 10–15 years. Prior to that time, very little sustained experimentation had been directed toward understanding the mechanisms that generate spatial organization. The reasons for this are multiple, complex, and subject to individual opinion. Our view maintains that two major factors worked against the development of this discipline. First of all, biologists are usually trained to recognize differences between things, and as a consequence, they tend to overlook the all-too-obvious similarities. It should not be forgotten that some of the most brilliant advances in biology have dealt with similarities, or universal principles, such as mechanisms of inheritance, the nature of the genetic code, and the theory of evolution and natural selection. Major advances in understanding pattern formation will, in our opinion, come from an appreciation for the generality of the rules that govern spatial organization during development.

Unfortunately, the discipline of pattern formation is still at a technically unsophisticated state, rather comparable to the study of genetics prior to Mendel. In fact, as will be clear from many of the chapters in this volume, a substantial amount of effort is currently devoted to uncovering the formal rules that will allow us to predict the behavior of cells under different experimental conditions. This effort necessarily involves much modeling and theory as well as experimentation. Accordingly, several chapters are devoted to modeling, while others emphasize experimentation. Some manage to combine both approaches.

A second major reason for the slow progress of the field is the pervasive belief that a greater understanding of pattern formation will be achieved when more information is available about smaller and smaller parts of the whole. Such a view has no applicability to pattern formation in its current state of sophistication. Meaningful knowledge at the molecular level will not be experimentally approachable until the general principles have been satisfactorily laid out at the cellular and/or tissue levels. Nevertheless, progress toward this goal is well underway in many different systems. By far the most useful approach to the inference of general principles has been the phenomenology of the developmental response following surgical intervention, just as the means to uncovering the principles of genetics lay in the phenomenology of abnormal or mutant organisms.

The dominant theme of the recent resurgence of interest in pattern formation is the concept of *positional information*. This concept was first explored at the turn of the century by Driesch on the basis of his studies on sea urchin embryos. He concluded from the fact that pattern regulation can occur that cells behave as if they had information about their position in the whole. However, the concept of positional information has become irrevocably tied to the name of Lewis Wolpert, who has galvanized a generation of developmental biologists into action through his articulate and compelling expositions of the problem. He has, almost single-handedly, made us aware of how intellectually impoverished the study of developmental biology had become by neglecting the central issue of pattern formation during the first 60 or so years of this century. Fortunately, Wolpert's research efforts were successful, as several contributors to this volume make clear in their chapters.

In this volume we have attempted to call attention to some of the scientists and several of the developing systems presently making major contributions to an overall understanding of pattern formation. The diversity of organisms being studied is an especially encouraging sign. Comparisons between organisms carrying out similar patterning processes has, in the past, proved extremely valuable in sorting out the important similarities from the more trivial differences. Presently, there is no one ideal system, and all those described in this volume have relative strengths and weaknesses. Some may fall by the wayside as experimentation enters the realm of molecular biology. But for now we should be careful not to overlook the importance of diversity of experimental systems for helping to simplify the major pattern formation issues.

We have come a long way in the last decade or so, in terms of both the quality of data and the sophistication of ideas. Yet, we still have a long way to go before we will be able to share fully the secrets of development with the embryos we study.

Susan V. Bryant

Contributing Authors

Kathryn V. Anderson

Friedrich-Miescher-Laboratorium, Max-Planck-Gesellschaft, Spemannstrasse 37–39, 7400 Tübingen, West Germany

Hans R. Bode

Developmental Biology Center and Department of Developmental and Cell Biology, University of California, Irvine, CA 92717

Patricia Macauley Bode

Developmental Biology Center and Department of Developmental nad Cell Biology, University of California, Irvine, CA 92717

Peter J. Bryant

Developmental Biology Center and Department of Developmental and Cell Biology, University of California, Irvine, CA 92717

Jonathan Cooke

Division of Developmental Biology, National Institute for Medical Research, Mill Hill, London NW7 1AA, United Kingdom

Danielle Dhouailly

Laboratoire de Zoologie et Biologie Animale, Université Scientifique et Médicale de Grenoble, B.P. 53 X, 38041 Grenoble, France

John F. Fallon

Department of Anatomy, University of Wisconsin, Madison, WI 53706

Joseph Frankel

Department of Zoology, University of Iowa, Iowa City, IA 52242

Vernon French

Department of Zoology, University of Edinburgh, West Mains Road, Edinburgh, EH9 3JT Scotland

Lewis I. Held, Jr.

Developmental Biology Center and Department of Developmental and Cell Biology, University of California, Irvine, CA 92717

Nigel Holder — Department of Anatomy, King's College London, Strand, London WC2R 2LS, United Kingdom

Lorette Javois — Developmental Biology Center and School of Biological Sciences, University of California, Irvine, CA 92717

Jane Karlsson — Genetics Laboratory, Department of Biochemistry, South Parks Road, Oxford OX1 3QU, United Kingdom

Stuart A. Kauffman — Department of Biochemistry and Biophysics, University of Pennsylvania School of Medicine, Philadelphia, PA 19104

Harry K. MacWilliams — Abteilung für Entwicklungsbiologie Zoologishes Institut der Universität Muenchen, Luisenstrasse 14 8000 Muenchen 2, West Germany

Malcolm Maden — Developmental Biology Division, National Institute for Medical Research, Mill Hill, London NW7 1AA, United Kingdom

George M. Malacinski — Program in Molecular, Cellular, and Developmental Biology, Indiana University, Bloomington, IN 47405

Carl N. McDaniel — Department of Biology, Rensselaer Polytechnic Institute, Troy, NY 12181

Hans Meinhardt — Max-Planck-Institut für Virusforschung, Spemannstrasse 35, 7400 Tübingen, West Germany

R. Scott Poethig — Department of Agronomy, Curtis Hall, University of Missouri, Columbia, MO 65211

Richard Nuccitelli — Department of Zoology, University of California, Davis, CA 95616

Christiane Nüsslein-Volhard — Friedrich-Miescher-Laboratorium, Max-Planck-Gesellschaft, Spemannstrasse 37–39, 7400 Tübingen, West Germany

Wolf-Ernst Reif — Institute and Museum of Geology and Paleontology, University of Tübingen, Sigwarstrasse 10, 7400 Tübingen, West Germany

Tsvi Sachs — Department of Botany, The Hebrew University of Jerusalem, 91904 Jerusalem, Israel

Klaus Sander — Institut für Biologie I (Zoologie) der Universität Albertstrasse 21a, 7800 Freiburg, West Germany

Jonathan M. W. Slack — Imperial Cancer Research Fund, Mill Hill Laboratories, Burtonhole Lane, London NW7 1AD, United Kingdom

Wilfred D. Stein — Department of Biological Chemistry, The Hebrew University of Jerusalem, 91904 Jerusalem, Israel

David L. Stocum — Department of Genetics and Development, University of Illinois, Urbana, Il 61801

Manfred Sumper — Institut für Biochemie, Genetik und Mikrobiologie, Lehrstuhl Biochemie I, Universität Regensburg, Universitätsstrasse 31, 8400 Regensburg, West Germany

Arthur T. Winfree — Department of Biological Sciences, Purdue University, West Lafayette, IN 47907

Lewis Wolpert — Department of Anatomy and Biology as Applied to Medicine, The Middlesex Hospital Medical School, London W1P 6DP, United Kingdom

Editor's Discussion with the Contributors

George M. Malacinski

During the development of this volume, I had an opportunity to engage the contributors in "off-the-record" conversations about several aspects of the discipline of pattern formation. Several of the more thought-provoking conversations are summarized here.

On the subject of positional information:

Q. Please explain what you feel would provide the best experimental test of the "positional information theory" which dominates so much of the thinking expressed by contributors to this volume.

A. Let me begin by explaining that the "positional information theory" is *not a theory* at all; rather, it is a *concept*. The distinction is an important one, for in the former instance an experimental test is called for. From the latter point of view, i.e., the conceptual one, a description or explanation, rather than a direct test, is warranted. Positional information is best viewed as a way of looking at a developmental phenomenon, rather than as a mechanistic way of interpreting one. Wolpert made that clear in his *J. Theor. Biol.* **25**:1–47 (1969) paper. He emphasized that within a "field" the position of individual cells is specified with regard to one or more points (i.e., coordinates) within the field. That "specification of position" is what he aptly termed "positional information." Most researchers would probably agree that such a concept is indeed a useful way of describing a developing system.

Now the "test" that you mentioned should be designed to determine what sort of mechanism is responsible for specifying the position of cells within a field. Wolpert originally postulated that gradients of morphogens are active in establishing positions. He emphasized, as do several contributors to this volume, that various types of morphogen gradients are present in developing systems and that cells "interpret" the gradient and in that manner establish their position. "Tests" should be directed therefore toward elucidating the characteristics of morphogen gradients. Alternatively, as in the case of cytoplasmic localizations such as the germplasm of amphibian and insect eggs, "position" is specified by inheriting—usually during early cleavage division—a unique cytoplasmic component. In those instances "tests" should be designed to learn about the mechanism whereby the cytoplasmic localization sets up a developmental program in the cells which contain it.

On the subject of model builders:

Q. Is there a gap between the rate at which model builders have constructed their models and experimentalists have collected data?

A. Yes, definitely. Many model builders have simply become too sophisticated! With their computers all they do every time a new piece of data emerges which is inconsistent with their model is to make a minor program change to accommodate the new data. Their models continually change, and eventually become albatrosses, that is, contraptions of the first order! Those contraptions often bear no relationship to reality. Some model builders have become completely absorbed in their endeavors and many have lost touch with the laboratory scientist. Many play no active role in the discipline of pattern formation. They don't suggest experimental designs for the laboratory scientist. All they do is *react* to the data as it is developed in experimental laboratories. Obviously, I feel strongly that a gap exists, and that a concerted effort should be made to bring the theorists into phase with the experimentalists (e.g., through workshops associated with symposia).

On the subject of morphogen gradients:

Q. Aren't there really *three* general issues that must be dealt with when morphogen gradients are invoked to explain pattern formation? *First*, a distinction should be made between whether the morphogen itself is small or large molecular weight. That is important, because small molecules could easily diffuse (or be transported) between cells. Gradients of small molecular weight substances could be active in whole embryos, or particular regions of embryos, or in special clusters of cells. Large molecules don't easily pass through cell membranes, so their activity gradients would necessarily be confined to single cells or groups of cells that are specialized for intercellular transport (e.g., the cells surrounding the developing insect oocyte).

Second, plausible models to account for the "shape" of the gradient should be formulated. Some gradients could be steep, while others could be relatively shallow. Several possible mechanisms could—in principle at least—formally be invoked to explain the shape of individual gradients.

Third, competence to respond to the morphogen must be developed in the cells that react to the gradient. Most models of morphogen gradients emphasize the geometry of the gradient. What about the capacity of cells to respond? Certainly it is not expected that all cells in a developing system share equally the ability to react to a specific morphogen.

A. I agree fully with your statement of some of the major issues. Regarding the *first,* I'd like to say that your statement is a bit of an oversimplification. Not only is molecular size important, but so are charge, solubility, etc. Those latter features may actually cancel out (or amplify) the importance of molecular size.

Concerning the *second* point, one must keep in mind that the story is more complicated than you have described. The presence of "inhibitor" gradients could easily modify the "biologically effective" shape of the morphogen gradient as much as any other property of the morphogen gradient.

Finally, regarding the *third* point, let me say that your thinking is a bit further ahead than both the present level of experimentation and attempts at designing gradients. Certainly the capacity to respond is as important and probably as complex as the construction of the morphogen gradient itself. However, "developmental competence" will probably be less easy to detect (i.e., bioassayed or measured) than the emergence of activator gradients.

Q. Isn't it somewhat of an embarrassment that despite the fact that "gradients of morphogens" have been aggressively promoted as key regulatory mechanisms in pattern formation, none has yet been *unequivocally* identified—in any experimental system?

A. No, not really. If one considers the rate at which progress has been made in most areas of embryology, the progress being made in morphogen identification is *not* extraordinarily slow. Consider, for example, the cell surface. It has been postulated to play a direct role in cellular determination for just as long. Yet no truly significant discoveries regarding the identification of cell surface regulatory molecules have been made.

Clearly, it is not fair to single out the morphogen idea as an example of a hypothesis that has failed to be substantiated by experimental fact. In the context of developing systems relatively few models have actually been proven right down to the level of the definitive isolation and characterization of the biologically active molecules.

On the subject of "spinning our wheels":

Q. Do you get the feeling that many of the recent papers in pattern formation represent repeats (with only minor variations) of the classical studies? That is, are we "spinning our wheels?"

A. Yes and no. In many instances we don't do adequate literature searches before embarking on "new" projects. It amazes me how often recent data are not new at all. Rather, it exists—in one form or another—in the old (e.g., pre-World War II, and in some cases—e.g., amphibian embryology—pre-World War I) literature.

Perhaps I'm being a bit harsh. The key phrase is "one form or another." The early workers often used different organisms and very inferior (by today's standards) measuring techniques. Likewise, they lacked the benefit of recent conceptual advances in molecular biology. So having each new generation of embryologists repeat the old experiments—within the context of a new conceptual framework—probably makes sense.

We must be careful, however, that we don't carry that approach to an extreme. Then we'd certainly be guilty of "spinning our wheels." Like probably most scientists I have personal opinions about which research areas have been overworked, but I'm not going to offend the sensitivities of some of my colleagues (and friends) by listing them for you!

On the subject of recombinant DNA technology:

Q. Should the discipline of pattern formation expect major advances to emerge from the field of recombinant DNA technology?

A. Immediately, I'd like to say that some of us are very anxiously awaiting the contributions recombinant DNA technology will provide. Some of us are optimistic and expect that several problems in pattern specification will benefit greatly from "in depth" knowledge of gene structure and function (e.g., transcription). Others are less optimistic. They feel that to the extent that pattern specification is highly coupled (temporally and spatially) to gene expression, great advances in knowledge will be provided by recombinant DNA technology. In other instances, especially those in which a pattern of development (e.g., axis of symmetry) is specified in transciptionally inactive cells (e.g., newly fertilized eggs of some animal species), less optimism is expressed. What is expected instead is that recombinant DNA technology will serve to define some of the traditional problems in pattern formation at a new level of organization—the primary structure of the gene. To me it is fascinating how often recombinant DNA technologists remark that they need more information at the "descriptive" level before they can design precise experiments at the "molecular" level. Particularly often I hear it emphasized that we need more *basic* information about such things as cell lineages, cell behavior patterns, etc. before a truly significant use can be found for some of the high-powered nucleic acid technology.

On the subject of growth industries, and sunset industries:

Q. Please provide a brief and very general overview (*in terms employed by industrial economists*) of the research areas of pattern formation that you consider to provide *growth* opportunities, to be relatively *mature*, and to have begun to *recede*.

A. My answer to that question will represent, naturally, more of a personal perspective than the broadly based general overview which you might like. I'll list a few research areas under each category (in alphabetical order), then I'll pretend I have some venture capital to invest in a highly speculative endeavor. Under "growth industries," I would list the characterization of electrical currents, *Drosophila* developmental genetics, morphogen identification, morpholaxis in amphibian embryos, plant pattern specification, protozoa cell patterning, and slime mold patterning. For "mature industries" I would select appendage regeneration and computer modeling. As an example of a sunset industry I'd single out "systematics." Specifically, taxonomic studies (e.g., of plants) which fail to elucidate underlying themes which may have guided the evolution of pattern formation mechanisms. Clearly, the area of pattern specification is, on balance, to be regarded—in my opinion—as a "growth industry."

Right now I'd invest my venture capital into a recombinant DNA technology firm (i.e., laboratory) which studies an organism with a relatively simple genome (e.g., *Drosophila*).

On the subject of "sleepers":

Q. Can you think of any biological systems that might provide unique opportunities for pattern formation research, but have been overlooked?

A. I'd suggest, without a substantial amount of forethought, the following: diatoms—they display considerable prepatterning, lens regeneration (does it follow the roles established for appendage regeneration?), and plants—in general they have been overlooked, despite the wide variety of pattern specification phenomena they display.

Clearly, those moments spent in conversation with the contributors were stimulating, for they dealt with several of the more important issues in pattern formation, many of which are not usually discussed in formal publications.

Glossary of Terms Frequently Used in Pattern Formation

determination   fixing of the developmental fate of a cell. Best employed as an "operational" definition—to interpret, for example, tissue grafting or cell transplantation data.

epimorphosis   the duplication or regeneration of structure that depends upon cell proliferation and tissue growth (e.g., limb regeneration).

morphogen   a regulatory molecule that is distinguished by its ability to promote morphogenesis (e.g., generate positional information). Usually considered to be small molecular weight and diffusible.

morphallaxis   the duplication or regeneration of a structure by the respecification of pre-existing cells. Cell proliferation is not required for the development of new structures (e.g., twinning of amphibian embryos).

pattern formation   the mechanisms that generate the spatial organization of a developing system. Followed by overt cell differentiation.

pattern regulation   the response of parts of a cell (e.g., surface structure in protozoans) or groups of cells (e.g., vertebrate appendage) to natural damage or experimental manipulation. Often results in either regeneration or duplication of the original pattern.

pattern specification   the mechanisms under normal circumstances that establish the original morphogenetic program or spatial organization of cells or tissues, for example, during embryogenesis.

polar coordinate model   a two-dimensional model that accounts for pattern regulation (e.g., regeneration). Assumes that cells possess information about their position on a radius and around a circumference. That information guides the appropriate morphogenetic or differentiation events.

positional information   a concept that proposes that cells are able to recognize their relative positions within a coordinate system. Those cells can recognize discontinuities in the coordinate system and respond by pattern regulation.

positional value   the information a cell is assumed to possess about its relative position within an organizational framework or structure.

We know what a masquerade all development is, and what effective shapes may be disguised in helpless embryos.—In fact, the world is full of hopeful analogies and handsome dubious eggs called possibilities.

George Eliot, in *Middlemarch*

SECTION I

General Articles

CHAPTER 1

Positional Information and Pattern Formation

Lewis Wolpert and Wilfred D. Stein

THE CONCEPT OF POSITIONAL INFORMATION was explicitly formulated by Driesch (11). He argued that all the cells in the early sea urchin embryo have the same inherent capacity for development and that the differences in their development were due to their position within the embryo; as if they were in a coordinate system that enabled each cell to know what to do. He derived this conclusion from his observation that half and quarter embryos develop normally. Because he could not imagine that the cells could re-establish the appropriate coordinate system no matter which parts of the embryo were removed, he decided that this re-establishment required a vital force, entelechy, outside the normal bounds of physics and chemistry. We now know that Driesch was wrong on two counts. First, there are differences among the cells of the sea urchin embryo (21). It is true that if the cell embryo is divided along the animal vegetal axis, both halves develop normally, but, if the division is such as to separate animal and vegetal halves, they develop quite differently. Second, it is possible to construct models which would enable the cells to know their position even when the system is grossly perturbed. For example, a Turing system (55) of diffusing and interacting chemicals does just this. It can set up a chemical concentration gradient that is self-organizing, and such a concentration gradient could effectively provide the cells with knowledge of their position.

The problem confronting Driesch was that of pattern formation and its regulation (65). What is it that specifies the three main regions of the early sea urchin embryo as it becomes divided along its main axis into mesenchyme, endoderm, and ectoderm in fixed proportions? Moreover, how are these proportions maintained over an eight-fold size range: that is, from quarter embryos to double embryos? This problem of pattern formation, concerned as it is with spatial organization, should be distinguished from two other processes: cellular differentiation and

3

changes in form. Cellular differentiation is the process of specifying those particular macromolecular components that characterize a cell type, such as haemoglobin in red blood cell differentiation. This process is different from pattern formation which is concerned with the spatial organization of cellular differentiation. For example, the cellular differentiation of cartilage is similar in the arm and leg, but its spatial organization is quite different. Pattern formation is also concerned with regulation, of the type noted by Driesch. In some developing and regenerating systems, it is as if there are field properties, the cells behaving in a cooperative and coherent way when parts are removed or added. Consider, for instance, the regeneration of hydra which can form a normally proportioned animal from a fragment 1/20 the original size (72). The other process—change in form—is brought about by the cells' generating forces which bring about such changes, for example, the folding of cell sheets and the migration of cells to different locations. A typical example, again with reference to the sea urchin, is gastrulation. Here the spherelike blastula is grossly transformed by the translocation of the future gut cells from the vegetal pole to the other side of the embryo so as to form a tube (17). Folding of sheets during development is a common process and is essentially a problem in mechanics. The pattern problem is that of specifying which cells will generate the forces and is always one of spatial organization.

Patterns of cellular differentiation can be generated during development by what appear to be several different mechanisms. The major distinction is probably between those in which cell-to-cell interactions are minimal and those in which such interactions are dominant. The pattern of cellular differentiation may, for example, reflect cytoplasmic localization within the egg. As the egg divides during cleavage, different cells can acquire different cytoplasms and the resulting pattern can reflect the original pattern of cytoplasmic localization in the egg. Ascidian development seems to proceed on these lines (61), as does the determination of germ cells in insects (23). Such a process need not involve cell-to-cell interactions. In the development of the nematode, cell lineages are highly determined and it is the cell lineage that in some manner specifies the pattern of cellular differentiation (26). Cell-to-cell interactions are involved, but these are not the dominant process. Instead the asymmetry generated at cell division appears to be the determining process. Such mechanisms for pattern formation must be contrasted with those where cell-to-cell interactions are involved. This later mechanism is most strikingly illustrated by the regulative properties of vertebrate and echinoderm embryos. In such embryos, at early stages, quite large portions can be removed, or the parts rearranged, yet normal development occurs (6). This regulation, as Driesch recognized, requires interactions between the cells. An interesting example of such interactions, which would seem completely to exclude any kind of lineage mechanism, comes from clonal analysis of mosaic insect embryos (14). It has been shown that about 10 founder cells give rise to the anterior portion—compartment—of the insect wing. Thus, one cell usually gives rise to about 1/10 of the anterior wing. However, by genetic techniques, it is possible to arrange for just one of the 10 cells to grow much faster than the others. This one cell can then form the whole of the anterior compartment which still has a completely normal pattern. It is very hard to explain this other than by invoking cell-to-cell interactions.

In this chapter, we are concerned only with the development of patterns in which cell-to-cell interactions play a major role. We believe that there are general principles governing this development. We believe, moreover, that when we understand pattern formation we shall be able to relate the different kinds of mechanisms to one another in a meaningful way. At present, the study of pattern formation is at quite a primitive stage. We have models that can account for many of the phenomena observed. However, because the molecular basis is quite unknown, at this stage we should try at least to understand the processes in terms of cellular activities.

The process of pattern formation usually occurs quite early in development. In vertebrates, the major patterning process is concerned with the laying down of the main body axis. Patterning thus occurs while the embryo is quite small. It is now clear that cell-to-cell interactions occur over distances of less than 1 mm and that changes take place within hours rather than minutes (65). This is one of the reasons for Crick (9) suggesting diffusion of a chemical as the basis of cell-to-cell communication. Since the embryo, or parts of the embryo, is so small when the pattern is laid down, most later development is growth and programmed very early (70). For example, the humerus in the arm is only about one-third of a millimeter long when it is laid down and it must grow about a thousand-fold. Thus a major feature of pattern formation is to program this growth. There is quite good evidence for the autonomy of development once the basic pattern has been laid down.

Spacing Patterns

There seem to be few, if any, random patterns formed during development. One might expect to find them when repeated structures such as hairs cover a surface, as in insects. However, it usually has been found that such patterns are not random, but ordered to the extent that there is a minimum distance between the repeated elements (25). Such order can be quantified and, in principle, ranges from the truly random to the completely ordered, as seen in the hexagonal pattern of feather follicles. Such patterns can be considered spacing patterns. Related to such spacing patterns are slightly more complex ones where the pattern is anisotropic, such as stripes. As Crick has pointed out, embryos are very fond of stripes (10).

A mechanism for spacing patterns that has been proposed for a number of systems is that based on inhibition (28, 63). Existing structures inhibit similar structures from forming close to them. One of the best examples is that of the alga *Anabena*. The filaments of *Anabena* are a single file of vegetative cells, with specialized cells, heterocysts, at regular intervals. As the vegetative cells divide, heterocysts differentiate from the vegetative cells, to maintain this pattern. The mechanism involves a diffusible heterocyst inhibitor, made by the heterocysts themselves, that prevents heterocyst formation unless the concentration of the inhibitor falls below a threshold concentration (63). On this simple mechanism, since the distance between heterocysts is increasing because of growth of the vegetative cells, cells midway between two heterocysts will begin to develop as heterocysts only when the inhibitor concentration falls below threshold level. Near the middle, more than one cell will begin to develop into a heterocyst. But the most advanced cell, by beginning to secrete heterocyst inhibitor, will prevent all less advanced cells from completing their differentiation. In order for the cell not to inhibit itself, it must have an autocatalytic mechanism to drive differentiation and to prevent autoinhibition. We can think of differentiation as involving the accumulation of a substance, an activator, that cannot diffuse from cell to cell. When the concentration of the activator reaches a critical or threshold value, the cell is determined to be a heterocyst cell. It will continue to develop and acquire all the characteristics of a heterocyst. At any time before the critical concentration of activator is reached, if the synthesis of the activator is inhibited, the cell will remain a vegetative cell. Such a mechanism will give rise to discrete single cells more or less uniformly spaced along a line, or, if extended to two dimensions, in a plane.

It should be emphasized that this mechanism requires a threshold for inhibition, a diffusible inhibitor whose production is dependent on an activator, and on an autocatalytic production of a nondiffusible activator, for these are the essential components of several different models for pattern formation that rely on reaction-diffusion mechanisms. In the case of spacing patterns, we see the three factors combined in their simplest form. If one now increases the complexity of the model system by allowing the activator to diffuse, one has a typical reaction-diffusion system of the type originally proposed by Turing (58) and which forms the basis of several related models (27, 34). Under appropriate conditions and by the choice of particular parameters, such systems will give rise to waves of chemical concentration. If structures only form at concentrations of the substance distributed in this way—the morphogen—only above a certain threshold that corresponds to the peak of the waves, a spacing pattern will result. Murray (37) has recently provided an invaluable analysis of such systems and the parameter space in which they operate. It is also of great interest that one can approximate a hexagonal pattern by such mechanisms (27).

One aspect of reaction-diffusion mechanisms requires particular mention, that is the process of branching or bifurcation. This is nicely illustrated by Lacalli (27). If a reaction-diffusion mechanism sets up a pattern with a single peak of concentration in a one-dimensional system and if the effective length is then doubled, for example, by, growth, the central maximum will break up—bifurcate—to produce two lateral peaks. The length can also be effectively doubled by an appropriate reduction of the diffusion constants (25).

The development of stripes requires a further modification of such mechanisms. This can be done in two ways. Either by making the geometry of the sheet of cells in which the reactions are taking place anisotropic (36), or by making the diffusion constant in one direction different from that in another (27).

These mechanisms generate waves in space by diffusion-reaction. It is possible to generate such waves by a completely different mechanism that is based on oscillations in time combined with a progress-zone type of model (50). The model thus combines some features of the Cooke-Zeeman (8) model for somite formation with that of the progress-zone model for limb development (52). In essence our model is based on the assumption that in a special region, the progress zone, the cells are coupled with respect to a chemical reaction which oscillates. That is, the concen-

Concentration

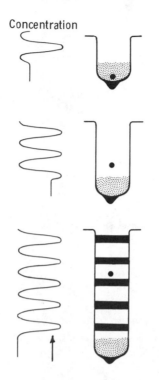

FIGURE 1. A model for generating repeated structures based on the progress zone. All the cells in the zone at the tip of a growing structure are coupled, and the concentration of a particular chemical changes in a periodic way. Since the cells in the progress zone are dividing, cells leave the progress zone, as shown by the black dot. When cells leave the progress zone the concentration of the chemical is frozen. This generates a sine wave. If structures only form when the concentration is above a threshold concentration, as indicated by the arrow, then stripes may develop. The diagram shows the system at three successive stages.

tration of a particular chemical changes in a periodic way in all cells at the same time. When cells leave the zone, the concentration of the oscillating chemical is frozen at the particular value it had as it left the zone (Fig. 1). This fixes the phase of the oscillation. Cells continuously leave the zone either because of migration or because cell division forces them out, and they therefore form a train of cells in which the concentration of the chemical—morphogen—is fixed in each cell but varies in a sinusoidal fashion along the train. Thus a chemical wave of concentration is generated. If the cells respond (as above) to some threshold mechanism then a repeated pattern will be generated.

All these mechanisms generate regular patterns of structures in which any formed structure is equivalent to any other—the peaks of the waves are all the same. One could not alter one structure without affecting all. This places a severe limitation on the class of patterns to which they can give rise. To bring about differences between the formed structures, it is necessary for them to be nonequivalent (32). This requires the organism to make use of positional information.

Positional Information

The concept of positional information (67, 69) suggests that cells have an intrinsic record of their position that effectively gives them an address. This record, the positional value, is used by the cells to determine cellular properties according to the cells' genetic constitution and developmental history, that is, to interpret positional value. It is as if the cells were in a coordinate system and the overt pattern emerges from the process of interpretation. Thus the pattern need not be isomorphic with the set of positional values which could be a set of graded concentrations. In principle, a specific cell type could arise sharply at any point. In contrast, a prepattern mechanism—such as for a spacing pattern—develops a distribution of morphogen that is isomorphic with the overt pattern of formed structures. For example, if a particular cell type forms only above a threshold concentration, its occurrence will correspond to the crest of the waves.

Positional information can be provided by the monotonic decreasing concentration of a morphogen as will be seen. That gradients play an important role in pattern formation is an old idea championed by many workers. However, discussion as to how such gradients were to be set up, maintained, or how they gave rise to overt patterns was, on the whole, neglected. Child (5) summarized numerous studies arguing for gradients in both development and regeneration. The classic studies of Horstadius (21) on sea urchin development invoked gradients in a more explicit manner as did the work of Pasteels (42) on early amphibian development. Of particular importance was the book by Huxley and de Beer (22) where many of the early ideas and results were codified. The central concept was that of the dominant region in relation to which other regions were specified. The concept of positional information (64) was very much along the same lines, but attempted a more rigorous and explicit formulation of the problem. It was also an attempt to revive interest in pattern formation.

The assigning of position in a coordinate system requires boundary regions with respect to which position is specified (like the origin in a coordinate system) a scalar that gives the distance from the boundary region, and a vectoral property that specifies the polarity of the system, that is, specifies in which direction to measure. All these requirements are satisfied by a gradient in concentration of a morphogen whose concentration is fixed at the boundary region and decreases monotonically with distance from the boundary. The boundary could be the source of the morphogen which diffuses away from it. If, in addition, it is broken down, then an exponential gradient of concentration will be set up. There is substantial evidence that what might be thought of as boundary or reference regions have special properties (65). For example, the polarizing region in the chick limb bud appears to provide the reference point for position along the antero-posterior axis (53). The micromeres in sea urchin development specify the vegetal pole of the animal-vegetal axis, and the hypostome of hydra similarly defines the head end. Such reference regions, when grafted to appropriate sites, can assign new positions that alter the emergent pattern so as to relate it to the new reference region. As regards the scalar, it is most likely to be the concentration of a chemical. We shall return to consider how this might be specified and interpreted. Here we wish only to emphasize that a monotonic gradient in concentration could specify position.

There is quite good evidence that polarity is determined by the sign of the gradient, that is, the direction in which positional values are increasing or decreasing. Polarity can manifest itself in regulating systems, such as, for example, *Hydra*, where a head will only be regenerated at the end closest to the original head in an isolated fragment (72). Or, it can show itself by the direction in which bristles or hairs point, as in insects (28). In both cases, reversal of polarity can be understood in terms of a reversal in the sign of a gradient.

It is of great interest that all the requirements for a coordinate system can be provided by a reaction-diffusion system. For example, Meinhardt and Gierer (34) have shown how their reaction-diffusion system mimics many of the properties required to account for the regeneration of hydra. The reaction-diffusion systems could also provide a means whereby a gradient in a diffusible morphogen can be set up by self-organization. It is not necessary to have a localized source of the morphogen.

It should not be thought that positional information must be either specified or recorded by a diffusible morphogen. Other mechanisms for specifying position can be based on the time spent in the progress zone (52) or membrane interactions (2). Positional information requires an ordered, but not necessarily a graded set of cell properties. The distinction between *ordered* and *graded* must be made. In a graded series, there is a quantitative variation as in a concentration gradient. In an ordered series, such as the telephone directory, there is a set of rules for placing the elements in a correct order. In general, it is not easy to design simple experiments to distinguish between graded and ordered systems. However, in the case where the signal is a diffusible morphogen, such a distinction becomes possible because of the specific properties of such a system: the signal can be attenuated (48), is additive (71), and has particular time/distance relationships (72).

The positional information mechanism for pattern formation has certain interesting characteristics. (1) The only cell-to-cell interactions that are required in principle are those involved in specifying the position. There need be no interactions between the overt developing structures themselves, since these arise from the interpretation of positional values. (2) The same set of positional values can be used for very different patterns, the difference lying in the interpretation. That is, the same signals and coordinate system can be used for quite different patterns. There is very good evidence that the positional fields in different imaginal discs are the same and use the same positional signals. In genetic mosaics of homeotic mutants, cells develop according to their positions and genetic constitution (44). Again, intercalation appropriate to equivalent positions occurs when fragments from different discs are combined (3). The positional signal from the polarizing region in the chick limb is the same in wing and leg, and is similar in all amniotes (12). Thus, mouse polarizing region grafted to the anterior margin of the chick limb bud specifies a further sequence of chick digits. It also seems that the positional field comprising the main body axis may be the same in different vertebrates. The induction of epidermal structures seems to be the same in the urodeles as in the anurans. This induction can be thought of as the transfer of positional information from mesoderm to overlying ectoderm followed by interpretation to form ectodermal structures (69). Anuran ectoderm grafted to the ventral region of a urodele embryo forms suckers in the appropriate anuran position even though urodeles do not form such structures (18). Finally, Rawles (44) has shown that pigment patterns in birds are characteristic of the donor melanocytes. Thus the patterns formed when the melanoblasts of pigmented birds are injected into embryos of white birds are characteristic of the donor. It is as if the melanoblasts read the local positional values and develop according to their own genetic program. (3) A key concept of positional information is nonequivalence (32). This concept emphasizes the differences between cells of the same differentiation class. For example, in the vertebrate limb cells with different positional values may differentiate as cartilage. While they are all cartilage cells, they are nonequivalent because they have different positional values. This nonequivalence can be expressed by the cells in terms of, for example, different growth or surface properties. Thus, for example, the cartilage in the wrist of the chick wing has growth properties quite different from those of the radius and ulna (51). These differences, it is suggested, reflect differences in positional value.

The Recording and Interpretation of Positional Information

If positional information is initially specified by the concentration of a chemical substance, which could be as simple as hydrogen ions, calcium, or cAMP, the problem is two-fold. How could these concentrations be converted into a stable record of position, and how could the cell interpret this record of positional value in order to determine its behavior? It is convenient at this stage to draw a distinction between two modes of positional information. The first, and perhaps more primitive, involves an isomorphism between the positional values and their interpretation. That is, the continuous gradient in positional information is expressed directly as a continuous gradient in some cellular property. If the positional value were recorded as the concentration of some substance, this concentration could then directly specify some cell characteristic, such as the number of adhesive sites on the membrane, or rate of a particular chemical reaction, or more generally, the concentration of any other substance. The various responses would then be graded and continuous. There are quite a few examples of graded cell properties that might well be generated by such a mechanism (Table 1), for instance by the gradient in adhesiveness in the insect segment (40) or the gradient in cell properties in the sea urchin embryo (17).

The second mode of positional information is that in which the continuous set of positional values is interpreted in a discontinuous manner. This interpretation must involve thresholds at particular positional values. This mode, depending as it does on multiple thresholds, almost certainly requires a more complex mechanism. At a particular threshold concentration, a particular gene or enzyme must be switched on, and remain on when the signal is removed. This probably requires a positive feedback loop for the maintenance of the state (31). The switched-on state characterizes all those cells whose positional value is above the threshold. In this state, a whole set of cellular attributes can be specifically attributed by what is essentially a cascade mechanism. In yet more complex models, the signal can turn on a gene only at a particular concentration of a morphogen (33). It is of interest that a recent model for specific gene activation is dependent on the local concentration of ions (16). This model, based on modern views of chromatin structure, involves the propagation of regions of single-stranded DNA and shows, at least in principle, how positional information could be recorded by bringing about a change in chro-

TABLE 1.

System	Graded Property
Sea urchin embryo	Adhesiveness and motility along animal and vegetal axis (17)
Nucleus laminaris of bird brain	Extension, number and branching pattern of dendrites (47)
Hydra	Time to regenerate a head (60)
Mesoderm of amphibian embryo	Time of somite formation (8)
Moth wing	Adhesiveness (38)
Insect segment	Adhesiveness (40)
Avian retina	Antigen (57)

matin structure. Another model, based on the development of the immune system, (20) proposes that the cells display different combinations of molecules on their surfaces as do antibody producing cells, the expression being controlled in part by DNA modification. While it is not clear how this could be linked to positional information, combinations of surface molecules could provide a very economic and powerful way of specifying positional value, just as the ZIP code specifies postal addresses. It is the implementation that is so difficult. How can a gradient be concentrated into a ZIP code and then interpreted in an appropriate manner? This is the central problem in positional information, and at present we have no idea how positional value is either recorded or interpreted.

In principle, patterns of unlimited complexity can be made with a mechanism based on positional information. All that is required is a position-specifying mechanism that is sufficiently fine grained, and, of crucial importance, a mechanism that provides for appropriate interpretation of position at every address. This, in effect, requires each cell to have a table of the appropriate interpretations at each cell position. This is not only very difficult to envisage, but implies an absence of constraints as to what can develop at any position, an absence that is not easily reconciled with what is observed. Such an extreme positional information mechanism could allow the formation of new cartilaginous elements or even liver cells at any point in the limb. This, together with the requirement of a large number of thresholds for interpretation, presents a formidable problem for models based on positional information. For this reason, we have recently put forward a model for pattern formation in limb development which, in some ways, reduces the complexity of the problem.

A Model for Chick Limb Development Based on Prepattern and Positional Information

A model for chick limb development based on positional information suggests that position along the two main axes, the proximo-distal and antero-posterior are specified by two different mechanisms (67, 68). For the proximo-distal axis, it is suggested that position depends on how long cells remain in a region at the tip of the limb bud, the progress zone (52). For the antero-posterior axis, it is suggested that there is a signal from a region at the posterior margin of the limb, the polarizing region (55). Both these mechanisms have aroused some controversy, but these controversies cannot be considered here (68, 24, 71). Of general importance is that such a model is based strictly on positional information being interpreted to give the observed pattern. This requires a large number of thresholds to be specified. For example, for each digit there need be two positional values, one at which cartilage formation is switched on at the digit's anterior margin, and another at which it is turned off at the posterior margin. This specification must be repeated for each digit. As pointed out, such complexities present problems for the interpretation of the positional information. Therefore, even though this model can account for a wide variety of results that have come from experimental manipulation of the limb, the model is unsatisfactory because of the multiple thresholds involved. A more serious objection is that there are experiments which argue against a strictly positional based model. The most important are those in which the cells of the limb bud are disaggregated, reaggregated, and then placed in a jacket of limb bud ectoderm, where they can develop quite well-formed digits even though a polarizing region is absent. This finding suggests that some type of spacing pattern-prepattern mechanism is involved in specifying the cartilaginous elements. Several prepattern models for the limb have been put forward (39, 62). We (50) have suggested a new model based on both prepattern and positional information, which one can use to illustrate both prepattern, positional information and nonequivalence.

On our new model, the patterning process is assumed to occur in the progress zone. In this zone, all the cells are assumed to be oscillating in unison; that is, the concentration of some chemical fluctuates in a regular, cyclic manner. This is a clock substance. The cells in the progress zone divide and cells then leave the progress zone. When the cells leave the progress zone, concentration of the clock substance in them is frozen. They thus bear a permanent record of the phase of the clock oscillator at which they left the progress zone. We assume also that they bear a record of the total number of oscillations that they experience while residing in the progress zone. This effectively provides positional information along the proximo-distal axis, and makes successive elements nonequivalent. The positional information is represented by a sine wave whose successive cycles are numbered (Fig. 2). The successive elements in the limb can be delimited by the troughs in this sine wave if cartilage does not form below a threshold corresponding to the troughs of this spacing pattern. Thus the successive elements in the limbs are laid down by successive cycles of the oscillating substance.

The number of elements at successive levels can be thought of as an increasing series along the proximo-distal axis. Thus there is one element (the humerus) in the upper arm, two in the forearm and (primitively at least) three and four in the wrist, followed by five digits. This pattern along the antero-posterior axis is assumed to be generated by a reaction-diffusion mechanism. The system displays a single peak in the first cycle that gives rise to the upper arm, two peaks for the forearm and so on. The increase in the number of peaks is determined by the discontinuous change in some cell property (possibly a diffusion constant of a morphogen) as the cells in the progress zone go through successive cycles. This model will allow a limb with only symmetrical elements all of the same size, and without antero-posterior asymmetry. Asymmetry along the antero-posterior axes comes, we assume,

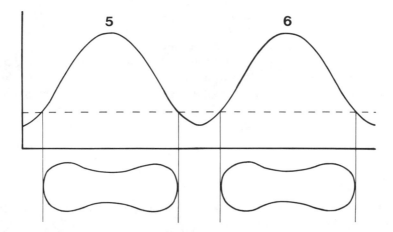

FIGURE 2. If position along the proximo distal axis is specified by the length of time the cells spend in the progress zone, then this might be based on a mechanism of the type illustrated in Fig. 1 together with a means for counting cell cycles. The figure shows just two such cycles, the fifth and sixth. If cartilage only forms above a threshold—the dashed line—joints begin to be accounted for. The concentration gradients, with high points in the center of the elements, may control aspects of cartilage morphogenesis, such as the wave of hypertrophy that spreads from the center and the mechanical properties of the perichondrium. This might partly account for the dog-bone-like form of the element if the perichondrium was a constraining sheath that created greater mechanical constraint in the center than at the ends. The difference in the number of cycles between the two elements could be used to specify different growth programs, since the elements would be nonequivalent.

from the action of the polarizing region, which is a region at the posterior margin of the bud, that is assumed to be the source of a diffusible morphogen (53). This morphogen provides a gradient in positional information across the antero-posterior axis. It effectively names the elements, such as the individual digits, that have developed from the prepattern. This specification affects both the width and the growth of the elements. The naming can be altered by grafting in additional polarizing regions. It is crucial to recognize that the naming is a discrete process and no digits of intermediate character are ever formed as would be expected if, instead of naming digits, the polarizing region were merely altering growth parameters in some continuous way. The specification enables, for instance, the metacarpal of digit 3 to be both the longest and widest of the three digits.

The characteristic form of each of the cartilaginous elements of the wing with their narrow central region (diaphysis) and broader ends (epiphyses) might be understood in terms of the gradient in the oscillating substance that decreases from the center of the element, both proximally and distally. If, for example, this substance affects the mechanical properties of the perichondrium, such that it is most strong in the diaphysis, then the overall form of the elements might, in part, be accounted for (1). In addition to this, as for the antero-posterior axis, detailed positional specification is required. For example, the proximal epiphyses may differ from distal ones, and most strikingly the distal end of the tibia captures the distal end of the fibula, a process that requires highly localized cellular changes (1). In the wrist, in particular, the structures have a complex and varied shape, specification of which, again, requires detailed positional information.

This model illustrates the possible interplay between prepattern and positional information. For a structure like a limb to have evolved, it seems necessary that a prepattern mechanism applied, since it is hard to conceive how the many thresholds that would be required for even a primitive limb, could have evolved sufficiently rapidly to provide even the most primitive limb.

Epimorphosis, Morphallaxis and Positional Signaling

Regeneration and regulation of patterns can be understood in terms of respecifying positional values or generating new ones (65). Regeneration provides the most direct evidence that cells have positional values that are separable from the structures that they form. Cells with the same differentiated phenotype can have quite different positional values. For example, the amphibian limb can regenerate when it is amputated and the structures regenerated reflect the level of amputation. Since the cell types are the same at different levels, regeneration requires that the cells are nonequivalent, that is, that they have different positional values. The most interesting development from the concept of positional information has been the polar coordinate model of French, Bryant, and French (13) which, while controversial (4) has been able to account for a wide variety of results on epimorphic regulation. The model suggests that cells have positional values which are specified in relation to polar coordinates and that simple rules describe how these positional values alter when parts of the system are removed or are joined up in different ways. The point relevant here is that the model relies on cells having positional values and that when normally noncontiguous regions are placed in contact with each other, new positional values are specified to smooth out the differences. Such experiments thus provide an assay for positional value.

The usual distinction between morphallaxis and epimorphosis is that the former does not involve growth whereas the latter does. In general terms, morphallaxis involves changing existing positional values. So, in the regeneration of the head end of hydra, the cells at the cut end acquire the positional values of the head end and new positional values are appropriately changed in the rest of the system. By contrast, in epimorphic regulation, as in amphibian limb regeneration, new positional values are generated by the growing cells at the cut end. However, it is not easy to draw a sharp distinction between morphallaxis and epimorphosis once the differences based on growth are recognized. In general terms, morphallaxis is considered to involve long-range interactions in which many of the positional values of the system are changed, whereas it is thought that in epimorphosis the original positional values are conserved and interactions are short range. But the distinction becomes blurred by the difficulty in deciding between short-range and long-range interactions. It is one of the surprises of work in recent years that interactions take place over rather small distances. It is hard to find good evidence of interactions over more than a few hundred micrometers, or more than about 30 cell diameters (62). We know of no work that suggests that in epimorphosis, interactions do not take place over such distances. There is also no evidence that in epimorphosis change in positional value is restricted to a region less than about 20 to 30 cell diameters in extent. The issues raised by the apparent differences between morphallaxis and epimorphosis must thus focus on the cellular activities involved. Over what distances are interactions taking place, what is the nature of these interactions, and which cells are changing their positional value?

Some of these issues are confronted in relation to the behavior of the chick limb bud when posterior tissue—the zone of polarizing activity—is grafted to more anterior positions. It is not disputed that this gives rise to additional limb structures. But what has been questioned is whether or not the polarizing region has special signaling properties, as has been suggested, and whether it is the source of a signal, possibly a diffusible morphogen whose range of action would be over several hundred micrometers (53). A possible alternative is that the results obtained are due to intercalation of positional values between the graft and host (24). It is not possible to review this issue here (71), but merely to use it to illustrate some implications for positional signaling. The most important issue is whether or not the polarizing region is the source of a signal. If it is such a source, one should be able to attenuate the signal and to obtain results consistent with such an attenuation. Attenuation of the signal leading to the predicted results has been obtained with γ-irradiation (48), uv-irradiation (19), and reduction in cell number (54). This attenuation is very difficult to account for in terms of intercalation, particularly in terms of a polar coordinate system.

It is commonly assumed that the main means of cell-to-cell interaction is via the channels provided by gap junctions. These channels allow direct cell-to-cell movement of molecules, having a molecular weight less than about 1,000 (66). It is the diffusion of such molecules that could provide a positional signal. But there is, at present, no direct evidence even for a diffusible molecule playing a role in pattern formation in the ways suggested in this chapter, and it must be emphasized that while there is a good correlation between the presence of such junctions and the requirement for a diffusible signal (66), there is no direct, or even very persuasive evidence, that the junctions provide the means of communication in pattern formation. We do not even know the pathways of cell-to-cell interaction in the chick limb bud. There is, however, evidence that the absence of such junctions may provide a means of defining separate positional fields or of establishing boundaries between them. Thus, Warner and Lawrence (59) have found that the cells at the boundary between insect segments allow the passage of ions but not fluorescein (MW300). That is, fluorescein will pass between cells of the same segment but will not cross the intersegmental boundary. This differential permeability could be the basis for restricting cell-to-cell communication.

We also remain largely ignorant of the chemical nature of the positional signal. Schaller (46) has indeed isolated a peptide from hydra—it is also present in mammalian brain—which she believes to be the activator molecule for head formation. Unfortunately, the peptide has rather a small effect, only increasing the rate of head formation and the number of tentacles by one or two. Nevertheless, this is an exciting development and raises the problem of the criteria to be met in order to establish that a substance is indeed a morphogen. These criteria have yet to be established and one cannot but be struck by the similarity between this problem and that of identifying putative neural transmitters. For example, the local application of retinoic acid to the chick limb bud can mimic the action of the polarizing region to a remarkable degree (56), but it in no way follows that retinoic acid is a morphogen. One would at least have to show that the distribution of retinoic acid in the bud had precisely the distribution required by the model.

Handedness

The problem of handedness has been largely ignored in recent years (see, however, 35). By handedness, we mean that a clear distinction can be drawn between a left-handed and a right-handed organism. In vertebrates, for example, handedness is very clearly shown by the position of the heart and viscera. In molluscs, handedness is manifested in the pattern of cleavage and coiling of the shell. Here we wish only to emphasize the special nature of the problem with reference to the main axes of development. The characteristic feature of vertebrate handedness is not just that there is bilateral asymmetry—the heart being on one side rather than the other—but that this asymmetry is consistently biased towards one side. It is this latter feature that presents special problems. Even though we do not understand how the axes of the developing embryo are established, in principle they require only the localization of a region that specifies an end. For example, in the development of the sea urchin embryo, one could imagine the animal-vegetal axis being determined by a gradient in a morphogen that was produced at the vegetal pole. This region, in turn, could be determined by some external asymmetry such as the site of attachment of the egg in the ovary. The next axis, the dorso-ventral, could arise at any point on the equator (as defined by the animal-vegetal axis) which could then set up bilateral symmetry with respect to the meridian that passes through it.

While we do not wish to minimize the problems associated with setting up these axes, they are quite different from specifying a consistent difference between the two sides defined by the axis of bilateral symmetry. Mechanisms similar to those used for defining the other axes could set up a gradient which would break the bilateral symmetry in a way that the high point of the gradient would be on either the left or right side. The difficulty lies in specifying that the gradient should always have, for example, its highpoint on the left side. How does the embryo distinguish left from right? We suggest that the consistent development of handedness requires asymmetric molecules to generate the handedness. There has to be a mechanism whereby such molecules can recognize the animal-vegetal axis and can then bias the gradient consistently towards one side or the other. It is thus of great interest that Layton (30) has described a mouse that carries a mutation that results in 50% of the offspring being morphologically left-handed and 50% right-handed.

Conclusions

Positional information can provide a framework for the discussion of both pattern formation and pattern regulation. The central concept is that of positional value with a cell property that records the cell's position with regard to some reference point. At this stage we have no idea either as to how positional value is specified in molecular terms, or how it is encoded within the cell. The problem is a difficult one and the field is at a stage where, while models at the phenomenological level can be quite elegant and persuasive, at the level of mechanism, we remain abysmally ignorant. While much discussion concerns diffusible morphogens, their existence has yet to be established. However attractive reaction-diffusion mechanisms may appear to be, they are still no more than attractive hypotheses. It would be encouraging to think that the field is at a stage comparable to that of genetics in its earliest days, when the physical basis of inheritance was totally unknown. Part of the problem is that we still understand too little about the biochemistry of both intracellular and intercellular processes. (We should remember, for example, that the mechanism whereby insulin controls cell activity, is still to be elucidated. This is a particularly sobering thought when we consider how crude our assays for putative morphogens are, in relation to pattern formation.) Even at the phenomenological level, much remains to be done. This is clear from the great impetus the polar coordinate model gave to experiments on regeneration. It is one of the virtues of positional information that it can lead to experiments and to the disproof of hypotheses. For example, Cooke (see Chapter 20) (7) has provided quite persuasive evidence that a model based on positional information has great difficulty in accounting for observations on the development of the dorso-ventral axis of the body in amphibians. This is particularly so since his evidence suggests that each region produces an inhibitor which inhibits further production of that region. His model, like that of Rose, (45) does not require the cells to have different positional values.

One of the problems with a model based purely on positional information is that a large number of different thresholds may be required. A multithreshold mechanism is not easy to handle at the present stage, particularly as we know so little about the process of interpretation. For this reason, we think it is likely that positional information may be used to modify prepatterns which initially, at least, require fewer thresholds. Nevertheless, the evidence for a graded set of discrete positional values is very strong. Finally, we have been silent on the question of genetic control of positional information and its interpretation. Undoubtedly, the *bithorax* system in insects is the best model at present for trying to understand this question. It appears that a gradient activates additional genes along the body axis so that in each segment, as one progresses anterio-posteriorly, an additional gene is activated (29). This might be regarded as recording, by gene activation, the positional value of each segment though we cannot exclude the possibility that positional value is encoded by some other means, and that the *bithorax* genes represent the first phase of interpretation. Even if we accept that the positional value of each segment is represented by particular combinations of the *bithorax* genes, we nonetheless have no idea as to how this leads to different spatial patterns. For example, mutations in the homeotic gene *aristapaedia* lead to antennal structures being replaced by leg ones (42). How can we account for this? Does the *aristapaedia* gene activate a whole set of other genes, and does this lead to the positional values in the antennal structure being interpreted differently and to the formation of antenna rather than leg? Or does the *aristapaedia* gene code for RNA molecules or proteins which modify the interpretation of leg so that it forms an antenna? We need to know what processes and molecules are held in common in interpreting leg and antennal structures. Are mutations that affect different structures in similar ways acting on the positional field? One such example may be *decapentaplegic* which seems to be involved in the elaboration of distal positional values in a number of different discs (49). The current investigation of the molecular basis of the action of the homeotic genes will have very important implications for pattern formation.

Acknowledgments

We are indebted to the SERC for financial support, to Dr. C. Tickle for her comments and to Miss M. Maloney for preparing the manuscript.

References

1. Archer, C., Rooney, P., Wolpert, L.: The early growth and morphogenesis of limb cartilage. In: *Limb Development and Regeneration*. Fallon, J.F. and Caplan, A.I. (eds.). pp. 267–278. New York: Alan R. Liss (1982).

2. Babloyantz, A.: Self-organization phenomena resulting from cell-cell contact. *J. Theoret. Biol.* **68**: 551–562 (1979).

3. Bryant, P.J.: Pattern formation, growth control and cell interactions in *Drosophila* imaginal discs. In: *Determinants of Spatial Organization*. Subtelny, S. and Konigsberg, J. (eds.). pp. 295–316. New York: Academic Press (1979).

4. Bryant, S.V., French, V., Bryant, P.J.: Distal regeneration and symmetry. *Science* **212**: 993–1002 (1982).

5. Child, C.M.: *Patterns and problems of development*. Chicago: University of Chicago Press (1941).

6. Cooke, J.: The emergence and regulation of spatial organization in early animal development. *Annu. Rev. Biophys. Bioeng.* **4**: 185–217 (1975).

7. Cooke, J.: Morphallaxis and early vertebrate development, this volume.

8. Cooke, J., Zeeman, E.C.: A clock and wavefront model for the control of the numbers of repeated structures during animal development. *J. Theoret. Biol.* **58**: 455–476 (1976).

9. Crick, F.H.C.: Diffusion in embryogenesis. *Nature* **225**: 420–422 (1970).

10. Crick, F.H.C.: Thinking about the brain. *Sci. Amer.* **241**: 181–188 (1979).

11. Driesch, H.: *Die organischen Regulationen*. Leipzig (1901).

12. Fallon, J.F., Crosby, G.M.: Polarizing zone activity in limb buds of amniotes. In: *Vertebrate Limb and Somite Morphogenesis*. Ede, D.A., Hinchliffe, J.R., Balls, M. (eds.). pp. 55–71. Cambridge, UK: Cambridge University Press (1977).

13. French, V., Bryant, P.J., Bryant, S. Pattern regulation in epimorphic fields. *Science* **193**: 969–990 (1976).

14. Garcia-Bellido, A., Ripoll, P., Morata, G.: Developmental compartmentalization in the dorsal mesothoracic disc of *Drosophila*. *Devel. Biol.* **48**: 132–147 (1976).

15. Gierer, A.: Biological features and physical concepts of pattern formation exemplified by hydra. *Curr. Top. Dev. Biol.* **11**: 17–60 (1977).

16. Graudine, M., Weintraub, H.: Propagation of DNA ase I—hypersensitive sites in absence of factors for induction: a possible mechanism for determination. *Cell.* **30**: 131–139 (1982).

17. Gustafson, T., Wolpert, L.: Cellular movement and contact in sea urchin morphogenesis. *Biol. Rev.* **42**: 442–498 (1967).

18. Holtfreter, J., Hamburger, V.: Amphibians. In: *Analysis of Development*. Willier, B.N., Weiss, P.A., Hamburger, V., (eds.). pp. 230–296. Philadelphia: Saunders (1955).

19. Honig, L.: Effects of ultraviolet light on the activity of the avian limb positional signalling region. *J. Embryol. Exp. Morph.* **71**: 223–232 (1982).

20. Hood, L., Huang, H.V., Dreyer, W.J.: The area-code hypothesis: the immune system provides clues to understanding the genetic and molecular basis of cell recognition during development. *J. Supramolec. Str.* **7**: 531–559 (1977).

21. Hörstadius, S.: *Experimental embryology of echinoderms*. Oxford: Clarendon (1973).

22. Huxley, J.S., de Beer, C.R.: *The elements of experimental embryology*. Cambridge, UK: Cambridge University Press (1934).

23. Illmensee, K.: Nuclear and cytoplasmic transplantation in *Drosophila*. In: *Insect Development*. Lawrence, P.A., (ed.). pp. 76–96. Oxford: Blackwells (1976).

24. Iten, L.E., Murphy, D.J.: Pattern regulation in the embryonic chick limb: supernumerary limb formation with anterior (non-ZPA) limb bud tissue. *Devel. Biol.* **75**: 373–385 (1980).

25. Kauffman, S.A., Shymko, R.M., Trabert, K.: Control of sequential compartment formation in *Drosophila*. *Science* **199**: 259–276 (1978).

26. Kimble, J.E.: Strategies for control of pattern formation in *Caenorhabditis elegans*. *Phil. Trans. Roy. Soc.* **295**: 539–551 (1981).

27. Lacalli, T.C.: Dissipative structures and morphogenetic pattern in unicellular algae. *Phil. Trans. Roy. Soc. B.* **294**: 547–585 (1981).

28. Lawrence, P.A.: Polarity and patterns in the postembryonic development of insects. *Adv. Insect. Physiol.* **7**: 197–266 (1970).

29. Lawrence, P.C.: The cellular basis of segmentation in insects. *Cell* **26**: 3–10 (1981).

30. Layton, W.M.: Random determination of a developmental process. Reversal of normal symmetry in the mouse. *J. Hered.* **67**: 336–338 (1976).

31. Lewis, J., Slack, J.M.W., Wolpert, L.: Thresholds in development. *J. Theoret. Biol.* **65**: 579–590 (1977).

32. Lewis, J.H., Wolpert, L.: The principle of non-equivalence in development. *J. Theoret. Biol.* **62**: 479–490 (1976).

33. Meinhardt, H.: Space-dependent cell determination under the control of a morphogen gradient. *J. Theoret. Biol.* **74**: 307–321 (1978).

34. Meinhardt, H., Gierer, A.: Applications of a theory of biological pattern formation based on lateral inhibition. *J. Cell Sci.* **15**: 321–346 (1974).

35. Morgan, M.J., Corballis, M.C.: On the biological basis of human laterality: II. The mechanism of inheritance. *Behav. Brain Sci.* **2**: 270–277 (1978).

36. Murray, J.D.: On pattern formation for lepidopteran wing patterns and mammalian coat markings. *Phil. Trans. Roy. Soc. B.* **295**: 473–496 (1981).

37. Murray, J.D.: Parameter space for Turing instability in reaction diffusion mechanism: a comparison of models. *J. Theoret. Biol.* **98**: 143–164 (1982).

38. Nardi, J.B., Kafatos, F.C.: Polarity and gradients in lepidopteran wing epidermis. II. The differential adhesiveness model: gradients of a non-diffusible cell surface parameter. *J. Embryol. Exp. Morph.* **36**: 489–512 (1976).

39. Newman, S.A., Frisch, H.L.: Dynamics of skeletal pattern formation in developing chick limb. *Science* **205**: 662–668 (1979).

40. Nübler-Jung, K.: Pattern stability in the insect segment. I. Pattern reconstitution by intercalary regeneration and cell sorting in *Dysdercus intermedius. Wilhelm Roux' Arch. W.* **183**: 17–40 (1977).

41. Odell, G.M., Oster, G., Alberch, P., Burnside, B.: The mechanical basis of morphogenesis. I. Epithelial folding and invagination. *Dev. Biol.* **85**: 446 (1981).

42. Pasteels, J.: Centre organisateur et potential morphogénétique chez les batraciens. *Bull. Soc. Zool. France.* **76**: 231–270 (1951).

43. Postlethwait, J.H., Schneiderman, H.A.: Developmental genetics of *Drosophila* imaginal discs. *Ann. Rev. Gent.* **7**: 381–433 (1974).

44. Rawles, M.E.: Origin of melanophores and their role in development of colour patterns in vertebrates. *Physiol. Rev.* **28**: 383–408 (1948).

45. Rose, S.M.: Cellular interaction during differentiation. *Biol. Rev.* **32**: 351–387 (1957).

46. Schaller, H.C., Bodenmuller, H.: Isolation and amino acid sequence of a morphogenetic peptide from hydra. *Proc. Nat. Acad. Sci.* **78**: 7000–7004 (1981).

47. Smith, D.J., Rubel, E.W.: Organization and development of brain stem. Auditory nuclei of the chicken: dendritic gradients in nucleus laminaris. *J. Comp. Neurol.* **186**: 213–240 (1979).

48. Smith, J.C., Tickle, C., Wolpert, L.: Attenuation of positional signalling in the chick limb by high doses of α-radiation. *Nature, Lond.* **272**: 612–613 (1978).

49. Spencer, F.A., Hoffman, F.M., Gelbart, W.M.: Decapentaplegic: a gene complex affecting morphogenesis in *Drosophila melanogaster. Cell* **28**: 451–461 (1982).

50. Stein, W.D., Wolpert, L.: A model for chick limb development using both a prepattern and positional information. *In preparation.*

51. Summerbell, D.: A descriptive study of the rate of elongation and differentiation of the skeleton of the developing chick wing. *J. Embryol. Exp. Morph.* **35**: 241–260 (1976).

52. Summerbell, D., Lewis, J.H., Wolpert, L.: Positional information in chick limb morphogenesis. *Nature* **244**: 492–496 (1973).

53. Tickle, C.: The polarizing region in limb development. In: *Development in Mammals. 4.* Johnson, M.H., (ed.). pp. 101–136. Amsterdam: Elsevier-North Holland (1980).

54. Tickle, C.: The number of polarizing region cells to specify additional digits in the chick wing. *Nature* **289**: 295–298 (1981).

55. Tickle, C., Summerbell, D., Wolpert, L.: Positional signalling and specification of digits in chick limb morphogenesis. *Nature, Lond.* **254**: 199–202 (1975).

56. Tickle, C., Alberts, B., Wolpert, L., Lee, J.: Local application of retinoic acid to the limb bud mimics the action of the polarizing region. *Nature, Lond.* **296**: 564–565 (1982).

57. Trisler, G.D., Schneider, M.D., Nirenberg, M.: A topographic gradient of molecules can be used to identify neuron position. *Proc. Nat. Acad. Sci.* **78**: 2145–2149 (1981).

58. Turing, A.M.: The chemical basis of morphogenesis. *Phil. Trans. Roy. Soc. B.* **641**: 37–72 (1952).

59. Warner, A.E., Lawrence, P.A.: Permeability of gap junctions at the segmental border in insect epidermis. *Cell* **28**: 243–252 (1982).

60. Webster, G., Wolpert, L.: Studies on pattern regulation in hydra. I. Regional differences in time required for hypostome determination. *J. Embryol. Exp. Morph.* **16**: 91–104 (1966).

61. Whittaker, J.R.: Cytoplasmic determinants of tissue differentiation in the ascidian egg. In: Determinants of Spatial Organization. Subtelny, S., Konigsberg, I.R., (eds.). pp. 29–52. New York: Academic Press (1979).

62. Wilby, O.K., Ede, D.A.: A model generating the pattern of cartilage skeletal elements in the embryonic chick limb. *J. Theoret. Biol.* **52**: 199–217 (1975).

63. Wilcox, M., Mitchison, C.J., Smith, R.J.: Pattern formation in the blue-green alga *Anabaena. J. Cell Sci.* **12**: 707–723 (1973).

64. Wolpert, L.: Positional information and the spatial pattern of cellular differentiation. *J. Theoret. Biol.* **25**: 1–47 (1969).

65. Wolpert, L.: Positional information and pattern formation. *Curr. Top. Devel. Biol.* **6**: 183–224 (1971).

66. Wolpert, L.: Gap junctions: channels for communication in development. In: *Intercellular junctions and synapses in development.* Feldman, J.D., Gilula, N.B., Pitts, J.D., (eds.). pp. 83–96. Chapman Hall (1978).

67. Wolpert, L.: Pattern formation in biological development. *Sci. Am.* **239**: 154–164 (1978).

68. Wolpert, L.: Pattern formation in limb morphogenesis. In: *Progress in Developmental Biology*, Sauer, H.W., (ed.). pp. 141–152. Stuttgart: Fisher (1981).

69. Wolpert, L.: Positional information and pattern formation. *Phil. Trans. Roy. Soc. B.* **295**: 441–450 (1981).

70. Wolpert, L.: Cellular basis of skeletal growth during development. *Brit. Med. Bull.* **37**: 215–220 (1981).

71. Wolpert, L., Hornbruch, A.: Positional signalling along the antero-posterior axis of the chick wing. The effect of multiple polarising region grafts. *J. Embryol. Exp. Morph.* **63**: 145–159 (1981).

72. Wolpert, L., Hornbruch, A., Clarke, M.R.B.: Positional information and positional signalling along hydra. *Amer. Zool.* **14**: 647–663 (1974).

CHAPTER 2

The Involvement of Transcellular Ion Currents and Electric Fields in Pattern Formation

Richard Nuccitelli

MOST ORGANISMS BEGIN development as rather patternless, spherical eggs. Many plant spores and eggs have no preformed developmental axis at this stage and most animal eggs exhibit only the animal-vegetal axis. The process of pattern formation is initiated as this spherical egg acquires polarity and establishes its main morphological axes. Often this axis establishment is reflected in a patterned redistribution of cytoplasmic and surface components such as mitochondria, chloroplasts, yolk platelets, and plasma membrane glycoproteins. For example, in the spore of the horsetail, *Equisetum*, chloroplasts move to the future rhizoidal end before germination; in the amphibian egg, yolk and pigment move to opposite ends of the animal-vegetal axis during oogenesis; and in the egg of the brown alga, *Pelvetia*, Ca^{2+} channels and pumps in the plasma membrane become concentrated at opposite ends of the rhizoid-thallus axis after fertilization. What mechanisms are involved in this acquisition of cell polarity?

There is much evidence that the cell's plasma membrane and cortex are actively involved in this process of axis establishment. Perhaps the best example of how an imposed change in the cell cortex can radically affect subsequent pattern development comes from the classical studies of Vance Tartar (40) on the unicellular ciliate, *Stentor*. The pattern of longitudinal surface stripes on *Stentor* is such that the stripe width increases with each stripe around the cell periphery so that there is only one region of discontinuity where a broad stripe region meets a narrow stripe region. The oral primordium always develops at that discontinuity. When a second such discontinuity is implanted, a second oral primordium is formed and a second oral opening appears at the next regeneration. Thus, the cell cortex can have a profound effect on cellular pattern formation. One might argue that this is a special case since *Stentor* exhibits cortical inheritance, but there are too many examples of this in other systems to dismiss it on these grounds. A second example of plasma membrane involvement in pattern formation is found in the Plant Kingdom where nearly every phylum has at least one cell type, usually a spore or egg, which is completely apolar. These cells optimize the position of their germination axis by detecting the pertinent environmental vectors such as light. There are many examples in which these light receptors lie in the plasma membrane, so the membrane must in turn act as a transducer to communicate information, such as the position of minimum light absorption, to the cellular machinery responsible for controlling the localized vesicle secretion necessary for germination. Here again the plasma membrane is directly involved in the establishment of pattern or polarity.

If the cell's plasma membrane is actively influencing cell polarity, one should be able to detect membrane asymmetries associated with the axis of polarity. This could be accomplished using many different approaches. For example, one might morphologically investigate the protein distribution in the membrane by using freeze-fracture electron microscopy or fluorescently labeled lectins. Another approach would be to map the location of those proteins involved in ion transport by electrophysiological methods. Since one of the main functions of the plasma membrane is the maintenance of ion concentration gradients of Na^+, K^+, Ca^{2+} and H^+ across itself by using ion pumps and channels, ion transport sites should be plentiful and might well be distributed asymmetrically. This chapter concentrates on studies using this last technique of electrophysiological detection of membrane asymmetry because much new information has been obtained with this technique during the past ten years.

The main concept behind this work lies in the spatial separation of ion leaks and pumps in the plasma membrane. All cells use either the Na^+–K^+ ATPase or H^+ ATPase to generate an ion concentration gradient across the plasma membrane. Selectively permeable ion channels allow these ions to leak across the membrane, generating a membrane potential. If these membrane proteins are uniformly distributed within the plasma membrane as diagrammed on the left in Fig. 1, many highly localized current loops would be generated. This current pattern provides no axis of asymmetry and can contribute little to the cell's overall polarity. However, by simply separating these same channel types in space, or equivalently by separating channels and electrogenic pumps, one could generate a transcellular ion flow, as shown in Fig. 1 on the right. This pattern of ion channels *will* result in an asymmetrical current flow through the cell which might indeed influence the cell's polarity.

This review summarizes the evidence for transcellular ion currents in developing systems. In most cases, the spatial current pattern coincides with the major growth axis of the cell, and in a few cases there is evidence that this current can actively influence the polarization process. These ion currents can have the most direct effect on the cell by either changing local ion concentrations or by generating voltage gradients both inside and outside of the cell's plasma membrane. Specific examples of each of these will be discussed, and include the germinating seaweed egg, the developing insect follicle and the growing neuron. Finally, the studies of transembryonic currents in developing multicellular embryos will be summarized.

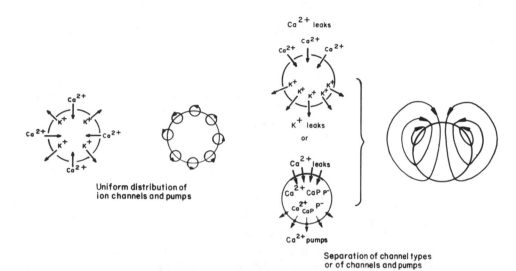

FIGURE 1. Possible distribution of ion channels and pumps in a cell's plasma membrane along with the resulting current pattern. Left: A uniform distribution of Ca^{2+} and K^+ channels will result in many highly localized current loops as shown. Right: Separation of the Ca^{2+} and K^+ channels (above) or Ca^{2+} channels and pumps (below) will result in the transcellular ion current pattern shown on the right.

The Detection of Transcellular Ion Currents

Steady-state ion fluxes have been measured in cells for decades by using radioactive ion tracers. When these measured flux values are combined with the known cell surface area, an estimate of the possible transcellular ion currents can be made. Such estimates suggest that possible current densities of 0.1 to 10 μA/cm^2 might pass through single cells. How can such small currents be detected?

As diagrammed in Fig. 2, such ion currents must flow through the extracellular medium of fixed resistivity and in so doing will generate a small voltage gradient. By measuring this voltage gradient, one could directly calculate the current density generating it. This can be seen clearly from Ohm's Law:

$$I = E/\rho = -(1/\rho)\, \vec{\nabla}\, V$$
$$= (1/\rho)\, \frac{\partial V}{\partial r}\, \hat{a}_r$$
$$\approx -(1/\rho)\, \frac{\triangle V}{\triangle r}\, \hat{a}_r$$

where I is current density, E is the electric field strength, ρ is the medium resistivity, ΔV is the voltage gradient measured over the small distance, Δr, and $\hat{a}_r$ is the unit vector in the radial direction. The smaller the Δr used, the better the approximation to the true current density. Therefore, the challenge is to measure extracellular voltage gradients generated by an ion flux of less than 1 μA/cm^2 over very small distances. The magnitude of this challenge can be appreciated when the calculation of the expected voltage gradient generated over a 20 μm displacement

FIGURE 2. Diagram of the transcellular current pattern in a single cell generating an extracellular voltage gradient. Two standard 3 *M* KCl-filled microelectrodes are positioned outside the cell attempting to measure this voltage gradient over the small displacement necessary to calculate the current density midway between them.

along a radial vector beginning 25 μm outside the cell's plasma membrane is made, resulting in a value of 6 nV in seawater and 30 nV in serum. Standard $3M$ KCl-filled microelectrodes have resistances on the order of 10^6 ohms and a resolution limited by their noise values of 10^{-5} volts. They are therefore not sensitive enough to detect these steady voltages of 10^{-8} volts. Lionel Jaffe and this author overcame this difficulty nine years ago by developing a much more sensitive electrode system called the vibrating probe (15). This was accomplished by lowering the electrode's resistance 1000-fold (by filling the glass microelectrode with metal and plating a platinum sphere at its tip as shown in Fig. 3) and by signal averaging and filtering using a phase-sensitive lock-in amplifier. The probe impedance is on the order of 10^3 ohms and the system's sensitivity is 10^{-8} volts. These changes resulted in a 100- to 1000-fold improvement in resolution over the standard microelectrode. This is comparable to the increase in resolution obtained when one goes from the light microscope to the electron microscope, so that now we can detect the nanovolt gradients generated by the steady ion fluxes flowing through single cells. Moreover, the spatial resolution of the vibrating probe technique is about 10 μm so that the spatial distribution of these membrane ion fluxes can now be revealed. A photograph of the vibrating probe is shown in Fig. 4. This instrument is now commercially available (Vibrating Probe Co., Davis, CA) and is being used in fifteen laboratories around the world.

Transcellular Ion Currents Are Found in Polarizing Cells

Investigation of developing cells in both the Plant and Animal Kingdoms with the vibrating probe has revealed transcellular ion currents that are closely correlated with each cell's axis of polarity. Figure 5 summarizes most of the single cell studies to date and each will be briefly described.

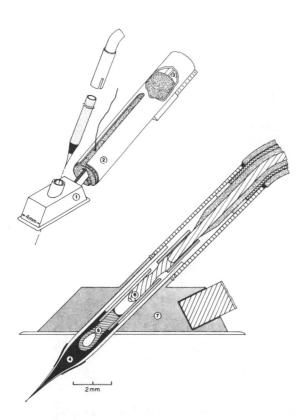

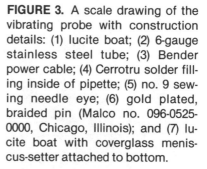

FIGURE 3. A scale drawing of the vibrating probe with construction details: (1) lucite boat; (2) 6-gauge stainless steel tube; (3) Bender power cable; (4) Cerrotru solder filling inside of pipette; (5) no. 9 sewing needle eye; (6) gold plated, braided pin (Malco no. 096-0525-0000, Chicago, Illinois); and (7) lucite boat with coverglass meniscus-setter attached to bottom.

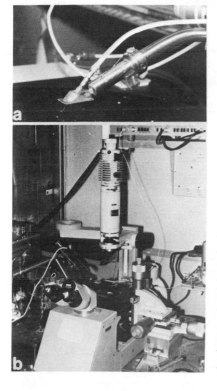

FIGURE 4. The vibrating probe. (a) Photograph of vibrating probe assembly. A piezoelectric element is housed in the stainless steel tube and attaches to the lucite probe holder to vibrate the probe between the desired two points in the extracellular medium. The coaxial signal cable shown carries the signal from the probe to a lock-in amplifier. (b) Photograph of the entire vibrating probe system including the inverted microscope, the Line Tool X-Y-Z micropositioner and the lock-in amplifier used to measure the probe signal.

PLANT KINGDOM

Water Mold   Brown Alga   Flowering Plants

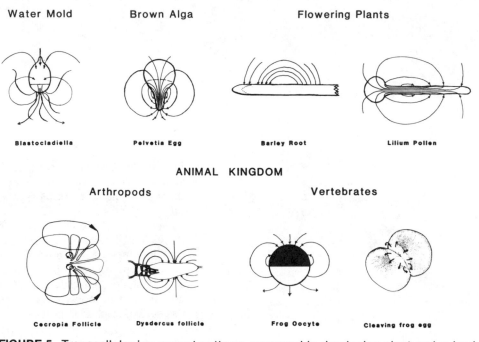

Blastocladiella   Pelvetia Egg   Barley Root   Lilium Pollen

ANIMAL KINGDOM

Arthropods   Vertebrates

Cecropia Follicle   Dysdercus follicle   Frog Oocyte   Cleaving frog egg

FIGURE 5. Transcellular ion current patterns measured in developing plant and animal cells.

Plant Cells

1. WATER MOLD

One of the lowest plant cells to be studied is the water mold, *Blastocladiella emersonii* (39). In growing cells, positive current on the order of 1 $\mu A/cm^2$ enters the rhizoid and leaves from the thallus. There is some evidence that H^+ carries part of this current. Sporulation is associated with a reversal of this current pattern, and positive current begins entering the thallus region from which the spores will be released. This cell, therefore, drives a transcellular current through itself along its axis of polarity, and reverses its direction according to the stage of the life cycle.

2. PELVETIA EGG

The egg of the brown alga, *Pelvetia fastigiata*, also drives a current along its axis of polarity (27) and evidence will be presented next that this current, which is carried by Ca^{2+} influx and Cl^- efflux, is a controlling factor in axis determination.

3. BARLEY ROOTS

Growing roots and root hairs of barley seedlings, *Hordeum vulgare L.*, drive steady currents through themselves. Current of about 2 $\mu A/cm^2$ in magnitude enters both the main elongation zone of the root and the growing tips of elongating root hairs (43). Both the inward and outward currents appear to be carried largely by H^+, OH^-, or HCO_3.

4. EASTER LILY POLLEN GRAIN

The higher flowering plant, *Lilium longiflorum*, exhibits transcellular currents in its germinating pollen grain (42). Positive current of about 4 μA/cm^2 enters the ungerminated grain's prospective growth site and leaves its opposite end. After the grain germinates to form the pollen tube, current enters along most of the tube and leaves the grain as diagrammed in Fig. 5. This is another example of a tip-growing plant cell that drives an ion current into the region of vesicle secretion and along its axis of polarity.

Animal Cells

Transcellular ion currents have been measured in a wide range of animal cells as well. As in most of the plant cells mentioned, the current pattern is closely correlated with the axis of polarity.

1. SILK MOTH FOLLICLE

The nurse cell-oocyte complex of the silk moth, *Hyalophora cecropia*, drives an ion current into the nurse cell end and out of the oocyte end (44, 19). This current is important for the polarized transport of protein along the cytoplasmic bridge connecting the nurse cells with the oocyte, and there is strong evidence discussed in detail next that an intercellular voltage gradient that might be generated by this current drives the electrophoretic transport of protein from nurse cells to oocyte.

2. COTTON BUG OVARIOLE

The morphology of the ovariole of the cotton bug, *Dysdercus intermedius*, is quite different from the silk moth ovariole because nurse strands between the trophic syncytium and oocytes can measure several millimeters in length. The basic current pattern, however, is analogous to that found in *Cecropia* with current entering over the tropharium and leaving the small follicles (6). This current is also about an order of magnitude larger than that measured in *Cecropia*. In another telotrophic hemipteran, *Rhodnius prolixus*, an intercellular voltage gradient of up to 10 mV with the oocyte more positive has been measured (41). The magnitude of this gradient is very sensitive to juvenile hormone. Therefore, in these three insect systems, the transcellular current pattern is strongly correlated with the axis of polarity and may be directly involved in the polarized transport of protein between cells.

3. Frog oocyte

The immature oocyte of the frog, *Xenopus laevis*, drives 1 μA/cm^2 into the animal hemisphere and out of the vegetal hemisphere prior to complete maturation (35). Here, again, the current pattern is along the main axis of polarity and the inward current exhibits Ca^{2+} influx and Cl$^-$ efflux components. This current pattern is found in all oocytes; however, when progesterone or any of several other maturation-producing agents is applied, the current decreases to nearly zero. Thus the mature egg appears to be in a more quiescent state waiting for fertilization to reactivate it.

4. Cleaving frog egg

The mature, unfertilized *Xenopus* egg has no detectable transcellular current until fertilization triggers an inward current pulse which enters the egg at the site of the sperm-egg fusion and spreads over the egg in a ring-shaped wave during about 3 min after fertilization. The fertilized egg then becomes quiescent until first cleavage. About 10 min after first cleavage has begun, new, unpigmented membrane begins to appear in the cleavage furrow. At this same time, outward current is measured near this unpigmented membrane (22). From this point on through at least the 32-cell stage, each blastomere drives a current through itself, into the old apical membrane and out of the newly inserted basal membrane. These blastomeres form an epithelium with the blastocoel on one side and the external medium on the other side, and the outward current region forms the basolateral margins of this epithelial layer. Therefore, a transcellular current is associated with the apical-basal polarity of these epithelial cells. Lionel Jaffe has proposed a model of epithelial cell polarity control by transcellular ion currents (13).

These investigations of the steady ion current patterns mainly around single cells have led to two generalizations:

1. Ion channels are generally not uniformly distributed in the cell's plasma membrane, but are spatially separated resulting in transcellular ion currents.
2. These ion current patterns are often closely correlated with the axis of cell polarity in both plant and animal cells.

Based on these steady-ion current measurements made in a wide variety of cell types from both lower and higher plants and animals, it is clear that transcellular ion currents are *signals* of cell polarity. However, the far more interesting question is whether these currents can affect, influence or control cell polarity. Does the plasma membrane directly influence cell function by separating ion channel types and driving these ion currents through cells?

Do These Transcellular Ion Currents Play a Causal Role in the Development of Cell Polarity?

Transcellular ion currents have the most direct effect on the cell by either changing local ion concentrations or by generating voltage gradients. Furthermore, these effects could act on the inside or outside of the cell's plasma membrane. The remainder of this section on single cells will concentrate on a few well-studied cases which indicate that the transcellular currents can affect cell polarity (13, 14).

Intracellular Ion Concentration Gradients

1. TIP-GROWING PLANT CELLS

These cells provide the strongest evidence that steady transcellular currents can generate ion concentration gradients which, in turn, can influence the cell's polarity. The most extensively studied system is the germinating egg of the brown alga, *Pelvetia*, and this current pattern is quite similar to that found in the germinating pollen grain of the higher plant, the Easter Lily.

The *Pelvetia* egg has no predetermined axis of polarity and can germinate a rhizoid at any point on its surface. This process involves the localized secretion of wall-softening enzymes and the increase in turgor pressure to 5 atmospheres which is accomplished by pumping in K^+ and Cl^-. The cell then bulges (germinates) at the weakest region where wall softening has occurred, the originally apolar zygote becomes differentiated into two different regions, and then into cells. The establishment of this axis of secretion can be influenced by a variety of environmental vectors including light, temperature, and pH (11). The response to unilateral light is to germinate on the dark side. This would be useful in nature since the rhizoid outgrowth will form the holdfast for the plant and should form opposite the sun. One can use the light response to orient the germination site while studying the transcellular ion current pattern associated with it using the vibrating probe, as shown in Fig. 6A (27). Currents are detected around this egg as early as 30 min after fertilization and tend to enter at the dark hemisphere. The early spatial current pattern is unstable and shifts position, often with more than one inward current region. However, current enters mainly on the side where germination will occur and is usually largest at the prospective cortical clearing region where the rhizoid forms. The current pattern observed during the two-hour period prior to germination is more stable and is shown in Fig. 6B. The site of inward current always predicts the germination site, even when the axis is reversed by light direction reversal.

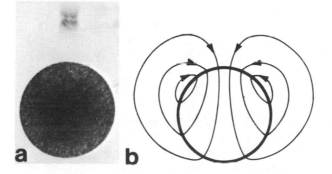

FIGURE 6. Current measurements around the *Pelvetia* egg. (a) Photomicrograph of the vibrating probe near a 3-hour-old *Pelvetia* egg. The egg's diameter is 100 μm. (b) The electrical current pattern generated by the germinating *Pelvetia* egg just before germination that will occur at the top of the figure.

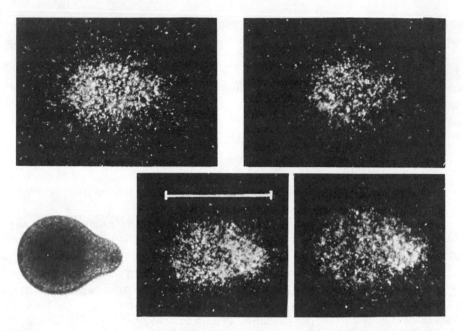

FIGURE 7. Low-temperature autoradiographs of 18-hour-old, germinated *Pelvetia* eggs. Made from deep-frozen median sections of an egg similar to the one shown on the lower left; top row: Controls grown in seawater labeled with $^{36}Cl^-$; bottom row: grown in seawater labeled with $^{45}Ca^{2+}$. Bar in center autoradiograph: 100 μm. Courtesy of John Gilkey and Lionel Jaffe.

To understand how this current might be affecting the egg, one must first know what ions are carrying the current. Therefore, these current measurements were next carried out in artificial sea waters in which various ion substitutions were made. These studies indicated that Ca^{2+} influx and Cl^- efflux carried the inward current with K^+ efflux as the probable outward current carrier. Robinson and Jaffe (37) directly measured the $^{45}Ca^{2+}$ tracer flux through the egg by separating the rhizoid and thallus ends by fitting eggs tightly into holes in a nickel screen, and measured 2 picoamps of Ca^{2+} current. This means that about 10% of the total transcellular current is carried by Ca^{2+} influx at this stage. While only a small fraction of the current is due to Ca^{2+} influx, the subsequent intracellular free ion concentration change is certainly much larger for Ca^{2+} than for the major current carrier, Cl^-. This is because the free Ca^{2+} concentration in all cells is extremely low (about $10^{-7}M$) compared to Cl^-_i, so that even a small Ca^{2+} influx will result in a very large concentration gradient. A first attempt to look for such a gradient was made by Gilkey while he was in Jaffe's lab using low-temperature autoradiography. Figure 7 shows the results of this study in germinated embryos indicating higher Ca^{2+} concentrations at the rhizoid tip than at the thallus end. This technique does not distinguish between free and bound Ca^{2+}, but since the overall Ca^{2+} concentration is higher at the rhizoid end, it is likely that the free concentration is higher there as well. It must also be noted that efforts to detect a Ca^{2+} gradient in the ungerminated egg have so far been unsuccessful.

Up to this point, we have a plasma membrane-driven, partly Ca^{2+} current entering the prospective germination region to generate a Ca^{2+} gradient within the cytoplasm. If such a Ca^{2+} gradient were important for the establishment of an axis of polarity, manipulations of the gradient should affect cell polarization. The best test for causality is to impose a Ca^{2+} gradient on the ungerminated embryo. This critical experiment was done by Robinson and Cone (36) using a gradient of the calcium ionophore, A23187. They found that 60 to 80% of the eggs lying within one egg diameter of a glass fiber coated with ionophore formed a rhizoid on the hemisphere nearer the fiber. This is strong evidence that imposed Ca^{2+} gradients can orient that axis of polarity and supports the hypothesis that the natural transcellular Ca^{2+} current is an effector of cell polarity in the *Pelvetia* egg. Moreover, a wide variety of other tip-growing plant cells exhibit high Ca^{2+} levels at their growing tips (10, 33, 34) and some drive an ion current into that region (42, 43).

The germinating pollen grain of *Lilium longiflorum* is the best example of a higher plant cell which exhibits many of the same membrane properties found in *Pelvetia*. The hydrated, ungerminated pollen grain also drives a current through itself, the entry region of which predicts the site of tube outgrowth very accurately. The elongating tube also shows a gradient of total Ca^{2+}, with Ca^{2+} higher in the tip region according to both low temperature autoradiography (10) and chlorotetracycline fluorescence (33). Therefore, it would appear that in lower and higher plants the plasma membrane is involved in the control of axis establishment by determining the site of Ca^{2+} influx and the transcellular current pattern.

2. ANIMAL CELL POLARITY

Animal cell polarity can also be influenced by Ca^{2+} gradients. Very recent work by Jeffery (21) indicates that the orange crescent in the egg of the tunicate, *Boltenia villosa*, forms on the high side of a calcium ionophore gradient formed with the glass fiber technique. This pigment and mitochondrial segregation normally marks the future dorsal side of the embryo and during normal development these organelles are further segregated into the muscle cells of the tadpole larva. However, when the egg is activated by the ionophore by placing it in the gradient, 82% of the activated eggs formed orange crescents and mitochondrial localizations with midpoints 45° or less from the point nearest the ionophore-coated fiber. Moreover, 10% of the eggs positioned between two coated fibers formed two orange crescents, each about half the normal size. This supports the hypothesis that there is no predetermined polarity of cytoplasmic localization in the ascidian egg, and suggests that cytoplasmic Ca^{2+} gradients can strongly influence this axis of polarity.

Intracellular Voltage Gradients

The second way that transcellular ion currents could directly affect cell function is through the intracellular voltage gradient which they might generate as they traverse the cytoplasm. In fact, there is a well-documented example of this very phenomenon in the insect oocyte-nurse cell syncytium which has been elegantly elucidated by Woodruff and Telfer (44, 45).

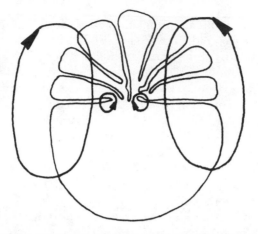

FIGURE 8. Diagram of the oocyte-nurse cell syncytium of the *Cecropia* moth. Three cytoplasmic bridges are evident and the transcellular ion current pattern is drawn in.

1. THE CECROPIA MOTH FOLLICLE

In many insects, the oocyte is connected to several nurse cells by cytoplasmic bridges which are formed by incomplete cytokinesis at cell division. A diagram of this oocyte-nurse cell complex or follicle is shown in Fig. 8. The main role of the nurse cells is to synthesize protein and RNA and to transport it into the larger oocyte. This transport occurs through the cytoplasmic bridges and is known to be unidirectional, that is, in a variety of insect types, RNA and protein are only observed to flow from the nurse cells into the oocyte and never in the opposite direction. In their study of this polarized intercellular transport, Woodruff and Telfer measured the membrane potential of both the oocyte and nurse cells and found a surprising 10 mV difference between them with the oocyte more positive than the nurse cells (44). This was surprising because these cells have conductive interiors and are joined by a continuous strand of cytoplasm. Such a conductor would be expected to have the same voltage throughout unless there were a steady current flowing through it. Such a current passing across the 35 μm wide intercellular bridges which have a resistance of 13 KΩ could generate a voltage difference between the two cell groups. They then investigated this possibility and found, using two independent techniques, that the follicle indeed drives a current through itself. The first method used was to increase the extracellular resistance by drawing the follicle into a tight-fitting capillary so that a larger extracellular voltage would be generated by the same transcellular current. The second method was to map the extracellular current pattern with the vibrating probe in collaboration with Jaffe (19). Both techniques detected current entering the nurse cell cap and leaving the oocyte's vegetal end, as diagrammed in Fig. 8. Up to 20 μA/cm^2 enters the anterior end of the nurse cells giving a total transfollicular current of about 100 nA. However, the measured intercellular voltage gradient implies a very large back current (order of 1000 nA) across the cytoplasmic bridge from oocyte to nurse cells in the opposite direction as the extracellular current pattern would suggest. One plausible explanation for this large cytoplasmic bridge back current that has been proposed

(16) is that it is driven by the furrow membranes. This would naturally occur if there were an animal-vegetal polarity in each of the nurse cells and in the oocyte such that the inward current ion channels were located in the animal hemisphere and the outward current channels or pumps were located in the vegetal region of each cell. This would result in the juxtaposition of outward current nurse cell membrane with inward current oocyte membrane at the furrow to drive current across the furrow from nurse cell to oocyte. This current would then leak back through the path of least resistance which is through the cytoplasmic bridge nearby as shown by the inner current loops in Fig. 8.

If this intercellular voltage gradient is involved in the polarized transport of protein, one should be able to demonstrate that the transport is dependent on the protein's net electrical charge. Woodruff and Telfer (45) showed this by microinjecting fluorescently labeled proteins with different pKs into either the nurse cells or oocyte and following their movement across the cytoplasmic bridge. Injected electronegative protein, serum globulin, only moves from nurse cell to oocyte, while electropositive lysozyme (pK 11.5) only moves from oocyte to nurse cell. Two nearly neutral proteins, haemoglobin and myoglobin, were able to move in both directions across the bridges. These observations suggest that the protein's charge is the main factor determining the direction of intercellular transport. Finally, this conclusion was even more convincingly supported by their demonstration that by simply reversing the charge on lysozyme by methylcarboxylation they could reverse its transport polarity. As shown in Fig. 9, essentially the same protein which now has a net negative charge will only move from nurse cell to oocyte. These data provide strong evidence in support of an intercellular electrophoresis transport mechanism in the developing *Cecropia* follicle, that is, a polarized transport of cytoplasmic components which is driven by transcellular ion currents.

2. Elongating Fungal Hyphae

Twenty years ago, Slayman and Slayman (38) reported a 100 mV voltage gradient along *Neurospora* hyphae with the elongating tip more positive than the center of the hypha. Since the membrane potential was smaller at the relatively fine cellular tips where electrode penetration injury is more likely, it is difficult to know how much of the measured potential difference is natural and how much is due to a lower membrane potential measurement at the tip where the injury leak could be greater. Nevertheless, recent unpublished vibrating probe measurements by Kropf in Caldwell's lab in Denver, indicate that current enters the elongating tip of another fungus, *Achlya bisexualis* (*unpublished results*). Perhaps this transcellular current is involved in the generation of the intracellular voltage gradient which could, in turn, electrophorese negatively charged secretory vesicles to the elongating tip where they are secreted.

The intracellular voltage gradient, which a given transcellular current generates, depends very much on the ion species carrying the current. For ions which exhibit little tendency to bind to fixed charges in the cytoplasm (such as K^+ and Cl^-), the voltage generated can be calculated from Ohm's law. However, ions which bind to fixed charges (such as Ca^{2+} and H^+), will create fixed charge gradients which can generate larger voltage gradients than expected by Ohm's law (20). This

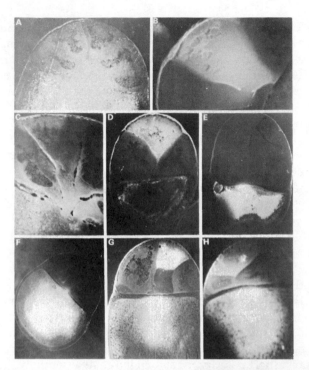

FIGURE 9. Fluorescence photomicrographs of follicles injected with proteins labeled with fluorescein isothiocyanate (FITC) by the method of Nairn. Micro-injection was carried out hydrostatically. The follicles were incubated in blood from the animal diluted 1:2 with dissecting solution. In dilutions as great as 1:10, blood protein concentration is still sufficient to maintain membrane permeability and steady-state potentials identical to those which occur in whole blood. Following injection and 1–2 hr incubation, follicles were fixed either by freeze substitution (a, c) or in Carnoy's acid-alcohol (b, d-h), embedded in paraffin, sectioned, and examined with a fluorescence microscope. Photographs were taken with Kodak Tri-X pan film (ASA 400). (a, c) Fluorescent lysozyme (FLy) micro-injected into oocytes; (b, d) FLy micro-injected into nurse cells. Note in (b) that protein has entered a co-bridged nurse cell (see Fig. 1), but failed to cross into the oocyte. (e, f) Methylcarboxylated FLy (MCFLy) micro-injected into oocytes gave no detectable fluorescence in the nurse cells. (g, h) MCFLy micro-injected into nurse cells. In all cases shown here, the width of the nurse cell cap adjacent to oocyte was about 450 µm. Courtesy of R.I. Woodruff and W.H. Telfer, reprinted by permission from *Nature* 286:84 Copyright (c) 1980 Macmillan Journals Limited.

is due to the Donnon Potential established by the displacement of K^+ to the end opposite the Ca^{2+} leak. The high K^+ at the low Ca^{2+} end will set up a balancing potential difference across the cytoplasm since it must be at a uniform electrochemical potential. Therefore, the Ca^{2+} leak end of the cytoplasm will become electropositive with respect to the opposite end, and this Donnon Potential can contribute substantially to the intracellular voltage gradient. For example, in the 100 µm diameter *Pelvetia* egg, 1 µA/cm² traversing a 200 Ω-cm cytoplasmic resistivity would generate an ohmic voltage difference of 2 µV across the cytoplasm, but the field due to the fixed charge gradient could be as large as 2 mV (20).

Extracellular Voltage Gradients

1. REDISTRIBUTION OF MEMBRANE PROTEINS BY WEAK EXTERNAL ELECTRIC FIELDS

Many of the externally protruding plasma membrane glycoproteins bear a high negative charge density. If the lateral mobility and charge heterogeneity among these membrane proteins are sufficient, a steady electric field should redistribute them within the plane of the plasma membrane by lateral electrophoresis. Jaffe introduced this idea and calculated that rather small extracellular electric fields could polarize the membrane protein distribution (12). For example, a population of 10 nm membrane particles would become one-tenth to one-half polarized by a steady voltage drop of only 0.8–4.0 mV across the cell in 3 hours.

This theoretical prediction was first confirmed by Poo and Robinson (32) using *Xenopus* muscle cells in culture. They showed that the Con A receptor distribution could be grossly polarized within 4 hours in a 400 mV/mm field (1.2 mV/cell), independent of cell metabolism. Since then, this lateral electrophoresis has been observed in a wide variety of cells including frog neurons (30), mouse macrophages (28), mouse fibroblasts (46), and sea urchin eggs (24), and the other membrane glycoproteins which have been redistributed by the imposed fields include acetylcholine receptors (29, 31), *Phaseolus vulgaris* lectin receptors (28) and ricin receptors (46) [See review by Poo (31).] Imposed fields also redistribute freeze-fracture membrane particles in *Micrasterias* (5). Surprisingly, the Con A and acetylcholine receptors and freeze-fracture particles move in the membrane toward the negative pole, opposite to the expected direction for negatively charged glycoproteins. Jaffe suggested that the explanation of this important anomaly lay in electro-osmosis. Since the negatively charged membrane proteins will have positive counterions in solution which are in turn surrounded by waters of hyrdation, the positive counterions pull the water which hydrodynamically drags the membrane proteins toward the negative pole. This concept has recently been tested by McLaughlin and Poo (25) who showed that the direction of Con A receptor migration in an imposed field could be reversed by changing the surface charge using a positively charged lipid (3,3' dioctadecylindocarbocyanine iodide) or by cleaving sialic acid residues with neuraminidase. Both of these treatments reduce the negative surface change, reduce the zeta potential and decrease the electro-osmotic flow parallel to the surface. Therefore, it is quite likely that electro-osmosis is responsible for moving the negatively charged glycoproteins to the negative pole.

2. EFFECTS OF WEAK ELECTRIC FIELDS ON CELL GROWTH

There is little double that weak electric fields on the order of 1 mV per cell diameter can redistribute membrane proteins. However, does this membrane polarization influence cell growth or behavior? The answer to this question is a definite yes. In a previous review, several plant and animal cell types exhibiting morphogenetic responses to weak external fields (0.1–6 mV/cell) were listed (16, Tables 6 and 7). Since then, four exciting reports of nerve galvanotropism have appeared (17, 8, 7, 30). All these reported enhanced growth toward the cathode and reduced

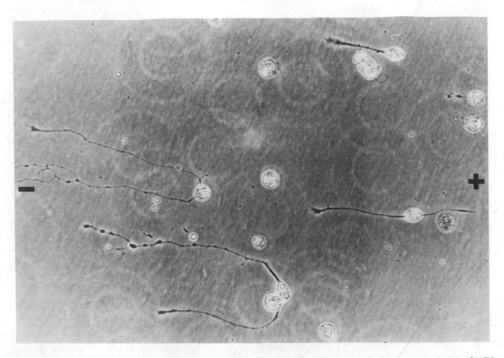

FIGURE 10. *Xenopus* neurons in culture in the presence of a steady electric field of 170 mV/mm. The poles of the field are indicated by +, −. The neuron at the bottom, left with a single neurite is 100 μm long. Courtesy of K.R. Robinson reprinted by permission from *J. Physiol.* 314:121 (1981).

growth towards the anode in small steady fields using a variety of neurons (chick dorsal root ganglia, frog neural tube and goldfish retina). Growing neurites were observed to bend toward the cathode, as shown in Fig. 10 with an extremely low threshold of 7 mV/mm (about 0.4 mV/cell) over a 16–20 hr period (8). The possible targets for such fields could be cytoplasmic, transmembrane, or perimembrane in nature, but the first two are less likely. The cytoplasmic voltage gradient across the interior of a 70 μm growth cone would be 10^{-5}–10^{-3} times the total gradient since most of the resistance is across the cell membrane (16), so one would expect 0.004–0.4 μV across the cytoplasm. Given a typical cytoplasmic electrophoretic mobility of 1 μm/sec per V/cm, cytoplasmic components would move at the most 0.2 μm/hr and would take 50 hours to cross a 10 μm growth cone!

The effect of these low fields on the transmembrane voltage or membrane potential is also very small. The typical neuron membrane potential is ⁻70 to ⁻90 mV, so the threshold field would perturb this by less than 1%. This is not likely to change ion transmembrane driving forces sufficiently to produce the reported polarization response 36% above controls (8, 16).

The third possibility, perimembrane or lateral electrophoresis of membrane proteins, is the most likely target of these small fields. In fact, Patel and Poo (30) showed that Con A receptors in *Xenopus* neurones accumulated on the cathodal side of the cell after 6 hours in a 500 mV/mm field (0.7 mV/cell). Furthermore, incubation in 40 μg/ml Con A during field application blocked Con A receptor migration and also blocked the galvanotropic response completely. This suggests that lateral membrane glycoprotein movements are important for galvanotropism, and supports the contention that the mechanism of action of these low fields is by lateral electrophoresis.

It should be evident from these examples that transcellular ion currents are closely associated with pattern formation in single cells and can exert their effect by establishing ion concentration gradients or voltage gradients. Does this also hold for multicellular organisms undergoing pattern formation?

Ion Currents and Voltages in Multicellular Embryos

In some respects, multicellular embryos can be quite similar to single cells undergoing pattern formation. For example, until gastrulation, most blastomeres are electrically coupled by gap junctions and can be considered as an electrical syncytium. The blastomere membrane at the blastula surface determines what passes into and out of the blastula much as the plasma membrane does for a single cell. Therefore this outer membrane could drive an ion current through the embryo at this stage. Our studies which were described earlier of isolated blastomeres from 32-cell stage *Xenopus* embryos support this notion since we find that each blastomere drives an apical-basal transcellular current through itself. Another example of the similarity to single cell pattern formation is found after gastrulation. Once the three germ layers have differentiated and the embryonic cells are no longer all electrically coupled, the outer epithelial layer might play a role in controlling transembryonic currents analogous to the plasma membrane's role in the single cell. In most embryos, the cells of this outer layer are electrically coupled and are actively pumping ions across the epithelial layer into the embryo. This in turn generates a positive voltage across the epithelium which will drive current out through epithelial regions of low resistance where cell junctions are leakier than elsewhere. These regions of junctional disruption would constitute a leak region analogous to an accumulation of ion channels in the plasma membrane of a single cell. However, exactly the opposite current flow will occur in the two systems. Single cells usually pump their most permeant ion (such as K^+) in, and as it leaks out through K^+ channels the inside becomes electronegative so that most cations (except perhaps K^+) will leak *into* the cell rather than outwards as in epithelial systems. Despite this difference in current direction, both single cells and epithelia appear to display their most active growth and pattern formation in regions of ion leaks. Thus, we find that in single cells current leaks into regions of plant cell elongation and into the animal hemisphere of immature *Xenopus* eggs, and, as we will soon see, in epithelial systems current leaks *out* along the main axis of symmetry or elongation in both chick and frog embryos.

Currents Through Chick Embryos

The electrical fields outside of developing chick embryos at the primitive streak stages 3 to 5 have been studied by Jaffe and Stern (18). They found that large current densities of the order of 100 μmA/cm^2 leave the whole streak and return else-

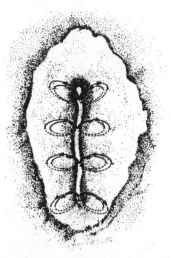

FIGURE 11. Current pattern through early chick embryo at primitive streak stage. Courtesy of Lionel Jaffe reprinted by permission from *Phil. Trans. R. Soc. Lond. B.* 295:427 Copyright (c) 1981 Cambridge University Press.

where through the epiblast, as shown in Fig. 11. This current is probably pumped into the intraembryonic space by the epiblast and then leaks out of the streak because it is a zone of junctional disruption. It will certainly generate a voltage gradient across some ingressing cells, and the magnitude of this gradient will depend on the resistivity of the embryonic spaces through which the current flows. The possibility exists that this current might exert a galvanotactic influence on cell migration.

Currents Through Frog Embryos

Similar large current densities leave the neural tube in *Xenopus* embryos, stages 10–20 (R. F. Stump and K. R. Robinson, *personal communication*). Again, these currents appear to be driven by a pumping epithelium. The potential difference across the skin of the adult frog has been extensively studied, and recently McCaig and Robinson (23) have shown that this transepidermal potential difference first begins to appear at neurulation. However, no gradients in skin potential within the embryo which might influence nerve or muscle growth have been detected thus far.

At early limb bud stage (stage 49), *Xenopus* embryos drive current into the gill or operculum and out of the hind limb bud (K. R. Robinson, *personal communication*). This is the only transembryonic current detected at this stage, and this current pattern is first detected at stage 43, a full day before limb bud appearance. Therefore, this outward current precedes and predicts the site of bud formation. Here the leak region of the outer epithelium is closely associated with the axis of limb development just as found in regenerating limbs as discussed next. In general, these embryonic ion currents are associated with the overall polarity of the embryo. By emerging from the prospective hind limb region, the current may stimulate growth in that region. There is as yet no evidence that the more detailed local pattern formation involved in the ensuing limb morphogenesis is influenced or predicted by the ion current pattern.

Ion Currents and Voltages Measured in Regenerating Systems

The regenerating systems that have been studied also drive currents through themselves using epithelial pumps. Long ago, Monroy (26) discovered that the distal surface of an amphibian limb is positive with respect to the proximal surface, and after amputation, this potential difference becomes larger. Recent measurements of current densities around regenerating limbs by Borgens, Vanable, and Jaffe (2) have confirmed Monroy's observation and indicate that a positive current of 10–100 $\mu A/cm^2$ leaves the cut surface of the regenerating limb stump. These large currents persist for up to 2 weeks and will produce a significant voltage gradient along the inner limb tissue which could exert a galvanotropic influence on the nerve cells in that region. In fact, Borgens has shown that when he supplements the natural currents in frog regenerating limbs with an imposed current, it greatly enhances the quantity of nerve which grows into the cathodally stimulated regenerates and results in greater limb regeneration. Since we know that nerve growth is enhanced toward the cathode in culture, it is quite likely that these imposed fields *in vivo* are also exerting a direct galvanotropic influence on the limb nerve. It is also worth noting here that a similar current pattern has been measured around regenerating human fingertips, and the regeneration is prevented if the skin is sutured shut over the finger tip, blocking the current flow (9).

These amphibian stump currents appear to help initiate limb regeneration. If they are inhibited using ion blockers such as amiloride, regeneration is greatly inhibited in naturally regenerating species. If these currents are enhanced by implanting batteries, greater regeneration results in nonregenerating species. Furthermore, the induction of supernumerary limb formation by chronic injury such as ligatures or insertion wounds may be due to the injury current that would leak out of these injured regions.

Another system in which imposed voltages enhance nerve growth is the regenerating spinal cord of the lamprey (4). When 10 μA is imposed across the completely severed spinal cord for 5 days, enhanced regeneration of reticulospinal neurons is observed. As in the frog limb, these imposed currents correspond in direction to the natural extracellular voltage gradient produced by injury-induced current which enters the cut face of the proximal cord segment (3). Therefore, this constitutes another example of *in vivo* nerve galvanotropism.

One unanswered question is whether these ion currents in developing and regenerating systems can generate a sufficient electric field inside the growing tissue to stimulate nerve galvanotropism. The magnitude of the field generated depends on the local resistance to the current flow. Assuming tissue resistances of 1000–5000 Ω-cm, typical limb current densities of 20 $\mu A/cm^2$ could generate a field of 2–10 mV/mm along the limb. This is very near the threshold field for nerve galvanotropism of 7 mV/mm (8). The most direct measurements of a tissue voltage gradient *in vivo* have recently been made beneath guinea pig skin during wound healing (1). When an incision is made through the glabrous epidermis of the guinea pig, a microampere flows through each millimeter of the cut's edge. These wound currents generate lateral, intraepidermal voltage gradients of about 100–200 mV/mm near the cut. Therefore, epithelial systems certainly can generate electric fields strong enough to exert a galvanotropic influence within the underlying tissue. These wound currents may also act on the epidermal cells to orient their motility and stimulate wound closure.

In these regenerating systems, the ion current appears to stimulate growth, and the evidence is quite strong that nerve galvanotropism is involved. Growth stimulation is an early step in pattern formation but there is not yet evidence that these ion currents predict the detailed patterns arising in the latter stages of limb morphogenesis. This area of investigation has just been thoroughly reviewed by Borgens (2).

Summary

This review has focused on the evidence that transcellular ion currents are commonly found in cells, and that the spatial pattern of these currents is intimately correlated with the cell's polarity. With the advent of the vibrating probe technique in 1974, it is now possible to detect the steady transcellular ion flow resulting from the separation of ion channel types and ion pumps in plant and animal cell plasma membranes. It appears that most cells do not have a uniform distribution of plasma membrane ion channels and pumps, but instead separate these transport sites, generating the transcellular currents. In most cases, these currents are signals or indicators of the cell's polarity, and in some cases there is evidence that these currents play a causal role in determining cell polarity.

The three ways that these transcellular ion currents can have the most direct affect on the cell are: (1) by changing local intracellular ion concentrations; (2) by generating cytoplasmic voltage gradients; (3) by generating extracellular voltage gradients. One or two examples of each of these effects is discussed in detail.

The tip-growing plant cells, the *Pelvetia* egg and lily pollen provide the strongest evidence that steady transcellular currents can generate ion concentration gradients which, in turn, influence the cell's polarity. Both of these cells drive a current carried partly by Ca^{2+} into their growing tips, and both exhibit intracellular Ca^{2+} concentration gradients. Furthermore, the *Pelvetia* egg's polarity can be manipulated by imposed Ca^{2+} ionophore gradients, suggesting that the Ca^{2+} gradient plays a causal role in polar axis establishment. The same Ca^{2+} ionophore gradient can also influence tunicate egg polarity, suggesting that animal cells may also use a Ca^{2+} gradient as part of their polarizing mechanism.

The second way that transcellular ion currents can directly affect cell function is by generating cytoplasmic voltage gradients. The best example of this is found in the insect oocyte-nurse cell syncytium that exhibits polarized transport from nurse cell to oocyte across cytoplasmic bridges. These cells drive a steady current through the bridge to generate a 10 mV difference between nurse cells and oocyte which electrophoreses negatively charged proteins and RNA into the oocyte. Protein transport across the cytoplasmic bridges is strictly a function of the protein's net charge, and the same protein can be made to reverse transport direction by simply reversing its net charge.

The third way that transcellular ion currents could directly affect cell function is by generating extracellular voltage gradients to laterally electrophorese membrane proteins. There is much evidence that such membrane protein redistributions can occur at very low field strengths on the order of 1 mV/cell and examples of galvanotropism at these low field strengths are discussed.

Developing and regenerating multicellular systems drive ion currents through themselves just as single cells do. These currents are usually concentrated in growth regions such as the area of the developing limb bud, and may act to stimulate growth in those regions. One likely target of the electric fields generated by these ion currents in tissues is the neuron which displays galvanotropism at physiological field strengths.

These ion currents are usually driven by epithelial layers in the embryo, and growth appears to be initiated by ion leaks where junctional disruption is occurring. Therefore, the role of ion currents in embryonic pattern formation appears to be one of growth stimulation and orientation at the appropriate positions as determined by the leaky regions in the epithelium. Our challenge is to uncover the mechanisms which determine the locations of these junctional disruptions.

The vibrating probe has opened up a new dimension of cellular organization that is providing many insights into the mechanisms of cellular and embryonic polarity control via transcellular ion currents and electric fields.

General References

JAFFE, L.F., NUCCITELLI, R.: Electrical controls of development. *Annu. Rev. Biophys. Bioeng.* 6:445–476 (1977).

JAFFE, L.F.: The role of ionic currents in establishing developmental pattern. *Phil. Trans. R. Soc. Lond. B.* 295:553–556 (1981).

BORGENS, R.B.: What is the role of naturally produced electric current in vertebrate regeneration and healing? *Int. Rev. Cytol.* 76:245–300 (1982).

References

1. Barker, A.T., Jaffe, L.F., Vanable, J.W., Jr.: The glabrous epidermis of cavies contains a powerful battery. *Am. J. Physiol.* **242**: R358–R366 (1982).

2. Borgens, R.B.: What is the role of naturally produced electric current in vertebrate regeneration and healing? *Int. Rev. Cytol.* **76**: 245–300 (1982).

3. Borgens, R.B., Jaffe, L.F., Cohen, M.J.: Large and persistent electrical currents enter the transected lamprey spinal cord. *Proc. Nat. Acad. Sci. (USA).* **77**: 1209–1213 (1980).

4. Borgens, R.B., Roederer, E., Cohen, M.J.: Enhanced spinal cord regeneration in lamprey by applied electric fields. *Science* **213**; 611–617 (1981).

5. Brower, D.L., Giddings, T.H.: The effects of applied electric fields on *Micrasterias*. II. The distribution of cytoplasmic and plasma membrane components. *J. Cell Sic.* **42**: 279–290 (1980).

6. Dittman, F., Ehni, R., Engles, W.: Bioelectric aspects of the Hemipteran Telcotrophic ovariole (*Dysdercus intermedius*). *Roux'Arch. Develop. Biol.* **190**: 221–225 (1981).

7. Freeman, J.A., Weiss, J.M., Snipes, G.J., Mayes, B., Norden, J.J.: Growth cones of goldfish retinal neurites generate DC currents and orient in an electric field. *Soc. Neurosci. Abstr.* **7**: 550 (1981).

8. Hinkle, L., McCaig, C.D., Robinson, K.R.: The direction of growth of differentiating neurones and myoblasts from frog embryos in an applied electric field. *J. Physiol.* **314**; 121–135 (1981).

9. Illingworth, C.M., Barker, A.T.: Measurement of electrical currents emerging during the regeneration of amputated finger tips in children. *Clin. Phys. Physiol. Meds.* **1**: 87–89 (1980).

10. Jaffe, L.A., Weisenseel, M.H., Jaffe, L.F.: Calcium accumulation within the growing tips of pollen tubes. *J. Cell Biol.* **67**: 488–492 (1975).

11. Jaffe, L.F.: Localization in the developing *Fucus* egg and the general role of localizing currents. *Advan. Morphogen.* **7**: 295–328 (1968).

12. Jaffe, L.F.: Electrophoresis along cell membranes. *Nature* **265**: 600–602 (1977).

13. Jaffe, L.F.: The role of ionic currents in establishing developmental pattern. *Phil. Trans. R. Soc. Lond. B.* **295**: 553–566 (1981).

14. Jaffe, L.F.: Developmental Currents, Voltages and Gradients. In: *Developmental Order: Its Origin and Regulation*. Subtelny, S., Green, P.B. (eds.). New York: Alan R. Liss (1982).

15. Jaffe, L.F., Nuccitelli, R.: An ultrasensitive vibrating probe for measuring steady extracellular currents. *J. Cell. Biol.* **63**: 614–628 (1974).

16. Jaffe, L.F., Nuccitelli, R.: Electrical controls of development. *Ann. Rev. Biophys. Bioeng.* **6**: 445–476 (1977).

17. Jaffe, L.F., Poo, M.-m.: Nuerites grow faster towards the cathode than the anode in a steady field. *J. Exp. Zool.* **209**: 115–128 (1979).

18. Jaffe, L.F., Stern, C.D.: Strong electrical currents leave the primitive streak of chick embryos. *Science* **206**: 569–571 (1979).

19. Jaffe, L.F., Woodruff, R.I.: Large electrical currents traverse developing *Cecropia* follicles. *Proc. Nat. Acad. Sci.* **76**: 1328–1332 (1979).

20. Jaffe, L.F., Robinson, K.R., Nuccitelli, R.: Local cation entry and self-electrophoresis as an intracellular localization mechanism. *Ann. N.Y. Acad. Sci.* **238**: 372–389 (1974).

21. Jeffery, W.R.: Calcium ionophore polarizes ooplasmic segregation in ascidian eggs. *Science* **216**: 545–547 (1982).

22. Kline, D., Robinson, K.R., Nuccitelli, R.: Ion currents and membrane domains in the cleaving *Xenopus* egg. *J. Cell Biol.* **97**: 1753–1761 (1983).

23. McCaig, C.D., Robinson, K.R.: The ontogeny of the trans-epidermal potential difference in frog embryos. *Develop. Biol.* **90**: 335–339 (1982).

24. McCaig, C.D., Robinson, K.R.: The distribution of lectin receptors on the plasma membrane of the fertilized sea urchin egg during first and second cleavage. *Develop. Biol.* **92**: 197–202 (1982).

25. McLaughlin. S., Poo, M.-m: The role of electro-osmosis in the electric field-induced movement of charged macromolecules on the surfaces of cells. *Biophy. J.* **34**: 85–93 (1981).

26. Monroy, A.: Ricerche sulle correnti elettriche dalla superficie del corpo di Tritoni adulti normali e durante la rigenerazione degli arti e della coda. *Pub. Stn. Zool. Napoli.* **18**: 265–281 (1941).

27. Nuccitelli, R.: Ooplasmic segregation and secretion in the *Pelvetia* egg is accompanied by membrane-generated electrical current. *Develop. Biol.* **62**: 13–33 (1978).

28. Orida, N.: Directional protrusive pseudopodial activity and motility in macrophages induced by extracellular electric fields. *Cell Motility* **2**: 243–255 (1982).

29. Orida, N., Poo, M.-m: On the developmental regulation of acetylcholine receptor mo-

bility in the *Xenopus* embryonic muscle membrane. *Exp. Cell Res.* **130**: 281–290 (1980).

30. Patel, N., Poo, M.-m.: Orientation of neurite growth by extracellular electric fields. *J. Neurosci.* **2**, 483–496 (1982).

31. Poo, M.-m.: *In situ* electrophoresis of membrane components. *Ann. Rev. Biophys. Bioeng.* **10**: 245–276 (1981).

32. Poo, M.-m., Robinson, K.R.: Electrophoresis of concanavalin A receptors along embryonic muscle cell membrane. *Nature* **265**: 602–605 (1977).

33. Reiss, H.D., Herth, W.: Visualization of the Ca^{2+} gradient in growing pollen tubes of *Lilium longiflorum* with chlorotetracycline fluorescence. *Protoplasma* **97**: 373–377 (1978).

34. Reiss, H.D., Herth, W.: Calcium gradients in tip growing plant cells visualized by chlorotetracycline fluorescence. *Planta* **146**: 615-621 (1979).

35. Robinson, K.R.: Electrical currents through full-grown and maturing *Xenopus* oocytes. *Proc. Nat. Acad. Sci. (USA)* **76**: 837–841 (1979).

36. Robinson, K.R., Cone, R.: Polarization of fucoid eggs by a calcium ionophore gradient. *Science* **207**: 77–78 (1980).

37. Robinson, K.R., Jaffe, L.F.: Polarizing fucoid eggs drive a calcium current through themselves. *Science* **187**: 70–72 (1975).

38. Slayman, C.L., Slayman, C.W.: Measurement of membrane potentials in *Neurospora*. *Science* **136**: 876–877 (1962).

39. Stump, R.F., Robinson, K.R., Harold, R.L., Harold, F.M.: Endogenous electrical currents in the water mold *Blastocladiella emersonii* during growth and sporulation. *Proc. Nat. Acad. Sci. (USA)* **77**: 6673–6677 (1980).

40. Tartar, V.: Pattern and substance in *Stentor*. In: *Cellular Mechanisms in Differentiation and Growth*. Rudnick, D. (ed.). Princeton: Princeton University Press (1956).

41. Telfer, W.J., Woodruff, R.I., Huebner, E.: Electrical polarity and cellular differentiation in Meroistic ovaries. *Am. Zool.* **21**: 675–686 (1981).

42. Weisenseel, M.H., Jaffe, L.F.: The major growth current through lily pollen tubes enters as K^+ and leaves as H^+. *Planta* **133**: 1–7 (1976).

43. Weisenseel, M.H., Dorn, A., Jaffe, L.F.: Natural H^+ currents traverse growing roots and root hairs of Barley (*Hordeum vulgare L.*). *Plant Physiol.* **64**: 512–518, (1976).

44. Woodruff, R.I., Telfer, W.H.: Electrical properties of ovarian cells linked by intercellular bridges. *Ann. N.Y. Acad. Sci.* **238**: 408–419 (1974).

45. Woodruff, R.I., Telfer, W.H.: Electrophoresis of proteins in intercellular bridges. *Nature* **286**: 84–86 (1980).

46. Zagyansky, Y.A., Jard, S.: Does lectin-receptor complex formation produce zones of restricted mobility within the membrane? *Nature* **280**: 591–593 (1979).

Questions for Discussion with the Editors

1. *What general types of mechanisms might be responsible for initially segregating or distributing the ion pumps and ion channels so that asymmetrical current flows result?*

At least four mechanisms could segregate ion pumps and channels: (1) Since these channels and pumps are integral membrane glycoproteins, they could be segregated by lateral electrophoresis. Imagine an apolar *Pelvetia* egg with an initially uniform distribution of ion channels whose plasma membrane-bound light receptors generated a Ca^{2+}

leak at the dark end. This inward current would generate an electric field across the cytoplasm that could electrophorese proteins along the inner leaflet of the plasma membrane. If more Ca^{2+} channels were attracted to the dark end in this way due to their electrical charge, the inward current would increase, attracting more channels increasing the current even more to attract yet more channels until a segregation of ion channels and pumps results. (2) Similarly, extracellular electric fields such as those found across epithelial cell sheets could segregate ion channels if charged groups protruded outside the outer leaflet. (3) Localized vesicle secretion could insert ion channels or pumps into a specific region such as observed in the cleaving *Xenopus* egg. (4) Cell-cell contact could generate ion channel asymmetries by the clearing of membrane regions of ion channels as they come in contact with one another. Such a cell-contact mediated effect on cell polarity has been reported in mouse blastomeres by Martin Johnson's laboratory.

2. *Would you speculate on the possibility that morphogen gradients of the type described in this volume by Meinhardt might be established or regulated by ion currents of the type you described for follicles or tip-growing plant cells?*

Meinhardt's morphogen gradient theory proposes the existence of a short range autocatalytic activator and a long ranging antagonistic inhibitor. It is quite intriguing that the Ca^{2+} current will perturb the intracellular Ca^{2+} concentration which has these same properties. Due to the extensive Ca^{2+} sequestration by mitochondria and the endoplasmic reticulum as well as the Ca^{2+} pump and Na^{2+}–Ca^{2+} exchanger at the plasma membrane, increases in free Ca^{2+} which could result from the inward current have a very short range. Moreover, Ca^{2+} is known to be autocatalytic since it can stimulate its own release from intracellular stores. The various Ca^{2+} pumps just mentioned are quite analogous to a long ranging antagonistic inhibitor. However, Ca^{2+} itself need not be the morphogen since it is well known to control the phosphorylation of many proteins via calmodulin. Thus the intracellular Ca^{2+} level might control the morphogen concentration through calmodulin by phosphorylating a "pro-morphogen" molecule to activate it. Another real possibility is that an intracellular voltage gradient generated by the transcellular ion current could electrophorese some already present morphogen activator to one region of the cell. A long ranging inhibitor could remain uniformly distributed. This would generate the proposed morphogen gradient as well.

CHAPTER 3

Models for Pattern Formation during Development of Higher Organisms

Hans Meinhardt

MOST INFORMATION about how development is controlled has been obtained from experiments in which normal development has been disturbed—for instance, by removal or transplantation of parts of a developing organism. However, from the regulation observed after such an interference, one cannot directly deduce the molecular mechanisms by which development is controlled (in the same way that one cannot deduce Newton's law of motion directly from the observation of planets). One can, however, make hypotheses about the molecular interactions on which pattern formation during development is based and see to what extent such models can account for the observed phenomena. Only if a model is provided in a precise mathematical form can one check whether the model is free of internal contradictions and whether it can account for the observations in a quantitative manner. If a model is simple, fewer parameters are involved which reduces the danger that a wrong mechanism will fit the observation by a convenient choice of the many parameters.

In recent years, we have developed a series of models for different developmental situations which provide a rather complete picture about how pattern formation could be achieved (20). In the present article, an outline of this work is presented and it is shown how these mechanisms could be linked with one another to account for the high reproducibility of embryonic development as a whole. Models for the following processes are discussed: (1) the primary generation of polarity and positional information in a fertilized egg with an explanation for the formation and for the regulatory properties of embryonic organizing regions; (2) the interpretation of positional information; interpretation leads to the activation of particular control genes in certain regions; the homoeotic transformations of the bithorax gene complex of *Drosophila* and the formation of metameric pattern will thereby become understandable; (3) the generation of secondary embryonic fields such as legs and wings in insects and vertebrates. The latter model accounts for the reproducible positioning of a limb as well as for its predictable handedness, and orientation in respect to the main body axis of the embryo. These models are discussed mainly in application to insect development since very detailed experimental observations are available for that system which challenges any model. However, some hints are given which indicate that the proposed mechanisms have more general applications.

Generation of Organizing Regions by Local Autocatalysis and Long-Ranging Inhibition

The final structure of an organism cannot already exist in a hidden form in the fertilized egg because an early separation into two parts frequently leads to a complete development of both halves. Therefore, the pattern must emerge in a self-regulating process, starting from more or less homogenous initial conditions. We have proposed (11, 10, 20) that primary pattern formation proceeds by local self-enhancement coupled to an antagonistic effect of longer range (lateral inhibition). A possible molecular realization of this general principle is shown in Fig. 1. A short-ranging substance—the activator—promotes its own production (autocatalysis) as well as that of its rapidly diffusing antagonist, the inhibitor. In an almost homogeneous initial situation, any local increase of the activator caused, for instance, by random fluctuations will increase further since the surplus of the antagonistic inhibitor, resulting from this local increase, diffuses rapidly into the surroundings and inhibits the activator production there while the local activator increase grows further (Fig. 1b). Therefore, a homogeneous activator-inhibitor distribution represents an unstable situation and any inhomogeneity would initiate the pattern. The stimulus can be very unspecific. Any asymmetry in oxygen supply, in the pH, or in temperature could be sufficient. Local disadvantages would shift the maximum toward an opposite area. According to the model, such an unspecific stimulus only orients the pattern. The pattern itself is generated by autocatalysis and long-range inhibition. It is fairly independent of the initial stimulus (Fig. 1c), but depends on diffusion and decay rates of the substances involved as well as on the geometry of the tissue.

An activator maximum has many properties of a classical organizer. If the size of the area in which pattern formation can take place is of the order of the activator range, the maximum appears at one side of the field since space is available for only one activator slope (Fig. 1c). This leads to monotonically graded (polar) distributions that are appropriate for supplying positional information. As mentioned, an activator maximum can emerge in an initially almost homogeneous situation and minute, unspecific influences can be decisive as to where it will appear. However, with progressing formation of the activator maximum, the position becomes more and more stable and can ultimately hardly be influenced by external manipulations. A biological example for this fading sensitivity in respect to external stimuli can be seen in amphibians. In many species the organizer—the dorsal

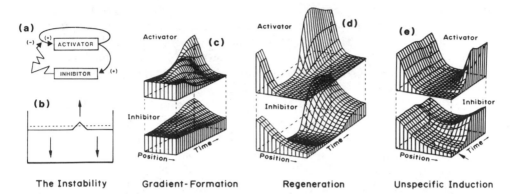

The Instability   Gradient-Formation   Regeneration   Unspecific Induction

FIGURE 1. Short-ranging autocatalysis and long-ranging inhibition—a mechanism for the generation of patterns in initially almost homogenious areas (11, 10, 20). (a) The assumed interaction: a slowly diffusing activator enhances its own production and simultaneously that of a rapidly diffusing inhibitor which, in turn, slows down the activator autocatalysis. (b) Such a system creates an unstable situation since any local activator increase grows further. The local increase cannot be compensated since the antagonistically acting inhibitor diffuses rapidly into the surroundings, inhibiting the activator production outside of the incipient activator maximum. (c–e) Pattern formation in a linear array of cells as a function of time. (c) Generation of a polar pattern. Initiated by random fluctuations, the maximum appears first in the center but moves laterally to the margin since space is available for only one activator slope. A graded concentration profile results. (d) Regeneration: after removal of the activated region, inhibitor is no longer produced. After the decay of the remaining inhibitor, a new activator maximum can appear via autocatalysis. (e) Unspecific induction: At the nonactivated side, the inhibitor concentration is low and any unspecific decrease (arrow) resulting, for instance, from a leakage through an injury or from destruction by uv-irradiation can trigger a second maximum. This leads to a symmetrical pattern.

lip—appears approximately opposite to the side of sperm entry. This positional cue creates only a weak preference that can be overridden, for instance, by changing the egg axis in relation to gravitation; i.e., by 90° rotation of the egg (9). However, as development progresses, egg rotations have less and less effect. Simultaneously, the dorsal cells become progressively more potent in their ability to induce a secondary dorsal side upon transplantation (17). According to the model, the higher the activator is evolved, the higher is the probability that, upon transplantation, a second maximum will be induced. Simultaneously the position of the maximum becomes stable due to the shielding cloud of inhibition.

The stability of an activator maximum results from the kinetics of the reaction and does not indicate that any special irreversible process has taken place. The area of high activator concentration can even be removed and a new maximum will "regenerate" since with the removal of the site of high activator concentration the site of inhibitor production is removed, too. The remnant inhibitor decays until a new activator maximum is triggered from a low level activator production in the remaining cells. The pattern is restored in a self-regulatory way (Fig. 1d).

After the discovery that the dorsal lip of an amphibian blastopore acts as an organizer (29) and that a second embryo can be induced after its transplantation to the ventral side, a great optimism arose that the substances that control development could be isolated. However, further investigations showed disappointingly that very unspecific stimuli can induce a secondary embryo. In an unspecific way, one can only lower a substance concentration, for example, by destruction of molecules due to irradiation or due to leakage through an injury. Unspecific induction is a straightforward consequence of the proposed activator-inhibitor scheme. At a distance from an existing activator maximum, the inhibitor concentration can become so low that even a very small additional decrease caused by the interference may result in the onset of autocatalysis there. A second activator maximum is developed whose form and peak height is independent of the triggering stimulus since the activator concentration increases until it is in equilibrium with the surrounding inhibition (Fig. 1e).

Generation of Positional Information in the Early Insect Embryo by an Activator-Inhibitor System

A characteristic feature of an organizer region is that it controls the development of a whole sequence of structures in space. An old concept in developmental biology is that of the morphogenetic gradient. Wolpert (34, 35) proposed that a graded distribution of a morphogenetic substance supplies positional information. The local concentration would determine which structure is to be formed at a particular location. A morphogen gradient could be generated by a localized source in conjunction with diffusion. However, the assumption of a local source as the origin of different determinations in space represents a circular argument as long as no mechanism is provided explaining how a local source can emerge. As we have seen, the activator-inhibitor mechanism is able to localize the synthesis of substances and generate graded distributions.

The developing early insect embryo is a very suitable organism for studying embryonic organization. The resulting embryo can be regarded as a good approximation to a linear array of structures (head lobe . . . thoracic and abdominal segments). Many experiments have been published that show that the basic segmental body pattern of the segments can be altered in a reproducible way by simple manipulations, such as uv-irradiation, ligation, or centrifugation (25, 26). Different species can react quite differently to similar experimental manipulations, but it is likely that the underlying mechanism is the same. We have shown (18) that the pattern regulation observed after such experimental interferences can be accounted for by the assumption that the sequence of segments is formed under the control of a single morphogen gradient. A narrow activator maximum is located at one egg pole which becomes in this way the posterior pole. The long-ranging inhibitor provides the positional information for the embryo. Thus a low inhibitor concentration would cause head formation—a high inhibitor concentration would lead to the formation of an abdomen. The activator-inhibitor model can be adapted to the different behavior of different species by minor changes of parameters.

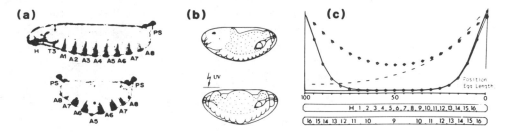

FIGURE 2. Double abdomen malformation in insects. A diversity of treatments can lead in different species to embryos in which a second abdomen is formed in the anterior half of the egg. Head and thoracic structures are missing. (a, b) Normal and symmetric embryo of (a) *Drosophila* (see Chapter 11) and (b) *Smittia* [drawn after Kalthoff(15)]. (c) Model calculation for *Smittia* (18): Activator (-----) and inhibitor (- - -, positional information) forming a normal and a symmetrical embryo. The anterior egg pole is normally the nonactivated side which carries low inhibitor concentration (signal for head formation). This pole is thus sensitive to unspecific induction of a second activator maximum (see Fig. 1e). A symmetric positional information (inhibitor, –o–o–o–) results. Both egg poles carry the signal to form an abdomen.

Unspecific Induction: Formation of a Second Set of Abdominal Structures Instead of a Head

From the many phenomena in insect embryogenesis that can be explained by this model in a quantitative manner, only one is mentioned here: The formation of symmetrical embryos in which a second abdomen is formed instead of a head. This observation provides the best evidence available that the segment pattern is under the control of a diffusible morphogen. Such "double abdomen" embryos (Fig. 2) are formed in many insects after quite diverse experimental treatments such as, for instance, uv-irradiation, punctuation, or temporal ligation of an anterior portion of an egg, as well as centrifugation of eggs (25, 26). Mutants are also known which show this phenotype (see Chapter 11). The variety of different triggers indicates a general instability at the anterior egg pole to form posterior structures. In terms of the model, the formation of an abdomen at an unusual position would indicate the formation of a new activator-inhibitor maximum. The anterior egg pole is the non-activated pole. Thus the inhibitor concentration is very low there and, for wide ranges of parameters, an additional decrease of the inhibitor concentration by experimental interference can trigger the onset of autocatalysis and thereby the transition from the polar to the symmetrical distribution (see Fig. 1e). In agreement with the experimental observation, this is an all-or-none effect; either the small activator increase induced by the treatment develops to the full peak height or it disappears completely.

An important feature of such double abdomen embryos is predicted by the proposed model: The segments in the center have a more posterior specification. In normal embryos of *Smittia* and *Drosophila*, thoracic structures are laid down in the center of the egg. In contrast, in double abdomen embryos, abdominal segments are formed at the center (the plane of symmetry). According to the model, in the latter case the morphogen diffuses from both poles toward the center. This leads to a morphogen increase and therefore to the formation of more posterior structures (Fig. 1c–e). Thus the changing of the central segment is a strong indication that a long-ranging substance controls the pattern.

Interpretation of Anterior-Posterior Positional Information Is a Stepwise Unidirectional Process

The assumption of a graded morphogen distribution as the controlling agent for the determination of a sequence of structures immediately raises the question as to how the local concentration can be precisely measured by the individual cells. From an analysis of ligation experiments during early insect embryogenesis, we have concluded that the cells do not measure the local concentration instantaneously but they switch stepwise from more anterior to more posterior determinations until the actually achieved determination corresponds to the external signal, the local morphogen concentration (Fig. 3a–c). In other words, an area exposed to a minimum threshold concentration becomes first determined to form structure 1. All cells exposed to a higher threshold concentration switch to a determination for structure 2, and so on (Fig. 3d–g). Each step appears to be essentially irreversible. A cell remains stable in a once obtained state of determination when the morphogen concentration declines; for example, in an anterior egg fragment after ligation. However, after a further increase of the morphogen concentration, the cell can achieve an even more posterior determination. The latter takes place in a posterior egg fragment. Thus this mode of cell response explains the formation of gaps in the segmentation after ligations (Fig. 3h).

This type of model—the absolute concentration of a morphogen determines the structures to be formed—appears to work only for segmented structures. Examples are the basic body pattern of insects as discussed above, the digits of vertebrates (see Chapter 1), and the segments of insect legs (see next). In contrast, pattern formation in nonsegmented structures such as the dorso-ventral dimension of vertebrates and insects, as well as the intrasegmental pattern in insects, follow different rules, as will be discussed next.

Segmented Structures: The Superposition of Sequential and Periodic Structures

During insect development, almost simultaneously with the determination of the segments during blastoderm formation, at least the thoracic segments become subdivided into anterior (A) and posterior (P) compartments (8, 30). Thus, simultaneously, two patterns arise: The periodic pattern of the alternating compartmental specifications (A-P-A . . .) and the pattern of segmental specifications (1, 2, 3 . . .). Such a dual (sequential/periodic) pattern is widespread in higher organisms. Other examples are the digits (see Chapter 1), the somites, and the teeth (see Chapter 26) of vertebrates. In insects, both patterns are precisely in register. With the precision of a single cell, the segment border between the meso- and a me-

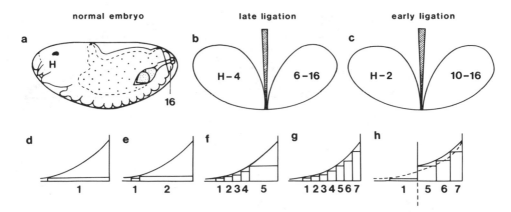

FIGURE 3. Ligation experiments indicating a stepwise interpretation of positional information. (a) The normal sequence of segments in a nonoperated *Smittia* embryo is termed H, 1 . . . 16. (b) After a ligation at the blastoderm stage with a blunt razor blade, the embryo behaves as mosaic. Both halves together form (almost) the complete sequence. (c) After a much earlier ligation (during nuclear cleavage stage), central segments are missing in both fragments. The terminal segments remain present (25, 26). (d–f) A quantitative explanation of this and many other ligation experiments, performed with different species at different stages and positions is possible by assuming that under the driving force of the morphogen (———), the cells attain more and more posterior determinations (1, 2 . . . 7) until the achieved determinations correspond to the local morphogen concentration (18). (g) After a ligation, the morphogen concentration drops in the anterior fragment. The segment formed indicates how far the stepwise advancement has already progressed. In the posterior portion, the morphogen concentration accumulates due to the diffusion barrier caused by the ligation (– – – morphogen distribution in the normal egg). Fewer and more posteriorly specified segments are formed. The gap in the segmentation after a ligation has different origins in both fragments and, therefore, it can be asymmetric.

tathoracic segment is always also a border between P and A. Both patterns must arise therefore in a coupled process; it is not sufficient that they are under the control of the same morphogen gradient. The question then is "Which is the primary event?" Either the periodic alternation between two (or three; see below) states controls the transition from one segmental specification to the next or, alternatively, each segment becomes secondarily subdivided into the compartments (Fig. 4).

The Formation of the Periodic Structure Is the Primary Event

As discovered by Lewis (16), a particular segmental specification can extend into a neighboring segment if *Drosophila* flies carry mutations of the bithorax gene complex (BX-C). For instance, in a fly carrying the mutation *bx* the mesothoracic (MS) specificity extends into the anterior metathorax (MT) (Fig. 4d). This indicates that the confrontation of two segmental specifications (e.g., MS-MT) is not the signal to induce a segment border, even if this border coincides with an A-P compartment

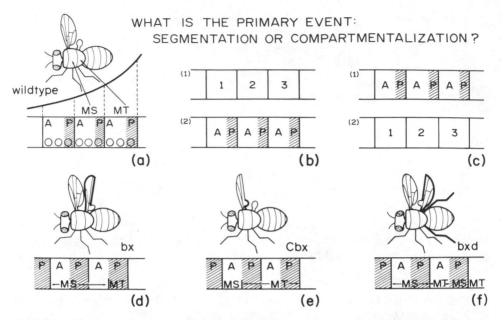

FIGURE 4. Evidence from mutations in the bithorax gene complex (BX-C) shows that the formation of the periodic pattern is the primary event. (a) In a normal *Drosophila*, the sequential pattern of segmental determinations (. . . meso- [MS] and metathorax [MT] . . .) are precisely in register with the periodic pattern of anterior (A) and posterior (P) compartments, indicating that both patterns arise in a coupled process. (b, c) Either the sequential (b) or the periodic pattern (c) is formed primarily and the other pattern appears in the frame of the first. (d–f) Some phenotypes of flies carrying a mutation in the BX-C (16); abnormal structures are drawn in heavy lines. A particular segmental specification can extend into a neighboring segment but, nevertheless, the periodic pattern /AP/AP/ . . . remains unchanged. Clearly, a particular segmental specification does not become subdivided into an A and P compartment.

border. If the segmental specifications can be changed without the periodic pattern of segments and A-P compartments being changed, we have to conclude that the formation of the periodic pattern is the primary event and that under its control the particular segmental specifications are achieved. The mutations of the bithorax gene complex obviously abolish the coupling between the (unchanged) A-P-A pattern and the segmental specifications. The same conclusion can be drawn from the phenotype in which the bithorax gene complex (BX-C) is completely deleted. In such a case, the MT and all abdominal segments become MS in character. However, the segmentation and even the total number of segments remain normal. Counting segments and giving them specific names (. . .MS, MT. . .) are two different processes, and any model that assumes that one abdominal segment is added after another until the last abdominal segment is formed is untenable.

The assumption that the periodic pattern is the primary event is also reasonable from an evolutionary point of view. Lower arthropods and annelids may possess many more, and more similar, segments.

The Gating of Control Genes by the Alternation Between Two (or Three) States and the Mutations of the Bithorax Complex

How could the periodic A-P pattern control the pattern of segmental specifications? This author has shown that the mutations of the BX-C can be explained under the following assumptions (20, 21): the cells can be in two states, either A or P.
Under the influence of the morphogen the cells begin to oscillate between A and P.
The total number of A-P changes depends on the local morphogen concentration.
With each P-A transition, a transition from one to the subsequent control gene occurs in a manner analogous to the way in which the periodic movement of the pen-
dulum and escapement controls the stepwise advancement of the hands of a clock.

The periodic alternations between two (or three) states allow a "gating" of
control genes. In the P state, an attempt is made to activate the control gene that
specifies the adjacent posterior segment, but the transition is blocked. After switching to the A state the transition is no longer blocked, but no activation of the next
control gene takes place. Thus a transition from one control gene to the next is
coupled to a P-A transition. This scheme accounts for the mutant phenotypes observed in the BX-C. For instance, if the block in the P-state does not work, the subsequent control gene is activated too early, namely in the P region of the preceding
segment. Corresponding mutants are *Cbx* (Fig. 4e) and *Hab*. Or the activation of
the next control gene does not work correctly. Then the same segmental specificity
would be repeated. For instance, in *bxd* the first abdominal segment maintains
thoracic character and a fourth pair of legs is formed (Fig. 4f). A third type of mutation is less obvious. Here the transition to the subsequent control gene takes place
but the preceding segmental character is not suppressed. The segment acquires a
similar character to the preceding one. Examples are the mutations *bx* and *pbx*.
This model allows the assignment of specific functions to the elements of the BX-C
and details such as the phenotypes of double mutations with different cis-trans arrangements are correctly described (for details, see 20, 21).

Segmentation Requires the Reiteration of Three Different States

We have seen how segmentation is *not* controlled: a segment border is neither
formed where A and P cells are juxtaposed nor where cells with different segmental
specifications are confronted. But, nevertheless, the periodic pattern must be the
primary event. What, then, is the signal for a segment border? A rule could be formulated that in a periodic APAPAP sequence when an AP-pair always forms a segment it does not define the grouping and the polarity of the segments since either
AP/AP/AP/ or A/PA/PA/P could result.

We have proposed (20) that a segment is primarily not only subdivided into
two (A and P), but into three different cell types (S, A, and P; S like "separation" or
"segment boundary"). If in a periodic sequence SAPSAP. . . , confrontations between P and S lead to the formation of segment boundaries, the positions of the
boundaries would be defined (. . P/SAP/SAP/ . .). In addition, each segment
would have an unambiguous polarity (Fig. 5a).

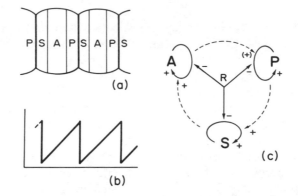

FIGURE 5. Molecular mechanisms proposed for primary intrasegmental pattern formation. (a) It is assumed that the initial step in the formation of segments is the formation of a periodic pattern of three different cell types: S, A, and P. A confrontation between S and P leads to the formation of a segment border. (b) It is not assumed, as in most other segmentation models, that the primary intrasegmental pattern consist of a sawtoothlike gradient with a concentration jump at the segment border. (But such a pattern would result secondarily if P acts as source and S as sink.) (c) A sequence of three (or more) structures in space can be generated if three autocatalytic feedback loops exist (S, A, P) that locally compete with each other (e.g., via a common repressor R) but support each other over a long range (22). A long-range self-inhibition has a similar effect as a long-range activation of neighboring structures since, in competing systems, a disadvantage of one structure is equivalent to an advantage of the others. (———) short-ranging substances, (– – –) long-ranging substances.

Evidence for the Threefold Segmental Subdivisions

The assumption of the periodic SAP-subdivision can account for several seemingly unrelated phenomena. As we shall see next, a confrontation between A and P cells is a precondition for leg and wing formation. Then, with an AP/AP/ subdivision, one would falsely expect legs and wings to be formed at the segment border. Bohn (4) has shown that indeed the confrontation of anterior and posterior tissue can induce legs but that each AP pair in one segment is located from the next by a "sclerite inducing membrane" (our S) at the most anterior part of each segment (therefore, /SAP/ and not /APS/). Removal of this isolating membrane leads to a P-A confrontation and, as expected, to the induction of an additional leg with reversed A-P polarity at the location of the former segment border (Fig. 9).

By grafting experiments with tissues from abdominal segments of a bug, Wright and Lawrence (36) have juxtaposed cells from the anterior and posterior parts of a segment and observed additional segment boundaries along the confrontations. In terms of the model, in these experiments S and P tissue is confronted and this leads to a new boundary.

In one of the segmentation mutants found by Nusslein-Volhard and Wieschaus (see Chapter 11), the number of segment boundaries is doubled. This would be expected on the basis of the proposed model if the A specification is hit by the mutation. A sequence . . . /S P/S P/S . . . would remain that leads to an additional segment border in the center of each segment. (So far the model does not provide an explanation for the even-skipped and odd-skipped mutants that they have found.)

How to Obtain a Periodic SAP-Pattern

A sequence of structures can be generated and maintained in a self-regulatory manner by a long-ranging activation of states that locally exclude each other (22). Imagine the following reaction scheme (Fig. 5c): There are three feedback loops, S, A, and P. They produce and react upon a common repressor. This creates an unstable situation in which one of the three loops dominates and represses the other two. Let us assume further that over a long range, the loop S activates loop A, A activates P, and P activates S (and possibly vice versa). Then at a particular location, a particular loop will dominate, but in the adjacent region long-range activation causes a neighboring feedback loop to dominate. Therefore, this mechanism has the possibility of generating sequences of stable states. A stripelike arrangement of the S, A, and P cells is especially favored since stripes—structures with a long extension in one direction but with a small extension perpendicular to the first—are separated by long borders. A very efficient mutual stabilization across such borders is possible. It is tempting to assume that the S, A, and P loops are control genes, but other realizations of the autocatalytic feedback loops cannot be ruled out at present.

To make this mechanism useful for the formation of segments we have to ensure that the patches of S, A, and P specification arise in an ordered way and not at random positions in the insect blastoderm. The obvious means of achieving a global control of the /SAP/SAP/ . . . pattern is the organizing morphogen gradient discussed previously.

How can a gradient control the SAP pattern? To maintain cells stably in a particular state of determination, the stabilizing influence of its neighbors is required. If only one cell type (let us say S) is present, the S loop is not supported by A (or P) neighbors. However, the S loop will activate the A loop and, after a certain time interval, a homogeneous cell population of S cells can switch to A, later from A to P, and again to S, and so on. As long as no stabilizing borders are present, the cells tend to oscillate between the alternative states. Borders can be formed at predictable positions by the organizing action of the morphogen gradient. Let us assume that initially all cells are in the P state (Fig. 6). Only those cells which are exposed to a certain threshold morphogen concentration switch to S. The first P/S border has arisen. All S cells at distance from this P/S border are not stablized by nearby P-cells and they switch to A, creating an S-A border. With each transition from one state to the next, a new stablizing border is formed. After each full cycle, a new stable /SAP/ sequence is added. As the process progresses, the region of stable SAP/SAP/ . . . pattern enlarges at the expense of cells which oscillate between S, A, and P. The borderline between stable and oscillating cells moves over the field in a wavelike manner from anterior to posterior (Fig. 6). The activation of the control genes for segment specification under S, A, and P control can be achieved in the same manner as described for the A-P-A oscillation.

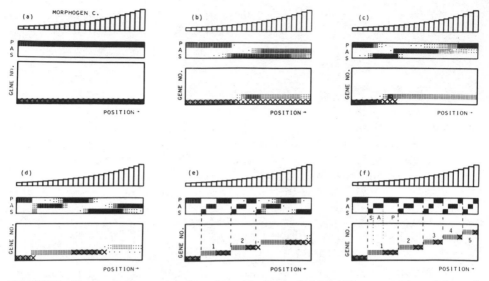

FIGURE 6. Stages in the generation of a periodic ... /SAP/SAP/ ... pattern and, in register, a sequential pattern of segmental specifications. The activity of the feedback loops (S, A, P) forming the intrasegmental pattern and of the genes controlling segmental specifications (1, 2..) are indicated by the density of dots. In the P state, a substance (XXX) is produced which induces the transition to the subsequent control gene. (a) Initially, all cells are assumed to be in the P state and gene O is active. (b) All cells exposed at least to a threshold morphogen concentration (top row in each subpicture) switch from P to S—the first segment border is formed. Since it is a P-S transition, also a switch from gene O to gene 1 occurs. (c) All S cells distant to the P-S border are not stabilized by nearby P cells and switch to A, forming a S-A border, and so on. (d–f) with each full SAP cycle, a new segment with a particular specification is added. Details for such computer simulations can be found in 20.

In principle, a single threshold would be sufficient to form a large segmented structure. The threshold would determine the first P-S border and all further segments would be sequentially added by the SAP mechanism. Not every structure has to be determined individually by a special threshold since the width of the segments can be controlled by the width of the S, A, and P stripes. In this way, far more structures can be determined by a single morphogen gradient than would appear reasonable under the assumption of a threshold-controlled interpretation. However, the reduced number of segments in symmetrical embryos (Fig. 2) and in ligated egg fragments (Fig. 3) indicate that the steepness of the gradient determines the width of segments. In the model, this would occur if the switch to a new control gene is accompanied by a small increase in the threshold of the next P-S transition.

Why has the expected boundary between S and A not yet been observed as a clonal border in *Drosophila*? In holometabolous insects, the ectodermal structures of the adult thorax are derived from imaginal discs. These discs are formed, as discussed later, around the A-P boundary. Thus it could be that the S region does not contribute to adult structures.

The proposed segmentation mechanism is capable of explaining several basic features. A repetitive pattern emerges and the units (/SAP/) exhibit from the beginning a polarity and a fine structure. Each of the elements can be different from the others since the periodic alteration between the three alternative states allows a reliable transition from one control gene to the next. That is, accounting operates on the DNA level by a stop-and-go mechanism. This enables a high spatial resolution. The morphogen gradient ensures that the S, A, and P stripes are added in an ordered fashion.

As we shall see, legs and wings are induced at A-P boundaries. Thus the primary subdivisions of the segments create the preconditions for the generation of the next finer substructures.

In some insects, the segments appear roughly simultaneously. In others the head lobe is first formed and the remaining segments are added in a sproutinglike process. The model allows for the sequential addition of new segments in a zone of growth. For instance, if a terminal S area becomes too large, some of the S cells switch to A, a too large A area will form P cells, and so on, and again a superposition of a sequential and a periodic pattern is formed during outgrowth.

The same general mechanism may be involved in the generation of other structures that consist of similar but not identical subunits. The somites are formed visibly in an anterior-posterior order as suggested by the model proposed. One somitic cleft is formed after another. In terms of the model, this could occur where P- and S-cells are juxtaposed (20). Or, as shown by Reif (see Chapter 26), the teeth of sharks are sequentially added in a medio-lateral sequence. The repetitive units, the teeth, frequently have a clearly defined polarity and they can be, but need not be, different from each other. The SAP-mechanism has regulatory features such as observed by Reif for the shark dentation, namely the formation of symmetric patterns after an interference or a subdivision into two segments if the available area becomes too large.

An Alternative Mode of Interpretation of Positional Information: Orientation of Self-Regulating Sequences under the Influence of a Morphogen Gradient

The loss of central segments in double abdomen embryos (Fig. 2) and the embryo's appearance after ligation (Fig. 3) provide a strong argument that the absolute concentration of a morphogen is decisive concerning which structure will be formed in the A-P pattern. The dorso-ventral dimension of insects has different regulatory features. This is especially evident in *Euscelis* embryos. After a ligation parallel to the long axis, a complete embryo can be formed from each fragment (26). Similarly, after an induction of a symmetrical D-V pattern in amphibians, two complete sequences of D-V structures are formed in a mirror-symmetrical arrangement. Each structure is much smaller because twice as many normal structures have to be fitted into the same space (see Chapter 20). The mechanism of lateral activation of structures that exclude each other locally, discussed for the SAP pattern, has pattern formation capabilities. A morphogen gradient (or any other slight asymmetry) can *orient* the pattern. The simulation in Fig. 7 shows that a sequence

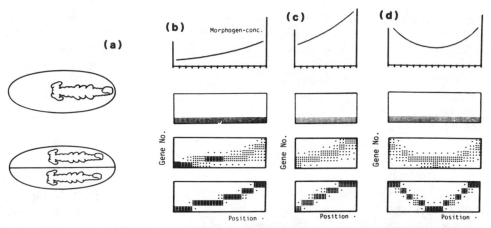

FIGURE 7. Orientation of a self-regulating sequence by a gradient. (a) Biological observation: the dorso-ventral pattern of an insect [*Euscelis*, see Sander (26)] can show very good size regulation. After a longitudinal ligation, a complete embryo is formed in each half (compare with the completely different regulatory behavior of the A-P dimension, Figs. 2 and 3). (b–d) Model: The formation of a sequence of structures by lateral activation of cell states is assumed [Fig. 5 (20, 22)]. From top to bottom in each subpicture: the orienting gradient; the initial, intermediate, and final patterns of activities of five feedback loops ("genes"). A higher morphogen concentration is assumed to provide a slight advantage for a higher gene (b, c). The resulting self-regulating pattern is, to a large extent, independent of the field size and of the absolute morphogen concentration (d). A complete symmetrical pattern can arise in the same space (compare with Fig. 2).

determined in this way is independent of the absolute value or of the slope of the morphogen distribution. A symmetrical pattern can be compressed into the same space. Thus we assume that the D-V pattern is generated in a similar way as the pattern within a segment, namely by mutual activation of states which locally exclude each other (20).

How to Keep the Antero-Posterior and the Dorso-Ventral Patterns Perpendicular to Each Other

For a developing organism, it is absolutely essential that the main body axes are not aligned parallel but, optimally, kept perpendicular to each other. This is not automatically the case if both patterns are generated by reaction-diffusion mechanisms as described previously. Orthogonality has to be enforced by an appropriate coupling of the two pattern-forming systems. How could such a coupling be achieved? Let us assume an initially more or less homogeneous tissue with a spherical shape. By a pattern-forming reaction as shown in Fig. 1, the sphere can be subdivided into a dorsal and a ventral (or in an animal-vegetal) half. If both dorsal and ventral cells are equally required to generate a high point of the A-P pattern, this high

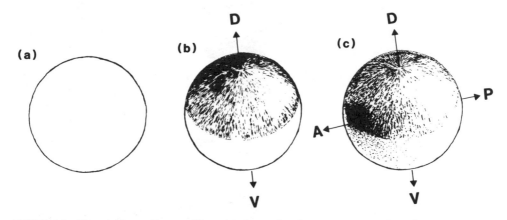

FIGURE 8. How to keep the positional information for two axes perpendicular to each other. (a, b) A spherical mass of cells (a) is—by a mechanism as shown in Fig. 1—first subdivided into a dorsal (D) and a ventral (V) hemisphere (b). If the condition for a high point in the anterio-posterior pattern (A–P) is a proximity of D *and* V cells, the high A point has to arise on the DV equator and both patterns are perpendicular to each other. This will occur if the autocatalysis required for the AP-pattern takes place in two steps, one in D-cells, the other in V-cells.

point can arise only at the "equator" and never within purely dorsal or ventral tissue. This would ensure that both coordinate systems are perpendicular to each other (Fig. 8). The particular location at which the high A or P point appears can be selected by autocatalysis and lateral inhibition, as described.

Evidence for such a D-V confrontation as a precondition for the A-P pattern has been obtained for planarians (7). During regeneration (or frequently even after an injury), a new head is formed only if in the process wound closure D and V tissue becomes juxtaposed.

Cell Determination Boundaries as Organizing Regions for Secondary Embryonic Fields

The final spatial complexity of an organism is much higher than could be achieved by the interpretation of only one (or two orthogonal) gradients. Pattern-formation mechanisms for secondary embryonic fields (subfields) are necessary. Among the best investigated subfields are the imaginal discs of insects (see Chapter 12) and the vertebrate limb (see Chapter 1). We have proposed (19, 20) that cell determination boundaries, resulting from primary pattern formation, can act as organizing regions for secondary embryonic fields. Imagine that a cooperation of two adjacent patches of cells with different determinations (e.g., two compartments) is necessary for the synthesis of a substance that controls pattern formation in the subfield. For instance, in one compartment a precursor of a morphogen is produced while it is processed into the final product in the other compartment. Morphogen produc-

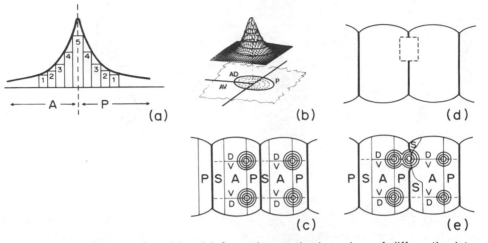

FIGURE 9. Generation of positional information at the boundary of differently determined cells. (a) If a cooperation of two cell types (A and P; e.g., compartments) is required for the synthesis of a morphogen, this synthesis can take place only in the region where both cell types are close to each other; i.e., in the boundary region. The local morphogen concentration is a measure for the distance of a cell from the boundary. (b) Three compartments (or two pairs of compartments) are close to each other only at a single point. A cone-shaped morphogen distribution results which is appropriate to determine the roughly circular disc and the concentrically arranged primordia of the leg segments within the disc. (c) Generation of leg fields in an insect. Schematic view of the ventral side. A subdivision of each segment into S, A, and P regions, as well as a subdivision into D and V regions in the dorso-ventral dimension, is assumed. Two leg fields are created within each segment at the two A-P and D-V intersections. Both legs have opposite handedness. A clockwise arrangement of the AD, P, and AV region corresponds to a left leg. (d, e) Formation of a supernumerary limb field. Bohn (4) has removed the "sclerite inducing membrane" (the most anterior part of a segment, our "S") or a cockroach (d) and observed a supernumerary leg at the former segment border. In terms of the model, this leads to new confrontation of AD, AV, and P tissues and thus to a new leg field (e). The A-P polarity and the handedness are reversed. Bohn has interpreted his result in a similar way.

tion would be possible only at the region close to the common boundary. A ridge-like morphogen distribution would result that is centered over the common boundary (Fig. 9). The local concentration would provide a measure for the distance from the border. This allows a reliable pattern formation in the course of development: The primary positional information leads to cells of different determinations separated by sharp borders. At these boundaries, in turn, new positional information is generated for the next finer subdivision, and so on. In this way, the boundary between differently determined cells becomes a boundary region in the sense used by Wolpert (34, 35); i.e., the high point of a positional information scheme. If only one of the two cell types is competent to respond to the morphogen, these cells are exposed to a nearly exponential gradient; a polar pattern results. The anterio-posterior pattern of vertebrate digits may be generated in this way (20).

Generation of Positional Information by Cooperation of Compartments in Insect Appendages

According to the mechanism we propose, the formation of the boundaries precedes the formation of subfields. This sequence of events is especially evident in the development of imaginal discs of *Drosophila* which form legs, wings, antennae, and other structures. Instead of homogeneous imaginal discs being initially formed and subsequently subdivided in anterior and posterior (A and P) compartments, the formation of A-P compartment borders occurs much earlier than the disc formation. It occurs almost simultaneously with the determination of the segments at the blastoderm stage (8, 30, 33). As discussed, the (S)AP subdivision is presumably the primary event.

How are imaginal discs formed in this process and how do they get their fine structure? The legs, for instance, are segmented structures as is the whole insect. It is tempting to assume that the sequence of leg segments is laid down in a similar way—namely under control of a morphogen gradient. The primordia of the leg segments are arranged in the disc as a series of concentric rings, the central ring forms the most distal structure, the tarsus, and the claws. The outer rings form the proximal leg segments and the thoracic structures (27). An appropriate morphogen distribution would have the shape of a cone since particular threshold concentrations then form a series of concentric contour lines. No additional positional information system would be required for the formation of the roughly circular disc. If a certain minimum concentration is required to initiate the primordium of an organ, a disc-shaped patch of cells would be recruited, whereas cells exposed to a subthreshold concentration would remain larval ectoderm.

But how could a cone-shaped morphogen distribution be generated within the embryo at a precise location? The A-P compartment borders are presumably lines which surround the embryo like belts. A second pattern forming event organizes the dorso-ventral (D-V) dimension. We assume that this leads to at least two pairs of D-V compartments—one pair for the future wing, the other for the future leg disc. Three different compartments (or two pairs of compartments) are close to each other only at one particular point. If a cooperation of the three or four compartments is required for morphogen synthesis, the point where all compartments are close to each other will be the very local source area. By diffusion and decay, a cone-shaped morphogen concentration is formed. The local concentration is then a measure of the distance from the intersection.

The wing disc surrounds the intersection of the A-P and D-V compartment border (8). Cells close to this intersection form the distal-most part, the wing tip. Surrounding cells form the wing blade. In terms of the model, each of the two borders can create a morphogen distribution that provides positional information for the A-P and for the D-V dimension. The cooperation of the AP- and DV-morphogen would lead to the cone-shaped distribution, allowing the distinction of cells forming either the thorax or wing blade as well as a distinction between disc cells and larval ectoderm. The color-pattern of the wings of butterflies frequently form stripes parallel to the outer wing margin. This can be interpreted in that the position of a cell is measured as distance from the wing margin (the D-V boundary).

The model makes several predictions: (i) The distal-most structures are formed where cells of all major compartments are close to each other; i.e., at the intersection of compartment borders. (ii) Distal transformation occurs whenever, after experimental interference, cells of all compartments come to lie close to each other. (iii) Distal transformation does not require a complete set of circumferential structures but a close proximity of cells of all compartmental specifications. (iv) Cell-nonautonomous mutants should exist in which the positional information but not the response of the cells is altered. Distally incomplete structures are expected in this case. (v) Due to the unidirectional interpretation of the positional information, no distal to proximal transformation of leg segments should occur. (vi) The arrangement of compartments determines the handedness of a structure. A clockwise arrangement of the AD, P, and AV compartment leads to a left leg.

Many experimental observations with disc fragments support this prediction (19). For instance, Schubiger and Schubiger (28) found that distal transformation requires the presence of the A-P compartment border. When a fragment which has been derived entirely from the anterior compartment undergoes distal transformation, this is always connected with a respecification of some anterior cells into posterior cells. However, the presence of the A-P border is not sufficient since fragments that contain the A-P border but lack cells of ventral specification do not show distal transformation.

More direct evidence for the involvement of compartments in the generation of the proximo-distal positional information can be derived from leg duplications and triplications. Such malformations can be induced in a heat-sensitive strain of *Drosophila* by heat shock (23, 24, 12, 13). The triplication follows the rules discovered by Bateson (1): (i) The three legs are formed in a plane (and not like a tripod). (ii) The two outer legs have the normally expected handedness while the central leg has the opposite handedness. (iii) Sometimes two of the three legs are fused and consistent; for instance, of two anterior halves of a limb. Such limbs are frequently distally incomplete.

These empirical rules can be explained by the proposed model in a straightforward manner. After induced cell death or after surgical interference, cells can change their compartmental specification (32). As mentioned, anterior cells are especially prone to become reprogrammed into posterior cells. According to the model, A and P cells stabilize each other mutually. If, for instance, killed cells—resulting from the heat shock—do not participate any more in this cell-cell communication, A-cells might be insufficiently supported by P-cells and switch to P specification (as we have assumed to take place during the original segmentation). The reverse transition seems to occur much less frequently. Perhaps a P-A (or P-S) transition requires the driving force of the morphogen. A new patch of posterior cells in the anterior compartment can lead to additional intersections (Fig. 10a, d) and thus to supernumerary legs. For instance, an island of posterior cells in the A compartment can cross twice the D-V boundary. Two additional legs (i.e., a triplication) is expected (Fig. 10d-f). They are formed in a plane since the intersections lie on the straight line of the dorso-ventral boundary. If the patch of posterior cells lies close to the ventral compartment but does not touch it, distally incomplete triplications are expected. In this case, the cooperation is restricted and the maximum morphogen concentration is not achieved (Fig. 10i). The model predicts that distally incomplete structures lack cells of the ventral compartment. This prediction is

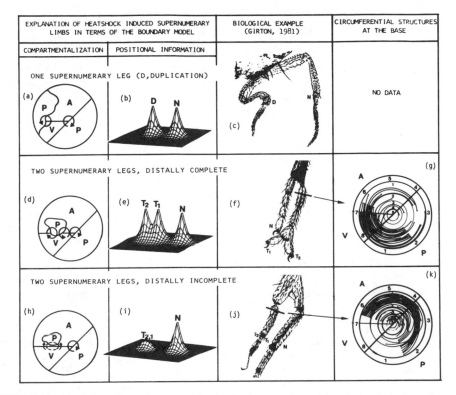

FIGURE 10. Model explanations and biological examples of heat-shock induced supernumerary legs. It is assumed that by the heat shock some anterior cells are respecified to posterior cells. (a–c) A new patch of posterior cells at one margin can lead to a second intersection and thus to a second (mirror-symmetrical) leg (D = duplication; N = normal leg). (d–g) A nonmarginal patch of new posterior cells can lead to two new intersections and thus to a triplicated leg (T_1, T_2). All distally complete triplication contain, at their base, cells of the ventral compartment. (h–k) If the new posterior cells are close but do not touch the ventral compartment, the cooperation of the three compartments is restricted, the maximum morphogen concentration is not achieved, and a distally incomplete leg consisting of two fused anterior halves (T_{12}) results. All distally incomplete legs lack cells of the ventral compartment (12, 13) providing strong support for proposed model (photographs kindly supplied by J. Girton).

fully supported by the results of Girton (12, 13). The distally incomplete supernumerary legs consistently lack a specific region, the cells of the ventral compartment, while distally complete legs contain ventral cells (Fig. 10k, g). This indicates that a complete set of compartments is required for distal transformation.

Interpretation of Positional Information in the Discs

The response of the disc cells to a particular morphogen concentration seems to proceed in the same way as described for the formation of body segments—in a stepwise manner. Examples for such a unidirectional distal transformation can be seen after a confrontation of proximal and distal fragments of wing discs (14) or evaginating leg discs (31). In both cases, a restricted distalization of some proximal structures is observed but never the reverse transformation.

Unidirectional promotion appears to be a necessary requirement if a sequence of structures, once determined under the influence of the diffusible morphogen, is to undergo overall growth. The slope of the gradient is determined by the diffusion rate and lifetime of the morphogen. Thus, in spite of continued growth, the slope of a gradient will remain the same as long as these parameters remain unchanged. Thus, during growth, cells at a distance from the source become exposed to lower and lower morphogen concentrations. Unless determination events are irreversible, this would lead to a loss of already achieved determinations.

The unidirectional mode of interpretation agrees with observations made on regenerating cockroach legs. Assuming that a cockroach leg is similarly compartmentalized as in *Drosophila*, the three compartments would form three long stripes along the proximo-distal axis of the tube-shaped leg. Only at the tip of the leg are the three different cell types close to each other. The gradient generated at this point has, however, only the extension of the order of the size of an imaginal disc. It is thus much smaller than the leg. Nevertheless, due to the irreversibility of interpretation, the determination of the proximal leg segments remains stable. The smallness of the gradient, however, enables a complete regeneration of a leg after removal of distal parts (Fig. 11). After wound closure, a new intersection of compartments is formed and a new steep gradient will emerge. In the vicinity of this new intersection, cells become exposed to a morphogen concentration which is higher than corresponds to their own determination. These cells become distally transformed. Thus, as observed by Bulliere (6), the missing parts will be specified on a very minute scale and grow out later to their final size. Regeneration of a cockroach leg is therefore assumed to proceed by a reprogramming of existing cells; i.e., by morphallaxis. The growth to the final size is a secondary event.

Another strange-appearing experimental observation becomes explicable under this assumption. While a stump of a cockroach leg regenerates completely, a removed femur-tibia joint is not replaced (2). Thus regrafting of distal structures onto a stump suppresses complete regeneration. In terms of the model, after such regrafting, the tube-shaped ectoderm is not closed but is kept as a tube (Fig. 11). Thus the graft prevents the formation of a new intersection of compartments. Since the morphogen distribution at the tip of the grafted leg has only a short range, the concentration at the host/graft juction is too low to enable any distal transformation and gap repair.

Cooperation of Compartments: A Molecularly Feasible Basis of the Polar Coordinate Model

The polar coordinate model (PCM) of French, Bryant, and Bryant (see Chapter 12) provides formal rules which have been very successful in describing pattern regulation in imaginal discs, insect legs, and vertebrate limbs. Our boundary model has regulatory features of the PCM since the requirement of two intersection boundaries (i.e., a meeting point between three sectors or four quadrants) is equiv-

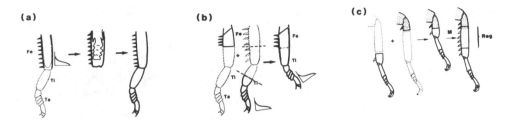

FIGURE 11. Regeneration and intercalation of cockroach legs. (a) Truncation leads to a complete regeneration of the removed distal leg parts. According to the model, during wound closure the three compartments come into contact. A new steep gradient is generated which leads to determination of the missing parts at a minute scale—as experimentally observed (6). (b) A removed femur-tibia (Fe, Ti) joint is not replaced by intercalary regeneration (2). In terms of the model, due to the graft the ectodermal tube is not closed and no new intersection of compartments occurs. The existing intersection at the tip of the grafted leg is too remote to allow any respecification. (c) In contrast, missing elements of the intrasegmental pattern are intercalated (3). The different results (b, c) indicate that both patterns, the sequence of segments, and the pattern within a segment are controlled by different mechanisms. Intrasegmental pattern is assumed to be achieved by direct mutual activation of cell states (22), while the sequence of segments results from the interpretation of a morphogen gradient as shown in Fig. 9.

alent to the requirement of a complete circle. Thus we propose that the complete circle rule of the PCM is in fact a "complete-set-of-compartments rule."

The polar coordinate model makes predictions only about how a disc or limb regulates after a particular experimental manipulation. No statement has been made about how a "complete circle" is initially formed during development. According to our model, the regulatory features of the PCM appear as the straightforward consequence of the mode of determination of these structures during normal embryogenesis. Compartmentalization, which is certainly the primary event in the formation of appendages, therefore plays an essential role in our model.

Predictions about the handedness of a structure are the same in the PCM and in our model since the handedness is independent of whether 3, 4, or 12 positional values are assumed. Since compartmentalization can be determined by independent methods, our model is inherently testable. The PCM is, in this respect, much less specific since the assumption of the 12 positional values and their assignment to real structures is somewhat arbitrary. Despite the similarities, our boundary model makes specific predictions not made by the PCM. The leg triplication discussed (Fig. 10) demonstrates the differences: According to our model, the three legs have to lie in a plane since they emerge along a line, the anterior-dorsal/anterior-ventral compartment border. This coplanarity is not predicted by the PCM. A more recent version of the PCM (5) predicts that distal transformation can be induced also in disc fragments that contain only a partially complete circumferential pattern. Nonetheless, a more complete circumferential pattern leads to a more complete distal transformation. However, as the results of Girton (Fig. 10g, k) show, a supernumerary leg may have more than half of the circumferential structures (bristle rows) and nevertheless remain distally incomplete. On the other hand, supernumerary legs with only a few bristle rows at the base can form distally complete legs if, and only if, bristles belonging to the three compartments are included, in agreement with our boundary model.

In the PCM model, the proximo-distal (P-D) sequence of structures is a continuous sequence of positional values. According to the model we propose, the P-D pattern is formed by the superposition of two patterns: One is the sequence of the segments determined by the morphogen gradient; the second is the pattern within each segment, generated by the mutual induction of neighboring states (see Fig. 5). We expect intercalation of missing structures within a segment, but not of missing segments. This is in agreement with the experimental observation. The failure of repair after removal of a femur-tibia joint has been mentioned—a result which is hard to understand assuming a continuous sequence of P-D positional values.

Intrasegmental Pattern Formation: Mutual Activation of Locally Exclusive States

In principle, such a mechanism could be used repetitively. Positional information leads to different determinations with sharp boundaries and, in turn, at these boundaries the positional information for the next finer subdivision is generated. At least in the insects, there are so far no indications that such a mechanism is used more than twice; a primary gradient specifies the segments and secondary gradients the proximo-distal pattern of the appendages. The next finer subdivision, the pattern formation within each segment or within each compartment around the circumference, follows different regulatory rules. Most important, intercalary regeneration of missing pattern elements is possible (Fig. 11c). The mechanism we envision for intrasegmental pattern formation—mutual activation of states that exclude each other locally—has already been discussed in connection with the SAP pattern during primary segmentation. This mechanism allows intercalation of missing structures if nonadjacent structures are juxtaposed, even if this intercalation is connected with a polarity reversal (20, 22). Such an intercalation may take place during normal development to insert many more structures into the original sequence of three structures which presumably form the primordial segment.

The mutual activation mechanism allows a much greater size of regulating morphogenetic fields. If a field has to be organized by a single diffusible substance, the field cannot be much larger than 1 mm, otherwise the time to set up the gradient would be too long (days instead of hours). If, however, the field consists of, let us say, ten structures that induce and stabilize their neighboring structures by ten diffusible substances, the field may be of the order of a centimeter. Indeed, systems which show intercalation of missing pattern elements (for instance, planarians or segments of cockroach legs) can be much larger than a millimeter.

Conclusion: How to Achieve Reproducible Pattern Formation During Embryonic Development

We have been able to construct mathematically precisely formulated models that provide a rather accurate description of many experimental observations (20). Each of these models postulates a basic principle by which a particular developmental step is achieved. According to these postulates, the linkage of the following steps would allow reproducible pattern formation in higher organisms.

1. The primary anterio-posterior pattern in a developing egg is achieved by local autocatalysis and lateral inhibition.
2. The positional information generated in this way controls the formation of two patterns which are superimposed on each other: A serially repeated pattern (for instance, the compartments), and a sequence of differing structures (e.g., *different* segments). The formation of the periodic pattern is the primary event. By an alternation between two or three alternative cell states, particular control genes determining individual segmental specifications become activated.
3. Nonsegmented structures such as the dorso-ventral pattern of an embryo or the pattern within segments seem to be generated by a direct long-range cross-activation of structures which exclude each other locally. Such a sequence is self-generating and self-stabilizing. A gradient may orient the sequence.
4. The dorso-ventral pattern can be made to arise perpendicular to the anterior-posterior pattern if a high point of the A-P pattern can only be generated at the border between cells of dorsal and of ventral determination.
5. Secondary embryonic fields that give rise to substructures such as limbs can arise at the boundaries between differently determined cells. Since these boundaries in turn result from the primary organization of the embryo, the secondary structures have necessarily the correct location, handedness, and orientation in relation to the main body axes of the developing embryo.
6. Periodic fine structures (for instance, hairs, bristles, leaves, feathers, or stomata) can be determined by a pattern which is generated by short-range autocatalysis and long-range inhibition. This mechanism ensures that these structures have a certain maximum and minimum distance from each other.

Thus, reliable pattern formation during development may consist of a chain of relatively simple biochemical interactions. Each step generates the precondition for the subsequent step. Some counter-intuitive properties of the expected pattern-forming reactions (20) are presumably the reason why these reactions have not yet been biochemically identified. We hope that a preconception of the possible mechanisms will be helpful in this respect.

During the elaboration of these models, many mechanisms initially taken into consideration failed, after mathematical formulation, to account for the experimental observations. It would be interesting to discuss which models *cannot* account for the experiments since this would show how restrictive the known experimental data really are for any model, but this would go far beyond the limits of this article. By computer simulations, I realized at which point my intuition was misleading. This improved my intuition and enabled the formulation of new or altered hypotheses, which were checked again. After a model consistent with the initially chosen experiments was found, such a model frequently could also account for phenomena for which the model was not originally designed. This is, of course, not a proof, but provides good support for a model.

Acknowledgment

Much of the work described in this article is the result of a very fruitful collaboration with Prof. Alfred Gierer for which I wish to express my sincere thanks. I thank Paul Whitington for a critical reading of the manuscript.

General References

MEINHARDT, H.: *Models of biological pattern formation*. New York: Academic Press (1982).
GIERER, A.: Biological features and physical concepts of pattern formation exemplified by hydra. *Curr. Tops. Devl. Biol.* **11**: 17–59 (1977).

References

1. Bateson, W.: *Materials for the study of variation*. London: Macmillan (1894).
2. Bohn, H.: Interkalare Regeneration und segmentale Gradienten bei den Extremeitäten von *Leucophaea*-Larven (Blattaria). I. Femur und Tibia. *Wilhelm Roux' Arch.* **165**: 303–341 (1970).
3. Bohn, H.: Interkalare Regeneration und segmentale Gradienten bei den Extremitäten von *Leucophaea*-Larven (Blattaria). III. Die Herkunft des interkalaren Regenerats. *Wilhelm Roux' Arch.* **167**: 209–221 (1971).
4. Bohn, H.: Extent and properties of the regeneration field in the larval legs of cockroaches (*Leucophaea maderae*). I. Extirpation experiments. *J. Embryol. Exp. Morph.* **31**, 3, 557–572 (1974).
5. Bryant, S.V., French, V., Bryant, P.J.: Distal regeneration and symmetry. *Science* **212**: 993–1002 (1981).
6. Bulliere, D.: Etude de la regeneration d'appendice chez un insecte: standes de la formation des regenerats et rapports avec le cycle de mue. *Ann. Embr. Morph.* **5**: 61–74 (1972).
7. Chandebois, R.: The dynamics of wound closure and its role in the programming of planarian regeneration. *Develop. Growth Differ.* **21**: 195–204 (1979).
8. Garcia-Bellido, A., Ripoll, P., Morata, G.: Developmental compartmentalization of the wing disk of *Drosophila*. *Nature New Biol.* **245**: 251–253 (1973).
9. Gerhart, J., Ubbels, G., Black, S., Hara, K., Kirschner, M.: A reinvestigation of the role of the grey crescent in axis formation of *Xenopus laevis*. *Nature* **292**: 511–516 (1981).
10. Gierer, A.: Generation of biological patterns and form: Some physical, mathematical, and logical aspects. *Prog. Biophys. Molec. Biol.* **37**: 1–47 (1981).
11. Gierer, A., Meinhardt, H.: A theory of biological pattern formation. *Kybernetik* **12**: 30–39 (1972).
12. Girton, J.R., Pattern triplication produced by a cell-lethal mutation in *Drosophila*. *Dev. Biol.* **84**: 164–172 (1981).
13. Girton, J.R.: Genetically induced abnormalities in *Drosophila*: Two or three patterns? *Am. Zool.* **22**: 65–77 (1982).

14. Haynie, J., Schubiger, G.: Absence of distal to proximal intercalary regeneration in the imaginal wing discs of *Drosophila melanogaster*. *Dev. Biol.* **68**: 151–161 (1979).

15. Kalthoff, K.: Specification of the antero-posterior body pattern in insect eggs. In: *Insect development*. Lawrence, P.A., (ed.). Oxford, London: Blackwell (1976).

16. Lewis, E.B.: A gene complex controlling segmentation in *Drosophila*. *Nature* **276**: 565–570 (1978).

17. Malacinski, G.M., Chung, H.M., Asashima, M.: The association of the primary embryonic organizer activity with the future dorsal side of amphibian eggs and early embryos. *Develop. Biol.* **77**: 449–462 (1980).

18. Meinhardt, H.: A model for pattern formation in insect embryogenesis. *J. Cell Sci.* **23**: 117–139 (1977).

19. Meinhardt, H.: Cooperation of compartments for the generation of positional information. *Z. Naturforsch.* **35c**: 1086–1091 (1980).

20. Meinhardt, H.: Models for biological pattern formation. London: Academic Press (1982a).

21. Meinhardt, H.: The role of compartmentalization in the activation of particular control genes and in the generation of proximo-distal positional information. *Am. Zool.* **22**: 209–220 (1982b).

22. Meinhardt, H., Gierer, A.: Generation and regeneration of sequences of structures during morphogenesis. *J. Theor. Biol.* **85**: 429–450 (1980).

23. Postlethwait, J. H.: Development of cuticular pattern in the legs of a cell lethal mutant in *Drosophila melanogaster*. *Wilhelm Roux' Arch.* **185**: 37–57 (1978).

24. Russell, M.A., Girton, J.R., Morgan, K.: Pattern formation in a ts-cell-lethal mutant of *Drosophila*: The range of phenotypes induced by larval heat treatment. *Wilhelm Roux' Arch.* **183**: 41–59 (1977).

25. Sander, K.: Pattern specification in the insect embryo. In: Cell patterning. *Ciba Foundation Symp.* **29**: 241–263. Amsterdam: Associated Scientific Publishers (1975).

26. Sander, K.: Formation of the basic body pattern in insect embryogenesis. *Adv. Insect Physiol.* **12**: 125–238 (1976).

27. Schubiger, G.: Anlageplan, Determinationszustand und Transdeterminationsleistungen der männlichen Vorderbein-scheibe von *Drosophila melanogaster*. *Willhelm Roux' Arch.* **160**: 9–40 (1968).

28. Schubiger, G., Schubiger, M.: Distal Transformation in *Drosophila* leg imaginal disc fragments. *Dev. Biol.* **67**: 286–296 (1978).

29. Spemann, H., Mangold, H.: Über Induktion von Embryonalanlagen durch Implantation artfremder Organisatoren. *Wilhelm Roux' Arch. Entw. Mech. Org.* **100**: 599–638 (1924).

30. Steiner, E.: Establishment of compartments in the developing leg discs of *Drosophila melanogaster*. *Wilhelm Roux' Arch.* **180**: 9–30 (1976).

31. Strub, S.: Leg Regeneration in insects: An experimental analysis in *Drosophila* and a new interpretation. *Dev. Biol.* **69**: 31–45 (1979).

32. Szabad, J., Simpson, P., Nöthiger, R.: Regeneration and compartments in *Drosophila*. *J. Embryol. Exp. Morph.* **49**: 229–241 (1979).

33. Wieschaus, E., Gehring, W.: Clonal analysis of primordial disc cells in the early embryo of *Drosophila melanogaster*. *Dev. Biol.* **50**: 249–263 (1976).

34. Wolpert, J.: Positional information and the spatial pattern of cellular differentiation. *J. Theoret. Biol.* **25**: 1–47 (1969).

35. Wolpert, L.: Positional information and pattern formation. *Curr. Top. Dev. Biol.* **6**: 183–224 (1971).

36. Wright, D.A., Lawrence, P.A.: Regeneration of segment boundary in *Oncopeltus. Dev. Biol.* **85**: 317–327 (1981).

Questions for Discussion with the Editors

1. *What kinds of experimental approaches do you feel will bridge the gap between your sophisticated reaction-diffusion models and our knowledge of how the genome is organized and expressed?*

All the models provide strong predictions which are testable. Of course, it would be nice to have the substances isolated in a test tube or to monitor directly their regulation during development. By uncovering the counterintuitive properties the pattern-forming reactions responsible for primary embryonic organization must have, the models provide a rational explanation for why these substances have not yet isolated and will help to avoid pitfalls in future attempts (22).

For secondary embryonic fields that become organized from boundaries of differently determined cells the best strategy would be to aggregate together dissociated cells of the different "compartments." This would lead to many new sites where cells of the different compartments become juxtaposed and thus to a greatly enhanced morphogen production. This could lead to its identification and isolation.

For the generation of segmental specifications, the model proposes a stop-and-go mechanism which allows a counting on the DNA-level. The Bithorax gene complex of *Drosophila* is a most obvious candidate of such a region on the chromosome. It has been sequenced to a large extent so that its regulatory features can be deduced from the sequence. The model will presumably help to interprete that sequence data.

2. *In Chapter 20, experiments are described in which the medio-lateral patterning of the amphibian mesoderm can be altered by the presence of already committed cells from an older embryo. Can that observation be accounted for by a reaction-diffusion model?*

Yes. The model (Fig. 7) predicts that size of a particular structure in the medio-lateral pattern is determined by mutual activation of the participating structures. This can be realized by a "self-limitation" of each structure, providing in this way an advantage for other, competing structures. For instance, implantation of presumptive somite mesoderm could lead to an emission of a somite-inhibitor from these cells and thus to a reduction of the area which newly formed somites can occupy, in agreement with Cooke's observation.

CHAPTER 4

Pattern Generation and Regeneration

Stuart A. Kauffman

THE PHILOSOPHER LUDWIG WITTGENSTEIN once remarked that avoidance of errors was significantly enhanced by consideration of more than one theory for the same phenomenon. In this chapter, I take his advice and discuss a number of alternative models of pattern generation and regeneration. The chapter falls into four sections: First I briefly review the concept of positional information, and describe some of the basic phenomenology of epimorphic pattern regulation. Second, I discuss the strengths and weaknesses of three alternative possible "coordinate systems" for that positional information—polar, Cartesian, and spherical—and show that none of these models can account for current data if restricted to account for pattern regeneration by processes of *intercalation* alone. This review leads in the third section to a more general critique and the suggestion that attempts to understand pattern generation should focus on theories which naturally link the actual geometry of a tissue to the profiles of positional fields within them. In the fourth section, I discuss the previously unrecognized capacity of the familiar reaction-diffusion class of models not only to generate and regenerate positional gradients but also to generate *two* dimensional positional information naturally reflecting tissue geometry.

Part One: Positional Information

This volume attests a renewal of deep interest in the problem of pattern formation in developmental biology. In large measure, the resurgence of enthusiasm coincides with Wolpert's reformulation of this fundamental problem in terms of the concept of positional information (34; and see Chapter 1). Prior to Wolpert's introduction of the positional information concept, the dominant theory available postulated the existence of developmental fields possessing "prepatterns" (29), nonuniform spatial distributions of biochemical substances in a tissue, whose local peaks would induce the formation of pattern elements such as specific digits, sensillae, or bristles.

In contrast to Stern (29), Wolpert proposed the more abstract idea that cells within a developmental field possess positional information about their locations with respect to the boundaries of the field, through access to an underlying positional coordinate system. The behavior of each cell in the field was assumed to be due to two independent processes. The cell *assesses* its position, then *interprets* its positional information according to the type of cell it is, and forms a specific structural element in the overall pattern.

The chief difference between Wolpert's positional information and Stern's prepattern, is that positional information is free of assumptions about the existence of specific morphogen peaks underlying the subsequent differentiation of specific pattern elements. This freedom in one sense makes the theory of positional information less predictive, yet allows for two important possibilities; first, that the positional information system in *all* the developmental fields of one organism are identical, and, more radically, that the positional information system in all organisms or all epimorphic fields is identical (34, 35).

The general success of the positional information concept led to a search for the coordinate system which supplies positional information. At present, polar (4, 9), spherical (24), and generalized Cartesian (7, 16, 18, 27, 33), models have been proposed. Differences among these alternatives are far from trivial. Although it is always possible mathematically to transform from one to another coordinate system, the "forces" or tissue properties which must be postulated to explain the observed features of pattern regulation differ sharply in the different models. While the morphogens have yet to be found, one task in this area of biology consists of efforts to discover the coordinate system and "forces" or requisite cell properties that most simply account for the observations, with the hope that the proper formulation will provide both macroscopic laws at the tissue level, and aid in discovery of the underlying molecular variables.

The Phenomena

To assess the relative success of the alternatives that have been proposed, it is necessary to review briefly at least some of the major phenomena of pattern generation and regeneration.

1. Intercalary regeneration of intervening structures

If an amphibian limb capable of regeneration is transected proximal and distal to the elbow, and the distal wrist fragment grafted to the proximal shoulder stump, cell proliferation forms in the wound area, followed eventually by regeneration of the missing elbow region. This basic phenomenon is fairly ubiquitous (27) and fundamental. Juxtaposition of normally nonadjacent tissues from a single develop-

FIGURE 1. Gradient in concentration of [S] provides proximal/distal positional information in amphibian limbs. Serial threshold levels specify pattern elements A, B, . . . I. Removal of midregion of limb, D, E, F, and grafting, creates discontinuity, stimulating cell
proliferation. Diffusive smoothing of gradient discontinuity regenerates missing gradient
levels (wavy lines), and structures *DEF*. Reprinted from (17), with permission.

mental field is generally followed by regeneration, in proper spatial order, of the
structures normally lying between the juxtaposed tissue edges. This notion of "betweenness" is necessarily central to any theory of pattern formation.

The simplest physical model to account for "betweenness" in intercalary regeneration postulates the existence of one or more chemical concentration gradients spanning the tissue, whose concentration levels specify the positional information of cells at each point in the domain. As shown in Fig. 1, in which a proximal/
distal gradient along an amphibian limb is envisioned, surgical removal of the elbow and grafting of wrist to shoulder creates a discontinuity in the gradient, [S], at
the graft junction. If one imagines that gradient concentrations remain fixed in the
"old" tissue fragments, while diffusion occurs in the new cells of the wound blastema, then simple diffusive averaging of the concentration discontinuity at the
graft junction smooths over the discontinuity, recreating all the intervening gradient values in proper spatial order.

For the remainder of this review, I shall adopt the postulate that position is
specified by graded scalar properties in tissues, although it is important to stress
that alternative discrete models based on concepts analogous to sequential inductions have been formulated. Given the postulate of positional gradients, a fundamental question is the extent to which the simple property of diffusive-like averaging of gradient discontinuities can account for the phenomena of pattern formation
and regeneration. This simple property turns out to be sufficiently powerful to account for much, but not all, of the available data. This failure points to additional
critical processes in pattern generation and regeneration.

2. SEQUENTIAL FORMATION OF POSITIONAL AXES IN DEVELOPMENT

In several systems, positional axes appear to be established sequentially during
development. In classical experiments, Harrison (11, 12) removed the right forelimb bud of *Ambystoma* and grafted in its stead the left forelimb bud. Such grafts
must invert either the anterior-posterior limb axis while keeping donor and host
dorsal and ventral axes aligned; or invert dorsal-ventral donor and host axes, while
keeping anterior and posterior axes aligned. Harrison found that if very early left
limb buds were grafted to the right, they developed into normal right limbs. If late
left limb buds were grafted, they formed normal left limbs with that axis inverted
with respect to the host that was inverted at surgery. But if left to right grafts were

made at an intermediate stage, the outcome depended upon which axis was inverted at the graft junction. If the anterior-posterior axis remained normally aligned and the dorsal-ventral axis was inverted, the donor left limb bud formed a right limb, while if the dorsal-ventral axis remained aligned and the anterior-posterior axis was inverted at the graft junction, the donor left limb bud formed a left limb which remained inverted at the host donor junction. Harrison interpreted his results to imply that the donor anterior-posterior axis becomes autonomously self sustaining prior to the dorsal-ventral axis. Similar data suggest that amphibian eye axes (14), as well as limb axes are established sequentially, although the status of data on the eye is in dispute.

3. DISTAL TRANSFORMATION

If an amphibian limb or tail is transected at the elbow, the proximal stump can form a regeneration blastema and regenerate the distal limb (11). If the digits of the transected distal fragment are implanted into a host flank to establish an adequate blood supply to the distal fragment, the cut surface at the elbow, which initially faced proximally, forms a regeneration blastema and regenerates second distal wrist and hand structures that are mirror symmetrical to the initial implanted distal limb (11, 12). In these experiments, both the proximal stump and the implanted distal limb fragment regenerate the *same* set of distal limb structures from the cut surfaces at the elbow, identified as a regenerate in the proximal stump and duplicate on the implanted distal limb fragment. The fact that both fragments form distal limb has been called the "rule of distal transformation" (23). Similar results have been found in many insect legs and in the imaginal discs of *Drosophila* (9).

4. SUPERNUMERARY LIMBS

Among the most striking observations in pattern regulation is the induction of extra or supernumerary limbs following grafting operations. After both the anterior-posterior and dorsal-ventral amphibian limb axes are fixed, transplantation of a left distal limb to a right proximal stump which reverses the anterior-posterior axis of donor relative to host but leaves the dorsal-ventral axis aligned, typically results in formation of two supernumerary limbs at the anterior and posterior margins of the donor host junction. If instead the anterior-posterior axes of host and graft are aligned, but dorsal-ventral axes are inverted, the two supernumeraries emerge from the dorsal and ventral margins of the host donor junction. These supernumerary limbs generally have the handedness of the proximal stump (5, 11, 12). Similar results have been obtained in transplantation of cockroach limbs (1, 6).

Rotation of a left distal limb by 180° and regrafting to its own proximal stump results in a more variable range of results. After such rotations, the donor may partially rotate back toward its normal alignment; sometimes zero, one, two, or more supernumerary limbs are formed at the graft site, with the same or opposite handedness (1, 5, 6, 9).

5. Duplication and regeneration by
complementary fragments of a tissue

Distal transformation by both proximal and distal limb fragments of amphibians is one example of duplication and regeneration by complementary fragments of a developmental field. The phenomenon is common, however, and has been studied in greatest detail in the imaginal discs of *Drosphila melanogaster*.

Drosophila *Imaginal Discs*

Drosophila melanogaster is a holometabolous insect with egg, larva, pupa, and adult stages. During metamorphosis, the larval ectoderm lyses, and the ectoderm of the adult is formed by the terminal differentiation of special larval organs called imaginal discs (10; and see Chapter 12). In the late third instar larva, each imaginal disc is a two-dimensional sheet of cells forming the surface of a hollow sphere. The columnar cells on one hemisphere form the imaginal disc proper, while the thin squamous cells on the remaining hemisphere form the peripodal membrane which is lost during metamorphosis (10). Imaginal discs are found as bilaterally symmetric pairs, each destined to form specific left and right regions of the adult ectoderm: the left and right first leg discs form the two prothoracic legs; the two wing-thorax discs form the left and right mesothoraces and wings, etc.

By injecting specific fragments of each disc into host larvae for metamorphosis, it has been possible to construct a fate map of each type of imaginal disc (3). For example, the fate map of the wing-thorax (hereafter wing) disc, Fig. 2, shows that the upper and lower margins of the disc along its longitudinal axis form ventral and dorsal thoracic structures, while the midregion of the disc forms wing structures. During metamorphosis, the wing disc folds along a bent arc running from the anterior to posterior disc margin, and apposes ventral and dorsal thorax areas, ventral and dorsal wing hinge areas, and ventral and dorsal wing blade areas, creating a "bag" that everts through the peripodal membrane. The center of the disc maps to the distal wing tip, while an arcing line from anterior to posterior disc edge through the disc center corresponds to the wing margin.

Grafting experiments like those in amphibian or cockroach limbs are not yet feasible in *Drosophila*. However, closely analogous experiments can be performed by cutting the wing disc into known fragments, and injecting each fragment into the abdomen of a fertilized adult female. In that environment, the disc fragment heals its cut edge, apposing tissue regions which are normally nonadjacent and new cells form in the wound area. Over a week in culture, the mass of a disc fragment typically doubles. After culture in adult abdomens, fragments may be recovered and injected into host larvae for metamorphosis, then recovered from the emerged adult host. By comparison of the patterns of hairs, sensillae and bristles which form when a known disc subfragment is injected directly into larvae for immediate metamorphosis, it is possible to characterize the pattern regulation which occurs in the cultured fragment.

The following are the dominant results:

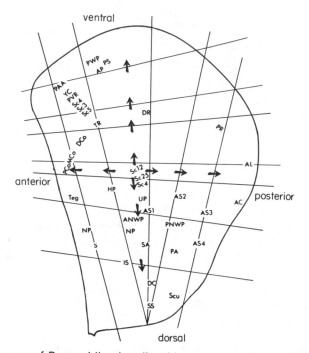

FIGURE 2. Fate map of *Drosophila* wing disc. Lines show positions of single cuts. Arrow across each line points from the fragment created by that cut which regenerates, into the fragment which duplicates (2). Abbreviations for wing disc pattern elements: NP, notopleural bristles; SA, supraalar bristles; PA, postalar bristles; DC, dorsocentral bristles; Scu, scutellar bristles; ANWP, PNWP, anterior and posterior notal wing processes; Teg, tegula; HP, humeral plate; UP, unnamed plate; AS1-4, first to fourth axillary sclerites; Pco, Mco, Dco, proximal, medial, and distal costa; TR, triple bristle row (anterior wing margin); DR, double bristle row (distal wing margin); PR, posterior row of hairs; Sc4, Sc25, Sc12, groups of sensilla campaniformia on the proximal dorsal radius; AL, alar lobe; AC, axillary cord; PAA, prealar apophysis; YC, yellow club; PVR, proximal ventral radius; PWP, pleural wing process; PS, pleural sclerite; AP, axillary pouch; Sc4, Sc3, Sc5, groups of sensilla campaniformia on the proximal ventral radius; anterior, posterior, ventral and dorsal disc margins. Reprinted from (17), with permission.

1. If the wing disc is cut in two fragments by a straight cut, the roughly smaller fragment duplicates some or all of its structures. In favorable cases, a mirror symmetrical duplicate is generated whose symmetry line lies near the cut on the wing disc; the larger complementary fragment regenerates structures normally formed by its smaller complement (2). Thus complementary fragments often exhibit complementary behavior, one duplicating, one regenerating. Consequently, it is possible to draw an arrow across each single straight cut on the disc, pointing from the fragment which regenerates into that which duplicates. As shown in Fig. 2, such arrows point radially outward from a small region in the interior of the disc. The polarity of regeneration and duplication reverses around this interior point, termed the "high point."

2. If the disc is cut into arbitrary ¾ and ¼ "pie" sectors, the ¾ fragment regenerates, the ¼ fragment duplicates.

3. If an interior "distal" circular region containing the "high point" is cut and cultured, it duplicates, the duplicate forming a mirror symmetric hemisphere to the original high point region. If the outer "proximal" annulus remaining after the high point region is removed is cultured, the outer annulus regenerates the central "distal" high point region (2).

4. If two narrow normally duplicating crescent fragments from opposite edges of the disc are mixed, they regenerate the intervening pattern elements spanning across the disc.

5. It is critical that complementary fragments do *not* strictly duplicate and regenerate. Under different experimental conditions and with varying frequencies, normally duplicating fragments can regenerate most of the pattern elements in the complementary fragment (8, 18, 25).

6. Anterior-posterior and ventral-dorsal asymmetries exist in the capacity of narrow duplicating fragments to regenerate: Narrow anterior margin and ventral thorax margin structures have been found by a number of workers to regenerate extensively, while narrow posterior and dorsal margin fragments do not (8, 18). The results in other imaginal discs are fundamentally similar although it should be stressed that an interior "high point" has not been demonstrated in other discs.

7. Mutants causing pattern duplication in different imaginal discs have been analyzed by a number of workers (e.g., Jürgens and Gateff (15) and will be discussed further.

As discussed in the next section, none of the current alternative proposed coordinate systems can account for the phenomena described based on smoothing of positional discontinuities alone.

Part Two: Models of Positional Coordinate Systems

The first major advance in predictive use of the positional information hypothesis lay in the formulation of the polar coordinate, or "clockface" model for pattern regulation in epimorphic fields by French *et al.* (9) and Bryant (3). The initial model was based on some of the results described in amphibian limbs, cockroach limbs, and imaginal discs, and is well illustrated by application to the wing disc of *Drosophila*.

The existence of a "high point" in the central (distal) region of the disc (Fig. 2) about which the direction of regeneration reverses suggested that cells might measure their distance in the tissue from this special point. This raised the possibility that the position of cells is specified in a polar coordinate system with the high point as its origin. Since the wing disc is a two-dimensional surface, an azimuthal angle must be measured. Were an angle specified by a single scalar variable, that variable would necessarily be discontinuous along some radial line in the tissue; for example along the radial line $\phi = 2\pi = 0$. But Bryant found that any ¼ pie-wedge fragment of the wing disc duplicates, while its ¾ complement regenerates. If an azimuthal discontinuity were present, the ¼ wedge fragment containing it should behave differently from the rest, and regenerate. Failure to find evidence of an angular discontinuity implies that, if cells measure an angle, they do so seamlessly. Therefore the model proposes that cells measure radial distance from a distal "high point" origin, and an angle seamlessly modulo 2π.

In order to account for the bulk of the data on epimorphic regeneration and duplication, the polar coordinate model initially proposed two rules of intercalary regeneration, and a third special rule for distal transformation.

Rule 1. If cells having different radial values are apposed, cell proliferation will be stimulated, and the missing intervening radial values will be restored back to a resting radial gradient, then proliferation will cease.

Rule 2a. If cells having different angular values are apposed, cell proliferation will be stimulated and the missing angular values will be intercalated back to a resting angular gradient.

Rule 2b. Since two angular arcs around a 2π circle of values join any two juxtaposed angular values, a choice rule is needed. The simplest postulates that angular intercalation occurs along the *shorter* arc.

Special Rule 3, the complete circle rule. If a complete circle of angular values at a proximal radial level is exposed, distal regeneration occurs. The special nature of Rule 3 will be discussed later.

The two intercalation Rules 1 and 2a, b, suffice to explain major features of epimorphic pattern regulation. For example, the classical example of grafting an amphibian hand to shoulder with regeneration of the missing elbow region is explained by Rule 1. Proximal radial values in the shoulder apposed to distal radial values in the hand lead to intercalation of the intervening radial values specifying elbow.

Duplication and regeneration by complementary fragments of the wing disc is explained by Rules 2a, b. Figure 3 shows a single, straight cut on the wing disc, yielding a narrow anterior fragment and a large posterior fragment. During culture, the narrow anterior fragment folds over, apposing the cut edge such that the ventral and dorsal thoracic regions heal together, ventral and dorsal wing blade regions also heal together. This healing juxtaposes cells with similar radial values, but discordant angular values. This discontinuity stimulates cell proliferation, and smoothing of the angular discontinuity along the shorter angular arc. Since this shorter angular arc is the arc present in the original narrow anterior fragment, the positional values in the new cells form a mirror symmetrical duplicate of those in the original anterior fragment, and the fragment duplicates.

The positional values present along the cut margin of the large posterior fragment are identical to those along the cut margin of the narrow anterior fragment. If the large fragment wound heals in a similar way, the same pairs of positional values must be apposed in the posterior fragment as in the anterior fragment. Therefore, the posterior fragment must intercalate, along the shorter angular arcs, the same intervening positional values as did the anterior fragment. Therefore the posterior fragment regenerates.

The polar model demonstrates a more general result. Whatever the coordinate system specifying position in a developmental field may be, the positional values along the two margins of a single cut are identical. If the two fragments heal in similar ways, both will appose essentially identical pairs of positional values.

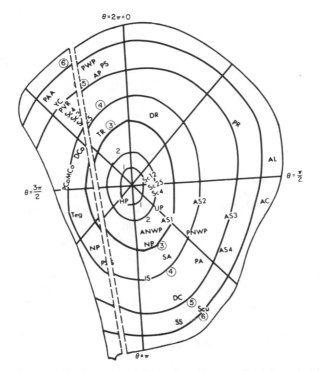

FIGURE 3. The polar coordinate model of the wing disc. Radial values (1–6) are measured from the "high point." Angle, Φ, is measured without discontinuity modulo 2π A straight cut anterior to the high point creates a narrow anterior fragment and wide posterior fragment. The latter heals by folding along its wound margin, apposing positions with equal radial but discordant angular values. Cell proliferation and smoothing of angular discontinuities along the short arc between apposed values leads to regeneration by the posterior fragment. The narrow anterior fragment folds and heals along its wound margin, juxtaposing the same positional values as did the posterior fragment. Short arc intercalation leads to duplication by the anterior fragment. Reprinted from (17), with permission.

Therefore, if subsequent pattern regulation is governed by diffusivelike averaging of positional disparities, both complementary fragments must reform the *same* set of structures. If one fragment duplicates, the second must regenerate. *The prediction of complementary behavior in complementary fragments is a coordinate free property* which follows from the postulates of graded positional value, and averaging of disparities.

Since the polar model is symmetric about the "high point," a narrow posterior fragment will duplicate, its complement will regenerate. Therefore the direction of regeneration will reverse about the high point. Further, any ¼ wedge section will heal its two cut margins and duplicate, while its ¾ complement will regenerate.

Intercalary Regeneration, Betweenness, and Convex Sets

A central feature of the postulates of intercalary smoothing is that only those positional values lying *between* the apposed values can be reformed. This leads to a critical restriction in the predictive consequences of any given coordinate system, since it implies that diffusivelike smoothing can *only* recreate positional values lying in the *convex set* bounded by the positional values in the apposed tissue edges. This restriction, in turn, implies that different coordinate systems may demand different special cellular behaviors beyond simple diffusivelike smoothing to account for the data.

The concept of a convex set, and the limitations it imposes can be brought out in the polar coordinate model. Radial positions can be visualized without loss of generality as a radially symmetric gradient whose peak is at the distal "high point." In Fig. 4, I show a wing disc from which the distal high point region has been removed, thus removing the radial gradient "peak." In the remaining outer proximal annulus, only lower values of the radial gradient are present. Therefore, no juxtaposition of tissue edges in the proximal annulus can lead to "diffusivelike" filling in of the missing high-point radial peak. In a polar model, if the region containing the origin is removed, that region does not lie "between" the positional values in the proximal annulus. That is, the region around the origin is not in the *convex set* of all those positional values derivable by averaging *any* pairs of positional values present along the cut margin of the proximal annulus.

The implication of this feature of any polar coordinate model is that distal re-

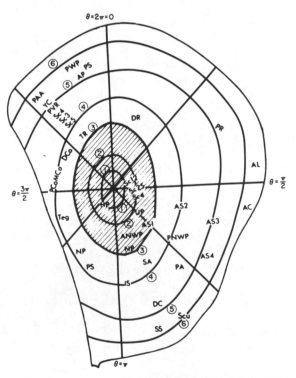

FIGURE 4. Polar coordinate system on wing disc, with distal wing "high point" region removed. No parts of proximal annulus have missing distal radial values. Therefore, no tissue contact and positional averaging can intercalate the missing distal radial values. Reprinted from (17), with permission.

generation by a proximal fragment cannot be obtained by diffusivelike averaging of positional discontinuities, and some special rule is needed. In the initial formulation of the polar coordinate model, special Rule 3, the complete circle rule was proposed. According to this rule, exposure of a complete circle of angular values at a proximal level leads to regeneration of distal radial values. With the assumption of this rule, the model accounts for the capacity of a truncated amphibian limb to undergo distal transformation, and regenerate distal structures. The same postulate accounts for distal regeneration by a proximal wing disc annulus. Equally, Rule 3 accounts for duplication of the cultured wing disc high point region from the exposed circle of angular values at a proximal radial level, and similarly explains duplication from the proximal cut surface by a newt hand whose digits are implanted into a host flank. Finally, Rule 3 accounts for the striking observation that grafting a left hand to a right stump yields two supernumerary limbs at the position of maximal discord of angular values. Such a graft creates two complete circles of angular values at the radial level of the graft. These undergo distal transformation and yield two supernumeraries, with the handedness of the host (9).

The successes of the polar model are considerable, and have led to testing several features of the model. I turn now to an evaluation of weaknesses in this model.

First, it should be stressed that special Rule 3, or any modified form of Rule 3, is formally equivalent to postulating a mechanism beyond diffusivelike averaging of positional values, to regenerate a missing radial gradient "peak." For example, in the current version of the polar coordinate model (4), its authors propose that short arc intercalation itself generates distal regeneration. While the postulate of a special mechanism to regenerate distal radial values is not a flaw, its status should be made explicit. As shown next, distal regeneration does not require special rules in other coordinate systems. If special mechanisms exist to recreate missing gradient values, then those mechanisms are of central importance. They are likely to play a role in the initial *establishment* of positional gradients as well as subsequent pattern regulation. The forms of special mechanisms which are suggested depend on the choice of coordinate system. Thus different coordinate systems suggest that different cellular properties beyond diffusive averaging are required to account for the data on pattern formation.

Second, as noted by Winfree (32, 33), in the vicinity of the origin, the gradient of the angular variable in the tissue becomes infinitely steep, since the origin is a singularity in the angular map on the plane. If cell proliferation is proposed to occur proportional to the gradient in radial or angular values, then proliferation near the origin should be unbounded. It is not. As Winfree points out, this implies a lack of independence of radial and angular gradients near the origin.

Third, anterior-posterior and dorsal-ventral developmental axes appear to be established sequentially. This is not deducible, hence not explained, by the proposal that position is measured in polar coordinates. For example, if a radial axis is established first, the amphibian limb bud should be angularly symmetric in transplantation between the time when the first axis and second axis are formed, in contravention to Harrison's classical observation (12). If angular values are established first, then reversal of anterior-posterior or dorsal-ventral areas should yield similar, not dissimilar results, at each stage, but do not (12).

Fourth, a considerable body of work testing the complete circle rule has shown it to be false in imaginal tissue, cockroach limbs, and amphibian limbs (18, 26, 28). It now appears that distal regeneration is roughly proportional to the proximal angular arc present. Confirmation of the complete circle rule would have been striking evidence in favor of a polar model. Its disconfirmation does not rule out such a model, but alternative coordinate systems have the deductive consequence that distal regeneration should be proportional to the proximal arc; hence the data disconfirming the complete circle rule leave a polar model with an *ad hoc* postulate as Rule 3, while alternatives deduce, hence explain, this phenomenon.

Fifth, accumulating evidence shows that normally duplicating imaginal disc fragments can with variable frequency regenerate some, most, or perhaps all of the complementary domain. Such results are now known in the wing, haltere, leg, and genital discs (8, 18, 25). No convincing comparable data are yet available on amphibian or cockroach limbs or other epimorphic systems, although *divergent regeneration* from symmetrical half limbs in amphibians (13) might be due to regeneration by normally duplicating fragments.

The polar coordinate model is constructed such that intercalation (Rules 1 and 2a, b) causes complementary fragments strictly to duplicate and regenerate (Fig. 3). The model is unable to account for regeneration by normally duplicating disc margin fragments by simple positional averaging. In fact, no proposed coordinate system, coupled with the assumptions of simple diffusive averaging of apposed positional values, can account for extensive regeneration by both the normally duplicating and normally regenerating fragments of a tissue. All coordinate systems share the property that both complementary fragments have identical positional values along their cut margins, hence if both fragments heal in similar ways, apposing similar pairs of positional values, both fragments must reform the same set of structures lying in the convex set bounded by the apposed positional values. Evidence that each complementary disc fragment can regenerate large domains of the other fragment is evidence that mechanisms beyond simple diffusive smoothing must operate during imaginal disc pattern regulation. The polar coordinate model can account for this phenomenon by weakening the short arc intercalation rule to allow long arc intercalation. Formally, this is equivalent to an active mechanism to regenerate angular values not in the convex set reached by diffusive smoothing.

Sixth, mutants in *Drosophila* are available which can produce completely duplicated legs (15). In order for a disc fragment to produce a complete duplicate appendage by apposing cut surfaces of the fragment and positional smoothing, the apposed surfaces would have to form a convex set containing the *entire* tissue image. Examination of the polar coordinate model (Fig. 3) shows no single fragment has this property. Within the frame of the polar model, a *partial* disc fragment can produce a mirror symmetric duplicate, but no fragment of a normal disc can produce both a normal leg, and a complete duplicate appendage.

Transverse Gradients, or Modified Cartesian Coordinate System

Recently several workers have independently suggested a modified form of a Cartesian coordinate model to account for the data on epimorphic pattern regulation (7, 16, 18, 27, 33). Suggestions by Winfree and myself are nearly identical. Figure

6a shows the *Drosophila* wing disc with roughly orthogonal monotonic anterior-posterior and ventral-dorsal gradients of two chemicals, X and Y. Lines of constant concentration are bowed outward on the disc, symmetrically about the "high point." Figure 5 shows the "image" of this cross gradient system in a "tissue specificity space," (18, 33) i.e., the XY morphogen space. In this "image," lines of constant X or Y chemical concentration, are straight lines, so a convex, bowed line of constant X in Figure 6a is a straight line in Figure 5. Similarly, a straight cut on the actual wing disc corresponds to a *concave* line in Figure 5 (see legends). The model assumes that a cut wing disc fragment heals its cut margin, apposing nonadjacent positional values, and that simple diffusion smooths discontinuities in X and Y, and fills in the convex set bounded by the apposed XY pairs along the cut margin. As shown in Figures 5 and 6a the bowing of lines of constant X and Y concentration on the disc implies that diffusive smoothing of X and Y discontinuities in new cells of a large posterior fragment of a single straight cut will fill the shaded convex set and *regenerate* anteriorly to the anteriormost value present along its cut margin. The complementary anterior fragment apposes the same pairs of discordant values, and duplicates to the same anteriormost value.

Symmetrical convex bowing of X and Y concentrations about the high point insures that the direction of regeneration reverses about the high point. A large anterior fragment from a straight cut will regenerate posteriorly, its posterior complement will duplicate. Similarly, any ¼ pie-wedge fragment which apposes its

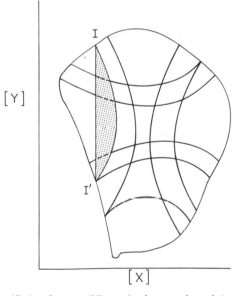

FIGURE 5. Tissue Specificity Space (33) assigning each point on wing disc a unique value of X and Y. *Bowed* concave lines are "images" in XY morphogen Tissue Specificity Space of *straight* cuts on the wing disc like those in Fig. 2. A single cut from I to I' on the actual disc lies along the concave arc from I to I' in morphogen space. Wound healing apposes the wound margin in the posterior fragment. Juxtaposition creates positional discontinuities that are smoothed by diffusion to fill the convex set, shaded area, bounded by cut margin. Reprinted from (17), with permission.

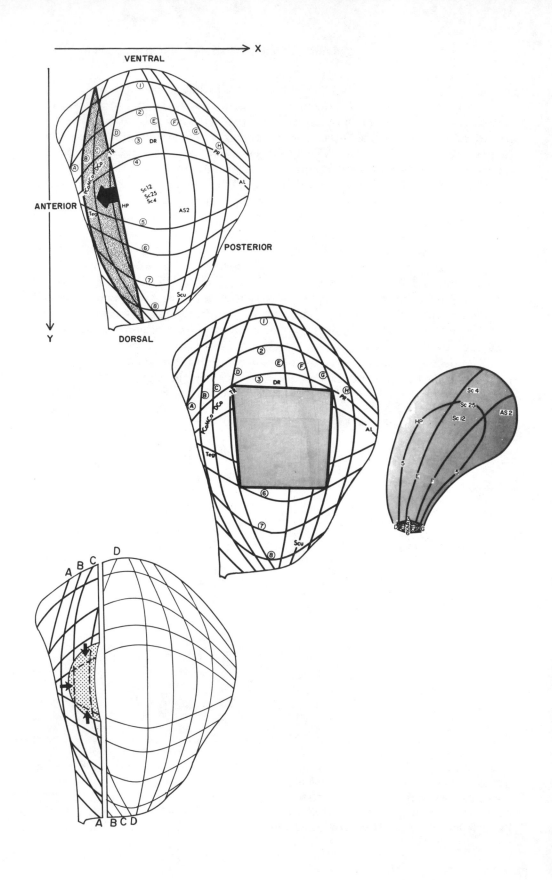

FIGURE 6. (A) A Cartesian coordinate system with two monotonic gradients, identical to Fig. 5, but showing the concentration profiles on the actual wing disc itself. The "Y" variable has ascending levels 0–9. The "X" variable has ascending levels A–H. A straight cut creates a large posterior fragment which wound heals by folding in half and juxtaposing opposite ends of the cut margin. For this particular cut, this leads to a discontinuity across the wound in Y values, but not in X values. The convexity in lines of constant concentration of each variable, *isocones*, assures that simple diffusive smoothing of X and Y variables yields regeneration out to the anteriormost X isocone, B, contained along the cut margin of the posterior fragment, stippled area. Similarly, the smaller anterior fragment wound heals the same way and duplicates to the B isocone. The symmetry of convexity in the X and Y isocones on the disc about a central region in the wing disc implies that the polarity of regeneration (heavy arrow) will alter about an apparent "high point" in this region, although cells do not measure position with reference to the high point. Regeneration by a normally duplicating fragment requires an active mechanism beyond simple diffusion to recreate the gradient extremum at the disc margin of the normally regenerating fragment. (B) Removal of a central (distal) region leaves a proximal outer annulus. Wound healing along the cut in the annulus juxtaposes tissues, creating discontinuities in the monotonic X and Y gradients, which are smoothed over by simple diffusivelike processes, leading to distal regeneration by the proximal annulus. The central square containing the high point itself purse-string closes juxtaposing tissues that again create X and Y gradient discontinuities across the wound healed zone. Smoothing of these discontinuities causes the positional values in the central square to be duplicated in the new cells of the wound zone, shown in the granular area. (C) Anterior disc fragment with a distal crescent removed, fills convex set, shaded area, and undergoes partial distal regeneration. Reprinted from (17), with permission.

two cut margins will duplicate, its ¾ complement will regenerate. Therefore, a transverse gradient Cartesian model can yield reversal of the direction of regeneration about the high point, without the assumption that the high point is a locus from which cells measure position.

Distal regeneration is a consequence of simple diffusive smoothing in a transverse gradient Cartesian model. As shown in Figure 6b, deletion of the high point region leaves a proximal annulus. Wound healing apposes tissue around the circular cut margin, thereby creating discontinuities in the X and Y gradients. Since the deleted high point region lies in the convex set reached by diffusive smoothing from the proximal annulus, the distal high point region is regenerated. Similarly, the high point region itself purse-string closes its circular wound margin, apposing the same pairs of positional values as did the proximal annulus, and therefore duplicates the X and Y high point values in the new cells of the wound area (Figure 6b). The same principle predicts that any proximal arc of tissue will undergo limited distal regeneration of structures lying in the convex set of the positional values along the cut margin (Figure 6c). Therefore, *limited* distal regeneration proportional to the proximal is present. It has recently been demonstrated in imaginal discs and amputated symmetric half limbs of amphibians and cockroaches, and is a deductive consequence of a Cartesian model.

As shown in Figure 7, a transverse gradient model explains the striking observations that grafts of distal left limbs to proximal right stumps can generate two supernumerary limbs with the handedness of the host. With minor distortion from symmetry, the Cartesian model also has the consequence that 180° graft rotations of the left hand onto the left stump can yield variable numbers of supernumeraries, including two supernumerary limbs of opposite handedness.

Finally, unlike the polar model, a transverse gradient Cartesian model is in direct accord with the widespread data suggesting that anterior-posterior axes are established prior to dorsal-ventral axes.

Any transverse gradient model suffers certain defects analogous to those of the polar model in which the "high point" region in the polar model lies outside the convex set of the proximal annulus. Its regeneration requires special mechanisms. In a transverse gradient Cartesian model, the extreme disc margins lie outside the convex sets of their complements. Thus, as shown in Figure 6a, the large posterior fragment only regenerates partially. It does not reach the anterior disc margin. Similarly, the narrow anterior fragment duplicates partially. Special mechanisms to recreate X and Y gradient maxima and minima are required to account for the observed capacity of disc fragments to regenerate or duplicate completely.

Like the polar model, no Cartesian model can explain the capacity of normally duplicating fragments to regenerate large domains of the complementary disc fragment by diffusive smoothing. Nor can a Cartesian model account by diffusive smoothing alone, for mutants that form complete limbs and complete duplicate limbs, since no subregion of a Cartesian coordinated tissue can have the entire

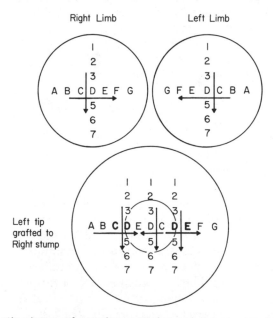

FIGURE 7. Schematic picture of anterior-posterior and dorsal-ventral Cartesian X, Y coordinates for left and right limbs, projected onto the plane. The position at the mid values of the two variables, 4, D, corresponds to the distal limb tip. Grafting a left distal limb to a right proximal stump leads to smoothing of discontinuities in X and Y (bold letters, CD and DE) and the formation of two supernumerary distal limb tips with the handedness of the host proximal stump, at the positions of maximal disparity of host and graft axes. Reprinted from (17), with permission.

tissue image in its convex set. The assumption of a transverse gradient model leads to the suggestion that cells in a tissue have special mechanisms capable of reforming X and Y gradient maxima and minima lying outside the convex sets of cultured fragments. Explicit mechanisms are described next.

Spherical Coordinate Model

Russell (24) proposed a spherical coordinate model using three orthogonal X, Y, and Z gradients to form a Tissue Specificity Space (32, 33). Position in a tissue is specified by a *solid* angle, Θ, Φ corresponding to latitude and longitude angles. Each positional value is a unique *ray* at a constant X:Y:Z ratio emanating from the origin. That is, each ray is a *line* of equivalent positional value in XYZ space. The longitudinal angle Φ, is defined by the ratios of X and Y in the equatorial plane, and the latitude angle Θ, is defined by the X:Z, or Y:Z ratio. The image in XYZ space of a two dimensional tissue is a two dimensional surface pierced by a set of solid angle rays emanating from the XYZ origin, which itself does not normally lie in the physical tissue (Figure 8).

The spherical model can account for almost all the available data using only the concept of diffusive smoothing. This is most easily visualized in the wing disc by remembering that the disc is topologically a spherical surface, the disc proper, backed by the peripodal membrane. Let this tissue spherical surface be embedded

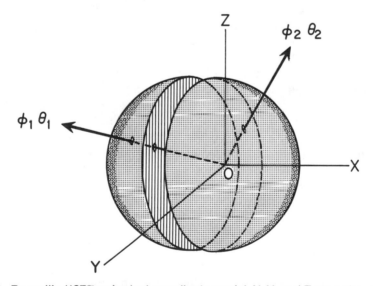

FIGURE 8. Russell's (1978) spherical coordinate model. X, Y, and Z are orthogonal gradients. Position is specified by ratio of X:Y:Z with respect to origin, 0; each ratio is a ray at a unique solid angle, Φ, Θ. The diagram shows a disc as a spherical surface (dotted region), with an anterior fragment cut off. Wound healing in the anterior fragment creates an additional new XYZ surface (parallel lines). Rays Φ,Θ, pierce both surfaces, hence the anterior fragment duplicates. The posterior fragment forms the same new XYZ surface and regenerates. Reprinted from (17), with permission.

in XYZ space such that each solid angle ray pierces the closed two dimensional surface of the image once, specifying the positional Φ, Θ value at that point. Therefore the image surrounds the origin, and each ray pierces a unique point on the tissue. If a narrow anterior fragment of the disc is cut, the margin purse string wound heals, and smoothing of X, Y, and Z discontinuities in the new cells forms a *second* surface in XYZ space whose edges join those of the original anterior fragment image (Fig. 8). Rays from the origin pierce both the new surface and the original anterior image surface (Fig. 8); hence the narrow fragment duplicates completely. The large posterior fragment wound heals similarly, and forms the same new image surface in XYZ space in the new cells of the wound area. These are pierced by the same rays that pierce the anterior fragment, hence the posterior fragment regenerates completely.

The model is spherically symmetric. Therefore not only do narrow anterior, posterior, dorsal or ventral fragments duplicate while their large complements regenerate, but a distal fragment containing the high point region will purse string close, creating a second image surface in the new cells in the wound area. This second surface will be pierced by the same rays that pierce the high point region, therefore it will duplicate the high point. Equally the proximal annulus will regenerate distally. Finally, this model directly explains distal regeneration proportional to the proximal arc cultured, and the incidence of supernumerary limbs.

Despite its strenths, the spherical model suffers weaknesses. First, the spherical model, like the polar model, has a singularity at the north pole high point, where all longitudinal angular values meet. The hypothesis that positional gradients regulate proliferation confronts infinitely steep longitudinal angular gradients at the "high point."

More importantly, while the spherical model is very appealing, it is unable to account for all the data by simple diffusive averaging of gradient discontinuities. Like the polar and Cartesian models, diffusion in the spherical model can yield only duplication and regeneration by complementary fragments. It cannot account for regeneration by normally duplicating fragments, nor the existence of mutants which form normal limbs, and complete duplicate limbs. To account for these phenomena, it is necessary to imagine cellular processes capable of recreating missing X, Y, or Z maxima and minima not in the convex set of the cultured fragment.

Part Three: General Critique

The review of the polar, Cartesian, and spherical coordinate models leads to several general comments.

I. What pattern regulation properties are coordinate-free, assuming only the fundamental process of intercalary smoothing of positional discontinuities?

1. Intercalation itself, in one sense is coordinate-free, occurring in any coordinate system with graded scalar variables. Yet it is not entirely free. In the polar model, for example, one might suppose *angular* intercalation occurs faster (or slower) than radial. In the former case, intercalary regeneration would initially spread around annular *rings*, in the latter, spread radially. In a Cartesian model, assumption of different X and Y intercalation rates leads to yet different directionalities to regeneration.

2. The sequential formation of anterior-posterior or dorsal-ventral axes is not coordinate-free.

3. Distal transformation using only positional smoothing is not coordinate-free for it does not arise in a polar model. Further, the arguments (21, 32, 33) that distal transformation can be coordinate-free implicity assumes embedding the tissue in a two variable coordinate system, i.e., in an R^2 space, with no singularities. This is equivalent to assuming *transversal gradients* of scalar variables. It is only coordinate-free in that the gradients need not be orthogonal.

4. Like distal transformation the generation of supernumerary limbs using only intercalation is not coordinate-free.

5. Duplication and regeneration by complementary fragments *is* coordinate-free in the restricted sense that both fragments must make the *same* new structures in pattern regulation. However, that any specific fragment should wholly duplicate or wholly regenerate is not coordinate-free, as seen in the specific "bowing" of X and Y concentrations in the Cartesian model (Fig. 6a, b).

II. The geometries of positional gradients do not identify the coordinate system. It might be supposed that determination of actual morphogen gradients in a tissue would suffice to determine the coordinate system used by the cells. That this is not so can be seen by reexamining the spherical coordinate model. Here the tissue is imbedded in an orthogonal X, Y, Z tissue specificity space. Ratios of these three chemicals specify *rays* of a solid angle emanating from the origin. A different use of the same three gradients provides a simple molecular interpretation of the polar model. Let the *concentration* of Z specify distance from the high point and the *ratio* of X and Y the azimuthal angle modulo 2π. Each *ray* in a plane of constant Z, emanating from the Z axis corresponds to a unique Φ angle at a unique Z radial distance from the high point. Simple diffusive smoothing of X, Y, Z discontinuities recovers the intercalation rules (1, 2a, 2b) of the polar model. Finally, position might be specified by the actual *concentrations* of X, Y and Z, not their ratios. Thus, the same three orthogonal gradients might be *used* by the cells in *diverse* coordinate systems, with differing predictions about the regulative properties achievable by positional intercalation alone.

III. There is no persuasive evidence of positional singularities. Although the same gradients might subserve different coordinate systems, one hope of distinguishing among them is to ask whether singularities exist as expected in the polar and spherical models. The data reviewed offer no persuasive evidence. Not only can "high point" behavior be generated in a Cartesian model without singularities, but behavioral high points, about which the direction of duplication and regeneration reverse have not been demonstrated in other discs. Indeed, if there is a "high point" in the leg disc, it lies at or near the edge of the upper medial quadrant, and certainly not at the distal tip.

IV. Since no coordinate system can account for regeneration by normally duplicating fragments through intercalation of positional discontinuities alone, further processes must occur in epimorphic pattern regulation. This implies that choice among potential coordinate systems cannot be made simply by excluding those, like the polar, that require active processes to explain large fractions of the data.

V. The conceptual framework of positional information itself, and of discussion of alternate coordinate systems is inadequate. The general problems are these:

1. Under the postulates of positional information, the pattern which results from *interpretation* of the information is *wholly arbitrary* with respect to the features of the positional field itself. Wolpert's French Flag can switch arbitrarily to a "Wolpert Flag" of Jackson Pollock complexity. No constraints are imposed by the field.

2. Further, even given the assumption of a coordinate system, whether polar, spherical, or Cartesian, its relation to the tissue domain is wholly arbitrary and free of the geometry of the tissue itself. For example, in a polar model, we have no grounds whatsoever to think that the high point will be located in the middle of the wing disc and at the upper medial leg disc margin. In short, beyond those properties that flow from smoothing positional discontinuity alone, the hypothesis of positional information is at best weakly predictive.

VI. The observations show that processes beyond smoothing of positional discontinuities in *pre-established* fields are involved in epimorphic pattern regeneration. This suggests that regeneration includes those processes initially involved in pattern generation. While alternative basic premises for considering pattern generation are possible, I suggest the following:

1. Pattern generation is fundamentally self-organizing. While this may not be true in mosaic eggs, the capacity for small duplicating leg disc fragments to regenerate entire legs is but one example of such self-organization.

2. The same example supports a preference for theories which do not postulate special (e.g., source or sink) cells.

3. Intercalation of positional gaps suggests position behaves as though it were spatially averagable. The simplest physical example is diffusion, which suggests that the Laplacian ∇^2 operator expressing such averaging is likely to be fundamental.

The implication of the assumption that diffusivelike averaging of positional variables occurs lies in the fact that such processes naturally couple the size and shape of a tissue to the profiles of gradients which occur in it. The central mathematical concept is an eigenfunction; intuitively this corresponds to a gradient profile having the property that, under the action of diffusion, the profile keeps its *shape*, but may change in amplitude. For example, in a long tube with sealed ends an initial cosine distribution of diffusible stain with maxima and minima at the ends would maintain the cosine distribution but relax to the uniform distribution. Even for modified forms of diffusive-like averaging of positional information, describable by modified operators, the eigenfunctions of such operators provides a natural link between the tissue geometries and the shapes of gradients of positional variables within it. Thematically, the tissue shape controls the profiles of the positional field, which in turn controls tissue geometries and internal inhomogeneities, providing one framework for a self-organizing theory of pattern generation. Further, if gradient profiles are naturally related to tissue shape, it also becomes reasonable to search for natural, rather than arbitrary relations between those profiles, and subsequent interpretation events. For example, "thresholds" might follow contours of constant concentration, rather than being placed arbitrarily.

Consistent with the hypotheses of self-organization, absence of special cells, and diffusivelike averaging, the task is to construct theories able to generate at least two-dimensional positional information. In the next section, I show that the familiar two variable reaction-diffusion model does just this is a particularly natural way.

Part Four: Sequential Generation of Two Positional Axes and Two-Dimensional Positional Information

It is now well known that model biochemical systems in which synthesis and degradation of chemical species are coupled, and their diffusive transport occurs, can lead to the formation of spatially inhomogeneous gradients of the constituents in the domain, (19, 22, 31). As a particular model, we postulate a single biochemical system of two components, with concentrations $X(r, t)$ and $Y(r, t)$ at position r, time t, which are being synthesized and destroyed at rates $F(X, Y)$ and $G(X, Y)$ and are diffusing throughout a tissue. The equations for this system are:

$$\frac{\partial x}{\partial t} = F(X, Y) + Dx \nabla^2 x$$
$$\frac{\partial y}{\partial t} = G(X, Y) + Dy \nabla^2 y \tag{1}$$

which are chosen to have a spatially homogeneous steady state X_o, Y_o at which $F(X_o, Y_o) = G(X_o, Y_o) = 0$.

Analysis of system (1) begins by linearizing about the spacially homogeneous steady state by substituting $X(r, t) = X_o + x(r,t), Y(r, t, = Y_o + y(r, t)$ and retaining only terms up to first orders in x and y in a Taylor expansion of $F(X, Y)$ and $G(X, Y)$. The resulting linear equations in deviations x and y from the homogeneous state are:

$$\frac{dx}{dt} = K_{11}x + K_{12}y + Dx \nabla^2 x$$
$$\frac{dy}{dt} = K_{21}x + K_{22}y + Dy \nabla^2 y \tag{2}$$

The stability of the spatially homogeneous steady state of the linearized equations to spatially inhomogeneous perturbations in the concentration of x or y may be analyzed by evaluating the determinant of the matrix

$$\begin{vmatrix} K_{11} - k_i^2 D_1 - \lambda & K_{12} \\ K_{21} & K_{22} - k_i^2 D_2 - \lambda \end{vmatrix} = 0 \tag{3}$$

yielding the dispersion relation between the two temporal eigenvalues

$$\lambda_i{}^{\pm} = \frac{1}{2}[K_{11} + K_{22} - k^2(D_1 + D_2)] \pm$$
$$\frac{1}{2}\sqrt{[K_{11} + K_{22} - k^2(D_1 + D_2)]^2 - 4[k^4D_1D_2 - k^2(D_2K_{11} + D_1K_{22}) + K_{21}K_{22} - K_{21}K_{12}]} \quad (4)$$

Here $k_i = \sqrt[2]{k_i^2}$ is inversely proportional to the wavelength of the perturbation l_i. For a given wavelength perturbation l_i, corresponding to a specific value of k_i, if both λ_i+ and λ_i- have negative real parts, that perturbation decays. In an unbounded domain, all perturbations with wavelengths corresponding to values of k_i for which the dispersion relation has at least one temporal eigenvalue with positive real part, will grow in amplitude in time and create a spatial pattern.

With appropriate constraints on the linearized reaction and diffusion constants (19), the dispersion relation $\lambda = f(k)$ is positive for a restricted range of k, $k_1 > k_i > k_2$, hence for a restricted range of wavelength 1, $l_1 < l_i < l_2$ In this case, system (1) acts as an amplifier, selecting and amplifying from thermal noise those wavelength components 1 in the range between l_1 and l_2.

If system (1) exists in a bounded spatial domain and no flux boundary conditions are imposed, then a chemical pattern will be established only if it simultaneously has a wavelength in the range $l_1 = l_2$, and also satisfies the boundary condition that the gradient have zero component normal to the boundaries. Assuming no flux boundary conditions, the shapes of chemical patterns which can "fit" on to the spatial domain must satisfy the Laplacian diffusion operator, and are an infinite series of specific *eigenfunction* patterns determined by the geometry of the domain. The first several eigenfunctions on a wing disc shape are shown in Fig. 9a–c (20).

Establishment of a First Positional Axis

As the spatial domain in which system (1) acts grows gradually larger, a succession of the discrete eigenfunctions or their superposition will sequentially become established in the domain (19). For any reaction-diffusion system which acts as an amplifier, perturbations below some minimal wavelength, l_1, will die away. Therefore, if the maximum physical length of the tissue is less than l_1, the spatially homogeneous state is stable. As the tissue grows in length past l_1, the first eigenfunction of the tissue domain will bifurcate from the homogeneous solution. A fundamental theorem of bifurcation theory guarantees that the first pattern will be a stable solution of the full nonlinear equations. Therefore, as a wing disc shape gradually enlarges, no chemical pattern forms for lengths below l_1, then eigenfunction 1, Fig. 9a, grows out of the homogeneous state, and forms a monotonic gradient along the long axis of the tissue.

It has been recognized for some time that this primary bifurcation provides a means to establish a first axis of positional information in a growing domain and supply one dimension of positional information.

The new question I wish to address is: Can a single reaction-diffusion system like (1) establish two transverse axes in a growing tissue?

At first glance, the answer appears to be "No." In the linearized equations of (1) or any similar system, each eigenfunction pattern which forms in the tissue is a

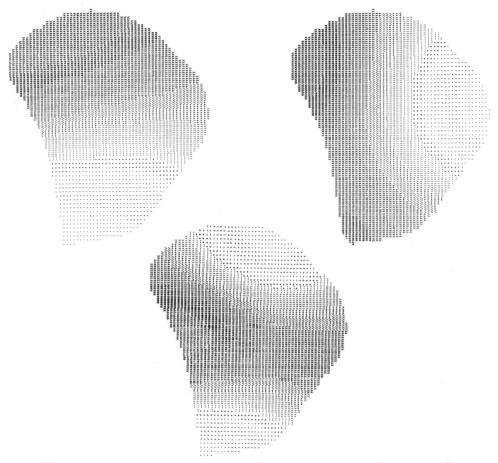

FIGURE 9. (A–C) First three eigenfunction patterns on wing disc as it enlarges. Concentration coded by visual density (20). Reprinted from (17), with permission.

spatially inhomogeneous distribution of a *fixed ratio* of the underlying chemical variables x and y, measured as deviations from the homogeneous steady state X_0Y_0. This property follows from the fact that the unstable wavelength l_i which is amplifying on the tissue, corresponds to a unique scaling factor k_i. Each unique choice of k_i specifies a particular set of values in matrix (3) having at most one positive temporal eigenvalue l_i; and a unique corresponding eigenvector $x_i{:}y_i$, which is a fixed ratio of the linearized deviations of x and y from X_0Y_0. This constraint implies that lines of constant concentration of X on the tissue parallel those of Y. Therefore although two chemical gradients are present in a two-dimensional domain, they supply at most one dimension of positional information. More generally, if N chemical species are in system (1) $N > 2$, all N occur as a fixed ratio when a single temporal eigenvalue λ_i is positive, hence lines of constant concentration of all N are parallel to one another and supply only one positional axis. Further, if any single eigenfunction pattern, e.g., Fig. 9c, is amplified alone in the tissue through a single positive eigenvalue, the ratio of all N species is again fixed, and at most one positional gradient is established.

Establishment of a Second Positional Axis

It has not been generally recognized that the familiar class of two-variable reaction-diffusion systems capable of establishing a first axis can *sequentially* establish transverse axes in growing two-dimensional domains. This can be exemplified on a wing disc shape. If the range of wavelengths l_1 and l_2 which are amplified by the linearized system (1) is made broad enough, then as the tissue domain enlarges past l_1, the first eigenfunction mode, Fig. 9a, forms. As the two-dimensional domain enlarges further, both the first ventral-dorsal monotonic pattern and the second anterior-posterior monotonic eigenfunction pattern Fig. 9b, satisfy the boundary conditions and may be amplified together. If the length and width of the tissue are unequal, each mode corresponds to a different specific wavelength l_i, l_j ($l_i > l_j$), and the corresponding scaling factors k^2 are unequal, $k^2i = k^2j$. Therefore, the eigenvector ratio of the chemical variables x and y, derived from matrix (3), differs for the two modes. The superposition of the two distinct modes, each amplifying a different ratio of X and Y, yields nonparallel (transverse) gradients of X and Y in the tissue.

As a second example, on a growing rectangle whose sides are of unequal length, $L_1 > L_2$, the linearized equations initially amplify the first eigenfunction creating parallel X and Y monotonic gradients in the L_1 direction, thereby creating a first axis of positional information. Then when the rectangle enlarges further the chemical system amplifies the *superposition* of the first two monotonic eigenfunctions in the L_1 and L_2 directions creating transverse gradients of X and Y, and two positional axes:

$$c_1 \begin{bmatrix} x_1 \\ y_1 \end{bmatrix} e^{\lambda_1 T} \cdot \cos \frac{\pi r_1}{L_1} \begin{bmatrix} x_1 \\ y_1 \end{bmatrix} + c_2 \begin{bmatrix} x_2 \\ y_2 \end{bmatrix} e^{\lambda_2 T} \cdot \cos \frac{\pi r_2}{L_2} \begin{vmatrix} x_2 \\ y_2 \end{vmatrix} \qquad (5)$$

The first major property of this class of models, therefore, is the prediction that two transverse positional axes should form *sequentially* as the tissue grows in size.

The linearized model of (1) is of limited value for several reasons. The amplitude of each of the two superimposed modes grows exponentially without bound. Whichever mode grows faster (λ_1 or λ_2) eventually dominates the superposition, therefore a single eigenvector $[x_1 : y_1]$ or $[x_2 : y_2]$ dominates the pattern and the underlying X and Y gradients become parallel. Second, the contributions of the first and second patterns to the superposition are given by arbitrary initial conditions, c_1, c_2. Finally, each eigenfunction pattern can occur in two opposite forms or branches, with maxima replacing minima. Which is established depends entirely on initial conditions. The conjunction of these defects implies that the relative orientation of X and Y gradients on the tissue is not uniquely fixed by tissue geometry in a linear theory.

These defects of a linear model are fully met by a fairly broad class of nonlinear reaction-diffusion systems in an *asymmetric* two dimensional tissue such as the wing disc. A number of workers have proposed nonlinear models which are known to bound the amplitudes of chemical patterns in one spatial dimension and two dimensions (19, 20). Generally, the establishment of the first pattern parallels the linear analysis, except that for large amplitudes, the pattern may deviate from the eigenfunction pattern. As the spatial domain enlarges, the second pattern commonly is gradually established through a secondary bifurcation from the primary spatially inhomogeneous monotonic pattern, yielding at each tissue size a fixed mixture or amplitude of distorted first and second modes. Finally, in a spatially asymmetric two-dimensional domain, the cube of the eigenfunction is almost always nonzero. From bifurcation theory this implies that, as the tissue enlarges, only one of the two branches (or forms) of the primary monotonic pattern emerges smoothly from the homogeneous state, while the other branch is stable but requires a macroscopic perturbation to be attained. For at least some nonlinear systems, only one branch of the subsequent secondary bifurcation emerges smoothly from that first mode (22). The novel implication is that the entire sequential establishment of a first positional axis, then two transverse positional gradients in two dimensional domains, their shapes and orientations, can be uniquely determined by the changing tissue size and geometry, and the kinetic parameters of the nonlinear model.

In Fig. 10a, b, I show the results of simulations of one nonlinear model given by:

$$\frac{\partial X}{\partial t} = -AX + \frac{BY^6}{1+Y^6} + D_1 \nabla^2 X$$
$$\frac{\partial Y}{\partial t} = -CX + \frac{D(Y^6+6)}{1+Y^6} + D_2 \nabla^2 X \tag{6}$$

with parameter values specified in the legend. As a wing disc shape enlarges, a first (Fig. 9a) ventral-dorsal monotonic gradient of X and parallel gradient of Y is established and supplies one-dimensional positional information along one positional axis. As the disc enlarges further and the underlying anterior-posterior second mode (Fig. 9b) begins to contribute to the pattern, the X and Y gradients both rotate slowly on the disc, Y faster than X, gradually establishing distinct nonparallel transverse, but not orthogonal, monotonic X and Y gradients spanning the tissue (Fig. 10b) forming a second positional axis. The model sequentially establishes transverse axes and two dimensions of positional values through two gradients whose forms and orientations are initially identical, but gradually come to differ as the tissue enlarges. The positional field is uniquely determined by the shape trajectory of the tissue, and the specific nonlinear model.

Novel Predictions

The inherent simplicity and minimality of this two-variable reaction-diffusion mechanism commends it as a serious model for the establishment of transverse positional axes and positional information in developmental fields. It may be that a

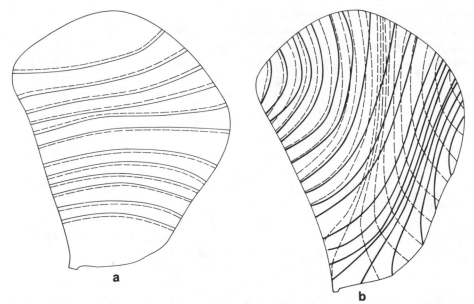

a

b

FIGURE 10. (A) Establishment of a first axis. First longitudinal monotonic spatially inhomogeneous steady state of nonlinear system (6), corresponding to first eigenfunction, Fig. 9A. Lines of constant X concentrations (solid lines) parallel lines of constant Y concentrations (dashed lines). Spacing between lines of constant X concentrations, or of Y concentrations are not strictly proportional to gradient slope, but are used to indicate shapes of X and Y gradients. (B) Establishment of Second Positional Axis. As wing disc grows, second eigenfunction, Fig. 9B, makes a contribution to mixed nonlinear steady state solution. X and Y gradients rotate away from positions in (A), and lines of constant X concentrations (solid lines) separate from lines of constant Y concentrations (dashed lines) creating transverse gradients and a second positional axis. Parameter values used for the nonlinear model (6). $A = 7.8$; $B = 15.6$; $C = 1$; $D = 1.87$; $b = .2$; $D_1 = 16$; $D_2 = 1$. Reprinted from (17), with permission.

more complex three variable model coupled with the superposition of three or more modes can yield three transverse gradients which might underly Russell's spherical coordinate model or a polar coordinate model. The minimal model has a number of predictive consequences.

1. Positional axes are established sequentially, as seen in several systems.
2. The longer axis in a field is established before the shorter.
3. During establishment of the second axis, the positional gradients may shift or rotate on the tissue. Duranceau (8) reports that positional values do appear to rotate on the developing wing disc.
4. The axes are transverse, but oblique, not orthogonal.
5. Since the monotonic gradients have values above and below the homogeneous steady state values X_0Y_0, that pair of values will typically occur at least once at an interior (hence, distal) point in the tissue.

The model predicts that very small tissue fragments cannot maintain gradients and will decay toward the steady state; therefore small fragments should drift toward the map position in the tissue corresponding to the steady state. Consistent with this prediction, Strub (30) in fact found that dissociated and reaggregated proximal leg or wing imaginal fragments from small regions transform distally.

Predictions of the two variable, nonlinear reaction-diffusion model with regard to duplication and regeneration are a combination of properties arising from simple passive diffusive averaging of positional discontinuities, and properties arising from the active dynamics and boundary conditions, and are described more fully elsewhere (17).

It is important to stress that the capacity of simple two-variable reaction-diffusion systems to generate transversal gradients supplying two-dimensional information is not restricted to a special subset of these models, but is rather a robust feature to be expected in essentially all members of this class of models. The general reason this is so is that these models generically amplify a finite *range* of wavelengths, $L_1 - L_2$. As the tissue in which such a reaction-diffusion system occurs grows larger, or other parameters alter, it must eventually reach a size at which N of the longer wavelengths "fit" on the tissue, while $N + 1$ of the shorter wavelengths simultaneously fit. Since both modes satisfy the boundary conditions, both will be amplified and the resulting (nonlinear) superposition of gradients will be locally transversal. A very substantial amount of mathematical work will be required to clarify the character of such patterns. I believe this class of models which can couple the shape of a tissue to the shape of the transverse gradients which arise in the tissue, warrants deeper exploration.

Important directions for further elaboration of this class of theories include:

1. Exploration of predictive rather than arbitrary relations of the positional field to differential cell behavior within it. Examples among these are thresholds as contours of constant concentration, and polarization of cells or alignment of division products by the local slope of the field. One past use of the successive eigenfunctions on the wing disc has been to account for the numbers, positions, two-fold symmetries and sequence of wing disc compartmentalization (19).

2. Analysis of the implications of non-monotonicity in positional gradients. The typical behavior of pattern generating mechanisms like reaction-diffusion systems with preferred wavelengths, as a tissue grows, is to shift from monotonic to nonmonotonic gradients. Positional information then becomes nonunique, with adjacent mirror symmetric domains, and repeated units of pairs of such domains. Maintaining unique positional information might be achieved by several alternatives and may have implications for the emergence of secondary fields in general.

Acknowledgements

This work was partially supported by current grants ACS CD-30 and NIH GM22341.

General References

Nicolis, G., Prigogine, I.: Self-organization in nonequilibrium systems. New York: Interscience (1977).

SUBTELNY, S. (ED.): Developmental order: its origin and regulation. New York: Alan R. Liss (1982).

References

1. Bohn, H. The Origin of the epidermis in the supernumerary regenerates of triple legs in cockroaches (*Blattaria*). *J. Embryol. Exp. Morphol.*28:185–208 (1972).
2. Bryant, P.J.: Pattern formation in the imaginal wing disc of *Drosophila melanogaster:* fate map, regeneration and duplication. *J. Exp. Biol.* 193:49–78 (1975).
3. Bryant, P.J.: this volume.
4. Bryant, S.V., French, V., Bryant, P.: Distal regeneration and symmetry. *Science* 212:993–1002 (1980).
5. Bryant, S.V., Iten, L.E.: Supernumerary limbs in amphibians: Experimental production in *Notophthalmus viridescens* and a new interpretation of their formation. *Dev. Biol.* 50:212–234 (1976).
6. Bulliere, D.: Interpretation des regenerats multiples chez les insectes. *J. Embryol. Exp. Morphol.* 23:337–357 (1970).
7. Cummins, F.W., Prothero, J.W.: A model of pattern formation in multicellular organisms. *Coll. Phenom.* 3:41–53 (1978).
8. Duranceau, C.: Control of growth and pattern formation in the imaginal wing disc of *Drosophila melanogaster*. Ph.D. thesis, Univ. Calif. Irvine (1977).
9. French, V., Bryant, P.J., Bryant, S.V.: Pattern regulation in epimorphic fields, *Science* 193:969–981 (1976).
10. Gehring, W., Nöthiger, R.: The imaginal discs of *Drosophila*, in: *Developmental systems: insects.* Counce S.K., Waddington C. (eds.). New York: Academic Press (1973).
11. Harrison, R.C. Experiments on the development of the forelimb of *Amblystoma*, a self-differentiating equipotential system. *J. Exp. Zool.* 25:413–461 (1918).
12. Harrison, R.G.: On relations of symmetry in transplanted limbs, *J. Exp. Zool.* 32:1–136 (1921).
13. Holder, N., Tank, P., Bryant, S.V.: Regeneration of symmetrical forelimbs in the axolotl, *Ambystoma mexicanum. Dev. Biol.* 74:302–314 (1980).
14. Hunt, R.K.: Developmental programming for retinotectal patterns, pp. 131–159, in: *Cell patterning.* Ciba Foundation Symposium 29. Amsterdam: Associated Scientific Publishers (1975).
15. Jürgens, G., Gateff, E.: Pattern specification in imaginal discs of *Drosophila melanogaster:* Developmental analysis of a temperature-sensitive mutant producing duplicated legs. *Wilhelm Roux' Arch.* 186:1–25 (1979).
16. Kauffman, S.A.: A Cartesian coordinate model of positional information in imaginal discs of *Drosophila.* 20th Annual *Drosophila* Conference (1978).
17. Kauffman, S.A.: Bifurcations in Insect Morphogenesis. I. and II, in: *Nonlinear phenomena in physics and biology*, p. 407, New York: Plenum (1981).
18. Kauffman, S.A., Ling, E.: Regeneration by complementary wing disc fragments of *Drosophila melanogaster*, *Dev. Biol.* 82:238–257 (1980).
19. Kaufman, S.A., Shymko, R., Trabert, K.: Control of sequential compartment formation in *Drosophila. Science* 199:259–270 (1978).
20. Kernevez, J.-P.: *Enzyme mathematics*, Amsterdam, North Holland (1980).

21. Lewis, J.: Continuity and discontinuity in pattern formation. In: *Developmental order: its origin and regulation*, p. 511–531, Subtelny S. (ed.), New York: Alan R. Liss (1982).

22. Nicolis, G., Prigogine, I.: *Self-organization in nonequilibrium systems*. New York: Interscience (1977).

23. Rose, S.M.: Tissue-arc control of regeneration in the amphibian limb. *Symp. Soc. Study Develop. Growth* **20**:153–176 (1962).

24. Russell, M.: A spherical coordinate model of positional information. 20th Annual *Drosophila* Conference (1978).

25. Schubiger, G., Karpen, G.: Extensive regulatory capabilities of a *Drosophila* imaginal disk blastema, *Nature* **294**:744–747 (1981).

26. Schubiger, G., Schubiger, M.: Distal transformation in *Drosophila* leg imaginal disc fragments, *Dev. Biol.* **67**:286–295 (1978).

27. Slack, J.: A serial threshold theory of regeneration, *J. Theor. Biol.* **82**:105–140 (1980).

28. Slack, J., Savage, S.: Regeneration of reduplicated limbs in contravention of the complete circle rule, *Nature* (London) **271**:760–761 (1978).

29. Stern, C.: Developmental genetics of pattern, in: *Genetic mosaics and other essays*, p. 135, Cambridge, Mass.: Harvard University Press (1968).

30. Strub, S.: Development potential of the cells of the male foreleg disc of *Drosophila*. *Wilhelm Roux' Arch.* **181**:309–320 (1977).

31. Turing, A.M.: *Philos. Trans. R. Soc. London*, Ser. B. **237**:37 (1952).

32. Winfree, A.T., this volume.

33. Winfree, A.T.: *The geometry of biological time*. Biomathematics Vol. 8, p. 345, Heidelberg, New York: Springer-Verlag, (1980).

34. Wolpert, L.: Positional information and pattern formation, in: *Current topics in developmental biology*, Vol. 6, p. 183, Moscana, A.A., and Monroy, A. (eds.). New York: Academic Press (1971).

35. Wolpert, L., this volume.

Questions for Discussion with the Editors

1. *In your introductory remarks, you adopted the postulate that graded scalar properties specify positional information. How do you propose to deal with mozaic localizations such as the insect egg's pole plasm, the snail embryo's polar lobe?*

The postulate that graded scalar properties specify positional information is attractive as perhaps the simplest account of the diffusivelike smoothing of positional discontinuities. While this chapter has focused on implications and use of this simplest hypothesis, it is critical to emphasize that the general problem is more complex. First, substantial evidence points to localizations of determinants in some eggs. It happens, in *Drosophila*, that evidence for localized determinants beyond pole plasm is weak. In its most extreme form, the hypothesis of localized determinants becomes a micromosaic model, with localized factors for each somatic commitment. Such prelocalization cannot, of course, account for pattern *regeneration*, and therefore, parsimony leads one to suppose that processes of pattern *generation* in the early embryo, perhaps using a few prelocalized determinants plus reaction-diffusion, or other mechanisms, is most generally to be expected. A stronger worry about the postulate of graded scalar "positional variables" is the full *independence* supposed to exist between the positional system and its interpretation. It seems useful to consider weakening this improbable idealization, either by coupling tissue geometry to gradient shapes and thence to commitments, as in this chapter's last section; or more radically, to consider the possibility that cellular commitments themselves generate and modify positional gradients. This area of unexplored theory raises questions about what it may mean for two tissues to "use" the "same" positional map.

2. *Your strategy seems to involve a search for models which are both* complete *and* universal. *Why? Couldn't a "polar coordinate" system regulate ⅔'s of the features of amphibian limb regeneration, a spherical coordinate system regulating the remaining properties? That is, in your search for the ultimate model, aren't you assuming that a* qualitatively uniform *set of signals generate pattern?*

This question touches a point of major difficulty in the application of theory in developmental biology. Of course a limb field might use a variety of mechanisms simultaneously in pattern generation and regeneration. At issue is whether we can find simple lawlike properties underlying the phenomena we see such that those phenomena are "naturally" explained. Success in physics and chemistry has been broad. In biology, we have acceded to the view that organisms are rather ad hoc accretions of mechanisms which have passed the evolutionary sieve; hence simple general laws are less expected than quixotic combinations of mechanisms. Under this view, shared properties derive from shared descent, and suggest shared mechanisms. In contrast to this deeply held view, consider the success of the polar coordinate model and its Cartesian and Spherical coordinate alternatives in explaining pattern regeneration in newts and cockroaches. It seems highly improbable that newts and cockroaches share a qualitatively uniform set of signals to generate pattern, due to common descent. Instead, as this chapter suggests, many of these phenomenologies are explained by almost any two dimensional positional map, and explained *regardless* of the detailed molecular mechanisms mediating that map, as topological properties of two dimensional maps in themselves. The very interesting possibility is that many properties of organisms, shared across distant phyla, may be consequences of ubiquitous underlying properties of self-organization rather than mere shared descent. The rational morphologists dream of "laws of form" may ultimately prove less foolish than premature.

CHAPTER 5

A Continuity Principle for Regeneration

Arthur T. Winfree

*One of the principal objects of theoretical research in any
department of knowledge is to find the point of view from
which the subject appears in its greatest simplicity.*

J. Willard Gibbs (5)

MANY KINDS of living organisms regrow appendages that are crushed
or torn off in the mishaps of an active life. Human beings have scarcely any abilities of this sort, a fact which contributes to their curiosity about the mechanisms of
regeneration in more resilient organisms. This curiosity runs deeper because regeneration in many ways resembles the initial normal development of an animal's
structures. Normal development plus regeneration, collectively called *morphogenesis*, presumably operates by some general rules that we might at least elucidate
empirically as a prelude to ferreting out deeper mechanisms. Yet for all the imaginative and meticulous efforts of at least four generations of developmental biologists, few general rules have stood the test of time. If principles of widespread applicability exist, they remain tantalizingly obscure.

In this situation, the whimsically dubbed *clockface* ("clock-phase") model of
regeneration created a stir among developmentalists of both experimental and theoretical bents. Its most fundamental and radical tenet is that positional information is stored in cells as a "phase" or an "angle" analogous to the hour-hand position
on a face of a clock, a quantity topologically unlike the conventional scalar variables of physiology. This has several startling implications which have been verified
experimentally in several contexts, so it seems to call for some interpretation, perhaps involving a rhythmic or somehow periodic process.

It also turns out that this model, unless adorned with special hypotheses to cover for exceptions, makes other predictions that are *not* observed. Moreover, a less original interpretation suffices to account for the remarkable regularities that gave the clockface model its first appeal. This relatively prosaic rendering assumes that positional information is encoded, not in a single angle or phase, but in a pair of transverse concentration gradients. This is a completely ordinary idea, but its implications were long overlooked. They include the essential qualitative facts about regeneration in epimorphic fields.

The Clockface Model

The clockface model was contrived as the vehicle for a brilliant resynthesis of observations by Vernon French, Susan Bryant, and Peter Bryant (1, 4). They draw on experiments with the fruitfly larva's "imaginal disks" (the precursors of the adult fly's various appendages), with the legs of cockroaches, and with amphibian limbs. Their original paper (4) is a *tour de force* of focused experimentation and organization of data. They come up with two principles which together account for an extraordinary diversity of peculiar experimental results. Let me first remark that what one takes to be "the principles" is presently subject to so much subjective interpretation that there are as many versions as there are people retelling this story. I will retell it based on four principles. The reader will want to examine the bibliography for other versions and for more detailed references to the original experimental papers. It should also be noted that the experiments under consideration here do not span the whole diversity of regenerative processes. We focus here on situations in which cells seem to have no intrinsic directional polarity, seem to retain their differentiated identity, and seem to impose on adjacent cells (possibly newly created from dividing populations) to adopt a similar identity.

The main point introduced by French et al. (4) is that each tiny patch of tissue is somehow labeled permanently with an unchanging intensive quantity—a tissue specificity—that behaves like time in a cycle or like a hue or like an angle in that any one of them denotes a point on some abstract circle of states. French et al. (4) placed the digits 1–12 as indicators of local tissue specificity around the circumference of a limb or other developmental field. Many of their diagrams thus resemble the face of a clock. Although this notation gives the model its name, it unfortunately also requires a numerical discontinuity (12/1) where none is intended biologically. A more apt analogy is provided by the seasonal states of an ecosystem: spring grades into summer grades into fall grades into winter grades into spring with no discontinuities. The numerical analogy is unfortunate, but the principle is clear and tantalizingly suggests something oscillatory, somehow periodic, underlying. Without committing ourselves to any such interpretation, it will still be convenient to accept the "clockface" metaphor and refer to the quantity denoted around its rim as a "phase." Independent of its circlelike topology, the import of a cell's neighborhood bearing this putative state label is that at an appropriate stage in regeneration, tissue will develop structures (sensory hairs, muscle attachments, color patches) corresponding to the local label. This is an application of Wolpert's (19) principle that tissue specificity, encoded in a local quantity of unknown origin, is conceptually distinct from the cell's interpretation of that "positional" speci-ficity and from other covert internal states such as "determination."

The following simple rules make use of the putatively *circular* label of cell states, and suffice to systematize a lot of otherwise very perplexing experimental results.

RULE 1:   Each little patch of cells is labeled with a phase which is part of a smooth phase gradient across the tissue. Thus there exists a smooth map from the tissue to an abstract ring of biochemical specificities. Once established, this map does not change. (Some independent second label, e.g., the proximo-distal level of cell specificity, is implicitly assumed, whereby to distinguish cell types on the two dimensions of an organism's surface.)

Rule 2 introduces an exceptional point near which this map is not smooth.

RULE 2:   In the normal limb, the phase values are supposed to run one full cycle around the limb axis (the azimuthal direction) as in Fig. 1. If a limb were an open-ended cylindrical surface, this would pose no problem. But besides having an inside, here discretely ignored, a limb has a foot or a hand at the end: its surface resembles the glove of Fig. 1, with phase increasing full cycle around the open end. Herein lies the tantalizing nucleus of this whole subject. If phase increases smoothly through a cycle around the perimeter of a two-dimensional patch of tis-sue, let us say that the patch has "winding number" $W = 1$ (or 2, should phase in-crease through *two* full cycles, etc.). It is an unfortunate fact of geometry (17) that only if the winding number is *zero* can a distinct phase value be attributed to every point inside. To convince yourself of this, try to sketch on the tissue a curve along which phase = 1, and an adjacent curve along which phase = 2, and so forth. All these curves start from the corresponding points along the tissue's boundary, and

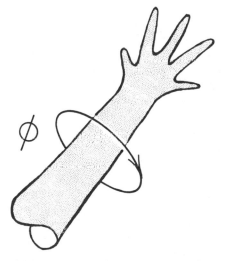

FIGURE 1. The circular dimension around an appendage is putatively encoded in the cells by a time-independent quantity that increases smoothly from an initial value and re-turns to it without ever decreasing. For the sake of graphic clarity, several diagrams here use humanlike appendages rather than the more diverse limbs peculiar to the actual ex-perimental subjects; humans, of course, do not regenerate much more than a fingertip.

extend inward. If the winding number of phase along that border is zero, then each curve that enters the patch from one point on the border can again exit the patch at another point where the phase value is the same. But if the winding number is not zero, then that integer number of full sets of phase contours enter and can't get back out. Wherever they converge, two points an arbitrarily small distance apart may have radically different assignments. This is a discontinuity. If $W = 1$ there must be a discontinuity in the assignment of phase values somewhere; the tissue could evade this discontinuity only by placing it in a hole (a region devoid of living, labelled cells). One might suppose that the discontinuity would appear along some "seam" running the length of the leg (the proximo-distal direction) like an International Date Line, but none was found (3). The discontinuity is apparently more compactly localized than that. It might be an isolated phase "singularity" as in Fig. 2. Alternatively, there could be several isolated phase singularities, e.g., three, of which two have opposite handedness and cancel out (Fig. 3).

Rule 3 corresponds to the Rule of Intercalation (4).

RULE 3: So long as the phase gradient is shallow enough (if adjacent cells are sufficiently similar in phase), cells divide only to replace those that happen to perish. But if cells with normally nonadjacent phases are juxtaposed (e.g., by cutting off a leg, rotating it, and sticking it back on), then those cells begin to divide in earnest. (Note that an exception apparently occurs at the singularity, where cells do not always divide, even though the phase gradient remains infinitely steep there. This point is especially developed by Sibatani (16).) As proliferation continues, the new cells take on phase values intermediate "between" their immediate neighbors. Thus the phase discontinuity is soon bridged through newly regenerated tissue. Proliferation continues until it has restored the initial shallowness of the phase gradient in space. (Note that some kinds of discontinuity cannot be smoothed over by intercalation of intermediates, viz. phase singularities such as are postulated (4) at each limb's distal tip.)

In some versions of the clockface model, intercalation of intermediate phase

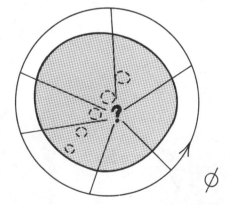

FIGURE 2. The epidermis in Fig. 1 is stretched flat. The dashed rings indicate fingernails. From five points on the abstract circle of phase values contour lines cross the epidermis, joining cells of like value. If the perimeter spans a full cycle, then these contour lines inevitably converge on the borders of an internal hole, or, if there is none, on a region of ambiguous phase (a "phase singularity").

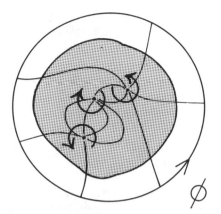

FIGURE 3. As in Fig. 2, but showing that any picture may be supplemented by any number of additional paired singularities of opposite handedness without violating the boundary conditions.

values might, without an additional rule, go either of two ways: there are *two* arcs of phase values "between" any pair of points of a circle. We defer this puzzle until it can be seen in its more general aspect.

Rule 4 corresponds to the second rule of (4), the Complete Circle Rule of distal outgrowth, revised in (1).

RULE 4: A new limb will grow out wherever there is a phase singularity. It is left- or right-handed according to whether the winding number of phase around the singularity is $+1$ or -1. Additional limbs of paired handedness are possible, and as we shall see, occur in fact. Within a patch of tissue whose border has winding number W around the phase ring, there will emerge R right limbs and L left limbs in such a way that $R - L = W$. For an exposition of this winding number formulation, see Glass (6).

In its original formulation (4), this rule was worded as though cells make a local decision based on fulfillment of a global criterion (nonzero integer winding number on a distant ring of tissue). This is implausible physiologically, and lends itself to the construction of paradoxes, besides. In the approach taken here, emphasizing strictly local principles of cell interaction, something like the Complete Circle Rule emerges as a mathematical artifact, valid in especially simple situations. These difficulties prompted several independent analyses that converged on substantially the same ideas as presented next (2, 10, 13, 16). It was not long before the contradictions implicit in the Complete Circle Rule made themselves felt in published experiments, too: an amended version of the clockface model subsequently appeared (1) in which a local rule replaces the former global rule.

Examples of the Foregoing

Applying the polar coordinate rules by way of illustration, consider the regeneration of a severed foot. A disorganized layer of vaguely differentiated tissue, a blastema, grows from the cylindrical edge to cover the stump. By Rules 1 and 3 and

Fig. 2, it must contain a phase singularity of the same handedness as the original foot. So by Rule 4 a new foot of the same sort replaces the old. In terms of winding numbers, $W = 1$ (or -1), so in the simplest case $R = 1$ and $L = 0$ (or $L = 1$ and $R = 0$).

Consider a second example in which a right foot is cut off and replaced by a left foot severed from the other leg. The cylinder of new tissue proliferating in the junction according to Rule 2 is bounded by two complete circles of phase, as indicated in Fig. 4. This two-part border has winding number $W = 2$ around the cylinder of skin it encloses. This is shown in Fig. 5 by slitting the stippled skin along the line AB and laying it flat : along path ABB'A'A the phase increases by $0 + 1 + 0 + 1$ cycle. Thus we expect two additional new right feet to emerge. In animals capable of regeneration, the actual result is indeed two additional right feet. A Chinese woman suffered this very operation in 1973, following piecemeal destruction of both limbs in a railroad accident (15). The present inability of humans to regenerate whole limbs presumably spared embarrassment to all concerned.

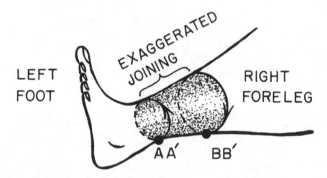

FIGURE 4. In the stippled cylinder of regenerating and healing tissue, the transition is made from the clockwise orientation of the left-handed graft (foot) to the anticlockwise orientation of the right-handed host (foreleg). (The human limb is intended to be generic.)

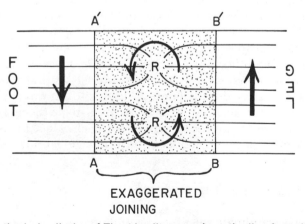

FIGURE 5. The stippled cylinder of Fig. 4 is slit open along the line from AA' to BB'. Lines A'B' and AB are the same. Contour lines and arrows show the smoothest way to join the oppositely oriented phase circles AA' and BB': two phase singularities of right-handed orientation are required.

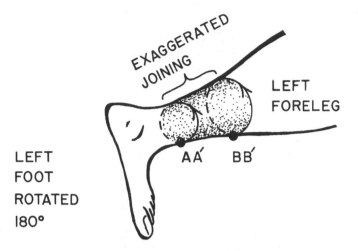

FIGURE 6. As in Fig. 4, but the left foot is rotated 180° rather than suffering removal to a right stump.

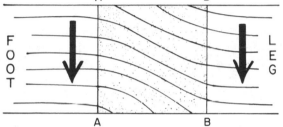

OR

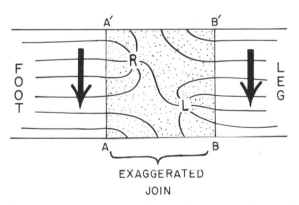

FIGURE 7. As in Fig. 5, but now the transition can be made without singularities, by just twisting all contour lines through 180° (top). An alternative construction (bottom) intercalates a superfluous pair of complementary singularities. Both outcomes are common.

As a last example consider a left foot cut off and replaced (Fig. 6) with a 180° rotation. By the same argument, we expect no supernumerary limbs, or else any number of left-right pairs (Fig. 7). A common result is a right foot and a left foot. The other common result is no new limbs.

An amusing implication of all this, incidentally, is that no organism capable of regeneration in this mode can have an odd number of asymmetric appendages, and its even number must occur in left-right pairs. A five-armed starfish, for example must be a covert tetrapod with a prehensile tail (a bilaterally fused appendage)—as it is in fact, the echinoderms being descended from bilaterally symmetric ancestors. I wonder what counterexamples—inevitable in the life sciences and always instructive—await recognition?

Alternative Formulations

Superficially, the data compiled by French et al. (4) would seem to argue forcibly that phase singularities do play a central role in the morphogenesis of higher animals, much as they do in wave-conducting sheets of oscillators such as the social amoebae, the ascomycete fungi, and certain chemical reactions (17).

This is a distinct possibility that I personally find very exciting, but it does have difficulties. These arise when we ask, as in the other organisms exhibiting phase singularities, what happens at the singularity? There is no abrupt discontinuity in cell type, no unique structure, and in apparent violation of Rule 2, cells do not keep proliferating indefinitely. All this suggests that the proximo-distal aspect of tissue specificity interacts in an essential way with the azimuthal tissue specificity represented on the phase circle. Although the original paper (4) does not dwell on this interaction, it is implicit in the polar coordinate diagram (Fig. 8; their Fig. 1) in which the proximo-distal aspect of tissue specificity is represented radially with the most distal types (making toes, fingers) at the center. Sibatani (16) gives a full explicit analysis of the peculiar implications of this diagram.

The landmark contribution that came with the clockface model consisted pri-

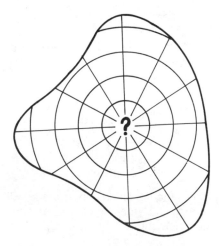

FIGURE 8. On a piece of generic tissue, level contours of proximo-distalness are drawn as rings concentric to the physical tip. Level contours of a complementary aspect of tissue specificity, locally transverse to the proximo-distal gradient, close in rings. The "?" denotes ambiguity in the phaselike quantity where all values ostensibly coexist without perpetual growth.

marily of a clear statement of correlated phenomena: There are some astonishing regularities spanning diverse phyla, never before perceived so lucidly. The model itself had less appeal, its own internal contradictions lying so near the surface. The inevitable flurry of counter-experiments and *ad hoc* amendments was well summarized (1), but the ostensible elegance of 1976 was irretrievable once the subject had been glimpsed from another perspective (2, 10, 13, 16). At present it appears that the reported phenomenology is best understood as a consequence of little more than the continuity properties of any topologically suitable representation of "positional information" or (to avoid confusion with physical position) "tissue specificity" underlying visible differentiation.

The most tantalizing feature of the clockface model was its use of a *circular* coordinate for measuring cell types. If cells have two interdependent aspects of tissue specificity then one must ask exactly what motivates the choice of one coordinate system over another on this two-dimensional state space. Of course, one would prefer a coordinate system natural to the topology of the state space. I find no reason to postulate any topology even as marginally exotic as the clockface model's cylinder (the product of a phase circle and a proximo-distal-ness axis) for the two-dimensional state space: A plain piece of paper seems to have the right topology.

One might prefer coordinates natural to a "dynamic" inherent in the system. The evidence for a "dynamic" would be that homogeneous patches of tissue change their specificity autonomously and predictably. I know of no certain evidence of such changes, except on a more prolonged time scale than concerns normal regeneration, after the earliest stages of growth. Also one might prefer a coordinate system that gives a preferred role to any unique tissue types (e.g., an indispensible organizing center such as the gray crescent of amphibian eggs or the pulsating tip of the slime mold slug). I am not certain, however, that special places or tissue types have been revealed by regeneration experiments. There is apparently no preferred origin for the coordinate system that we must choose to conveniently identify neighborhoods on the surface of an organism. In fact, since we know nothing of the biochemical basis, we might abandon coordinate systems altogether. By dealing in a coordinate-free representation, we isolate the essential facts necessary for understanding the empirical results. For two alternative and considerably more up-to-date executions of this approach, see Lewis (10) and Mittenthal (13).

An Inverse Fate Map

A simple coordinate-free representation might be constructed as follows. Let us imagine a tissue-specificity space (TSS). We endow it with enough dimensions (two) to distinguish cell types on a two-dimensional surface such as the surface of a leg. We depict as a *place* in TSS the latent tissue specificity of a cell or a patch of neighboring cells. To each region in TSS there corresponds a type of structure that cells in that region will make when they mature. Map the creature, organ, or tissue into TSS. There is nothing new in this procedure. This is only drawing a fate map, inside out as it were: instead of drawing the organism in the real world and writing tissue names on an overlay, we write the tissue names at fixed places in tissue-specificity space and then draw the organism as an overlay, distorted as required to put places on their corresponding names. For example, this "inverse fate map" of any bilaterally symmetric organ is folded so that two patches of cells symmetrically disposed to the left and right of a mirror plane both map to the same place in TSS. The mirror axis maps to a fold line (Fig. 9).

Now consider two pieces of an organism, normally not adjacent in the intact, mature stable organism. These appear in TSS as two islands of tissue. If they are now physically juxtaposed, tissue specificities are not initially affected, so fine lines must be used to connect the cells that are quite separate in TSS though physically adjacent along the surface of contact (Fig. 10). In experiments on the crayfish, Mittenthal (13) shows that these "in-between-ness" connections *cannot* always be made *straight* by suitable rubber-sheet rearrangement of TSS, contrary to my conjecture of 1977 (17). This may be a consequence of nothing more exotic than *curvature* of TSS: if TSS were like a steep hill on an otherwise flat plane, then between any two points astride the hill there would be *two* shortest paths (left and right

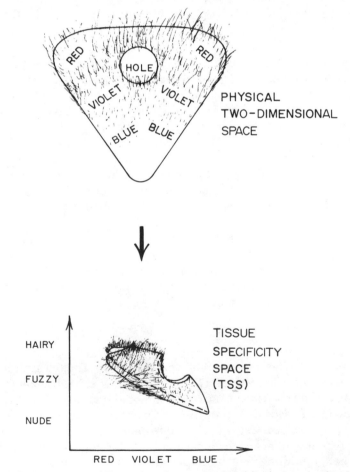

FIGURE 9. An imaginary two-dimensional organism and its image in tissue specificity space.

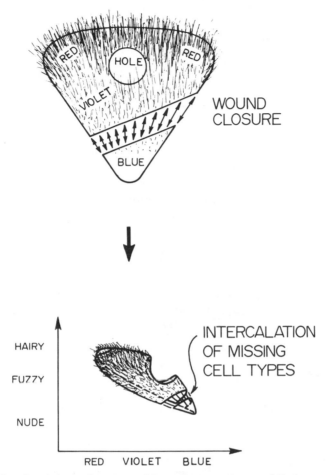

FIGURE 10. A strip of tissue is removed from the organism and its image without affecting adjacent tissues. Physical apposition of the wound edges is indicated by fine lines, pulling the edges into contact in the upper picture, but not altering tissue specificities (i.e., leaving the pieces unmoved, only bridging the range of states presently unpopulated by cells) in the lower image. New cells adopt tissue specificities intermediate between their neighbors', thus populating the fine lines and restoring the image, if not necessarily the physical geometry of the organism.

along the hill's flanks) from one to the other through "in-between" cell states. Any representation that purports to straighten all such curves would thereby inadvertently superimpose these two, though they consist of distinct cell types. So a two-dimensional TSS cannot always be so defined that the linking curves appear straight. Nonetheless, straight-line connections in two-dimensional-space appear to suffice for the purposes of this Chapter, obviating the need to adopt a special "shortest arc" rule and the implicit metric to decide which path to take in a one-dimensional space.

We rewrite the rules now in three parts using this geometric language:

RULE 1: Each little patch of cells is labeled with a state which is part of a smooth gradient of states across the tissue. This state is a point in a two-dimensional space of biochemical specificities topologically equivalent to a plane. Once established, this smooth map does not change. The two components of this "state" specification might be, e.g., the concentrations of two kinds of adhesive site on the outer cell membrane.

RULE 2:   No exceptional point or state exists.

RULE 3:   (which, in this formulation, does almost all the work): Cells proliferate at a rate determined by the separation of adjacent cells in TSS (the unlikeness of surface properties, for example). If they are initially quite different in tissue specificity (e.g., cells on the interface, connected by fine lines in Fig. 10), then they begin mitosis. The tissue expands in physical space while its image in TSS remains unmoved but becomes denser with cells. Proliferation stops when the density of cells throughout the image has risen to a local norm, i.e., when the local gradient of tissue specificity has diminished to normal for each kind of tissue. There is no problem about endless proliferation at an unremovable phase singularity: no singularities occur. Just as in the original Rule 3, new cells intercalate tissue specificities intermediate between their physical neighbors. (Note that we have yet to specify exactly what is "intermediate" between cell types that are not normally adjacent. This puzzle is pushed much closer to solution in Mittenthal's *Rule of Normal Neighbors* (13)).

Applying the TSS Image Rules

No separate rule is needed for induction of a limb. The image in TSS of any ring of tissue is necessarily a ring, and by Rule 3 the tissue inside the physical ring must acquire (at least) the tissue specificities inside the ring's image. At an appropriate stage in development, cells will develop the structures by which their positions in TSS were named. If those should happen to include the distal parts of a limb (fingers, for example), then such structures (among others) arise when tissue specificities are interpreted biochemically, and we have a "limb."

If the most distal structures are normally internal to more proximal structures in TSS, then amputation corresponds to ablation of a disk in TSS, leaving a ring. Blastema formation followed by wound healing corresponds to stitching across the empty disk fine lines along which new cells take up the missing tissue specificities, restoring the more distal organs (Fig. 11).

Replacing a left hand (L) by a right hand (R) on a left stump corresponds to cutting out the disk in TSS and replacing it with an identical disk. But remember that because of the inevitable L–R mismatch in physical space, tissue connections from the (R) central disk reach across to the opposite side of the (L) hole (Fig. 12). We thus have a three-layered map across the distal core of TSS. The three layers are the existing hand and the cylinder of wound along the wrist that joins hand to forearm. The two new layers (the cylinder of wound proliferation) have the same orientation, so that we get two replicate left hands.

Or imagine that the arm is amputated at both ends. Both the proximal and distal blastemas must then span the middle ("hand") region of TSS. So mirror image hands must emerge (and do) at each end. This is the simple geometric essence

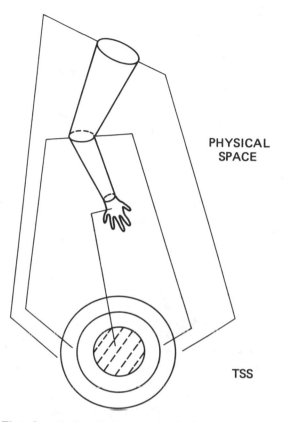

FIGURE 11. As in Figs. 9 and 10, a limb maps to its image in tissue specificity space. More distal structures are believed to map concentrically interior to more proximal structures. Amputation at any level thus deletes a central disk. Contact of "wrist" tissues through the regeneration blastema is indicated by dashed lines: they span the missing (more distal) tissue specificities. This is the geometrical interpretation for the empirical generalization that mutilated tissues grow only more distal structures from the exposed edges.

of the rule of distal outgrowth. According to this rule, more distal structures regenerate from any stump even at the proximal end where, if regeneration were naively expected, a shoulder should grow.

This picture in TSS is the same (*not* inverted) for right or left limbs. Thus replacing a left hand (L) by a right hand (R) on a left stump corresponds to cutting out the (left-hand, L) disk in TSS and replacing it with another (right-hand, R) disk. The new disk looks exactly the same as the old in TSS: it covers all the same regions of the "inverse fate map," since it has all the same cell types in the same neighborly arrangement. But the reversal of handedness corresponds to a reversal in the correspondence between cell types and points in real physical space. This only becomes important at the boundary where the graft and host are trying to fit together: in real physical space, that ring of cell types in TSS corresponds to a clockwise ring of host tissue, but to an anticlockwise ring of graft tissue.

Because of this inevitable L–R mismatch in physical space, tissue connections

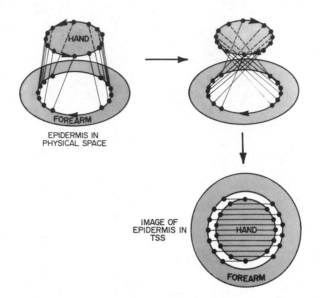

EPIDERMIS IN
PHYSICAL SPACE

IMAGE OF
EPIDERMIS IN
TSS

FIGURE 12. The experiment of Figs. 4 and 5 is represented in the terms of Figs. 9–11, rather than in terms of phase contours. In the upper left panel, a hand with clockwise polarity is inserted in a forearm with anticlockwise polarity. The separation of hand from forearm in a third dimension here is only for convenience of manipulation in the transitional second panel: the actual epidermis and its image in TSS are, of course, two-dimensional as in Figs. 9–11. In the lower right panel, the hand has been flipped to its standard orientation in tissue specificity space. Regenerating tissue, being suspended between mismatched left-handed and right-handed boundaries, is now seen to span the "hand" region twice, inverted relative to the graft hand. Two supernumerary hands equivalent to the one removed are thus inevitably induced by the requirement of continuity alone.

from the (R) central disk must reach across to the opposite side of the (L) hole (Fig. 12). Note that this is not quite crazy; to switch handedness, one must invert an organ through some axis, commonly the medio-lateral axis as in Fig. 12, but often (and with comparable effects) the antero-posterior axis or any intermediate. Each cell type at the junction finds itself adjacent to cells of the same antero-posterior-ness, but opposite medio-lateral-ness. The new cells arising in the junction between graft and host must accordingly cover the range of medio-lateral positional values (tissue specificities) between opposite extremes. This range is spanned by the new cells on each side (and is of course already spanned within the graft itself). We thus have a three-layered map across the distal core of TSS. The three layers are the existing hand and the cylinder of wound along the wrist that joins hand to forearm. The two new layers (the cylinder of wound proliferation) have the same orientation, so that we get two replicate hands. Since that orientation is upside-down relative to the (R) graft, they are of opposite (L) type, the same as the host. The parity-conserving final result is a left limb with a left hand, but with a complementary L–R pair in addition. This strange, and (originally) unexpected experimental result derives from nothing more than an assumption of continuity(!).

Equivalent Metaphors and a Coat of Many Colors

Another way to articulate the TSS rules uses the language of *color*. The diagrams in every paper about the clockface model would be much simplified, and both design and interpretation of experiments would be made transparent if only we could publish airbrushed color diagrams as follows. Let the phaselike, circular aspect of tissue specificity be represented not in digits 1–12, nor in contour lines, but in *hues* taken from a color wheel. On a color wheel, red grades into orange grades into yellow grades into green grades into blue grades into indigo grades into violet grades into red grades into orange, and so on: without backtracking, one can progress smoothly and unidirectionally through a cycle of tissue states, as required. Proximo-distal-ness may then be represented by *saturation* of the hue, the most distal tissues being least saturated (gray) so that all hues nearby are compatible without discontinuity. Our labelling of tissue types then consists only of assigning to each tissue type (and so to each place on the organism) a color: a hue and its saturation. Every organism, in this view, wears a coat of many colors. The color is permanently imbued in each cell at mitosis, in cases of strict epimorphosis. Surgical interventions consist of abutting together colors that were previously separated by gradations of color. If the colors now "run" and blend (as in morphallaxis), or if new fabric intercalates itself, adopting hybrid coloration compatible with its boundaries (as in epimorphosis), then we end up with arrangements of color that satisfy all the phenomenology of regenerating and duplicating tissues, and, incidentally, the mathematical phenomenology of phases 1–12 (hues) and winding numbers thereof (complete color wheels) and singularities (gray spots).

The phase circles of Glass (6) may be traced through full cycles of hue, ignoring saturation. They can be constructed by drawing circles on the physical tissue and then following the circles' images (rings) in TSS. The number of singularities corresponds to the winding number of the image about an arbitrarily chosen origin for the angular coordinate system. In the coat-of-many-colors metaphor, the usual choice of origin would be gray, but this is an arbitrary convenience dependent on ambient lighting. Biologically, the usual choice of origin is the distal-most tissue, but I see no biological reason to prefer such a choice to any other. From my point of view, the arguments of French et al. (4) and of Glass (6) about circles, phase maps, and winding numbers amount to using a circle embossed on the organism as a means of book-keeping the folds and rotations of the regenerating tissue's image in TSS. Glass's calculational procedure is very useful here, because until one has a little practice, it seems awkward to visualize handedness in TSS maps such as Fig. 12. But so far as I can tell, the predictions all come out the same with any origin for the polar coordinate system or (as done here) with no origin at all.

The essential feature common to all these metaphors is the mapping from a physically distributed tissue to an abstract state space of two dimensions with continuity properties like ordinary two-dimensional Euclidean space. The most general underpinning for such a mapping is a set of chemical substances arranged in the tissue along transverse concentration gradients. To take but one example, the local concentrations of any three distinct pigments define a three-dimensional state space. Any surface colored by mixtures of those pigments is thereby mapped into that space. If the total amount of pigmentation (the sum of the three concentrations) is limited to a fixed quantity, then the surface is confined to the two-dimensional triangular plane of standard colorimetric charts. Color mixtures happen much as in Fig. 12 and 13. If the "pigments" (or cell-surface antigens, etc.) should happen to mutually influence their own rates of synthesis and degradation then, as is well known from the classical theory of reaction/diffusion structures, the chemical gradients may become self-stablizing. The only experimentally demonstrable reaction/diffusion structure at present is the "rotor" in Belousov–Zhabotinsky reagent. Its structure, dynamics, and reaction to trauma may be expounded exactly in the manner of this chapter (17); its development into three dimensions resembles nothing so much as the embryology of simple organisms (18).

From the literature surrounding the clockface model (1), one gains the impression that proper understanding of the phenomena of regeneration somehow repudiates (in favor of the more modern preoccupation with cell surface chemistry) the classical and apparently sterile notion that chemical concentration gradients and reaction/diffusion structures underlie it all. Perhaps it is indeed time for a fresh perspective on these unanswered riddles, as old as the science of biology. But to my mind, nothing more resembles the phenomenology of the (known and nonlinear, as

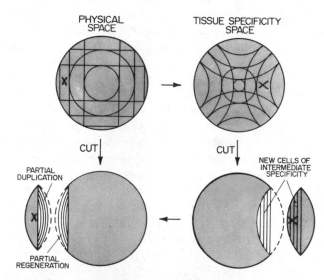

FIGURE 13. (Legend: above = before; below = after; left = physical space; right = tissue specificity space.) The imaginal disk of a fly's appendage is painted with grid lines and with concentric contours of proximo-distalness, then mapped (right) into TSS. The "puckered" mapping is essential for what follows. Cutting off a chord in physical space (lower left) excommunicates that chord's lens-shaped image in TSS (lower right; the lens is drawn as though displaced from its initial tissue specificities only for graphic clarity). Supposing that each piece folds to close the cut edge on itself, new cells arise with intermediate specificities indicated by fine vertical lines: each edge grows back about half of the cell types in the severed chord. More severe "puckering" of the map would improve this partial regeneration and reduplication, but to guarantee completion we must cease to ignore the *dynamics* of specificity in TSS.

opposed to theorized and linear) reaction/diffusion structure than contemporary reports on regeneration in limb fields of comparable size.

Experiments Needed

False facts are highly injurious to the progress of science, for they often endure long; but false views, if supported by some evidence, do little harm, for everyone takes a salutary pleasure in proving their falseness.

Charles Darwin

Certain implications of the geometric viewpoint adopted above may allow critical testing:

1. This style of mapping physical media or organisms into a state space or a TSS differs from its better-established prototypes (17) in that no "dynamic" is postulated in state space. Applied to nondividing or relatively slowly dividing tissue, this is the assumption of epimorphosis. But dynamics may play an essential role in more rapidly dividing young or regenerating tissue. As Goodwin (7, p. 175) was first to emphasize, this is a deficiency that needs attention. For example, consider the following:

a. Although French et al. (4) address themselves exclusively to epimorphosis (old cells retaining their specificity), some amount of morphallaxis (respecification) may commonly occur in regenerating limbs (1). According to Maden's interpretation (12), this fact is assimilated by simply allowing our Rule 3 to govern wounded tissues even prior to cell division. This scarcely alters our diagrams. However experiments to elucidate the following two points are potentially more subversive.

b. Every tissue starts as a tiny patch of indistinguishable cells which, as they grow, acquire divergent specificities. Developmental fields originate, like the "big-bang" model of the expanding universe, from a singularity in TSS. At least at that early stage of development a "dynamic" *dominates* TSS. Kauffman et al. (9; Chapter 4, this volume) provide an alluring linear theory of this "dynamic."

c. Observations on transected imaginal disks of the fruitfly argue for a strong tendency of tissue specificity to evolve more proximally. The qualitative observation is that both edges of the cuts produce the same new structures: one piece exactly duplicates and the complementary piece regenerates completely. The knife line in Fig. 13, though straight in the real world, is generally curved on the image of a disk in TSS, TSS being defined in such a way that inbetweenness follows straight lines so far as possible. (For *caveats*, see (13) as mentioned previously.) No matter how the cut edge heals onto itself, the proliferating cells will adopt tissue specificities intermediate (parallel lines in Fig. 13) between points on the knife cut's image in TSS. These intermediates cannot include the whole domain occupied by the excised piece if the disk's image is convex in this TSS—and it must be, or else intercalation between protrusive extremes of the image would produce tissue types outside the image, i.e., not normally present. So far as I know, this does not happen. So developmental fields are *convex* and cutting the edge off one leaves each part completely without access to some region of TSS occupied by the other part

(Fig. 13). Without a "dynamic," one piece must then fail to *fully* regenerate while the other fails to *fully* reduplicate. Transection experiments may thus provide the means to verify the existence of a "dynamic" and to quantitate the rates at which tissue specificity can change during normal growth and during regeneration.

Without a proximalizing "dynamic," there must be parts of any organism that *cannot* be regenerated: the adjacent most proximal parts of neighboring developmental fields, located on the common rim of their convex images in TSS. By the same token, the kind of regeneration here considered cannot proceed beyond any symmetry plane: a bilaterally symmetric organism, for example, could regenerate at most half of its complete body. How do these limitations contrast with the facts of regeneration in radially symmetric organisms and in morphallactically flexible organisms such as *Planaria*?

Recent experiments using artifically constructed bilaterally symmetric limbs (summarized in (1)) are particularly intriguing in this context. An unexpected progressive loss or gain occurs in circumferential positional values (hues, clock-phase values, TSS regions occupied). This may betray a slow underlying dynamic such as we inferred previously.

As this writing the "facts" about disk regeneration seem to me so much less certain than in 1977 that I have not attempted to update discussion *vis-à-vis* the experimental literature, much as it may be needed. In my view the first priority question still remains: What tissues *are* juxtaposed in wound closure? Consistently the same? Does it matter?

2. According to this coordinate-free representation, ablation of a feature on the symmetry plane of a bilaterally symmetrical organism (a tongue, a nose, a penis, a tail, the genital disk of a fly) corresponds to cutting a hole out of the fold of the two-layered image in TSS (Fig. 9). Such a hole can heal in two quite distinct ways (Fig. 14):

a. Closing horizontally (top-to-bottom contacts), the hole in TSS would be filled again. Thus we should anticipate complete regeneration.

b. In contrast, a vertical closure (side-to-side contacts) would join mirror image tissues. Thus, no proliferation should ensue. Even if it did, the missing tissue specificities would not be recovered.

The results of diagonal closure must vary between the extremal results a and b according to the angle of the diagonal. This series of experiments may help to distinguish between geometric interpretations of tissues specificity.

For discussion of experiments recently undertaken along these lines, consult (1). One source of ambiguity at present is the lingering uncertainty about the actual (uncontrolled and unobserved) geometry of wound closure.

3. If one had to guess before observing, it might reasonably be supposed that an organism would respond to surgical challenges by intercalating positional values between given boundaries in the smoothest way. This surmise comes from the observation that paired phase singularities (left-right pairs of supernumerary appendages) are uncommon where uniformity of phase might alternatively prevail.

Why then does the experiment of Fig. 6 commonly result in a left-right pair of limbs, adopting the bottom solution rather than the top solution in Fig. 7? The top

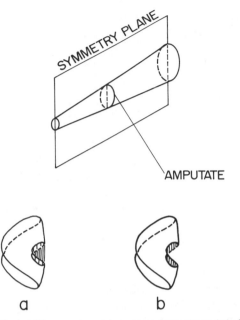

FIGURE 14. As in Figs. 9 and 10, a symmetric appendage maps into TSS as a folded disk, its corresponding left and right parts *twice* covering each occupied specificity. Amputation deletes a central disk *straddling the fold line*. In contrast to the inevitable distal regeneration of asymmetric structures, this one may regenerate to any degree, depending on the exact manner of tissue appositions during wound closure. The extremes are shown in (a) and (b).

solution is plainly smoother in terms of phase alone, but it turns out that the bottom solution can be the smoother if the tissue's state can vary not only in phase but also in some conjugate quantity such as the radius on a polar coordinate diagram. It therefore seems important to ferret out the sources of variability in the outcome of such experiments.

4. It seems not entirely proven by experiment that apposition of dissimilar tissues is required to elicit regeneration. Is it possible that the external medium may under some circumstances serve as a "tissue type"? Where is its point in TSS, if so?

5. It would be of interest to treat a strictly nonliving chemical "organizing center" (18) as though it were, for example, a limb field, repeating many of the physical manipulations of classical embryology. Some of the practical difficulties are similar: Accurate separation and recombination of microscopic chunks of wet gel that tend to adhere through surface tension or to float apart when submerged; accurate observation of outcomes that consist of garbled and sometimes subtle patterns, constantly changing; limited reproducibility due to uncertainty of initial conditions. The outcomes might also bear sufficient formal resemblance to provoke stricter attention to strategic issues: What questions are we asking? How will the answers aid understanding and control? Since the morphogenesis of organizing centers in Belousov–Zhabotinsky reagent can be understood in exact physical-

chemical detail, it might be illuminating to know how nearly it really does resemble cytochemical morphogenesis.

Conclusion

This chapter opened with Gibb's injunction to seek simplicity, an aesthetic criterion here interpreted quite subjectively as *geometric* simplicity. To end this discussion, I recall the amendment by Alfred North Whitehead, ". . . to seek simplicity *and distrust it.*" For the present, it appears to me that the existing experimental results reveal little more than that developmental fields map continuously into a suitable descriptive space, wherein the images of new cells lie close to the images of their progenitors. I hope that my posing the problem in that format may help provoke the execution of critical experiments to resolve present-day ambiguity about the topology of that space, about its metric, and about the dynamic flow that presumably guides cells to their final positions in it.

The present version of this chapter provides no more basis than the drafts that circulated in 1977 for using the marshalled facts (1, 4) to distinguish quantitatively among alternative models. Still awaiting suitably quantitative data, I finally included these remarks as a provocative chapter *against* phase singularities in a book celebrating the same (17). That similar notions occurred to Cummings and Prothero (2), to Lewis (10), to Mittenthal (13), and to Sibatani (16), and undoubtedly to many others who have not yet published their ideas, suggests that what remains here is still qualitative and may merit quantitative attention (as in developing theme 5, immediately above). Further refinements and experimental tests by Mittenthal (14) go far in this direction. But the "facts" are today so much less certain than they seemed years ago that my attempt to seriously update this essay foundered in a sea of scholarly contradictions (e.g., see 1, 8, 16). Rather than take more space for these simple ideas, I have only enriched them by a few comments and citation of more recent reviews, and herewith offer them for target practice.

Acknowledgment

I am indebted to Jack Dunne for urging upon me preprint (4), over which this fantasy was drafted, to Jay Mittenthal for repeated perceptive critique, to the Medical University of South Carolina for hospitality throughout 1977, to the West German Cancer Research Institute of Heidelberg for support to lecture on this topic in summer 1979 (portions here included), and to Walter Kaufmann–Buhler and Springer–Verlag (copyright holder, 1980) for permission to use portions of Chapters 2, 10, and 16 of *The Geometry of Biological Time.*

General References

BRENNER, S., MURRAY, J.D., WOLPERT, L. (ed): Theories of biological pattern formation, *Phil. Trans. R. Soc. Lond. B.* **295**:425–617 (1981).

WINFREE, A.T: The geometry of biological time. New York: Springer–Verlag (1980).

References

1. Bryant, S.V., French, V., Bryant. P.J.: Distal regeneration and symmetry. *Science* **212**:993–1002 (1981).

2. Cummings, P., Prothero, J.W.: A model of pattern formation in multicellular organisms. *Collective Phenomena* 3:41–53 (1978).

3. French, V.: Intercalary regeneration around the circumference of the cockroach leg. *J. Embryol. Exp. Morph.* **47**: 53–84 (1978).

4. French, V., Bryant, P., Bryant, S.V.: Pattern regulation in epimorphic fields. *Science* **193**: 969–981 (1976).

5. Gibbs, J.W.: quoted by L.P. Wheeler in *Josiah Willard Gibbs—The history of a great mind.* New Haven: Yale University Press (1962).

6. Glass, L.: Patterns of supernumerary limb regeneration. *Science* **198**: 321–22 (1977).

7. Goodwin, B.C.: *Analytical physiology of cells and developing organisms.* New York: Academic (1976).

8. Karlsson, J., Smith, R.J.: Regeneration from duplicating fragments of the *Drosophila* wing disc. *J. Embryol. Exp. Morph.* **66**: 117–126 (1981).

9. Kauffman, S.A., Shymko, R.M., Trabert, K.: Control of sequential compartment formation in *Drosophila. Science* **199**: 259–266 (1978).

10. Lewis, J.: Simpler rules for epimorphic regeneration: the polar coordinate model without polar coordinates. *J. Theor. Biol.* **88**: 371–392 (1981).

11. Littlefield, C.L., Bryant, P.J.: Prospective fates and regulative capacities of the female genital disk of *Drosophila melanogaster. Dev. Biol.* **70**: 127–149 (1979).

12. Maden, M.: The regeneration of positional information in the amphibian limb. *J. Theor. Biol.* **69**: 735–753 (1977).

13. Mittenthal, J.E.: The rule of normal neighbors: A hypothesis for morphogenetic pattern regulation. *Dev. Biol.* **88**: 15–26 (1981).

14. Mittenthal, J.E.: Preprints awaiting proper citation 1982–1983.

15. Mobile Medical Team of Wuhan Fourth Hospital.: Autotransplantation of severed limbs. *Chin. Med. J.* **2**: 417–422 (1976).

16. Sibatani, A.: The polar coordinate model for pattern regulation in epimorphic fields: A critical appraisal. *J. Theor. Biol.* **93**: 433–489 (1981).

17. Winfree, A.: *The geometry of biological time.* New York: Springer–Verlag (1980).

18. Winfree, A., Strogatz, S.H.: Singular filaments organize chemical waves in three dimensions, parts I, II, III submitted, *Physica-D*.

19. Wolpert L.: Positional information and the spatial pattern of cellular differentiation. *J. Theor. Biol.* **25**: 1–47 (1969).

Questions for Discussion with the Editors

1. *How does Wolffian lens regeneration, which involves a truly remarkable transformation of cell type (pigmented iris into lens), relate to your models for pattern specification in regenerating limbs and/or imaginal disks?*

This wonderful phenomenon draws attention back to items 2 and 4 under "Experiments Needed." The bilaterally-symmetric lens that normally develops from the iris' dorsal margin seems to inhibit the margin from further lens-making activity. Immediate cell-surface contact is apparently not required for this inhibition: the margin also fails to proliferate if merely exposed to cell-free liquid containing the "scent" of nearby lens. This much is not incompatible with the viewpoint adopted above, and might even illuminate the medium of "dashed-line" interactions in the figures. However, a critical experiment suggests itself. Suppose the de-lensed iris is sutured shut side-to-side. Margin cells then contact only their exact left-right counterparts, and should not proliferate (presumably because the inhibitor is autonomously present in dorsal margin cells, even if not in such massively diffusible quantities as in lens). If they do, then the notions of this chapter are irrelevant to lens regeneration.

2. *Your concluding remarks concerning the reduction of regeneration data to little more than a set of field maps with progenitor cells generating images of themselves are refreshingly provocative! However, can you provide any examples of pattern specification phenomena which are indeed presently* more fully *understood than either appendage regeneration of disc duplication?*

Unfortunately, I don't have a good understanding of the mechanism of *any* pattern-specification phenomenon.

SECTION II

Simple Systems

CHAPTER 6

Cell-Type Ratio and Shape in Slugs of the Cellular Slime Molds

Harry K. MacWilliams

CELLULAR SLIME MOLDS are "social amebae" that proliferate as free-living cells but aggregate upon starvation into multicellular masses. The aggregates then undergo a sequence of morphogenetic transitions (Fig. 1), culminating in the formation of a fruiting body made up of stalk cell and spores; the stalk cells die while the spores are released to reinitiate the unicellular portion of the life cycle (Fig. 1). Under some circumstances, the multicellular masses migrate over the substrate for a period before culmination. The migrating masses are elongate in the direction of movement and are fancifully called "slugs." Cellular slime mold slugs display several different sorts of spatial organization, and their length (about 0.2–2.0 mm) falls within the province of "Wolpert's law" (36) that biological systems at the stage of pattern formation are of size appropriate for organization by concentration gradients of diffusing substances.

Cellular slime molds are highly convenient and manipulable organisms in which multicellular developmental problems can be studied genetically (24), biochemically and physiologically (18), and by the classical embryological method of transplantation (9, 19). This combination of methods, which is not available in any other organism, may allow us to identify the diffusible factors controlling development. Several substances have indeed been isolated or partially purified from slime molds which may be involved in the regulation of pattern and/or morphogenesis. These factors include the differentiation-inducing factor DIF, (15) that appears to stimulate cells to enter the stalk and "prestalk" pathway; ammonia (32), which inhibits fruiting and acts as an antagonist to DIF; the slug-turning factor STF, (11) that appears to control phototaxis and thermotaxis in slugs, and may act by influencing the position of the slug "tip," a region having "organizer" properties in the slug. Cyclic AMP also has several effects during slime mold development (18). At

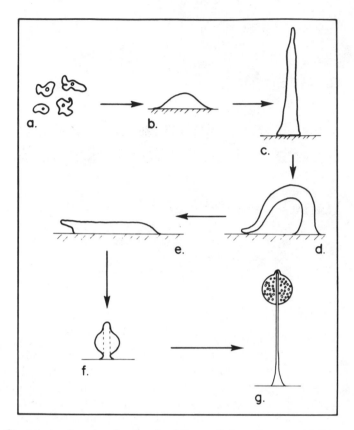

FIGURE 1. Stages in the development of the cellular slime mold *Dictyostelium discoideum*. Free living amebae (a) aggregate to form a hemispherical mound (b) which elongates vertically (c) and transforms (d) into a "slug" which then migrates over the substrate (e). The slug eventually rounds up and transforms (f) into a fruiting body (g) consisting of stalk cells and spores.

least one other "morphogen" may exist: experiments of Sternfeld and David (31) suggest that prestalk cells produce a factor that inhibits the transformation of other cells into prestalk cells.

In this chapter we consider the principles of multicellular organization in the slug stage of cellular slime molds. Slime mold morphogen research has not yet contributed significantly to our ideas as to what these principles are, and our approach will instead be based on a relatively small set of formal studies which will be discussed first. We will then consider three purely theoretical models of various aspects of slime models development.

The first of the three models is a simple scheme to account for the control of the cell-type ratio in slugs; it explains the phenomenon of "sorting" in slime molds and certain features of mutants in which the ratio of cell types is altered. The second model is a model of slug shape which is based on differential chemotaxis and can be viewed as an extension of the cell-type ratio model; it has morphogenetic power and accounts for the results of transplantation experiments as well as the modification of transplantation phenomena in cell-type ratio mutants. Unfortunately, however, the model makes several incorrect predictions about slug shape. The third model is an alternative model of slug shape based on adhesive interactions; this model explains all known facts about shape. The adhesive model can be combined with the chemotactic model to account for the pattern of cell differentiation in slugs as well as the results of transplantation experiments and the properties of cell-type ratio mutants. With additional assumptions, the adhesive model can also reproduce the desirable feature of the chemotactic model through adhesive interactions alone.

Although the last of the three models has, in principle, a chance of being "correct" it is more realistic to assume that all of the models are wrong at least in their details. The models, however, should serve to stimulate thinking about the *kinds of systems* which may be involved in the spatial integration of slime molds; the models also suggest what sorts of roles the slime mold "morphogens" substances may fill. The discussion of models, furthermore, should highlight some of the enormous variety of opportunities for exciting experimental and theoretical work that presently exists in the slime mold system.

Experimental Studies of Spatial Organization Mechanisms in Slugs

The Prestalk-Prespore Pattern

Cellular slime mold slugs are spatially organized, in that they contain two types of cells, *prestalk cells* and *prespores*, which are ordinarily located in distinct *prestalk and prespore zones*. The prestalk zone occupies the anterior portion of the slug and normally contains about one-fifth of the slug's cells; the prespore zone occupies the posterior four-fifths of the slug (Fig. 2).

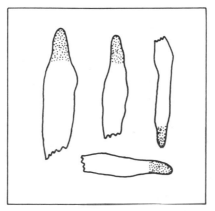

FIGURE 2. The prestalk/prespore pattern in four *Dictyostelium discoideum* slugs, strain Hs1, as revealed by neutral red staining.

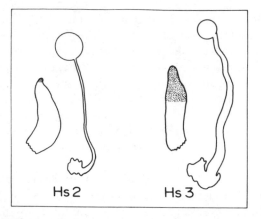

FIGURE 3. Slugs and fruiting bodies of the proportioning mutants Hs2 and Hs3. Hs2 shows a reduced prestalk zone in the slug and a decreased stalk volume in the fruiting body; in Hs3 the prestalk zone and the stalk volume are enlarged. Fruiting bodies of Hs2 and Hs3 average 8% and 41% stalk, respectively; the common parent Hs1 (whose slugs are shown in Fig. 1) makes fruiting bodies which are 22% stalk.

RELATION TO STALK CELLS AND SPORES

Many observations support the idea of a close relationship between the "pre-pattern" in the slug and the final pattern of stalk cells and spores. It appears that in normal differentiation the stalk is formed by cells from the anterior portion of the slug, where the prestalk cells ordinarily reside (27, 3). Prestalk and prespore cells differ ultrastructurally, biochemically, and physiologically, and in some cases the differences are clearly relevant to their normal differentiation fates (for recent data on cell-type specific polypeptides see Borth and Ratner (6), for general review see (18, 20). In some situations, prestalk and prespore cells tend to differentiate according to their "predispositions" even when removed from their normal surroundings (30, 34). The ratio of stalk cells to spores is influenced in the same direction as the prestalk/prespore ratio by temperature and nutritional manipulation (13, 12); two mutants with alterations in their prestalk/prespore ratio show corresponding alterations in the final differentiation pattern (Fig. 3).

REGULATION OF THE PRESTALK/PRESPORE RATIO

Prestalk and prespore cells, however, are not "determined" to form stalk cells and spores, respectively. Rather, the prestalk-prespore pattern is a regulative one. Thus when the two zones of the slug are separated from one another, cell-type conversion occurs within each of the isolated fragments, tending to restore the original prestalk/prespore ratio (5, 14). [Regulation in isolated posteriors appears to involve an additional process, the mobilization of "anterior-like" cells (31); these are cells which resemble the prestalk cells (28) but are found distributed throughout the prespore zone.] Regulation does not depend on these cells, however, as it also occurs in isolates of prespore cells from which the anteriorlike cells have been removed (28).

Regulation can explain several results which otherwise might be taken to suggest that the prepattern and final pattern are unrelated. Thus (i) all cells of the slug can be made to differentiate as stalk, or as spore, by appropriate mechanical manipulations (10); (ii) two mutant strains with drastic alterations in the proportion of stalk cells to spores—one which forms all stalk cells, and the other which forms virtually none—do not show corresponding changes in the prestalk/prespore ratio (23). One can explain this kind of result by assuming that in each case the final differentiation of one of the cell types is blocked while the differentiation of the other can proceed. After differentiation in the allowed direction one would expect regulation among the remaining undifferentiated cells, followed by further differentiation in the allowed direction. For example, in Morrissey's "stalky" mutants, one can imagine that the fundamental lesion is a block in spore formation. Once the available prestalk cells had differentiated as stalk, some of the undifferentiated prespore cells could convert to prestalk; these could then differentiate into stalk, and the cycle would continue until all of the cells of the aggregate had been exhausted.

Cell Sorting and the Prestalk-Prespore Pattern

A number of attempts have been made to explain the formation of the prestalk-prespore pattern with "traditional" pattern formation concepts (such as "positional information" or diffusion-reaction schemes) in which cells in particular positions are directed to differentiate in the prestalk or prespore directions (see (20) for review). It now appears that these attempts were misdirected; it seems more likely that prestalk and prespore cells initially appear in random positions in slime mold aggregates, and that the prestalk/prespore pattern is formed when these cells migrate through the aggregate and coalesce into separate zones.

PATTERN FORMATION BY SORTING

It has been known for some time that cells from the anterior and posterior regions of slugs, when mixed, return roughly to their initial positions. In early reports, it appeared that patterns produced by sorting were diffuse, resembling cell-type gradients; these reports lead to the suggestion (20) that sorting in slime molds was not good enough to produce the discrete-zone pattern of prestalk and prespore cells. Recent experiments using vital dyes as cell markers have produced sorting patterns that are much more distinct (31), with the bulk of the prestalk zone formed by the prestalk donor and the bulk of the prespore zone by the prespore donor.

Even the best patterns produced by sorting, however, contain a zone of mixed cell origin at the prestalk-prespore boundary. The existence of this zone suggests that some cells switch differentiation state during the sorting procedure. Are these cells caused to switch states because they are removed from the "home" zones and mixed with cells of the opposite type? If so, this would at least suggest that the prestalk and prespore zones stabilize the differentiation states of their component cells, which would be a positional effect of a sort. One can pose the question by comparing a sorting experiment in which, for example, marked prestalk cells and unmarked prespore cells might be stirred together, with a parallel experiment in which a marked prestalk zone is transferred intact to the front of an intact prespore zone. In the transplantation experiment, the pattern of marked and unmarked cells initially shows no mixing zone at all. After the few hours required, however, for the stirred cells to reorganize into a slug, a mixing zone develops in the transplanted slug as well, and the transplanted and sorted slugs are indistinguishable (MacWilliams, *in preparation*). Cell-type switching occurs just as quickly in an intact prestalk/prespore pattern as in a mixed slug, and there is no evidence for positional stabilization of cell type.

Is sorting a "positional memory"?

An advocate of position-dependent cell differentiation might propose that the sorting properties of cells are initially determined by a position-dependent process: prestalk and prespore cells would differentiate in response to positional cues during slug development, and would thereafter have the ability to return to their zones of origin. It turns out, however, that sorting occurs even among cells which are entering the slug stage for the first time. A well-studied example is the sorting of cells grown under different nutritional conditions: (17, Table 3) found that when cells grown axenically in the presence of glucose ("G cells"), were mixed with cells grown without sugar ("NS cells"), 80% of the prestalk cells were NS cells and 80% of the prespores were G. Since the experiments involved mixing vegatative cells, this sorting cannot have been due to "positional memory." One might of course argue that a positional memory operates in slugs which are not mixtures of G and NS cells, but its biochemical basis is such that the prespore and prestalk states can be imitated by growth in G and NS media. If this is true, however, the positional memory of slug cells must often be dominated by nutritional variables, and one could just as well suppose, from the beginning, that cells differentiate according to nutritional state.

A continuum of sorting types

Leach et al. (14) in fact studied cells of four different nutritional states, and found that the four states could be ranked by pairwise mixing experiments into a linear hierarchy of sorting preference; in any mixture the lower-ranking member would be recovered preferentially in the spores. This suggests that sorting may be controlled, not by a two-state system (with "prestalk" and "prespore" values) but by a continuously variable parameter. This idea in turn suggests that there might be sorting *within* the prestalk and prespore zones. In recent experiments (MacWilliams, *in preparation*), I have found two kinds of evidence for sorting in the prestalk zone. One kind of experiment involved mixing cells from the front-half prestalk zone of a genetically marked donor with the rear-half prestalk zone and prespore zone of a wild-type host; slugs were allowed to reform and cell samples were then taken from the two halves of the prestalk zone. (The marker used, methanol resistance (*acrA*) does not appear to affect sorting and the experiment was also performed with the genetic marker in the host.) In these experiments, 65% of the cells consistently returned to the half of the prestalk zone that they came from.

In a second kind of experiment, two strains which are known to sort from one another were mixed and samples taken from various positions in the prestalk zones of the mixed slugs. In such experiments using Hs1 (a wild-type) and Hs3 (a prestalk-sorter discussed below), I regularly saw a cell-type gradient in the prestalk zone; in a 50:50 mixture the front third of the prestalk zone was at least 95% Hs3, the rear third 60% Hs3, and the middle third in between. It is clear that the prestalk zone is not homogeneous, but contains an internal sorting gradient.

CORRELATION OF SORTING TENDENCY AND PRESTALK/PRESPORE RATIO

Strains that sort out from one another tend to differ in the proportions of prestalk and prespore cells they make when developing by themselves. Thus "NS" cells (see above) that are relative prestalk sorters, make slugs containing about 20% prestalk cells, while "G" cells, relative prespore sorters, make 10% prestalk (12). Two mutants with altered proportions of prestalk cells (19) conform to this pattern; thus Hs3, the prestalk-sorter mentioned above, makes 45% prestalk cells, and was in fact isolated as a prestalk/prespore ratio mutant; Hs2, a ratio mutant with 3% prestalk cells, is a prespore sorter under at least some conditions. The strains NP422 and NP429, which are "slugger" mutants (mutants which fail to fruit; (25)) are also both ratio mutants and sorters (MacWilliams, unpublished). In all cases the direction of the effect is the same: *relative prestalk sorters* show an *increased fraction of prestalk cells; relative prespore sorters* show a *decreased prestalk fraction.*

Spatial Organization Inherent in the Shape of the Slug

In addition to the division into prestalk and prespore zones, cellular slime mold slugs are spatially organized in another, subtler way: they are elongate in shape. The shape of the slime mold slug is regular and is independent of slug volume over a greater than hundred-fold volume range: large slugs and small slugs show, on the average, the same ratio of length to width (2); see Fig. 4. The regular elongate form of slugs does not necessarily follow from the presence of prestalk and prespore zones (which could exist in an aggregate of any shape), and might be considered a "pattern" in its own right. It may seem artificial to consider the shaping of a slug as a pattern formation process (involving the "differentiation" of "spatially indifferent" cells into ones occupying positions one slug radius from the front, two slug radii from the front, etc.) but the organization of an initially spherical aggregate into an elongate shape is, in any case, a clear instance of morphogenesis. Morphogenesis is, after all, the process which students of pattern formation ultimately seek to explain, so it seems perfectly suitable to consider it here.

FACTORS CONTROLLING SLUG SHAPE

The slug shape problem has received almost no attention, and the known facts can be listed quickly.

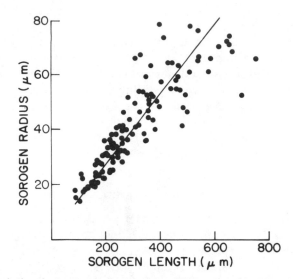

FIGURE 4. The relation between radius and length in a population of *Polysphondylium pallidum* slugs varying about eightfold in length. The measurements are reasonably well fitted by a straight line, suggesting that the average ratio of length to width is constant. [Figure courtesy of E.C. Cox; reprinted from (2).]

1. The elongation of slugs does not appear to require migration over the substrate. The initial elongation process (when the slug is formed) occurs vertically, into the air; migration commences only when the fully elongate structure falls over onto the substrate. Elongation is also known in suspension-culture aggregates freely suspended on fluid medium (Fig. 5).
2. Slug shape does not appear to be based on a permanent difference between front and rear cells. Thus if slugs are cut into anterior and posterior halves,

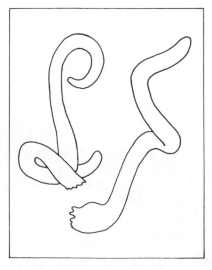

FIGURE 5. Elongate aggregates of Hs2 cells formed in suspension culture. Substrate contact is evidently not required for elongation.

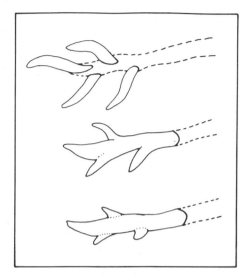

FIGURE 6. When slug tips are implanted into the side of a migrating slug, each tip "organizes" a fraction of the host tissue, and the slug breaks up into as many slugs as tips are present. Figure redrawn from Raper, 1940).

both of the resulting smaller slugs elongate to attain a length-to-width ratio similar to that of the original slug (MacWilliams, *in preparation*).

3. Elongation appears to occur under control of the slug *tip*. The tip is a nipple-shaped protuberance, more or less prominent, on the anterior end of the slug. Two observations suggest its role in the control of slug elongation.

   (a) When a tip is transplanted into the side of a host slug, the tissue subjacent to it forms a new slug axis, which then separates from the host as an independent slug (Fig. 6).

   (b) The proportioning and sorting mutant Hs3 makes slugs which are characteristically shorter and fatter than wild-type slugs of the same volume. When tips are interchanged between mutant and wild-type slugs, the chimeras assume a length-to-width ratio characteristic of the tip donor (Fig. 7).

FACTORS CONTROLLING TIP FORMATION

If the tip is responsible for slug shape, what is responsible for the formation of the tip? The first visible step in the process of tip formation is a clustering of pre-stalk cells at the site at which the tip will appear. cluster of pre-stalk cells may be a sufficient condition for tip formation: when prestalk-cell clusters are created artificially, either by coalescense of prestalk cells around an electrode emitting cyclic AMP (9) or microsyringe transplantation from the prestalk zone (17), tip formation often follows.

Tip inhibition gradient. Tip formation in such microsyringe transplantation experiments is more likely when transplants are made to sites far from the host tip, and is more likely when the tip of the host slug is removed (19). This suggests that

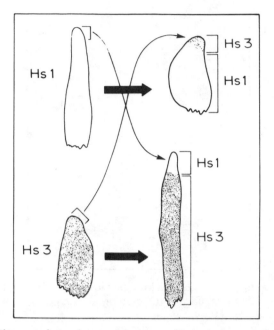

FIGURE 7. When tips are interchanged between Hs1 slugs and the characteristically short and fat slugs of strain Hs3, the Hs1 slug with a Hs3 tip becomes short and fat while the Hs3 slug with an Hs1 tip assumes a wild-type length/width ratio. Slug shape appears to be controlled by the slug tip.

the tip inhibits tip formation; one could imagine that the inhibition is due to a substance, made in the tip, which diffuses into and is broken down in the rest of the slug. Consistent with this idea is the observation that the *level* of the tip inhibition depends on the absolute, rather than the relative distance from the host tip; the rear of a short slug has the same inhibition level as the middle of a slug twice as long. This is expected in simple diffusion models since the slope of a concentration gradient generated by a local production, diffusion and breakdown depends primarily on the ratio of the diffusion and breakdown constants and is thus independent of slug size ((19) and *in preparation*).

Tip activation gradient. Prestalk cells from the anterior of the prestalk zone induce tip formation more efficiently than prestalk cells from the rear of the zone, and there is a continuous gradation of inductive efficiency in between (19).

Tip activation and inhibition in prestalk/prespore ratio mutants. Experiments with prestalk/prespore ratio mutants reveal alterations in transplantation properties. The strains Hs2, NP422 and NP429 (mentioned above as strains with reduced prestalk zones) all show significantly reduced levels of *both* tip activation and of tip inhibition. In the mutant HS3, in which the proportion of prestalk cells is increased, the tip activation level is normal but the tip inhibition gradient is steeper—the tip inhibition level drops off twice as fast as normal with increasing distance from the tip. A possible explanation for these changes will be given in the model of prestalk/prespore proportioning discussed in the next section.

A Model for Dividing a Population of Cells into Prestalk and Prespore Types

I will now discuss a model specifically designed to explain the observations (discussed above) that (1) various types of cells can be ranked linearly according to their tendencies to become spores in mixtures, and (2) that relative stalk-preferring cells tend to make slugs with enlarged prestalk zones, while spore-preferrers make slugs in which the prestalk zone is reduced. The model also turns out to provide at least a partial explanation for the transplantation properties of slugs discussed previously, including the changes in properties encountered in prestalk/prespore ratio mutants.

Principle of the Proportioning Model

INHIBITOR AND THRESHOLD

The model assumes that prestalk and prespore cells can interconvert freely, and that this interconversion is controlled by a diffusible substance. Specifically, imagine that the prestalk cells make a *prestalk inhibitor*, and that this substance is broken down by all cells, so that its steady state level, to a first approximation, is proportional to the fraction of prestalk cells present. We will initially ignore the inhibitor concentration gradient which is expected to result from the spatial separation of prestalk and prespore cells, and will speak of a single inhibitor level. Assume that individual cells show a threshold response to the prestalk inhibitor: when the inhibitor level is above threshold, a cell behaves as prespore; when the inhibitor level is below threshold, a cell behaves as prestalk. The system thus contains a stabilizing negative feedback loop: prestalk cells make the inhibitor, and at high inhibitor levels the cells convert to prespore.

"SORTING" ACCORDING TO THRESHOLD LEVEL

Assume that the inhibition threshold can be different in different cells. Assume that some of the variability is random, so that any "homogeneous" population of cells in fact has a continuous threshold distribution. But assume that the average value of the threshold can also vary among strains, and in particular can depend on nutritional state. Differences in average threshold will produce "sorting." To see this, imagine two populations of cells with different, nonoverlapping threshold distributions; in a mixture of the two types the threshold distribution will be bimodal. Imagine that the inhibitor level comes to lie between the two peaks of the distribution. All of the cells of the strain with the lower average threshold all then have threshold values lower than the inhibitor level, and will become prespores; the cells with the higher average threshold all have thresholds above the inhibitor level and will all become prestalk. This situation corresponds to perfect sorting. Sorting will be less than perfect, but will still occur, if the threshold distributions overlap or if the inhibitor level does not lie exactly between them.

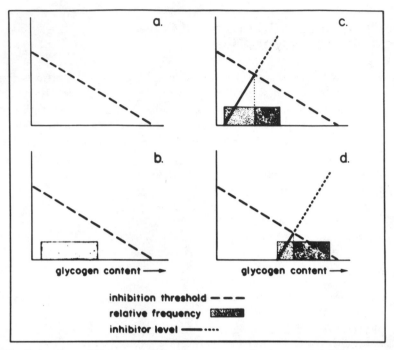

FIGURE 8. Operation of the prestalk/prespore proportioning model. The model assumes that cells can interconvert freely between prestalk and prespore states and that this conversion is regulated by an "inhibitor" which is made by the prestalk cells and broken down by all of the cells of an aggregate. Each cell has a threshold response to the inhibitor: it is prestalk at sub- and prespore at supra-threshold inhibitor levels. (a) One assumes that the threshold is different in different cells; in this figure it is assumed to be inversely correlated with cellular glycogen content. (b) In a given aggregate, there will be some particular distribution of thresholds, represented here by shading. (c) The actual inhibitor level in an aggregate depends on the fraction of cells in the prestalk state. Assume that all cells are initially prespore, and that the cells convert to prestalk in order of thresholds, high thresholds (cell on the left) first. Conversion, which causes the inhibitor level to rise, can continue until the inhibitor level is equal to the threshold of the prespore cell whose turn has come to convert. At this point, which is given by the intersection of the threshold curve and the "inhibitor level" curve, the system is stable. Coarse shading indicates cells that are stably prestalk cells; fine shading prespore cells. (d) In an aggregate of lower-threshold cells, the same mechanism gives a smaller fraction of prestalk cells. The inhibitor level in the stable state is also lower.

DETERMINATION OF THE INHIBITOR LEVEL

In the above argument, the inhibitor level was assumed. To see how this level would actually be determined, imagine a population of cells with a given threshold distribution (this time unimodal). Imagine further that all the cells are in the prespore state, and that the prestalk inhibitor level is therefore zero. Imagine that it becomes possible for the cells to convert to prestalk, and assume (to simplify the argument; this assumption is not logically required) that the cells convert to prestalk in order of their prestalk inhibition thresholds, the cells with highest thresholds converting first. As conversion proceeds two things will happen (Fig. 8): (i) the inhibitor level will rise; (ii) as high-threshold cells convert to prestalk and are thus withdrawn from the prespore population, the threshold level among the remaining prespore cells will decrease. In particular, the threshold of the prespore cell at the point of converting to prestalk (the "marginal threshold level") will become progressively lower. Eventually a point will be reached at which the rising inhibitor level equals the threshold of the cell whose turn has just come to convert. The inhibitor will prevent this cell, and all other remaining prespore cells, from converting to prestalk. All prestalk cells will now have thresholds greater than the inhibitor level, while all prespore cells have lower thesholds; the system will then be in a stable state, and the inhibitor level will cease to change.

[The proportioning mechanism would also work if all cells began in the prestalk state; in this case one would imagine that the inhibitor level was initially maximum and the cells deconverted in order. The model is thus a general model of the regulation of the prestalk/prespore ratio. If cells did not convert or deconvert in order, the inhibitor level would still stop changing at the equilibrium value because at this point net conversion or deconversion would cease. A period of interconversion would follow, during which all cells in the "incorrect" state would switch. The ultimate result would be the same as if the cells had converted in order.]

Average threshold and the prestalk/prespore ratio

By comparing Fig. 8c and 8d, one can see that the final proportion of prestalk and prespore cells is influenced by the threshold distribution. If thresholds are generally high, many cells have to convert to prestalk before the inhibitor curve and the threshold curve intersect; if threshold are low, only a few prestalk cells must be present before the inhibitor level reaches the marginal threshold. The model thus accounts for the observation that relative stalk-preferring strains tend to make higher proportions of prestalk cells, while in relative spore-preferrers the proportion of prestalk cells is reduced.

Spatial Effects and Transplantation Properties

Inhibition gradient

After the two cell types are formed, we assume that they migrate to separate zones to form the prestalk/prespore pattern. We will not now attempt to explain this migration, but will consider an important consequence, that can be used to account for the transplantation properties of slime mold slugs. Since the prestalk inhibitor is assumed to be made in the prestalk cells but broken down throughout a slug, one expects there to be a steady-state inhibitor concentration gradient, with highest inhibitor levels in the prestalk zone and lowest levels at the rear of the slug. This gradient is reminiscent of the observed gradation in the level of the *tip* inhibition—high in the front, and low in the rear; one could thus account for the tip inhibition gradient by assuming that tip inhibition and prestalk inhibition are the same. This assumption seems reasonable inasmuch as the first step in tip formation appears to be a clustering of prestalk cells at the site of the incipient tip; the prestalk inhibitor could presumably prevent this by converting the prestalk cells to prespore.

ACTIVATION GRADIENT

The gradient of prestalk inhibitor in the slug will "induce" a parallel gradient in prestalk inhibition threshold. At least two steps are required to show this. First: Since the inhibitor level is graded, note that there can be only one position in the slug at which the inhibitor level is exactly equal to the marginal threshold (the threshold of the most prestalk-preferring cell which has remained prespore). Let us call this position the *crossover point*.

Second, consider the positions of two cells, a prespore cell whose threshold is equal to the marginal threshold, and a prestalk cell whose threshold is infintesimally higher. The prespore cell must be located at or to the front of the crossover point; otherwise it would become prestalk. The prestalk cell must be located at or to the rear of the crossover point; otherwise it would become prespore. But the prestalk cell must be located at or to the front of the prespore cell (because the prestalk zone is anterior). Both cells thus lie at the crossover point, and this point is at the prestalk/prespore boundary.

In all of the prestalk zone, the inhibition level will be greater than or equal to the level at the crossover point, and will thus be greater than or equal to the marginal threshold. Imagine a prestalk cell with a threshold just slightly greater than this value. This cell can only remain prestalk if it occupies a position near the prestalk/prespore boundary. If it travels further forward in the prestalk zone, it will encounter suprathreshold inhibition and will switch to prespore, and should migrate backward to its "allowed" position. In a position somewhat further forward, a cell is required to have a threshold somewhat higher than the marginal value; in the front of the slug, a cell may be required to have a threshold considerably higher than the marginal value. In general terms, a gradient of inhibition threshold must exist in the slug which is at least as steep as the gradient of prestalk inhibition. If one assumes that the prestalk inhibition and the tip inhibition are the same, this gradient should also be a gradient of tip inhibition.

TRANSPLANTATION PROPERTIES IN MUTANTS

How can one explain the alterations in transplantation properties which one encounters in prestalk/prespore ratio mutants? For the mutants Hs2, NP422, and NP429, which have reduced prestalk zones, a straightforward explanation is possible. In all of these mutants, transplantation experiments disclose a reduced resistance to tip inhibition and a reduced tip inhibition level. Assuming that the tip inhibition and the prestalk inhibition are the same, the reduced resistance to tip inhibition indicates a reduced prestalk inhibition threshold; the proportioning model (Fig. 8c,d) thus predicts the observed reduction in the fraction of prestalk cells. The proportioning model also leads one to expect a lower steady-state inhibitor level; this accounts—again assuming that prestalk inhibition and tip inhibition are the same—for the reduced tip inhibition level.

For the mutant Hs3, which has an expanded prestalk zone, a different sort of explanation is necessary. Experiments suggest that the tip inhibition gradient in Hs3 is about twice as steep as normal, a change which could result from an increase in the rate of tip inhibitor breakdown. Increased breakdown of the prestalk inhibitor would mean a lower steady-state inhibitor level for any given production rate; the effect would appear in formal diagrams such as Fig. 8 as a flattening of the "potential inhibitor level" curve. Flattening this curve would postpone its intersection with the threshold curve, so that the steady-state fraction of prestalk cells would be increased.

A PROBLEM WITH Hs3

There is also an alternative explanation for Hs3; this is based on the observation that the strain is a strong prestalk-sorter, and is expected on this basis alone to have increased prestalk inhibition threshold and an increased fraction of prestalk cells. Unfortunately, this new explanation makes trouble for our accounts of the other mutants. For if Hs3 indeed has an increased prestalk inhibition threshold, it should, according to our scheme, have increased resistance to prestalk inhibition as well as increased tip inhibition level. Neither of these changes are observed. The only clear way out of this problem may seem a semantic solution at best; this is to assume that prestalk inhibition and tip inhibition are not identical but "correlated" on account of some biochemical link. In most instances we expect the correlation to be observed, but in Hs3 a mutation has presumably affected the biochemical link itself, with the result that the correlation is lost. There is circumstantial evidence for this idea in the observation that other proportioning mutants (as well as in G and NS cells) the alteration in proportioning is accompanied by an alteration in glycogen content. In Hs3, however, the glycogen content is unchanged. Hs3 may thus have a sustantially different biochemical basis than the other mutants (19).

A Model of Slug Shape Based on Differential Chemotaxis

The remainder of this chapter is devoted to the question of the control of slug shape. Can simple models explain how the tip of a slime mold aggregate "organizes" the remaining cells into an elongate cylinder, with a particular length-to-width ratio? The slug tip first of all appears to be a source of cyclic AMP (4, 29) which can act as a chemotactic attractant for slug cells (21). It is thus tempting to try to explain the organization of the slug as a chemotactic phenomenon. It seems clear, however, that the simple assumption that all cells are attracted to the tip to the same degree will not do; this would only produce a sphere of cells surrounding the tip. A more promising approach would seem to be to assume that cells are differentially chemotactic, some more strongly attracted to the tip than others.

How could a difference in chemotactic response be translated into a specific three-dimensional form? One possibility is suggested by the observation (16) that isolated prestalk zones migrate with more motive force than prespore zones. One might imagine that a migrating aggregate assumes an elongate shape because the actively chemotactic anterior cells provide the force for migration, while the less motile posterior cells are simply dragged along. This kind of model could at least explain why the prestalk cells assume a position at the anterior end of a migrating slug. One could attempt to elaborate the idea into a theory of the length-to-width ratio by assuming a continuous distribution of motive force among slug cells. During migration the faster cells would come to the front, as in a heterogeneous party of hikers, and a gradient of motive force would be established. If the distribution were narrow, so that all cells had close to the same motive force, one might imagine a short fat slug would result. A population of cells with a wide range of values of motive force might give long thin slugs.

There is an important argument, however, against this sort of model, which is that transverse cuts made partially through a slug show no obvious tendency to gape, and when transverse cuts are made entirely through slugs the fragments show no tendency to spring apart. There is no suggestion of mechanical stress within slugs. It seems that elongation in slime mold aggregates must be caused by a force other than substrate drag, a force that arises entirely within the cell aggregate.

Positive and Negative Chemotaxis

If the elongating force is to be supplied by chemotaxis, but must be generated entirely within the aggregate, one obvious possibility is that there are positively and negatively chemotactic cells that actively move apart from one another as an aggregate elongates. To set a limit to this elongation, one might assume that cells can switch between positive and negative chemotaxis, and that the switch is controlled by a substance whose level can reflect slug length. In this section I will discuss a model in which cells can respond either positively or negatively to a chemotactic agent diffusing from the aggregate tip; the cells switch between positive and negative responses on the basis of the local concentration of the chemotactic agent.

Assume that a substance diffuses from a point source in an aggregate and is broken down in all of the cells; this will create, in the steady state, a concentration gradient. Assume that each cell has a response threshold to this substance, such that the cell migrates up the concentration gradient (towards the source) if the level of the agent is below the threshold of the cell but migrates away from the source (down the gradient) if the agent's level exceeds the threshold. The effect of these assumptions is that a cell will "seek" a position in the aggregate at which the concentration of the agent exactly equals its threshold. If it can find such a position, assume that the cell will be "chemotactically neutral" and will remain there.

Assume further that the thresholds of individual cells differ, and that any cell population can be characterized by a distribution of threshold values. Imagine (for the moment only!) that the maximum and minimum concentrations of the chemotactic agent are fixed, so that one can speak of the range of gradient values to be found in a given aggregate. Those cells with response thresholds lying within the available range of gradient values will find stable positions within the aggregate. The chemotactic mechanism will in effect line these cells up in order of response threshold, creating a gradient of threshold whose level and slope matches the gradient in chemotactic factor. The chemotactic mechanism can thus arrange at least some of the cells of an aggregate into an ordered configuration.

The chemotactic movements of cells, furthermore, should be able to affect the range of gradient values prevailing in an aggregate and under appropriate conditions should be able to adjust the gradient range to fit the distribution of response thresholds of the aggregate's cells. Consider a situation in which the range of gradient values is considerably greater than the range of response thresholds. Make the additional assumption that the average level of the chemotactic agent is "appropriate" to the aggregate; under simple conditions this means that the gradient value at the midpoint of the aggregate equals the median response threshold (Fig. 9 bot-

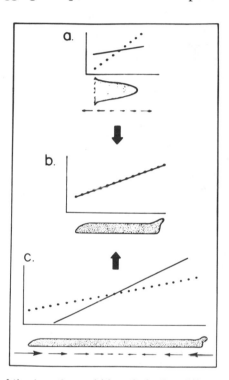

FIGURE 9. Regulation of the length-to-width ratio in the differential chemotactic model. The model assumes that an aggregate tip makes a "chemotactic agent" which diffuses into the aggregate and establishes a concentration gradient (———). Cells respond positively or negatively to this agent depending on whether its level lies below or above their "chemotactic response thresholds". High-threshold cells thus tend to move to the front and low-threshold cells to the rear, creating a gradient in chemotactic response threshold (· · · ·). In an aggregate which has not yet elongated (A), the concentration gradient of chemotactic agent is necessarily shallow; the range of gradient levels is narrower than the distribution of response thresholds. The cells in the front experience subthreshold levels and migrate forward; the cells in the rear experience suprathreshold levels and migrate backwards. As a result the aggregate elongates. Elongation causes the range of gradient levels to increase until the system reaches a stable configuration (B) with each cell in a position at which the level of the chemotactic agent is equal to its response threshold. If the aggregate should become too long (C), the range of gradient levels exceeds the range of response thresholds; this gives rise to positive chemotaxis in the rear and negative chemotaxis in the front, as a result of which the aggregate will shorten until it reaches the stable configuration (B).

tom). Under these circumstances cells in the rear of the aggregate will be positively chemotactic and will move forward, while cells in the front will be negatively chemotactic and will move backward. The overall effect will be to rearrange the aggregate into a shorter and thicker form. Since this rearrangement increases the cross-sectional area perpendicular to the concentration gradient, the concentration gradient should flatten, its values falling in a portion of the aggregate containing the source and rising in more remote positions. If the average level of the chemotactic agent remains appropriate, it should be possible for cell rearrangement and the resulting flattening of the concentration gradient to continue until the range of gradient values exactly corresponds to the distribution of response thresholds. At this point the entire system will be stable.

Consider now a situation in which the gradient range is too short, but again the average concentration of chemotactic factor is appropriate (Fig. 9, top) so that the gradient value and the response threshold at the midpoint of the aggregate are equal. In this case the cells in the rear of the aggregate will be negatively chemotactic, while the cells in the front chemotax positively; the result will be that the slug will stretch and become narrower. The cross-sectional area available for diffusion will decrease, and the concentration gradient should steepen. Again, if the average concentration remains appropriate, the aggregate will continue to elongate until the range of gradient values exactly matches the distribution of response thresholds.

In these examples, I have assumed a block-distribution of response thresholds, but it seems at least plausible that a differential chemotactic system could accommodate itself to threshold distributions of complex shape, producing in each case a stable structure. For instance, imagine a distribution of thresholds that was flat except for a central peak. All the cells within the peak would have similar threshold values, and all would "wish" to be at about the same distance from the chemotactic source. This would create a wide spot in the slug. Within the wide section, however, the concentration gradient of chemotactic factor should be reduced (because of the increased cross-sectional area), and there would thus be "room" in the gradient for the extra cells (the configuration would be stable). Consider next a threshold distribution with a central valley. In this case, few cells would wish to lie at a given distance from the source, which would create a thin spot in the slug. Within this "wasp waist," however, the concentration gradient would become steeper; the system would thus manage with the reduced number of cells.

Negative chemotaxis in the differential chemotactic model would not necessarily have to be realized by the active movement of cells away from the source of chemotactic agent. Another possibility is that cells that are not positively chemotactic to a certain minimum degree are carried away from the source by a constant flow of extracellular material, produced at or near the chemotactic source. This material could be identical with that making up the slime sheath of the slug. In this case the chemotactically neutral state would be one in which cells migrated forward at a rate sufficient to just "stay even" with the current of extracellular material; faster or slower chemotactic movement would appear as positive or negative chemotaxis relative to the coordinate system of the slug. One can then account readily for the forward migration of the slug as a whole: the sheath material is normally stationary with respect to the substrate, and what appears to individual cells as a "swimming against the current" appears to an outside observer as net forward movement. This process could occur just as easily in suspension culture aggregates as in aggregates in contact with a substrate. In suspension culture, one in fact often sees accumulations of amorphous material attached to the end opposite the tip.

The discussion so far suggests that the differential chemotactic model has morphogenetic power and that its assumptions are reasonable. The model seems particularly relevant to slime molds because it posits that the slug is "organized" by a point source of morphogen that one could easily identify with the slug tip. The discussion so far also suggests that the shape of aggregates is at least in part controlled by the distribution of chemotactic response thresholds. This brings up an important question: just what sort of influence can the tip have over aggregate shape? Is the model capable of explaining the observation, described previously, that the length-to-width ratio of chimeric slugs is determined by their tips?

Incorrect Predictions about Slug Shape

MORPHOGENETIC IMPOTENCE OF THE TIP

It unfortunately appears that the answer to this question is no. In the differential chemotactic model, two tips can differ from one another only in the amount of the agent that they produce. In the discussions so far we have always assumed that the average level of the chemotactic agent was "appropriate" for the cells in the aggregate. This suggests that the strength of the morphogen source may be dictated by the chemotactic response thresholds of the available cells. This can in fact be demonstrated rigorously in models in which one assumes that the chemotactic agent is lost from the slug by first-order breakdown, a common assumption in diffusion models. In this case, each cell breaks down the agent at a rate that is determined only by the concentration of agent it experiences. But recall that each cell comes to lie at a position at which the concentration of chemotactic agent equals its own response threshold. The total rate of breakdown is therefore fixed by the breakdown constant and the distribution of response thresholds. The rate of production is fixed in turn by the fact that production and breakdown must be equal in the steady state of the system. If the system is to be in its stable state, there is only one possible strength for the chemotactic source. This conclusion also applies if breakdown is other than a first-order process, provided only that breakdown is mediated by cells, and each cell breaks the agent down at a rate determined only by its local concentration.

What actually happens if the source has the wrong strength? In this case, the system cannot achieve steady state. If the source is stronger than the "appropriate" value, the concentration of the chemotactic agent would be too high for all cells to assume chemotactically neutral behavior. Some cells would remain negative-chemotactic under all conditions, and these cells would presumably be lost from the aggregate. It seems likely that the concentration of the chemotactic agent in the remaining tissue would then rise, leading to the loss of more cells; the slug would ultimately disintegrate. If the source were weaker than "correct," the concentration of the chemotactic agent would be too low to maintain all cells in a chemotactically neutral state; some or all of the cells would be continuously positive-chemotactic. These cells would form a sphere surrounding the morphogen source. The system would at best produce a sphere with an elongate tail, a shape that does not occur, to my knowledge, in slime molds. It is clear that changing the strength of the source of chemotactic agent will not modulate the length-to-width ratio of the slug; the differential chemotactic model gives no explanation for the observation that the length-to-width ratio of a slug may be controlled by its tip.

LARGE AND SMALL SLUGS ASSIGNED THE SAME LENGTH

What does the differential chemotactic model predict about slugs of different sizes, drawn from the same population of cells? Consider the "thought experiment" of cutting a differential chemotactic slug lengthwise into equal left and right halves. Assume that the original slug was in steady state, with each cell at a position at which the gradient value exactly equalled its chemotactic response threshold and the total production of chemotactic agent exactly equal to the total breakdown. Since the system has axial symmetry (arising from its one-dimensional organization), there was not originally any net movement of chemotactic agent across the plane separating the right and left halves; each half was originally self-sufficient. Bisecting the slug will thus have no effect on the gradient values and should assign half of the source, as well as half of the breakdown, to each half-slug. Both half-slugs will then be in their stable states without any rearrangement of their cells.

The two half-slugs produced by this operation do not have the same shape as the original slug; they have the same length. Similar arguments could show that two identical slugs fused lengthwise would form a stable slug which is thicker than the original but still has the same length. It thus appears that the differential chemotactic model predicts that all slugs drawn from a given population of cells will have the same length. As mentioned above, observations show that slugs of different sizes have the same shapes, but differ in length.

One can also consider the prediction of the differential chemotactic model if a slug is cut transversely in half. Assuming that the cells of the original slug were arranged in order of their chemotactic response thresholds, the threshold distributions in the two half-slugs should be narrower than the distribution prevailing in the original slug. One would then expect the two slugs to rearrange themselves only minimally, with the resulting slugs both having a length-to-width ratio considerably less than that of the original slug. This result also contradicts observation; as mentioned above, experiments show that the two half-slugs rearrange themselves to the same shape as the original slug.

The Prestalk-Prespore Pattern in the Differential Chemotactic Model

We have seen that the differential chemotactic model fails to account for several important features of slug shape. It thus seems unlikely that the model is "correct" in slime molds. One nevertheless hesitates to discard the model, since it is not entirely arbitrary. At least some of its features are similar to the features of the pre-stalk/prespore proportioning model, a model with considerable explanatory power. We will now see that a unified chemotactic/proportioning model has further explanatory power, and is attractive enough to motivate an attempt to "save" the differential chemotactic model.

Factors suggesting a unified chemotactic/proportioning model

Differential Chemotaxis in the Proportioning Model. In the differential chemotactic model, we saw that positive and negative chemotactic responses to an agent diffusing from the front of the slug lead to the formation of a gradient of chemotactic response threshold. In the prestalk/prespore proportioning model, we explained a gradient of prestalk inhibition threshold by assuming that cells migrate forward when their prestalk inhibition thresholds exceed the level of prestalk inhibition, and backward if the relationship between inhibition and threshold is reversed. This process is formally identical with differential chemotaxis.

Cyclic AMP as a Modulator of Chemotaxis and Proportioning. Cyclic AMP, furthermore, which is the most likely candidate for the chemotactic agent, also seems to have the properties of the prestalk inhibitor; in two laboratories it has been found to suppress the conversion of isolated prespore cells into prestalk (Durston, Takeuchi, *personal communications*). This clearly suggests a unification of the chemotactic and proportioning models in which cyclic AMP modulates both chemotaxis and cell-type interconversion.

Source Strength in a Unified Chemotactic Proportioning Model. In a unified model, an important problem of the differential chemotactic model would be solved: how the source of chemotactic agent could be adjusted to the "appropriate strength" for a given slime mold aggregate. Recall that the differential chemotactic model is stable when every cell experiences a concentration of chemotactic agent equal to its chemotactic response threshold. A stable state can only be achieved, however, if the rate of production of the chemotactic agent is equal to the rate at which the agent would be broken down in this situation. If one assumes that the chemotactic agent is identical to the prestalk inhibitor, this state becomes self-stabilizing; if the production rate were too low, cells would uniformly be exposed to subthreshold levels of prestalk inhibitor, new prestalk cells would form, and the production rate would increase; excessive production would bring about a decrease in the number of prestalk cells and a decrease in production rate.

Transplantation Properties in a Unified Model. Another intriguing feature of a unified chemotactic/proportioning model arises from making prestalk cells a source of the chemotactic agent: one automatically gives a cluster of prestalk cells the property of an aggregate tip. One expects such clusters to behave as tips, and one thus obtains a basic explanation for tip formation in transplantation experiments. It remains easy to understand the gradients disclosed by transplantation; factors that influence the chances that transplanted cells will remain in the prestalk state should influence the chances of tip formation; gradients of tip inhibition and tip activation should arise from the gradients of prestalk inhibition and prestalk inhibition threshold.

Within the unified chemotactic/proportioning model, furthermore, a second explanation of the tip inhibition gradient is possible. A group of prestalk cells transplanted into a host slug could in principle either be attracted to each other or to the tip of the host; in the first case they would form a new tip, while in the second case they would not. The chemotactic signal from the host tip would thus inhibit tip formation. Since the chances for inhibition would be greater for transplants to the front of the host, where the signal emanating from the host tip would be stronger, the chemotactic signal would produce an inhibition gradient.

One could imagine that tip inhibition in slugs is entirely due to this second mechanism. If so, one could rethink the explanation for the tip activation and the tip activation gradient. "Resistance" of transplanted cells to the chemotactic signal of the host would have to arise from their own production of chemotactic agent. The activation gradient within the prestalk zone would then suggest a gradient in the rate of production of this agent, and could ultimately suggest a correlation between prestalk inhibition threshold and chemotactic agent production rate. This correlation might account for the association (in the short prestalk mutants Hs2, NP422, and NP429 discussed above) of changes in prestalk inhibition threshold and changes in tip activation. The idea that the prestalk inhibition threshold and the production of chemotactic agent are correlated but not identical allows one to account for the long prestalk mutant Hs3 in which the prestalk inhibition threshold is increased, but there is no change in transplantation properties.

A NEW PROBLEM IN THE UNIFIED MODEL, AND A POSSIBLE SOLUTION

All these considerations show that the unified chemotactic/proportioning model is an interesting one, and it is distressing to recall that the model cannot be correct. A basic problem with the model is of course that as a differential chemotactic model it makes incorrect predictions about slug shape. A second problem should also be mentioned; this arises from the fact that in differential chemotactic models each cell seeks a position at which the inhibitor concentration equals its threshold. Why, having reached such positions, would cells be stably prestalk or prespore? Instead of forming two distinct cell types, it seems that the model would tend to bring all cells to a neutral state, exactly balanced between the two possibilities.

One way out of this dilemma is to assume that although the differential chemotactic mechanism "tries" to operate in slime mold slugs, there are other stronger forces that control slug shape. One could then imagine, in particular, that slugs were normally kept shorter than the differential chemotactic length. This in turn would mean that the range of inhibition gradient values would be less than the width of distribution of prestalk inhibition thresholds. In the bulk of the prespore zone, one would then expect that the cells would have thresholds below the inhibition level; while in most of the prestalk zone, the thresholds would be above; both sorts of cells would be stable.

A mechanism which might control the length-to-width ratio of slugs independently of the differential chemotactic mechanism is discussed in the final section of this chapter.

A Model of Slug Organization Based on an Adhesive Gradient

The tip is sometimes described as the "organizer" of the slug (29), a term which suggests an analogy with the "organizer" regions of other developing systems. This suggests an interesting line of thought: the activity of one noted "organizer," the dorsal lip of the blastopore in amphibian embryos, has sometimes been explained as a result of differences in adhesiveness (33). Recently, Mittenthal and Mazo (22) have suggested a model for evagination based on adhesive gradients; the idea is that a gradation in adhesiveness can create a force that tends to make tissue elongate. Cyclic AMP, which is probably produced in the tip of the slug, seems to be able to modulate the adhesiveness of slug cells (Durston, *personal communication*); this suggests that the tip could create an adhesive gradient within slugs. Yabuno (37) has found that cells from the front of the slug are more adhesive than posterior cells. Can one use the idea to explain the ability of the slug tip to "command" the cells of an aggregate to arrange oneself into an elongate form?

Principle of Morphogenesis in the Adhesive-Gradient Model

To address this question, let us first consider how an adhesive gradient could cause tissue elongation. We will start by assuming that an adhesive gradient is present in a tissue mass, and will defer considering how the gradient might be generated until later. Let us assume a cylindrically shaped block of tissue, not necessarily very elongate, along which a gradient of cell adhesiveness exists. Consider that the adhesiveness arises from adhesive molecules on the surfaces of the cells. For simplicity, assume that the adhesiveness is homotypic, that is, that there is only one sort of sticky molecule involved, and that it is stochiometric, that is, each adhesive molecule binds to exactly one similar molecule on another cell. Further assume that the binding reaction is energetically favorable, so that the system tends to a state in which the maximum possible number of adhesive molecules are bound to others.

Assume that the adhesive gradient is due to a graded variation in the number of adhesive molecules per unit area of cell surface. Assume finally that in any given cell the molecules occupy random positions, that is, there is no tendency for the adhesive molecules to cluster on one side of the cell, even if this would allow a greater number of adhesive molecules to be bound. This last assumption may be difficult to accept, and it probably does not have to be made in such a severe form; the properties of the model would not be qualitatively changed, if many of the adhesive molecules do cluster so long as a fraction are positioned randomly. For the sake of simplicity, assume here that all the adhesive molecules are distributed randomly over the cell surface.

ADHESIVE DISCONTINUITIES GENERATE AN ELONGATING FORCE

If cell adhesiveness changes with position, and adhesive molecules are uniformly distributed over individual cells, there must be adhesive discontinuities at

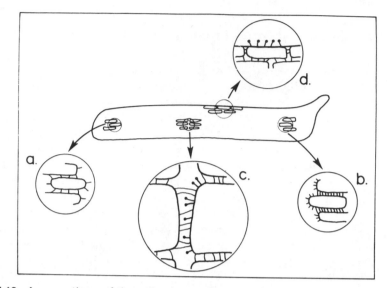

FIGURE 10. Assumptions of the adhesive-gradient model. (a) Each slug cell is uniformly covered with adhesive molecules which bind to similar molecules on neighboring cells. (b) Cells from the front have higher densities of adhesive molecules than cells from the rear; adhesiveness is graded in between. (c) At anterior-posterior cell interfaces, some of the adhesive molecules on the anterior cell are necessarily unbound. (d) At the slug free surface adhesive molecules are also unbound. Calculations show that to minimize the number of unbound adhesive molecules the system must elongate to a specific degree and then stop.

cell interfaces; the posterior aspect of a given cell n will have a greater density of adhesive molecules than the anterior aspect of the next cell $n + 1$. This means that, at such interfaces, some of the adhesive molecules cannot find partners on the opposing cell and necessarily remain unbound. The number of unbound adhesive molecules will depend on the slope of the adhesive profile: with a steep gradient there will be many unbound molecules, with a shallow gradient there will be few.

Such a system can decrease the number of unsatisfied adhesive molecules, and thus decrease its potential energy, by rearranging its cells to increase its length. Consider, for example, an aggregate 100 cells long and 10 cells wide, with an adhesive value of 100 at the front and 0 at the back. Assuming the adhesive gradient is linear, there will be an adhesive discontinuity of 1 at each anterior/posterior interface; since there are $99 \times 10 = 990$ interfaces there are a total of 990×1 units of potential energy in the form of unpaired adhesive molecules. Now assume that the aggregate has been rearranged into a shape 200 cells long and 5 cells wide. The discontinuity is now 0.5 at each interface, and although there are now $199 \times 5 = 995$ interfaces, the potential energy (995×0.5) has decreased considerably.

"SURFACE TENSION" RESISTS ELONGATION

This calculation suggests that there would be a further energy decrease by rearranging the aggregate into a still narrower form; the energy would continue to decrease at least until the point at which the aggregate had become just one cell wide. This is of course not observed in slime mold slugs, which seem to elongate to some specific extent and then stop. In order to predict realistic behavior, adhesive-gradient models must incorporate a mechanism that resists indefinite elongation. In Mittenthal and Mazo's (22) adhesive-gradient model of the insect leg, the problem is solved by considering a second component of the potential energy of the system, namely the energy required to bend the leg epithlium into a cylinder. As the leg elongates, the radius decreases, and this strain energy increases. Ultimately the system reaches a point at which the adhesive energy released by further elongation is less than the energy required for further bending of the epithelium; at this point, the shape of the system is stable.

Since slime mold slugs are solid, rather than epithelial, strain energy cannot be used to resist elongation in our model. A second force, however, that can resist elongation is surface tension. It is well known that the surface area increases if a droplet of constant volume elongates from a spherical to an ovoid shape and beyond; if energy is associated with the free surface of the droplet, then energy will be required for elongation, and the droplet will show a tendency to relax spontaneously into a spherical form. A surface energy arises naturally from our assumptions about cell adhesiveness; the energy arises from the adhesive molecules on the free surface of the aggregate, which are necessarily unable to find partners. Elongation increases the free surface of the aggregate and thus increases the number of unsatisfied adhesive molecules associated with the surface.

Length and Width of the Slug Shape with Minimum Adhesive Potential Energy

In the differential-adhesion model, therefore, elongation of the aggregate causes two sorts of changes in the number of unsatisfied adhesive molecules: a decrease in the number of molecules unsatisfied on account of adhesive discontinuities and an increase in the number of molecules unsatisfied because they have come to lie on the surface of the slug. A further analysis of the behavior of the model requires quantitative calculation of these two effects. In this calculation we will take into account a fact not previously discussed, namely that adhesion at anterior-posterior interfaces appears to involve different molecules than adhesion between lateral surfaces (1). We will assume here that only the anterior-posterior adhesiveness is graded with position. This assumption is helpful in maintaining the cylindrical shape of the slug: if the surface energy were higher in the anterior of the slug than in the posterior, the system could decrease its potential energy by constricting its circumference in the anterior and increasing the circumference in the posterior; it would thus tend to assume a conical shape.

POTENTIAL ENERGY ASSOCIATED WITH THE ADHESIVE GRADIENT

Assume an aggregate of cells with adhesiveness (of anterior/posterior interfaces) A_F at the front and the lower value A_B at the rear; the units of adhesiveness are adhesive molecules per unit area of cell surface. Assume that the adhesiveness is linearly graded between the fixed values at the front and the rear. Each cell thus has an adhesiveness which depends on its anterior-posterior position; there is an adhesive discontinuity at each anterior-posterior position; there is an adhesive discontinuity at each anterior-posterior interface. The potential energy associated with adhesive molecules that are unpaired due to the adhesive gradient.

E_G = (Adhesive discontinuity) (number of interfaces) (area per interface)

The adhesive discontinuity can be expressed as the adhesive difference between front and rear $(A_F - A_R)$ divided by the length of the slug in cells. The length in cells can in turn be expressed as the total length divided by the length of a cell, while the number of cells can be expressed as the total volume divided by the volume of a cell. This ultimately yields the expression:

$$E_G = (A_F - A_R)\frac{L_C}{L}\frac{V}{V_C}A_C \tag{1}$$

where L and V are the length and volume of the slug while L_c and V_c are the length and volume of a cell and A_c is the area of an anterior-posterior interface. The expression can be simplified by elimination of $\frac{L_c A_c}{V_c}$, which should equal 1 for cylindrical cells, to give

$$E_G = (A_F - A_R)\frac{V}{L} \tag{2}$$

POTENTIAL ENERGY ASSOCIATED WITH THE SLUG SURFACE

The surface energy is given by:

$$E_S = \text{(adhesiveness of lateral cell surfaces) (surface area)}$$
$$E_S = A_L \pi W L \tag{3}$$

where A_L is the adhesiveness of lateral cell surfaces, W is the width (i.e., diameter) of the slug, the slug is assumed to be cylindrical, and the surface area of the ends is neglected. W is related to L through the expression for the volume of the slug:

$$V = \left(\frac{W}{2}\right)^2 \pi L$$

which can be restated:

$$W = \left[\frac{4V}{\pi L}\right]^{1/2}$$

which in turn allows one to restate (3) as

$$E_S = A_L \pi L \left[\frac{4V}{\pi L}\right]^{1/2}$$

$$E_S = 2A_L[\pi V L]^{T} \tag{4}$$

MINIMIZING THE TOTAL POTENTIAL ENERGY OF THE SLUG

Equations (1) and (4) show that when slug cells are rearranged in such a way as to change the length, the gradient energy changes with the inverse of the length while the surface energy changes with the square root of the length. The inverse $(1/L)$ become infinite for $L = 0$ but is finite for larger values, while the square root of L is infinite for infinite L but finite for smaller values. This implies that the sum of the two energies has a minimum at some finite value of L. Conventional calculus can be used to show that the minimum occurs at:

$$L = \left[\frac{A_F - A_R}{A_L}\right]^{2/3}\left[\frac{V}{\pi}\right]^{3/8} \tag{5}$$

$$L/W = \frac{A_F - A_R}{2A_L} \tag{6}$$

Examining these expressions shows that the ratio of length to width in the minimum-energy configuration is independent of the volume of the slug. The model thus accounts for the fact that slug shape is independent of slug volume (Fig. 4).

Generation of an Adhesive Gradient with Fixed Front and Rear Values

NECESSITY OF AN ADHESIVE GRADIENT WITH FIXED ENDS

In the preceding discussion, we assumed that the adhesiveness values at the front and rear of the slug were fixed, independent of slug length. Consider what would happen if the adhesive gradient kept a constant slope, rather than constant extremes, when the length of the slug was varied. The potential energy associated with an adhesive gradient would then be independent of length, and in particular would not decrease with elongation. There would be no force tending to elongate the slug and the minimum energy configuration would be governed only by the necessity of minimizing surface; a spherical aggregate would result.

Even less natural behavior is predicted if we explicitly assume that the adhesive gradient is generated by a substance produced at a point source in the tip and broken down in the rest of the slug. In this kind of system, the slope of the concentration gradient would actually decrease if the cells are rearranged to make the slug shorter and fatter, since the cross-sectional area available for diffusion will increase. One would then also expect a decrease in the slope of the adhesive gradient. In this kind of system, minimization of the adhesive free energy would lead to the opposite of elongation; the system would tend to a "squat" form in which the anterior-posterior axis was shorter than the slug width.

These considerations should make two things clear. One is that "fixed ends" or some reasonable approximation are necessary to produce elongation in the adhesive gradient model of the slime mold slug. The other is that getting these "fixed ends" is tricky; it isn't sufficient to assume a point source of morphogen and diffuse breakdown.

Othmer–Pate Model. A scheme for producing "fixed ends" has been suggested more recently by Othmer and Pate (26); this scheme is very different from the earlier "source-sink model" of Crick (7) in that it would not require any special activities in the cells of the slug rear. A key component of the Othmer–Pate model is the control substance, a substance that is assumed to be produced by the cells throughout the slug but lost or destroyed at one end. In the simplest version of this model, one might assume that the substance diffuses very rapidly, so that its concentration is near-uniform throughout a slug. If one then assumes that the rate of destruction or loss is first-order in concentration of the control substance, the system will tend to a steady state in which the concentration of the substance is proportional to the slug's length. The control substance thus effectively measures the length of the slug.

Othmer and Pate envision that the control substance acts by modulating the diffusion constant of the morphogens which control the organization of the slug. In particular, the authors envision that high levels of the control substance (encountered in long slugs) increase the diffusion constants of the morphogens and thereby make the slug "seem shorter" to these molecules; low levels of the control substance decrease the morphogen diffusion constants and make the slug "seem longer." Othmer and Pate have found that under some simple conditions the diffusion constant can be adjusted in exactly the manner required to cancel out the effect of length changes on the morphogen pattern. Under these circumstances, even a simple gradient-generating mechanism (such as a morphogen produced at a point source and broken down throughout the slug) can produce a morphogen concentration pattern that is totally scale-invariant, and in particular has fixed ends.

It might seem that modulating diffusion constants would have to involve a complex mechanism, but this is not necessarily true. An attractive idea is that the control substance affects the concentration of immobile receptors that bind the morphogen reversibly; the morphogen would then be partitioned between a mobile and a stationary phase, and its movement would be slowed, rather like the movement of a solute can be retarded in column chromotography. In slime mold slugs, one would presumably have to vary the unbound fraction over a 40-fold range to compensate for the roughly sixfold length range over which slug shape has been shown to be constant (Fig. 4). This would call for about a 40-fold change in receptor concentration, which seems achievable.

It is interesting to ask what effects an Othmer–Pate control substance might have if one isolated it and added it back to slime mold slugs. Presumably, the concentration of the substance within the slugs would be increased. Since a high concentration of the substance is ordinarily associated with a long slug, one can infer that the slugs would be "deceived" into believing that they were longer than they actually were, and would thus respond by shortening. In mechanistic terms, the substance would lead to an increase in the diffusion constant of the morphogen whose gradient sets up the adhesive gradient. The difference $(A_f - A_r)$ would then decrease, which would decrease the steady-state length-to-width ratio for the slugs. This could be a relatively easy effect to see.

Influence of the Tip on Slug Shape

It is rather easy to explain the effect of the tip on slug shape in the adhesive gradient model if one assumes that the adhesive gradient is generated by diffusion of a morphogen from a small source in the tip. Imagine one had a mutant in which the production of the morphogen was decreased, although the diffusion of morphogen away from the tip was normal. The morphogen concentration would be shallower and the result would be a shorter and fatter slug.

Imagine now the consequence of interchanging the tips between wild-type and mutant slugs. The length-to-width ratio should be unaffected by the type of cells making up the body of the slug, since in these cells the morphogen only diffuses, and this occurs in identical fashion in mutant and wild-type tissue. The shape will thus be controlled only by the activity of the morphogen source, and in tip-interchange experiments will depend only on the genotype of the tip. The adhesive-gradient model thus appears to explain the results illustrated in Fig. 7.

Sorting and the Prestalk/Prespore Pattern in the Adhesive-Gradient Model

Unlike the differential chemotaxis model, the adhesive-gradient model explains slug shape without in any way relying on the heterogenity of slime mold cells; the model could work without difficulty in an aggregate of identical cells, since the required adhesive differences are directly induced by a morphogen diffusing from the tip. Slime mold cells do appear, of course, to be heterogeneous, and they do appear to sort out from one another in slugs. One can imagine dealing with this fact within the adhesive-gradient model in two fundamentally different ways. One is to assume that a differential chemotactic mechanism operates within the adhesive-gradient slug to create its cell-type and sorting patterns. It is also possible to explain sorting-out in purely adhesive terms by introducing the assumption that the adhesiveness of a given cell depends not only on the morphogen concentration but also on intrinsic properties of the cell.

Hybrid Model with Coexisting Adhesive and Chemotactic Mechanisms

Systems in which both differential chemotactic and adhesive-gradient mechanisms operate have interesting properties if one assumes that the adhesive mechanism is strong enough to dominate in the control of slug length. In such a system, the adhesive mechanism would not prevent chemotactic sorting, but would prevent slugs from attaining the steady state in which each cell lies at an inhibitor level equal to its chemotactic response threshold. It was suggested previously that this could stabilize the prestalk/prespore pattern in short slugs. The range of prestalk inhibition levels in such slugs will be narrower than the distribution of prestalk inhibition thresholds, and most prestalk cells will experience inhibitor levels that will hold them in their differentiation states.

Destabilization of Long Slugs. In large slugs, it seems there would be an opposite effect; the adhesive-gradient mechanism would make them longer than the chemotactic steady-state length, and the range of inhibition gradient values would exceed the width of the distribution of inhibition thresholds. In the rear, in particular, one could have inhibition levels too low to block prestalk differentiation in any cells. The prespore cells in the rear would accordingly convert to prestalk, but would be unable to migrate forward in mass because this would shorten the slug and steepen its adhesive gradient. It is easy to imagine that these cells would tend to form tips in the rear of the slug, which would then separate from the host, carrying with them masses of the "excessive" host tissue. The combination of adhesive and chemotactic mechanisms could thus impose an upper limit on the length of slugs. This is a possible explanation for the apparent upper length limit described by Byrne *et al.* (2), a limit which in fact appears to be enforced by the formation of tips in the rear of long slugs and the subsequent separation of excessive tissue.

Adhesive Sorting Within an Adhesive Gradient

A rather different approach to the problem of obtaining sorting and prestalk/prespore patterning within an adhesive-gradient slug arises from the idea that adhesive differences alone can produce cell sorting. Consider an adhesive gradient containing a single cell whose adhesiveness does not match that of its neighbors; assume that the cell is free to migrate within the tissue. If the cell is more adhesive than neighbors, it will have many unsatisfied adhesive molecules on its surface, some of which can be satisfied if the cell moves to a region of higher adhesiveness. If the cell is less adhesive than its neighbors, the neighbors will have unsatisfied adhesive molecules, and the potential energy of the system will decrease if the cell moves to a region of lower adhesiveness. The cell will thus in either case tend to move until it reaches a region in which its own adhesiveness matches the adhesiveness of the surrounding cells. "Sorting" could thus proceed by an adhesive mechanism; the mechanism will resemble chemotaxis toward the source of the adhesiveness-inducing morphogen, but will not require directed cellular actions.

Correlation of Adhesiveness with Prestalk Inhibition Threshold. This principle could be used in two different ways to explain the formation of the prestalk/prespore pattern. In one way, cells sort directly according to their prestalk inhibition thresholds. Imagine that the adhesiveness of slug cells is the sum of two components, one determined by a morphogen diffusing from the tip, and another proportional to the prestalk inhibition threshold. Consider a random mixture of cells which has just come under the influence of a tip. Morphogen diffusing from the tip will establish an adhesive gradient in the aggregate, but the gradient will initially be "bumpy" on account of the random variation in prestalk inhibition threshold. Individual cells will then tend to move within the gradient to points at which their own adhesiveness matches the average adhesiveness of their neighbors. This movement will smooth the adhesive gradient and generate a gradient in prestalk inhibition threshold.

One could explain the formation of the prestalk/prespore pattern in this sort of model by assuming that an independent prestalk/prespore proportioning mechanism operates to convert those cells with highest inhibition thresholds—already those in the front of the aggregate—to the prestalk state. There would be no prestalk/prespore sorting *per se*, and in this respect the model resembles the differential chemotactic model. The model improves on the chemotactic model, however, in one feature: Since the linear ordering of cells would proceed by an adhesive mechanism independent of the prestalk inhibition gradient, it could be efficient even in short slugs where the inhibition gradient range was far less than the width of the threshold distribution. One thus expects good sorting at the ends of the slug, even when the slugs are short. In long slugs, the model would predict the same sort of instability as the chemotactic/adhesive model discussed previously: the threshold gradient would be shallower than inhibition gradient, with the result that cells in the rear of the slug could become prestalk and form tips.

Increase in Adhesiveness upon Prestalk Differentiation. A second way to explain the prestalk/prespore pattern in adhesive terms is to assume, as before, that adhesiveness is the sum of two components, one controlled by a morphogen diffusing from the tip, but in this case to assume that the second component is conferred on any cell by prestalk differentiation. Prestalk cells would then be more adhesive than others and would tend to sort to the front of slugs. This idea is attractive in part because there is evidence for specific cell-surface glycoproteins in prestalk cells (35). The prestalk zone, furthermore, is often thinner than the prespore zone of a slug; it seems likely that one could explain this by assuming an additional adhesive mechanism.

This model would differ from the previous adhesive-sorting model in that sorting would occur in response to the gradient in prestalk inhibition. The model would be quite similar to the adhesive/chemotactic model, with adhesiveness providing the mechanism for chemotactic movement. As in the adhesive/chemotactic model, cell sorting at the front and rear of short slugs would not be efficient. The model has one aspect, however, which does not have a parallel in the adhesive/chemotactic scheme: the prestalk-prespore boundary represents a real discontinuity. Thus the final adhesive gradient, after sorting, should contain a step at the boundary; the step would be equal in height to the increase in adhesiveness that occurs on prestalk differentiation, and could thus be considerably greater than the discontinuity which normally prevails at anterior/posterior cell interfaces. Since unpaired adhesive molecules are expected at this discontinuity in proportion to its height, the system should have a strong tendency to minimize the area of contact between prestalk and prespore cells. The net effect would be to sharpen the prestalk-prespore boundary; this is an attractive feature of the model, as the prestalk-prespore boundary is indeed rather sharp, often seeming no more than one cell wide.

One might imagine that one could produce a similar boundary sharpening in the differential chemotaxis model by assuming an elevation of chemotactic response threshold in prestalk cells. There is a problem in this idea, however, as it seems that a chemotactic discontinuity would lead to a "wasp waist" (see discussion of the differential chemotactic model) so thin that it would contain no cells. The slug would then split into prestalk zone and prespore zone rather than migrating as a unified whole. Splitting is only predicted at an adhesive discontinuity if there are more unbound molecules in an area of prestalk-prespore boundary than occur in the same area of slug-free surface.

It would of course be possible to get both efficient sorting and boundary sharpening in a three-component adhesive model, in which adhesiveness depended on (i) a morphogen diffusing from the tip, (ii) the prestalk inhibition threshold, and (iii) the prestalk differentiation state.

TRANSPLANTATION PROPERTIES IN ADHESIVE MODELS

In models that assume coexisting adhesive and chemotactic mechanisms, one can explain the phenomena of tip formation, tip activation, and tip inhibition in the same way that one does in chemotactic models without the adhesive mechanism. In adhesive models without a genuine chemotactic component, one can accomplish much the same thing by assuming that prestalk cells are ordinarily the source of the morphogen that induces adhesiveness in slime mold cells. A cluster of prestalk cells should then produce an adhesive gradient in the surrounding tissue; the gradient should produce elongation so that a slug "axis" is formed with the prestalk cells at its tip. The inhibition in this sort of model can be explained, as in the differential chemotactic model, in two different ways; either as prestalk inhibition (which exists in both of the pure-adhesive models discussed here) or as a competitive effect of the adhesive gradient originating in the host tip. In the former case, as in the differential-chemotaxis model, a tip activation gradient is expected to arise from the gradient in prestalk inhibition threshold, and mutants with altered prestalk inhibition threshold are expected to show activation and inhibition changes. In the latter case, parallel with the chemotactic model, the activation gradient suggests that prestalk inhibition threshold is correlated with production of the adhesiveness-inducing morphogen; one can use this correlation to explain the observed reduction in tip activation in mutants with reduced prestalk zones, and one can explain Hs3 as a case in which the corelation mechanism has broken down.

Conclusions

The models discussed in this chapter demonstrate that three rather different processes—the cell type interconversion, chemotaxis, and adhesiveness—could all contribute to morphogenesis and patterning in important ways. It seems rather likely that more than one of these processes is involved in the actual spatial integration of slime mold slugs, and it is certainly conceivable that all three processes play important roles. Other mechanisms, such as the propagating waves that organize slime mold aggregation (15), might also be significant.

There is at least preliminary evidence linking a single molecule—cyclic AMP—to all of the processes discussed in this chapter (and to aggregation waves as well). This suggests that one might construct a grand unified adhesive/chemotactic/proportioning model in which a concentration gradient of cyclic AMP plays a central role in each of the three regulatory subsystems. At this point there appears to be one problem with this idea, which is that the tip inhibition gradient and the hypothesized gradient regulating adhesiveness have different formal properties. Experiments show that the tip inhibition gradient has a fixed slope in slugs of different length; the adhesive gradient is required to have fixed extremes. It seems at least possible that one could get around this problem by supposing that the morphogen gradient which regulates adhesiveness in fact has fixed slope, and the Othmer–Pate control substance, which we previously supposed to modulate the morphogen diffusion constant, thereby maintaining the extremes constant, instead modulates the relationship between the morphogen gradient and the adhesiveness of slug cells. This possibility has not, to my knowledge, been investigated mathematically, and is purely speculative at this point.

One might of course maintain that theoretical efforts in this direction would be misdirected; there seem to be at least four morphogenetically active substances in slime molds, and there is therefore no real need to construct a slime model based on a single morphogen. The major theoretical challenge may in fact be to build models in which all of the substances play meaningful and realistic roles.

Another theoretical problem at this stage is to explain the properties of Hs3. As of yet we have no good explanation why three of its characteristics—prestalk sorting, decreased length-to-width ratio, and increased inhibition gradient slope—should coexist in a single mutant. One might imagine that the strain contains multiple mutations; another strain, however, with at least the first two characteristics has been isolated recently in my laboratory.

In addition to these theoretical problems, there are several clear opportunities for new experimental work relevant to patterning in slime molds. In particular the shape problem is almost completely open. Are there substances that affect the length-to-width ratio of slugs? Could one isolate mutants with altered length-to-width ratios, and classify them into various physiological types? How do the various slime mold morphogens, singly and in combinations, affect cellular adhesiveness? Significant insights might emerge from any of these lines of work.

The cellular slime mold system is unique among the experimental systems studied today in that formal pattern formation models suggest immediate, feasible experiments on the biochemical level. There is, therefore, an excellent prospect for gaining a molecular understanding of spatial integration in slime molds in the next few years.

General References

Gross, J.D., Town, C.D., Brookman, J.J., Jermyn, K.A., Peacy, M.J., Kay, R.R.: Cell patterning in *Dictyostelium*. *Phil Trans. Roy. Soc. Lond.* B295: 497–508 (1981).

MacWilliams, H. Transplantation experiments and pattern mutants in cellular slime mold slugs. In: *Developmental order: its origin and regulation*. 40th Symposium of the Society for Developmental Biology, Subtelny, S., Green, P. (eds.). New York: Alan R. Liss (1982).

References

1. Beug, H., Katz, F.E., Gerisch, G.: Dynamics of antigenic membrane sites relating to cell aggregation in *Dictyostelium discoideum*. *J. Cell Biol.* 56: 647–658 (1973).
2. Byrne, G.W., Trujillo, J., Cox, E. Pattern formation and tip inhibition in the cellular slime mold *Polysphondylium pallidum*. *Differentiation, in press* (1983).

3. Bonner, J.T.: A descriptive study of the development of the slime mold *Dictyostelium discoideum*. *Amer. J. Botany* **31**: 175–182 (1944).

4. Bonner, J.T.: The demonstration of acrasin in the later states of the development of the slime mold *Dictyostelium discoideum*. *J. Exp. Zool.* **110**: 259–271 (1949).

5. Bonner, J.T., Chiquoine, A.D., Kolderie, R.O. A histochemical study of differentiation in the cellular slime molds. *J. Exp. Zool.* **130**: 133–158 (1955).

6. Borth, W., Ratner, D.: Different synthetic profiles and developmental fates of prespore versus prestalk proteins in *Dictyostelium discoideum*. *Differentiation, in press* (1983).

7. Crick, F.: Diffusion in embryogenesis. *Nature* (London) **225**: 420–422 (1971).

8. Durston, A.: Tip formation is regulated by an inhibitory gradient in the *Dictyostelium discoideum* slug. *Nature (London)* **263**: 126–129 (1976).

9. Durston, A., Vork, F.: A cinematographical study of the development of vitally stained *Dictyostelium discoideum*. *J. Cell Sci.* **36**: 261–279 (1979).

10. Farnsworth, P.: Experimentally induced abberations in the pattern of differentiation in the cellular slime mould *Dictyostelium discoideum*. *J. Embryol. Exp. Morph.* **31**: 435–451 (1974).

11. Fisher, P.R., Smith, E., Williams, K.L.: An extracellular chemical signal controlling phototactic behavior by *D. discoideum* slugs. *Cell* **23**: 799–807 (1981).

12. Forman, D., Garrod, D.: Pattern formation in *Dictyostelium discoideum*. I. Development of prespore cells and its relationship to the pattern of the fruiting body. *J. Embryol. Exp. Morph.* **40**: 215–228 (1977).

13. Garrod, D., Ashworth, J.: Effect of growth conditions on the development of the cellular slime mold, *Dictyostelium discoideum*. *J. Embryol. Exp. Morph.* **28.**, 463–479 (1972).

14. Gregg, J., Karp, G.: Patterns of cell differentiation revealed by L – [³H] fucose incorporation in *Dictyostelium*. *Exp. Cell Res.* **112**: 31–46 (1978).

15. Gross, J.D., Town, C.D., Brookman, J.J., Jermyn, K.A., Peacy, M.J., Kay, R.R.: Cell patterning in *Dictyostelium*. *Phil. Trans. Roy. Soc. Lond. B* **295**: 497–508 (1981).

16. Inouye, K., Takeuchi, I.: Motive force of the migrating pseudoplasmodium of the cellular slime mould *Dictyostelium discoideum*. *J. Cell Sci.* **41**: 53–64 (1980).

17. Leach, C.K., Ashworth, J.M., Garrod, D.R.: Cell sorting out during the differentiation of mixtures of metabolically distinct populations of *Dictyostelium discoideum*. *J. Embryol. Exp. Morph.* **29**: 647–661 (1973).

18. Loomis, W.F. (ed.): *The development of Dictyostelium discoideum*. Academic Press (1982).

19. MacWilliams, H.: Transplantation experiments and pattern mutants in cellular slime mold slugs. In: *Developmental order: its origin and regulation*. 40th Symposium of the Society for Developmental Biology, Subtelny, S., Green, P. (eds.). New York: Alan R. Liss, Inc. (1982).

20. MacWilliams, H. Bonner, J.T.: The prestalk-prespore pattern in cellular slime molds. *Differentiation* **14**: 1–22 (1979).

21. Maeda, Y.: Role of cyclic AMP in the polarized movement of the migrating pseudoplasmodium of *Dictyostelium discoideum*. *Develop., Growth and Differ.* **19**: 201–205 (1977).

22. Mittenthal, J., Mazo, R.M.: A model for shape generation by strain and cell-cell adhesion in the epithelium of an arthropod leg segment. *J. Theoret. Biol.* **100**: *in press* (1983).

23. Morrissey, J., Farnsworth, P., Loomis, W.: Pattern formation in *Dictyostelium discoideum*: An analysis of mutants altered in cell proportioning. *Developmental Biology* **83**: 1–8 (1981).

24. Newell, P.: Genetics of the cellular slime molds. *Ann Rev. Genet.* **12**: 69–93 (1978).

25. Newell, P., Ross, F.: Genetic analysis of the slug stage of *Dictyostelium discoideum*. *J. Gen. Microbiol.* **128**: 1639–1652 (1982).

26. Othmer, H., Pate, E.: Scale-invariance in reaction-diffusion models of spatial pattern formation. *Proc. Natl. Acad. Sci.* (USA) **77**: 4180–4184 (1980).

27. Raper, K.: Pseudoplasmodium formation and organization in *Dictyostelium discoideum*. *J. Elisha Mitchell Sci. Soc.* **56**: 241–282 (1940).

28. Ratner, D., Borth, W.: Comparison of differentiating *Dictyostelium discoideum* cell types separated by an improved method of density gradient centrifugation. *Exp. Cell Res.* **142**: *in press* (1982).

29. Rubin, J., Robertson, A.: The tip of the *Dictyostelium discoideum pseudoplasmodium* as an organizer. *J. Embryol. Exp. Morphol.* **33**: 227–241 (1975).

30. Sampson, J.: Cell patterning in migratory slugs of *Dictyostelium discoideum*. *J. Embryol. Exp. Morph.* **36**: 663–668 (1976).

31. Sternfeld, J., David, C.: Cell sorting during pattern formation in *Dictyostelium discoideum*. *Differentiation* **20**: 10–21 (1981).

32. Sussman, M., Schindler, J.: A possible mechanism of morphogenetic regulation in *Dictyostelium discoideum*. *Differentiation* **10**: 1–5 (1978).

33. Trinkhaus, J.P.: "Mechanisms of Morphogenetic Movements". In: *Organogenesis*, DeHaan, R.L., Ursprung, H. (eds.). Holt, Rhinehart and Winston (1965).

34. Tsang, A., Bradbury, M.: Separation and properties of prestalk and prespore cells of *Dictyostelium discoideum*. *Exp. Cell. Res.* **132**: 433–441 (1981).

35. West, C.M., McMahon, D.: The involvement of a class of cell-surface glycoconjugates in pseudoplasmodial morphogenesis in *Dictyostelium discadeum*. *Differentiation* **20**: 61–64 (1981).

36. Wolpert, L.: Positional information and pattern formation. *Current Topics in Developmental Biology* **6**: 183–224 (1971).

37. Yabuno, K.: Changes in cellular adhesiveness during the development of the slime mold *Dictyostelium discoideum*. *Develop., Growth and Differ.* **13**: 181–190 (1971).

Questions for Discussion with the Editors

1. *The models you have proposed are—in a formal sense—intellectually very satisfying. But what do you say to the experimentalist who asks, (a) what are the prospects for experimentally verifying the existency of the "response thresholds" which are important to your chemotaxis models, and (b) how might the small differences in cell adhesive properties required of your models be measured?*

(a) The best approach to the question of response thresholds would probably be straightforward cell physiology. Assuming that the chemotactic agent is cyclic AMP, one could measure the chemotactic velocity of slug cells in artificial gradients of varying slope and level. The chemotactic model would predict a velocity decrease at high cyclic AMP concentrations, but not necessarily negative chemotaxis; a cell that migrated forward, but more slowly than the slug was migrating, would, in the coordinate system of the slug, be negatively chemotactic. Once the apparatus was set up and working, one would also want to look for differences in chemotactic response between cells from the front and rear of slugs and between cells from strains which characteristically sort out from one another.

(b) Cell adhesiveness can be measured sensitively in "aggregometers" in which a cell suspension is exposed to controlled agitation; adhesiveness is measured as the amount of shear necessary to hold the cell suspension at a given degree of aggregation (corresponding to a given optical density). Anthony Durston (Hubrecht Lab, Utrecht, Holland) has used this kind of system to show that cyclic AMP modulates the adhesiveness of slug cells.

2. *Can you suggest any pattern formation phenomena in higher eucaryotic organisms which resemble those in slime mold morphogenesis for which you have proposed models?*

This depends on what you mean by "pattern formation." Up to now, pattern formation workers have devoted most of their attention to developmental events in which a sheet or block of cells is divided up into regions with well-defined boundaries (in which cells then differentiate in particular ways). The differential chemotactic and adhesive-gradient models are intended for rather different situations, namely developmental events in which (1) morphogenesis involves cell rearrangement, rather than position-dependent changes of cell type or shape, and (2) the behavior of the individual cells may be chaotic, order being evident only in the behavior of the system as a whole. The motion of cells through the primitive streak is one example of this *kind* of process, and might well be approachable with the kinds of ideas which are used in the differential-chemotactic and adhesive-gradient models.

CHAPTER 7

Pattern Formation in Ciliated Protozoa

Joseph Frankel

As ORGANISMS EVOLVED to become large and complex, they generally also became multicellular. However, a high degree of pattern complexity has also been achieved by members of diverse groups of unicellular eukaryotic organisms. Outstanding examples include the well-known giant green alga *Acetabularia*, the desmid *Micrasterias* with its multiple cytoplasmic axes, the remarkably complex flagellates that inhabit the hindgut of termites, and the ciliates. The ciliates are probably pre-eminent among these in contributions to the understanding of intracellular patterning. In part, this is a consequence of experimental advantages related to ease of cultivation and of microsurgical analysis and genetic manipulation. More important, in my opinion, are the advantages derived from the intrinsic features of patterning in the organisms: first, that there exists a complex array of structures with definite locations; second, that most of these structures are located at or near the cell surface, allowing analysis of patterns in two rather than three dimensions; third, that cells grow primarily along one of these two dimensions and become tandemly subdivided, making possible study of repeated expressions of patterning within what is effectively a single clonal axis; and fourth, that during the sexual process (conjugation), gamete nuclei are exchanged between cells that retain their structural individuality, so that one can distinguish the separate contribution of nuclear genes and of pre-existing patterns to newly arising patterns.

This review will not cover all aspects of patterning in ciliates. It will instead confine itself to a consideration of questions concerning the *location* at which structures develop, and the *geometry* of these structures. Within this deliberately restricted scope, the presentation of experimental results will be organized around four major ideas. The first is the concept of *hierarchical organization*, the old idea that different processes might be at work at different levels of the biological hierarchy and that one therefore cannot simply extrapolate mechanisms from a smaller to a larger biological realm. The second idea is that of a *direct continuity* of ciliate patterns along the longitudinal axis of clonal growth, which provides a means for inheritance of cellular differences at a supramolecular level. The third idea, derived from classical experimental embryology, is that of a *field organization* of systems controlling the location and organization of cell surface structures at the level of large-scale patterning. The final idea is that independent and labile *directional axes* underlie this field organization. A consideration of how one of these axes might reverse itself will lead to a tentative application to ciliates of one model that was first proposed to explain axial reversals in multicellular organisms. This experimentally and conceptually oriented presentation will follow a brief exposition of the basic structural layout of the organisms on which relevant experiments have been performed.

The Organisms

Ciliate Organization

Although very diverse in details of form and pattern, all ciliates are built on the same basic organizational plan. Some rudiments of this general plan will be presented first, followed by a brief characterization of the three experimental organisms that we will be "using" in most of the remainder of this essay. Exposition of structural detail will be confined to that necessary for understanding the experiments, and terminology will be simplified wherever possible, meaning that the terms used here will not always be the same as those encountered in the ciliatological literature.

Ciliates are characterized by nuclear dimorphism, with a diploid, germinal *micronucleus* and a compound, somatic *macronucleus*. Their locations differ among ciliate species and sometimes even within a species at different stages in the cell cycle. In some ciliates, nuclei are located directly underneath the cell surface (as depicted in Fig. 1A) and, as we shall see, are positioned in specific relations to the overlying cell-surface pattern.

Another characteristic ciliate organelle is the *contractile vacuole*, involved in osmoregulation and ion balance. It is generally located directly underneath the cell surface, and opens out on the surface at a *contractile vacuole pore* (Fig. 1A). The location of the contractile vacuole and its pore are also controlled in large part by the cell-surface pattern.

The cell is surrounded by two unit-membrane systems, under which is a fibrogranular layer [for more information about these, see (2)]. Ciliary basal bodies are embedded within the fibrogranular layer. These basal bodies are the central elements of *ciliary units*, which are the major expressions of cell-surface organization. The geometry of a simple ciliary unit is illustrated in Fig. 1B [for details and variations, see (29)]. A basal body is at its center. A cilium frequently protrudes outward from the basal body. Two sets of microtubule bands are organized in specific spatial relations relative to the basal body; in addition, a striated rootlet, when present, extends to the anterior-right[1] of the basal body. The crucial point of this

---

[1]Throughout this review, *right* and *left* refer to the right and left of an observer who stands inside the cell so that his antero-posterior axis coincides with that of the cell, and who keeps turning around his own longitudinal axis to face the surface of the cell. All figures in this chapter, other than polar views and cross sections, are drawn as seen from the outside of the cell, with the anterior end uppermost. Hence, when inspecting these figures, your left is the cell's right and *vice versa*.

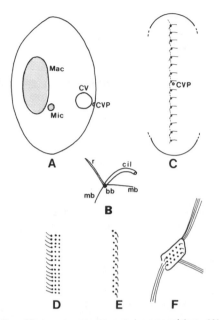

FIGURE 1. Characteristic ciliate structures and ensembles. (A) Internal structures: the macronucleus (Mac), micronucleus (Mic), contractile vacuole (CV) and contractile vacuole pore (CVP). (B) A ciliary unit as observed in the light microscope. The central structure is the basal body (bb) from which the cilium (cil) projects. Common accessory structures include left and right microtubule bands (mb), and a striated ciliary rootlet (r), which when presented is normally oriented anteriorly and to the right. (C) Organization of a ciliary row, in which ciliary units are lined up end to end. A contractile vacuole pore (CVP) is shown adjacent to one of the ciliary units. (D) A membranelle. (E) A paroral membrane. (F) A cirrus.

description is that the ciliary unit is inherently asymmetric. One could imagine a mirror-image "enantiomorph" to the unit of Fig. 1B, in which (for example) the striated rootlet is located to the anterior-left of the basal body rather than the anterior-right. Such mirror-image ciliary units have never been observed in any ciliate.

In ciliates, ciliary units rarely occur singly. The typical assemblage of ciliary units is the *ciliary row*, indicated diagrammatically in Fig. 1C. All of the ciliary units in the ciliary row are arranged in a single linear series, and all are oriented in the same manner. Characteristically, a ciliary row extends longitudinally over the body surface of a ciliate, from the anterior to the posterior end of the cell, as shown for a single row in Fig. 1C. Most ciliate species have a characteristic stable number, or range of numbers, of similarly oriented ciliary rows.

In addition to these "simple" ciliary rows, many ciliate species, including all of those that we shall consider, also have compound rows that are grouped together in various ways, three of which we shall describe. Two kinds of such assemblages make up the ciliature of the oral apparatus. These are the *membranelle* and the *paroral membrane*. Membranelles (Fig. 1D) consist of several (frequently three) rows of closely spaced ciliary units arranged in either hexagonal or square packing. All of the units are ciliated, but only some have accessory microtubule bands. The paroral membrane (Fig. 1E) is made up of two rows of ciliary units in a staggered array, with a characteristic disposition of ciliature (in general, only the right-most row is ciliated) and of accessory microtubule structures. Ciliate oral apparatuses frequently include three or more parallel membranelles on their left, and one or two paroral membranes on their right.

One important group of ciliates, the hypotrichs, also possesses a third type of compound ciliary structure, the *cirrus* (Fig. 1F). Each cirrus consists of several parallel rows of ciliated basal bodies packed hexagonally. Only certain rows bear accessory microtubule bands, some of which contribute to long and prominent fibers that extend from the cirrus in definite directions (Fig. 1F). Cirri are either located singly at definite positions on the ventral surface of hypotrichs, or are organized into longitudinal *cirral rows*, which are compounded versions of ciliary rows (see Fig. 2C).

Finally, certain cell-surface structures are not parts of ciliary units but are positioned in definite orientations with respect to these. Such structures include the previously mentioned contractile-vacuole pores, which are situated close to certain basal bodies of particular ciliary rows (Fig. 1C).

The detailed organization of these structures can be described only with the aid of the transmission electron microscope. However, the *locations* of these structures, including even microtubule bands associated with single basal bodies, can be discerned by light microscopy of cells stained with protein-silver (protargol). Use of this stain allows precise assessment of the consequences of genetical and microsurgical intervention. Many of the demonstrations that will subsequently be outlined depend upon this combination of experimentation and cytological analysis.

Three Experimental Organisms

There are about ten ciliate types (species or groups of closely related species) in which pattern formation has been extensively analyzed. We cannot consider all of these, yet no one suffices for exposition of all of the basic principles. So we will compromise with three, *Tetrahymena*, *Stentor*, and hypotrichs of the *Oxytricha-Stylonychia* group. Of these, *Tetrahymena* is the smallest and simplest ciliate, most amenable to genetic approaches and least to conventional microsurgery. Its cell surface organization is sketched in Fig. 2A-1. There are normally 18 to 21 longitudinal ciliary rows and an oral apparatus (OA) made up of one paroral membrane (known to tetrahymenologists as the undulating membrane) and three membranelles near the anterior end. Contractile vacuole pores (CVPs) are situated near the posterior end of two ciliary rows located about ¼ of the cell circumference to the right of the OA; one of these is shown in Fig. 2A. The first sign of impending cell division that is visible on the cell surface is the formation of an *oral primordium*, initially as a field of basal bodies proliferated to the left of the mid-region of the right-postoral ciliary row (i.e., the right-most of the two ciliary rows situated posterior to the old oral apparatus) (Fig. 2A-2). Membranelles and the paroral membrane later differentiate within this field. A fission zone then appears as an equatorial discontinuity in the ciliary rows (Fig. 2A-3). At about the same time, the CVP(s) for the anterior daughter cell are formed adjacent to the appropriate ciliary row(s) just anterior to the fission zone. The cell then cleaves along the fission zone, while the final modeling of the new oral apparatus (and remodeling of the old one) takes place.

Stentor (Fig. 2B) is enormously larger than *Tetrahymena* (about a millimeter in length, compared to 50 μm for *Tetrahymena*), and is the classic object for ciliate microsurgery. It has many ciliary rows, and numerous units within each row (not shown individually in Fig. 2B). The rows are also not evenly spaced around the cell circumference; rather, there is a gradient in spacing of ciliary rows, proceeding from the most narrowly spaced rows clockwise around the cell to more and more widely spaced rows. This arrangement necessarily implies a seam at which the most narrowly spaced rows abut on the most widely spaced. This region of abutment, generally obliquely oriented, is known as the *contrast zone* (Fig. 2B-1). The feeding system of the cell consists primarily of a long band of membranelles wrapped around an apical disc, and terminating in an *oral spiral* which is anterior to the contrast zone. The nodulated macronucleus is located directly underneath relatively closely spaced ciliary rows to the right of the contrast zone, while the contractile vacuole and its pore are situated underneath the widely spaced rows on the other side of the contrast zone (Fig. 2B-1). An oral primordium is initially formed by proliferation of basal bodies adjacent to several of the narrowly spaced ciliary rows in the contrast zone; the field then coalesces within a region of breaks in those rows (Fig. 2B-2). Membranelles then differentiate, and an oral spiral forms (Fig. 2B-3). A fission zone appears, and cytokinesis then proceeds with accompanying complicated morphogenetic maneuvers to generate the new posterior tip and anterior apical disc (Fig. 2B-4) of the respective daughter cells. The macronucleus goes through maneuvers of its own, first condensing, then elongating, and finally renodulating after being pinched in two by the constricting cell (not shown).

Hypotrichs of the *Oxytricha/Stylonychia* group are intermediate in size between *Tetrahymena* and *Stentor* and manifest a patterned complexity that exceeds that of either of these. The cells are flattened, with highly differentiated dorsal and ventral surfaces. The dorsal surface (not shown) possesses six longitudinal rows of short cilia. The ventral surface (Fig. 2C-1), on the other hand, has no rows made up of single cilia; the locomotory structures instead are cirri (cf. Fig. 1F) which are deployed in two longitudinal rows at the respective margins of the cell (the *marginal cirri*) plus three groups of somewhat larger cirri on the ventral surface in between (the frontal, ventral, and transverse cirri, collectively the *FVT* cirri). The oral apparatus includes a band of about 50 parallel membranelles plus two paroral membranes. During the development of the ventral surface, an oral primordium forms initially as a small group of basal bodies anterior to a particular transverse cirrus (Fig. 2C-1), which enlarges by addition of more basal bodies as it migrates anteriorly (Fig. 2C-2), and then differentiates membranelles and paroral membranes (Fig. 2C-3). The cirri are all replaced, in a truly remarkable manner. First two sets of frontal-ventral-transverse (FVT) cirral primordia appear as parallel arrays of short longitudinal ciliary rows (Fig. 2C-3), many of which develop by an unusual process that involves a dedifferentiation of certain frontal and ventral cirri followed by a repositioning of their constituent basal bodies. Two sets of new marginal-row primordia develop similarly, each from two or three adjacent marginal cirri (Fig. 2C-3). These new rows then undergo further ordered basal-body prolif-

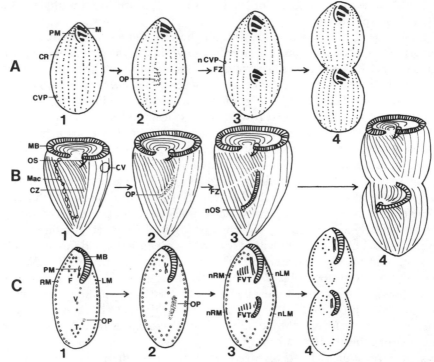

FIGURE 2. Cell-surface organization of three ciliates extensively employed in analysis of pattern formation. Each of these is illustrated in a sequence of four diagrams, beginning with the non-developing cell (1), proceeding through early (2) and late (3) stages of cell-surface development, and ending with cytokinesis (4). (A) *Tetrahymena*. Frame 1: Eight ventral ciliary rows (CR) are shown, with each dot representing a ciliary unit. The oral apparatus, near the anterior end, includes a paroral membrane (PM) and three membranelles (M). A contractile vacuole pore (CVP) is located near the posterior end (adjacent to a ciliary row that is invisible around the margin of the cell). An early oral primordium (OP) is shown in frame 2, which in frame 3 has completed development of membranelles and a paroral membrane. A fission zone (FZ) and new CVP (nCVP) appear at this time. Furrowing later takes place along the fission zone (frame 4). (B) *Stentor*. Frame 1: A large anterior membranelle band (MB) includes many membranelles and terminates in an oral spiral (OS). The ciliary rows are here depicted as continuous lines, with a striking discontinuity at the contrast zone (CZ). The macronucleus (Mac) is made up of a chain of nodes situated directly underneath the cell surface in the region of fairly closely spaced ciliary rows to the right of the contrast-zone, while the contractile vacuole (CV) is located in the region of widely spaced rows to the left of the contrast-zone. Frame 2 shows the oral primordium (OP) appearing within the contrast-zone, frame 3 the development of membranelles, a new oral spiral (nOS), and formation of the fission zone (FZ). The characteristic shape of the cell poles is molded during cytokinesis (frame 4). (C) The ventral surface of a hypotrich (*Oxytricha*). Frame 1: The oral apparatus is made up of a membranelle band (MB) and two paroral membranes (PM). Cirri, shown as open circles, include right marginal (RM) and left marginal (LM) rows, as well as frontal (F), ventral (V), and transverse (T) groups. In this case, the cell in frame 1 is in the earliest stage of cell-surface development, with a small oral primordium (OP) adjacent to the left-most transverse cirrus. Frame 2 shows the enlarged OP that has moved anteriorly; frame 3 illustrates the differentiation of mem-

branelles in the OP and the formation of the ciliary rows that constitute the frontal-ven-tral-transverse (FVT) and new right (nRM) and left (nLM) marginal cirral primordia. The cirri that later differentiate from these cirral primordia are shown as filled circles in frame 4; the old cirri are all ultimately resorbed (some that are still present during cytokinesis are not shown here).

eration and finally break up to form complete sets of new cirri for the two daughter cells (Fig. 2C-4). All of the old cirri remaining from the preceding cell generation (except for those that were used up in forming the primordial ciliary rows) are then resorbed.

Despite the diversity of detail in the organization and development of these three different ciliates, some important similarities foreshadow crucial results of experimental analysis. One is the existence of an organizational continuity along the longitudinal cell axis. Even in hypotrichs, where cirri are totally replaced as in-dividuals, cirral rows retain continuity as ensembles because the rudiments of the new rows develop within the old ones. A complementary generalization is that of tandem subdivision. In each ciliate, two new cell surface organizations are formed within what was originally one, with the two in tandem alignment along the an-tero-posterior axis. This segmental reorganization is essentially complete *before* cy-tokinesis begins. Hence, while the positional values of cellular longitudes might re-main the same, the values of latitudes *must* change dramatically: the "waist" of a cell in generation n must transform itself in a juxtaposed "head" and "tail" as it de-velops toward generation $n + 1$.

Hierarchical Organization

Two Systems of Polarity in Tetrahymena

The concept of an intracellular spatial hierachy can be explained through two ex-amples. The first concerns two distinct types of polarity that are both expressed with ciliary rows. This story begins historically with *Paramecium*, in which Beis-son and Sonneborn (4) were able to induce cells to *rotate* a few of their ciliary rows 180°, following which the paramecia actively maintained and propagated these rows in their inverted state. This demonstration was repeated in *Tetrahymena* over a decade later (40). Normally oriented and inverted ciliary rows of *Tetrahymena* are shown diagrammatically in Fig. 3A and 3B. In normally oriented ciliary rows, new basal bodies are first formed *anterior* to pre-existing basal bodies, then acquire the associated fibrillar structures in geometrical conformity with the analogous pre-existing structures of neighboring basal bodies, and finally become ciliated (1). In inverted ciliary rows (Fig. 3B), all of these features have been rotated 180°. The geometry of appearance of new basal bodies and associated fibrillar structures re-mains completely normal in relation to the internal architecture of the ciliary row, but when assessed with respect to the remainder of the cell new basal bodies appear immediately posterior rather than immediately anterior to old ones (40).

Now we will focus our attention on the appearance of new cilia. Nearly half of

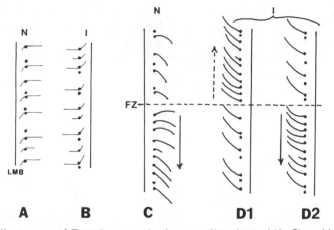

FIGURE 3. Ciliary rows of *Tetrahymena*, both normally-oriented (A, C) and inverted (B, D), in nondividing (A, B) and dividing (C, D) cells. A and B: Completed ciliary units are depicted as in Fig. 1, except that the striated rootlet is not shown because it is unstained in the preparations from which these diagrams are derived, and cilia are omitted for the sake of clarity. A continuous longitudinal microtubule band (LMB) is associated with each ciliary row of *Tetrahymena*. (A) A normally oriented ciliary row (N). New basal bodies are situated anterior to old ones. (B) Inverted (180°-rotated) ciliary row (I). Note that new basal bodies are now located posterior to mature old ones. C and D: Here cilia are depicted as curved lines emerging from the basal bodies, and all other structures except for the longitudinal microtubule bands are omitted. (C) A normally oriented ciliary row. The ciliation wave (arrow) proceeds posteriorly from the fission zone (FZ). (D) Inverted ciliary rows, showing two hypotheses concerning control of the direction of the ciliation wave. D1: control intrinsic to the ciliary row. D2: control extrinsic.

the basal bodies of ciliary rows of *Tetrahymena* are unciliated (34). This implies a long lag between the formation and ciliation of basal bodies. But ciliation is not uniform in space: anterior basal bodies are almost all ciliated, while those in the middle and posterior regions of the cell are often unciliated (27, 34). This unequal ciliation is generated just prior to and during cell division by what appears to be a wave of ciliary outgrowth that moves posteriorly from the fission zone, resulting in ciliation of all microscopically resolvable basal bodies in its path (Fig. 3C) (12). This apparent wave is polarized, since it propagates in a posterior direction from the fission zone, thus ciliating the basal bodies of the anterior region of the prospective posterior daughter cell; the region anterior to the fission zone, which will become the posterior region of the future anterior daughter cell, is not affected by the ciliation wave and includes many nonciliated basal bodies interspersed among ciliated ones.

What controls the polarity of the ciliation wave? Is it the intrinsic polarity of the ciliary row, or is it some other polarity superimposed upon the ciliary row? A decisive answer can be obtained simply by studying the ciliation wave in cells bearing some inverted ciliary rows. If the polarity of the ciliation wave is an expression of the internal polarity of the ciliary row, then when the row is inverted the ciliation wave should be inverted also; one would expect it to propagate anteriorly from the fission zone within inverted rows at the same time that it propagates posteriorly in normally oriented rows (Fig. 3D-1). But this is not what happens; instead, the ciliation wave is propagated posteriorly from the fission zone in inverted ciliary rows exactly as it is in the neighboring normally oriented rows (Fig. 3D-2). Hence there exist two superimposed systems of polarity: a local one internal to the ciliary rows governing (for example) the direction of nucleation of new basal bodies, and a second large-scale cellular polarity that is independent of the orientation of ciliary rows but nonetheless is able to impose its influence by controlling the time and place of formation of new ciliary structures.

Independence of Patterning and Assembly in Hypotrichs

In the previous example, we saw how a large-scale polarity can remain normal while a locally controlled polarity is inverted. Now we will consider an example that is roughly the converse: a locally controlled asymmetry that remains normal in the face of the reversal of large-scale asymmetry. Appreciation of this story requires some background on the nature of ciliate doublets. It is relatively easy to generate parabiotic ciliate doublets, which are essentially two complete cells fused back-to-back with their antero-posterior axes aligned. The two ventral surfaces of such parabiotic doublets are of identical polarity and asymmetry. These are the common ciliate doublets. We will here consider another, more uncommon type. Two decades ago some Chinese investigators reported a remarkable type of micro-surgically generated hypotrich doublet, in which the two ventral surfaces are located on the same plane rather than in the usually back-to-back configuration (56). The two oral systems of these doublets were found to be arranged in mirror-image configurations (54), with a normal and functional oral apparatus on the left side and a nonfunctional oral apparatus of reversed asymmetry on the right side (Fig. 4A); the arrangement of cirri is more irregular but also shows signs of mirror-image patterning. Such hypotrich doublets have been generated more recently in two other laboratories as well (21, 24), and their characteristics are always the same.

The mirror-symmetrical arrangement of ciliature of the flat doublets is a consequence of a similar mirror symmetry in the sites of appearance and patterns of development of ciliary primordia (54). One example is the site of initial appearance of the oral primordium. In the left component, the rudiment of the oral primordium appears just anterior to the cirrus that is at the left-anterior apex of the "V" pattern formed by the set of transverse cirri (Fig. 4A-1OP); this location corresponds to that observed in normal singlet cells (Fig. 2C-1). In the right component, the oral primordium appears immediately anterior to the cirrus at the right-anterior apex of the V-formation (Fig. 4A-rOP) (Pang, *personal communication*; Grimes, *personal communication*). The two locations are precise mirror-images.

So far, we have demonstrated mirror asymmetry for the locations of mature ciliary structures and of ciliary primordia. What about the internal organization of these ciliary structures? We would expect it to be normal on the left side, and it is; the normal organizations of membranelles and cirri are shown in Fig. 4B. Now what about the right side? If the ciliary organelles on the right side were in all respects the mirror images of those on the left, one would expect the membranelles and cirri to be organized as depicted in Fig. 4C-1. But that is not what was found.

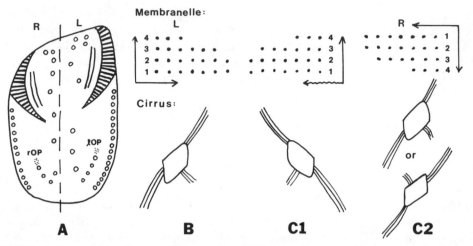

FIGURE 4. Schematic diagrams of aspects of the mirror-image pattern reversal of hypotrich ciliates (*Stylonychia* and *Pleurotricha*). (A) Organization of the ventral surface of "flat doublets". L indicates the normally organized left half, R the reversed right half, and the vertical dashed line indicates the plane of mirror-symmetry of arrangement of cell-surface elements. Both the diagrammatic conventions and the stage of development are the same as in Fig. 2C-1. The oral primordium of the left component (lOP) is located as in a normal cell, while that of the right component (rOP) is generated at the mirror-image location with respect to the cirral pattern (see text). (B) Organization of a membranelle (above) and cirrus (below) of the left component. The dots here represent basal bodies (the left end of the membranelle is not shown). The numbers and arrows shown adjacent to the membranelle designate arbitrary conventions for assessing the internal organization of membranelles. (C) Possible organizations of structures of the right component: (1) Hypothetical mirror-image internal organization of membranelles and cirrus. The wavy line shows the reversed axis of the hypothetical membranelle. (2) The orientations actually observed: inverted for the membranelle, normal or occasionally inverted for cirri. Frame A is drawn from Fig. 1b of (21) and Fig. 1 of (54), somewhat idealized and with information added concerning the site of origin of the oral primordium (Grimes, *personal communication*; Pang, *personal communication*); frames B and C are revised from Fig. 2 of (21), with addition of an inverted cirrus in frame C2 (Grimes, *personal communication*).

Instead, the cirri of the right component are individually normal (except that cirri of marginal rows may be inverted), while the membranelles are inverted, i.e., they are normal membranelles rotated 180° (21, 24).

We must pause to appreciate the meaning of this demonstration. At the level of large-scale organization, the right half of the doublet is a true mirror image of the left half. There is no way in which one can cut out an image of the right-half pattern, either of mature cells or of any developmental stage, and superimpose it on the left-half pattern while keeping it in the same plane. But when one ceases to look at the cell's pattern as a whole but studies the individual components of that pattern, the structures expressed in the right half are always superimposable on those expressed in the left half, although sometimes one must rotate them within the plane of the cell surface to achieve that superimposition. This implies that the local systems that link basal bodies to one another cannot respond to a large-scale enantiomorphic pattern by creating a local enantiomorph, presumably because the ciliary units themselves are of an unchangeable internal asymmetry and polarity. Yet, and this is the truly important point: the very fact that a large-scale pattern reversal can be achieved despite the absence of a local pattern reversal indicates that the large-scale pattern is indeed distinct from the local one (17).

The Developmental Role of Ciliary Units

The above experimental demonstrations indicate that there are at least two systems of polarity and asymmetry, and that one is not a simple spatial extension of the other. The geometry of the local patterning system appears to be controlled by ciliary units, whereas the large-scale system is superimposed on the ciliary units. There are two ways in which this distinction has been further borne out. One involves analysis of different ways in which the two systems influence the positioning of a structure that is not itself a part of the ciliary row. The clearest case is that of the contractile vacuole pore (CVP) of *Tetrahymena*. This structure normally develops very close to a ciliary unit, specifically to its posterior-left (38). But when a ciliary row is rotated 180°, the CVP develops at exactly the same position with respect to the intrinsic geometry of the ciliary unit, thus now to its anterior-right with respect to the cell's coordinates. This "fine-positioning" is a strictly autonomous property of a ciliary row, since if one of the two CVP-rows is normally oriented and the other inverted, the two CVPs develop on the same sides of the two rows with reference to the row's geometry, and hence on opposite sides with reference to the cell's geometry (37). But CVPs do not develop adjacent to all ciliary rows; which row is "chosen" to be a CVP-row is primarily a function of the large-scale cellular geometry, as we will see in a subsequent section.

The conclusion that some but not all patterns depend directly upon pre-existing ciliary organization is further supported by analyses of the persistence of variant patterns through an encysted stage in various hypotrichs. Grimes (14) examined starvation-induced cysts of the hypotrich *Oxytricha* by serial-section electron microscopy and found that all ciliary structures disappear upon encystment, and become reformed *de novo* during excystment. Hence, those structural variants whose maintenance depends upon the persistence of ultrastructurally visible ciliary organization would be expected to disappear after passage through a cyst, while variants that require only some other organization that is separable from the ciliature might persist through the cyst stage. The first result was obtained for extra marginal cirral rows (19), the second for doublet configurations either of the ordinary parabiotic kind (15) or of the flat mirror-image variety (17). These divergent results are consistent with a duality of patterning mechanisms, one tied closely to the pre-existing ciliary organization and the other dissociable from that organization (17).

Cytotactic Propagation

Ever since Weissmann's formulation of the germ-plasm hypothesis (59), we have been accustomed to think of characters of organisms as expressions of determinants that are separable from the characters themselves, determinants that we now know to be embodied in an intramolecular code. Sonneborn (45), however, noticed that ciliates provide a remarkable exception to this generalization. *Paramecium* cells that have acquired inverted ciliary rows as a consequence of developmental accidents (essentially involving insertion of an upside-down ciliary row) can propagate such inversions for hundreds of generations (4). Beisson and Sonneborn also demonstrated that the phenotypic difference between cells that carry inverted ciliary rows and those that lack such rows cannot be ascribed to any genic difference. Sonneborn (45) coined the term *cytotaxis* for "This ordering and arranging of new cell structure under the influence of pre-existing cell structure." Since the crucial criterion for cytotaxis is propagation of a structural variant in the absence of any relevant genic difference, cytotaxis represents a true exception to the Weissmannian view of the biological world.

Before considering further examples of cytotaxis in ciliates, we should be clear as to what this important concept does entail and what it does not. It does entail a causation of meaningful cellular differences by circumstances independent of genic differences. One may object that there is no internal difference between an inverted ciliary row and a normally oriented row: an inverted row is simply a normal row turned upside-down; if one examines a transmission electron micrograph of the two next to one another one cannot tell which row is inverted and which is normally oriented without the aid of other cellular reference markers (41). The factual basis for the objection is correct but the objection itself is beside the point. At the cellular level, there is a major and indisputable structural (and also functional) difference between a cell that bears one or more inverted rows and a cell that bears none. The fact that the difference is evident only at the organizational level of the whole cell simply re-emphasizes the importance of taking level of organization into account when seeking to understand biological phenomena.

Although cytotaxis entails real and important differences that are inheritable at a level of biological organization more complex than sequences of nucleotide base pairs, it does not require a unique underlying mechanism. The classic example of cytotaxis, the inheritance of the polarity and asymmetry of the ciliary row of *Paramecium*, can be thought of in terms of specification of sites of origin and of orientation of new ciliary units in the close neighborhood of old ones (46). Such a mechanism might not differ in principle from the dependence of each step of viral assembly upon the existence of a properly primed structural product of the previous step. The important difference is that whereas in the virus the whole structure is discarded and then built anew in each infectious cycle, in the ciliate the structural ensemble maintains its continuity across cell generations. The crucial special feature that makes cytotaxis possible is a mode of cell growth and division that encourages structural continuity across cell generations. Cytotactic propagation is likely in other kinds of cells also, to the degree that they can maintain such structural continuity [for examples, see (2) pp. 294–5].

The argument just presented would lead one to expect that evidence for cytotaxis should be obtained easily in ciliates, and for more than just ciliary-row inversions. This is indeed true, although before describing some examples, I should caution that in these additional cases the demonstrations of cytotaxis include evidence for propagation of new phenotypes initiated by microsurgery or its equivalent but often lack a true genetic test; in such cases it is assumed that the effect of the microsurgery is to alter structural configurations and not to induce genic mutations.

Although discovered in *Paramecium*, cytotaxis has been demonstrated in other ciliates as well. In *Tetrahymena*, this includes both a repeat of the classic demonstration of propagation of a ciliary-row inversion (40) and also demonstrations of the inheritance of differences in number of ciliary rows (31, 8). Most important, however, cytotaxis is observed at different levels of intracellular organization. Cytotactically propagated variants include, in order of ascending organizational complexity: (a) extra longitudinal microtubule bands in *Tetrahymena* (39); (b) the ciliary row variants already discussed; (c) supernumerary marginal cirral rows in the hypotrich *Oxytricha* (16); (d) parabiotic doublets in *Paramecium* (44), *Tetrahymena* (6), *Stentor* (49), and hypotrichs such as *Oxytricha* (15); and finally (e) the flat doublets with mirror-image patterns in the hypotrich *Stylonychia* (56, 21, 24). It should be noted that these examples span the two levels of hierarchical organization considered in the previous section. Examples (a) to (c) are configurations propagated by short-range interactions, while examples (d) and (e) are of large-scale organization which have been shown in hypotrichs to be maintained independently of pre-existing ciliary structures. The independence of the inheritability of variant patterns from their organizational level suggests that cytotaxis is compatible with more than one mechanism of structural propagation. Indeed, any structural ensemble that can propagate itself longitudinally and that can maintain both its individuality and its internal geometry (polarity and asymmetry) during this propagation can be cytotactically perpetuated during the growth of the ciliate clonal-cylinder.

In winding up the consideration of cytotaxis, we will need to anticipate the conclusion of the next section, namely that large-scale patterns of ciliates are organized as developmental fields. I will argue that the property of cytotactic propagation does not *in itself* imply that ciliate developmental fields are fundamentally different from those expressed, for example, in early embryos. There is no known case of cytotactic propagation of fields expressed during conventional embryonic and post-embryonic development, but this may be because embryos are made anew in each generation from tiny parts of the parental organism. Except in certain special cases, ciliates do not form embryos, but some multicellular organisms do reproduce like ciliates. Consider for a moment the turbellarian flatworm, *Stenostomum*. This worm can reproduce itself by elongation followed by tandem subdivision (essentially a segmentation in which the segments pinch apart to form complete worms). Long ago Sonneborn (43) generated parabiotic doublet worms, and found that these reproduce asexually true to type. That is, these worms perpetuate their doublet organization cytotactically. A worm that grows longitudinally and repeatedly segments itself into daughter worms as a ciliate does in daughter cells can be thought of as equivalent to a single organism with unrestrained longitudinal growth. Whatever it is that orients the underlying developmental fields then simply grows along with the organism.

Developmental Fields

Relational Determination of the Site of Oral Development

Stentor is the classic organism for demonstrating field organization in ciliates. Recall that the primordium of the new feeding structures of the posterior division product arises in a contrast zone, within a region of closely spaced ciliary rows adjacent to more widely spaced rows (Fig. 2B-2). In a classic analysis, Tartar (49) demonstrated that it is a relational feature, the "stripe contrast," that is paramount in specifying the position at which an oral primordium appears. In one particularly revealing experiment, Tartar implanted a region of widely spaced ciliary rows from one *Stentor* into the middle of the narrowly spaced region of another (Fig. 5A-1). This results in three contrast zones—the usual one plus two additional ones at the borders of the implant. When regeneration is stimulated by removal of the old oral spiral, oral primordia appear within the narrowly spaced rows at all three loci of contrast (Fig. 5A-2). This result neatly demonstrates that formation of new oral primordium is not dependent upon a unique local determinant. The same experiment also shows that a relational aspect of intracellular organization is crucial in calling forth the development of the new primordium; it is as if a region of more widely spaced ciliary rows induces oral primordium formation in adjacent narrowly spaced rows, provided of course that development has been triggered appropriately (in this case, by the removal of the oral spiral).

Many other experiments also support the above conclusion, but we will consider only one, to which we will later return. This involves the construction of a *Stentor* with a single contrast zone of reversed asymmetry. The operation is bizarre: first the anterior and posterior ends of the cell are excised, and then "the anterior half of the cell is rotated on the posterior, and after a longitudinal cut the left half is rotated 180° on the right" (52, p. 129) (Fig. 5B-1). The way that some cells respond to this drastic rearrangement is to regenerate a single contrast zone in which the juxtaposition of stripes is backwards: narrowly spaced rows now abut widely spaced rows to their right rather than the usual left. An oral primordium then forms, as always, in the region of narrowly spaced rows adjacent to the widely spaced rows. The spatial disposition of the oral primordium is a mirror image of the normal one (Fig. 5B-2). But as the membranelles complete their differentiation and move anteriorly to replace the excised feeding system, a flaw becomes apparent: the cilia of the membranelles, instead of beating posteriorly toward the oral spiral, beat anteriorly, *away* from the oral spiral (Fig. 5B-3). The consequent starvation presumably generates a signal for formation of a new oral primordium, which forms at the same place and in the same manner as the first (Fig. 5B-3). The cell repeats this useless performance as many as 12 times, and gradually starves to death (Fig. 5B-5). This result demonstrates more clearly than any other that the necessary and sufficient conditions for oral development—in this case, the pattern juxtaposition combined with the signal that calls forth development—operate autonomously regardless of whether or not the cell can benefit from the result.

One other aspect of Tartar's pattern reversal deserves notice. Not only is the oral primordium reversed, but so is the location of the contractile vacuole pore and the macronucleus (Fig. 5B-3). The former now is located under the region of widely spaced rows, the latter under the closely spaced rows, just as in normal cells, but the spatial arrangement is a mirror image of the usual one. This suggests that the cell-surface pattern is indeed determinative of the cellular location of these internal structures. This conclusion was clearly demonstrated, for the macronucleus, by de Terra (5). She showed that *Stentor* cells acquire a number of macronu-

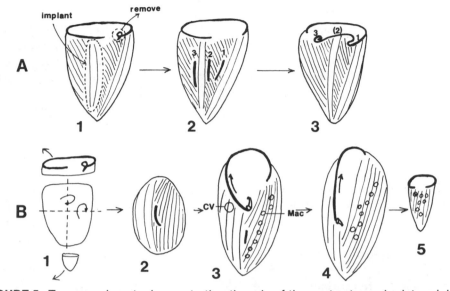

FIGURE 5. Two experiments demonstrating the role of the contrast-zone in determining the site of oral primordium formation in *Stentor*. (A) Consequences of implantation of a nonnucleated sector of widely-spaced ciliary rows from one stentor into a narrowly spaced region of another. (1) The implant is inserted and the oral spiral is removed to stimulate formation of new oral primordia. (2) Formation of an oral primordium at the usual site (1) plus simultaneous appearance of additional oral primordia within the narrowly spaced rows on both sides of the implant (2, 3). (3) Completion of regeneration: the three newly formed bands join to the remnant of the old membranelle band and to each other. Primordia (1) and (3) form oral spirals. (B) "Quartering" of stentors to stimulate formation of reversed contrast-zones. (1). The operation: both ends are removed. Then the anterior half of the remainder is rotated 180° relative to the posterior half along the horizontal dashed plane. Finally, the left half is rotated 180° with respect to the right half about the vertical plane, yielding a heteropolar complex with horizontal dislocations. (2) Formation of a reversed contrast-zone and oral primordium following healing and reorganization of the "quartered" *Stentor* fragment. (3) Differentiation of an incomplete oral spiral, and partial reconstitution of cell form. The arrow indicates the direction of ciliary beat, away from the oral spiral. The positions of the contractile vacuole (CV) and macronucleus (Mac) are indicated. A second oral primordium forms. (4) The consequence of a repeated reorganization. (5) A tiny incomplete *Stentor* resulting from numerous unsuccessful reorganizations. Panels A redrawn with slight modification from Fig. 4 in (49) and Fig. 46D in (51); Panels B redrawn from Fig. 13A in (50); Fig. 49 in (51); and Fig. 14 in (52), with an oral primordium added in frame 3 and arrows in frames 3 and 4 on the basis of text descriptions in the above sources.

clei equal to the number of contrast zones. Thus, if a singlet *Stentor* is converted into a parabiotic doublet by grafting an implant containing a contrast zone (but lacking a macronucleus) onto the dorsal surface, a portion of the extant macronucleus seems to be attracted to the region of narrow spacing near the supernumerary contrast zone, and before long the artificially constructed doublet has two macronuclei (5). The number of appropriate cell-surface sites controls the number of macronuclei, rather than the opposite.

What is the relationship between the contrast zone and the remainder of the cell surface? Working independently of Tartar, Uhlig (57) confirmed Tartar's basic conclusions, refining and extending many of them. However, his general formulation differed from Tartar's in a subtle but important way. Instead of viewing the contrast zone as an essentially local relational feature, Uhlig considered it to be an expression of the extremes of a circular gradient wrapped around the cell—the + + + to − − − gradient illustrated schematically in Fig. 6. The experimental justification for this formulation will be indicated only sketchily here; for details, see Uhlig's German original (57) or my summary of some of his crucial experiments (7). Uhlig observed that when contrast zones are created artificially at sites comparable to site 3 in Fig. 5A-2, an original moderate pattern contrast is enhanced by subdivision and branching of rows prior to formation of the oral primordium, thus changing a + / − − − contrast (for example) to a + + + / − − − contrast. This suggested a dynamic basis for primordium formation, with regulation of a gradient system at its foundation. Uhlig also investigated interactions among contrast zones of varying "strength," showing, for example, that a + / − − contrast could regulate to a + + + / − − − contrast and then form an oral primordium if—and only if—there is no pre-existing + + + / − − − contrast zone nearby. Uhlig's analysis is reminiscent of rather similar observations and formulations in hydroids, in which intact hypostomes and tentacle rings appears to inhibit the induction of new ones nearby (58).

Contractile Vacuole Pore Cytogeometry

While Uhlig's experimental results with *Stentor* met the classical criteria for a morphogenetic gradient, the demonstration was, as Sonneborn (47) emphasized, based

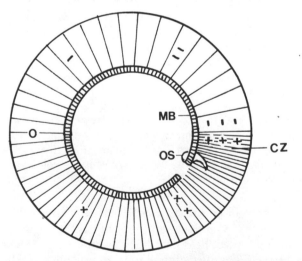

FIGURE 6. Uhlig's "circular gradient" model of the organization of the *Stentor* cell surface. The cell is seen from above, with the membranellar band (MB) contracted so that the organization of ciliary rows on the body posterior to it can also be seen. Note the juxtaposition of extreme values in the contrast zone (CZ), posterior to the oral spiral (OS). Redrawn from Abb. 6 in (57).

phogenetic gradient, the demonstration was, as Sonneborn (47) emphasized, based largely upon analysis of interactions involving apparent inhibition or competition among preformed sites. Demonstration of a positional gradient that controlled the locations of structures within its realm would require that new subsidiary structures be positioned proportionally with respect to the system as a whole. The locations of the contractile vacuole and macronucleus of *Stentor* are qualitatively where one would expect, but in the absence of a detailed quantitative analysis it cannot be decided whether there is true proportional positioning consistent with a positional gradient model, or whether positioning is by measurement of a fixed distance from the contrast locus, as postulated by models, such as Sonneborn's (47) sequential-nearest-neighbor-interactions, that are based on extensions of local allosteric relationships.

The paradigm of relational positioning is Nanney's quantitative analysis of contractile-vacuole-pore (CVP) cytogeometry in *Tetrahymena* (30). Recall that new CVPs develop just anterior to the fission zone, in a specific geometrical relationship to certain ciliary rows located to the right of the oral meridian. Which ciliary rows? Nanney's crucial discovery was that the cell's choice of CVP-row depends upon a relative distance from the oral meridian; the greater the total number of ciliary rows, the greater also is the number of rows between the oral meridian and the CVPs. The rough proportionality between the longitude of CVP formation and the cellular circumference, both assessed in terms of number of ciliary rows, was summarized diagrammatically in the form of a constant central angle (Fig. 7A and 7B). However, Nanney (30) also pointed out that the central angle formulation in its literal form breaks down as soon as one considers parabiotic doublets with two oral meridians and two CVP sets, in which the central angle is reduced to one-half of what it is in singlets (Fig. 7C). Nanney concluded that "The cell in some way measures a fraction of the distance between one stomatogenic meridian and the next (or possibly between other cortical features) and regulates the field size in relation to that distance" (30, p. 316).

We will now attempt to reformulate Nanney's fundamental insight by combining four ideas. The first is his basic concept of measurement of a fractional distance, as quoted. The second is the assumption that the stomatogenic meridian of *Tetrahymena* acquires its stomatogenic capacity as a consequence of some underlying relational pattern contrast, as in *Stentor*, despite the absence in *Tetrayhymena* of any visible contrast in arrangement of cell-surface elements. Although this assumption is unproven, the underlying idea that the site of oral development is specified by a spatial relation rather than by fixed information inherited by a particular ciliary row is strongly supported by Nanney's demonstration (32) that any ciliary row can be stomatogenic if it is in the right cellular location. The third idea is Uhlig's concept of a circular gradient, transposed to *Tetrahymena*. The fourth idea is that this is a positional gradient, so that the formation of a particular structure can be determined by the reading of a specific level of the gradient (48, Abb. 12). The combination of these four ideas converts the central angle diagrams (shown inside the cells in Fig. 7) into positional-information diagrams (shown outside the same cells). It is important to note that the gradients drawn in these diagrams need not imply graded concentrations of a diffusible underlying morphogen, but could refer to any *condition* that can be varied continuously and that can be "interpreted" or implemented to bring about the formation of a discrete structure. This representation is in the general spirit of Wolpert's positional information formula-

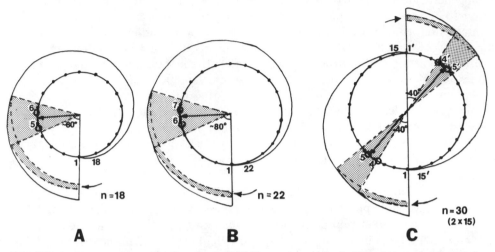

FIGURE 7. Three schematic diagrams of CVP cytogeometry in *Tetrahymena*. The circles studded with large dots represent cross-sections of *Tetrahymena* cells, the dots indicating ciliary rows. Frames A and B depict singlet cells with 18 and 22 ciliary rows respectively. The right-postoral row is numbered 1 and the numbering proceeds clockwise around the cell. Frame C represents a parabiotic doublet with oral areas directly opposite one another, and hence two right-postoral ciliary rows, numbered 1 and 1' respectively. Nanney's central-angle formulation is shown inside the cells, while a reformulation in terms of a postulated cell-surface positional gradient is shown outside. The reference meridian for both is the right-postoral ciliary row(s). The arrows show the central angle of the inductive field (inside) and the corresponding gradient-contour (outside). The shaded regions indicate the width of the inductive field. Basal bodies of rows within the inductive field can generate CVPs. The "inside" portions of these diagrams were adapted from Figs. 2 and 4 in (30) with minor modifications.

tion (60), although it differs from it in the crucial respect that interpretation here involves assembly processes taking place in specific parts of cells rather than selective gene activation resulting in specific protein syntheses by whole cells. This alternative idea was asserted as a possibility by Stumpf (48).

Alternative Sequences and Developmental Assessment

We will pause for a moment to summarize the concepts developed thus far, before considering some observations that supplement these concepts. Ciliates appear to possess a reference meridian along which oral structures develop. This meridian is specified by a localized relationship rather than a unique localized substance. Conditions that work over long intracellular distances influence the cell's capacity to establish oral meridians elsewhere and also determine the location of other structures. In a few cases such positioning has been demonstrated to be spatially proportional.[2] It is possible, then, to think of the large-scale organization of all elements of

[2]It should be emphasized here that evidence for true proportionality is available only for the sites at which structures make their initial appearance. The relative sizes and internal architecture of complex ciliary primordia generally do not show such clear-cut proportionality (see 3 and references cited therein).

the ciliate surface as an expression of fields that can be propagated in a clone by virtue of the geometry of ciliate growth. Such propagated fields are superimposed upon local and nonregulative systems of fine-positioning that are strictly dependent upon the spatial organizational of ciliary units.

There is more, however, to large-scale ciliate pattern formation than the organization and expression of fields in relation to reference meridians. Ciliates also have means of assessing what pre-existing structures are present and adjusting their developmental responses to the structural configurations that are thus assessed. We will consider examples of alternative developmental sequences and developmental assessment (17), first in hypotrichs and then in *Tetrahymena*.

Alternative means of generating ciliary primordia must exist in hypotrichs capable of cystment. Normally, most or all new ventral ciliary primordia arise in close juxtaposition to certain pre-existing structures: the oral primordium near a particular transverse cirrus, the frontal-ventral-transverse (FVT) cirral primordia from particular old frontal and ventral cirri, and marginal cirral primordia from specific old marginal cirri (Fig. 2C). But when cells emerge from cysts, they generate all of their ciliary primordia *de novo*. Hence the participation of old ciliary structures in the formation of new ones is usual but not essential.

The conditions that influence whether or not pre-existing ciliary structures are "used" in the creation of new ones have been explored in microsurgical studies involving ablation of particular regions. For example, Jerka–Dziadosz (23) observed that when a cell-margin of the hypotrich *Paraurostyla* is excised, removing a pre-existing marginal cirral row, a new marginal cirral row is formed nonetheless, and it differentiates at the reconstituted cell margin of the truncated cell. Even more dramatically, pre-existing ciliary structures may not be used to generate new ones if these structures are at inappropriate cellular locations. The clearest demonstration of this is for development near the anterior end of *Paraurostyla* (25). This organism possesses six prominent frontal cirri (Fig. 8A-1). During development, the anterior-most three of these cirri remain quiescent (and are ultimately resorbed), while the posterior three disaggregate to form the modified ciliary rows that will later differentiate into several FVT cirri (Fig. 8A-2). If, however, the anterior tip of the cell is first cut off (Fig. 8B-1), the cirri that would otherwise be employed in ciliary-row development now assume an extreme-anterior location and are not made use of; instead, ciliary rows develop more posteriorly, at a distance from the new anterior tip that is similar to the distance of a normal FVT primordium from the normal tip of an unoperated cell (Fig. 8B-2). This experiment demonstrates that at least in some cases relative intracellular location is decisive in determining where ciliary primordia develop, and that the cell makes use of pre-existing ciliary structures only when they are conveniently available at the right places.

In these studies on the hypotrich *Paraurostyla*, it was assumed that ciliary primordia of truncated cells develop *de novo* at their definitive site; e.g., that following lateral truncation the new marginal cirral primordium originates at the cell-margin (23). Grimes and Adler (18) challenged this conclusion on the basis of a re-analysis of early stages of development of ciliary primordia in laterally truncated cells, using scanning electron microscopy. They found, in brief, that the oral primordium, rather than remaining fairly compact and developing primarily into oral structures as is normally the case, appears to form extensions directed toward the affected sites; the appropriate cirral primordia then develop in direct continu-

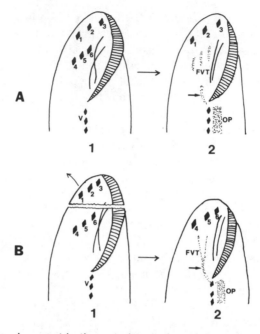

FIGURE 8. Ciliary development in the anterior portion of the ventral surface of the hypotrich *Paraurostyla* in intact cells (A) and following amputation of the anterior tip (B). Frames (1) show the cell prior to the onset of oral development, with the six large frontal cirri (numbered) and one of the ventral cirral rows (V) indicated. Frames (2) indicate early stages in formation of FVT ciliary rows. In (A), three FVT rows develop from disaggregation of frontal cirri 4, 5, and 6; a small ciliary field is also seen more posteriorly (arrow). The oral primordium (OP) had formed earlier close to a ventral cirral row (V). In sequence (B), there are (in this cell) two FVT ciliary rows, formed in close spatial association with a ciliary field (arrow) that extends from the vicinity of the oral primordium (OP). Frame A2 drawn from Fig. 14, Frame B2 from Fig. 28 in (25), with certain structures that are out of the focal plane in those photographs (OP in A2, frontal cirri in B2) added.

ity with these extensions, possibly by their further extension and reorientation. The apparent *in situ* origin of primordia of ablated structures that had been described earlier was thus viewed as a misinterpretation resulting from a failure to observe the earliest stages of the primordia in question. There is, for example, evidence of possible extensions of the oral primordium in photographs of the formation of FVT ciliary rows in *Paraurostyla* cells with truncated anterior tips (25; Fig. 8B-2). However, there is also evidence for the *in situ* origin of an FVT ciliary row following the selective destruction of frontal cirrus no. 6 by an ultraviolet microbeam (22). The "sources" of ciliary primordia are frequently not completely clear and may well vary in different species of hypotrichs and/or under different circumstances. Such variations are not surprising if one thinks of the cell as being capable of assessing the presence and location of all of its structural elements, and then utilizing this assessment in determining the means of initiating ciliary primordia as well as the fate of these primordia. From this perspective, the source of a ciliary primordium is less important than the position at which the primordium comes to organize itself, and it is expected that the former might be more variable than the latter.

In *Tetrahymena*, a developmental sequence exists that is truly optional, in the sense that it may take place or may not under a defined set of external conditions, depending only on the outcome of an assessment of a specific aspect of pre-existing cellular pattern. This sequence involves the contribution of the oral primordium to the formation of ciliary rows. Normally, each ciliary row perpetuates itself, and the oral primordium differentiates only into new oral structures. But there is one exceptional situation in which the oral primordium seems to give rise to a new ciliary row as well. When *Tetrahymena* cells are starved by a nutritional shift-down, they undergo a transformation that produces a long and thin "rapid-swimmer" form (35). During the course of this transformation, the pre-existing oral apparatus is replaced by a new one generated from an oral primordium formed in close proximity to it. In concluding phases of this oral replacement process, couplets of basal bodies may be peeled off from the posterior end of the developing undulating membrane and inserted between the two postoral ciliary rows (36). Eventually, this chain of couplets is organized into a new ciliary row, increasing the total number of ciliary rows by one. What is most interesting about this process is that it occurs under a very restricted set of conditions: starvation, oral replacement, and—most important of all—a number of pre-existing rows that does not exceed 19. Under identical conditions, the majority of 18-rowed cells, some 19-rowed cells, and few or no 20- or 21-rowed cells go through this row-insertion process. This threshold is near the upper end of the *stability range* (8) which defines the presumed optimal ciliary row numbers for this species. Thus, the cell somehow "knows" its ciliary row number, and "assesses" that number when deciding whether or not to use its oral primordium to insert a new ciliary row when an appropriate opportunity arises. The fascinating question, of course, is what is being assessed. In this case we know that it is *not* the total number of basal bodies on the cell surface, because the number of basal bodies in all ciliary rows taken together is roughly constant, with fewer basal bodies per row as the number of rows increases (33). The cell must have some way of assessing number of rows *per se*; at present we do not know how.

Axial Reversals and Mirror-Image Patterns

We have seen earlier that local patterns, dependent upon the organization of visible ciliary units, always preserve their original internal geometry, while large-scale patterns, which lack such dependence, can undergo mirror-image reversals. We will now inquire into how such mirror-image reversals can be achieved. Formally, the answer is simple: flip over one axis while preserving the others. For example, imagine that the dorsal-ventral axis of your right hand were reversed while the other two axes remained unchanged: the back of your hand would become a palm and *vice versa*, and your right hand would now be transformed into a left hand. I will argue that generation of mirror-image patterns in ciliates is a consequence of such reversals of single axes, which can be accomplished in one of two ways, depending (in part, at least) upon whether the reversal is initiated by a physical rotation of one part of a cell relative to another.

Mirror-Imaging by Reversal of Antero-Posterior Polarity

We have already seen, in Figs. 4 and 5, two examples of surgically generated mirror-image reversals, and we will now look in more detail into how the reversals were actually achieved. In the case of the *Stentor* reversal shown in Fig. 5B, the operation was sufficiently complex and the followup insufficiently detailed so that we cannot be certain of exactly how the pattern-reversal was generated. However, results of other, simpler, operations on *Stentor* allow us to make a reasonable guess. Tartar and Uhlig both were able to demonstrate a role of the cell's posterior region in inducing the posterior end of the oral apparatus, the oral spiral (Fig. 2B-3). If the posterior region of one *Stentor* is grafted adjacent to the midregion of another developing *Stentor*, a supernumerary though abortive oral spiral develops in a portion of the oral primordium that is closest to the graft (Fig. 9A) (51, pp. 175-6). Hence, there is something about the posterior cell region that promotes development of an oral spiral in neighboring parts of the oral primordium. This juxtaposition may also be effective when a segment of a *Stentor* including a contrast-zone is rotated 180° *in situ*; in such cases an oral spiral may develop both at the original posterior end of the oral primordium near a tail-pole constructed by the implant (Fig. 9B, t1), and also at the original *anterior* end of the oral primordium near the tail-pole of the host cell (Fig. 9B, t2) (51, pp. 174-5). Detailed studies by others, notably Uhlig (57), confirmed this lability of the antero-posterior axis within the oral primordium and the influence of the posterior region of the cell on determining this polarity. Uhlig (57) postulated the existence of an antero-posterior gradient that controls, among other things, the differentiation of the oral spiral within the oral primordium. If this gradient were reversed while the circular gradient orthogonal to it remained unchanged, then a mirror-image reversal of large-scale pattern would result. A glance at Fig. 5B-1 will show that the complex operation that resulted in a reversed contrast zone included a 180° rotation of the left half of the cell relative to the right half. The preceding horizontal rotation of the anterior half on the posterior created further dislocations that probably forced the cell to undergo a major reconstruction of its contrast zone. It is likely that a major portion of this zone was first mechanically inverted (as in Fig. 9B), and perhaps then underwent an *in situ* reversal of its inherent antero-posterior axis to generate a mirror image at the organizational level of large-scale pattern.

Flat hypotrich doublets with mirror-image ciliary patterns (Fig. 4A) have been obtained in three hypotrich genera (*Stylonychia*, *Pleuroticha*, and *Parauros-tyla*) by three different research groups (21, 24, 56). The mode of origin of these doublets has in no case been fully clarified, since even the operations that generate these doublets most efficiently do so in only a minority of the operated cells. The case that is closest to being understood is the outcome of the curious operation by which the Chinese workers have repeatedly succeeded in eliciting these doublets (53). Developing cells, at the stage shown in Fig. 2C-3, are chosen, and a central region is isolated. This region includes the condensed macronucleus in the center and the posterior portion of the oral apparatus and the anterior part of the oral primordium on its ventral surface (Fig. 10A & B). Following healing of the edges of the wound, the remnant of the oral primordium sometimes undergoes a clockwise

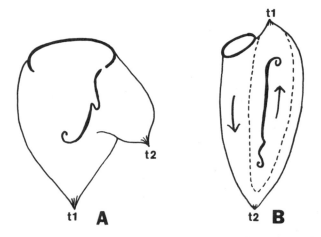

FIGURE 9. Two experiments indicating the importance of the posterior region of the *Stentor* cell in induction of an oral spiral. (A) Induction of an abortive oral spiral in the center of the oral primordium following lateral implantation of an extra posterior portion of a *Stentor*. t1 indicates the tail-pole of the host, t2 that of the implant. (B) Formation of an oral spiral at both ends of the oral primordium following excision of a nonnucleated segment from the region of the contrast-zone and reimplantation following 180° rotation about the antero-posterior axis. The arrows show the intrinsic polarity of the host and of the implant. An oral primordium develops within the implant, with one oral spiral forming at its intrinsic posterior end near the tail pole reconstituted from the implant (t1), and a second oral spiral formed at its intrinsic anterior end near the tail-pole of the host. Panel A redrawn from Fig. 44C, panel B from Fig. 44A, in (51).

rotation, resulting in an inverted configuration relative to the original antero-posterior axis (Fig. 10C & D) (55). During this period the ciliature of the ventral surface is undergoing successive reorganizations, as indicated by the presence of FVT cirral primordia (Fig. 10C & D). At some time during the course of these reorganizations the antero-posterior axis of the right half of the cell appears to be reversed to conform to that of the adjacent left half, so that the original anterior end of the rotated oral structures now becomes a posterior end (Fig. 10, compare D and E), with a clearly marked though abnormal oral cavity (54). Reversal of the antero-posterior axis of the right component will generate a mirror-imaged large-scale pattern, while the individual membranelles of that component preserve the inverted organization reminiscent of their original configuration.[3]

The operation depicted in Fig. 10 is not the only way in which mirror-image flat doublets can be obtained. Jerka–Dziadosz (24) has reported that the same configuration can be generated in *Paraurostyla* after fission-arrest followed by a forward sliding of the posterior daughter cell, so that the two presumptive division products come to lie almost side-by-side. In this case, it is possible (though not es-

[3]The most complete documentation of the origin of a mirror-image configuration by selective reversal of the antero-posterior cell axis was provided by Grimes and L'Hernault (20) in their study of the reorganization of the ciliature of the hypotrich *Stylonychia* following the counter-clockwise folding-over of the posterior half of a microsurgically isolated right longitudinal half-cell. This configuration could not yield self-propagating doublets because it failed to construct a normal cell-mouth.

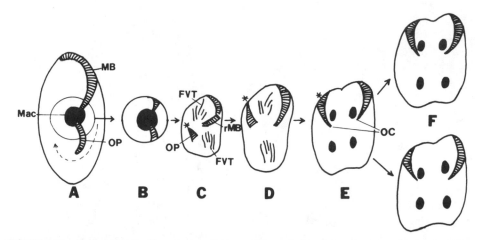

FIGURE 10. A diagrammatic representation of the origin of flat-doublets in the hypotrich *Stylonychia*. (A) A late-prefission stage (corresponding to Fig. 2C-3) showing the macronucleus (Mac), old membranelle band (MB), and oral primordium (OP). The dashed arrow indicates the clockwise direction of cytoplasmic flow observed at this stage. The solid circular outline indicates the region that is microsurgically removed. (B) The flat central disc that has been removed from the pre-dividing cell (the remainder is discarded). (C) Reorganization of the cell derived from the central disc. The anterior oral apparatus is being supplemented by a newly formed reorganization membranelle band (rMB). The fragment of the original oral primordium (OP) is being moved to the cell's right and simultaneously rotated nearly 180°; its original posterior end is indicated by an asterisk. Two sets of fronto-ventral-transverse cirral primordia (FVT) are also present. (D) A later stage, at which the inverted OP has now been moved to the anterior-right margin of the partially reconstituted cell. The form of this OP is still unaltered, with its original posterior end (asterisk) directed anteriorly. (E) The oral apparatus on the right side of the cell has now undergone a reversal of its large-scale polarity; the original posterior end (asterisk) now has the form characteristic of an anterior end, while an oral cavity (OC) is present at the original anterior end. The cell is now a doublet, with two sets of FVT cirri (cf. Fig. 4) and four macronuclei (a nondividing singlet cell has two tandemly arranged macronuclei). (F) The flat doublet can now divide as a heritable side-by-side twin configuration. Redrawn from a portion of Fig. 7 in (55).

tablished) that the right-left axis rather than the antero-posterior axis might undergo reversal, as in the *Tetrahymena* examples to be considered in the next section.

Mirror-Imaging by Reversal of Right-Left Asymmetry

Mirror-image patterns can arise without microsurgically generated polar reversals. The clearest cases have been observed in *Blepharisma* (related to *Stentor*) and in *Tetrahymena*. We will consider the *Tetrahymena* examples in some detail. There are two such situations, one involving an unusual pattern juxtaposition in wild-type cells, the other the action of a mutant gene. I will describe the wild-type case first because I suspect that the mutant allele is uncovering a developmental mechanism that can also be exhibited by wild type cells in an unusual geometrical situation.

The unusual situation is the "asymmetrical" regulation of a parabiotic doublet. A typical parabiotic doublet formed by back-to-back fusion of two normal cells is illustrated as a polar projection in Fig. 11A. The doublet shown in Fig. 11A is called "symmetrical" because the two oral systems, with their associated CVP sets, are directly opposite one another (placing the expressions "symmetrical" and "asymmetrical" within quotation marks will indicate reference to the relative placement of structures rather than their internal organization). Such doublets regulate to singlets by a process involving loss of ciliary rows and eventual "suppression" of oral development along one of the two reference meridians. When the two oral systems remain "symmetrically" placed during this regulation, this conversion does not involve any unexpected intermediates (Fig. 11, pathway 1). However, ciliary rows are sometimes lost unequally on the two sides of the cell (pathway 2), so that the two oral-CVP systems come to be "asymmetrically" placed (Fig. 11B). One oral system may then be lost and the cell revert to the singlet state, still in a relatively uncomplicated fashion (pathway 3). But occasionally something exceedingly strange happens (pathway 4). Fauré-Fremiet (6) noticed that a third oral apparatus of somewhat reduced dimensions may develop between the two "asymmetrically" placed major oral apparatuses. E. M. Nelsen (*personal communication*) rediscovered this phenomenon, and then studied it systematically. He found that such supernumerary oral apparatuses appear in some cells within all clones of parabiotic doublets in which the original oral meridians come to be "asymmetrically" placed. The position of the supernumerary oral apparatus is generally exactly as shown in Fauré-Fremiet's one drawing of such cells (Fig. 8 in Ref. 6): on the short arc between the two major oral systems, and generally closer to the oral apparatus on its left than to the one on its right (Fig. 11C). Nelsen's new observation was that these supernumerary oral structures are often of reversed asymmetry, particularly during their development (Fig. 11C); in fact, they are good phenocopies of the partly reversed oral configuration observed in cells homozygous for the *janus* symmetry-reversal mutant (see below). The ciliary rows of such cells are always normally oriented and never inverted, indicating that there has been an *in situ* reversal of right-left asymmetry without any preceding folding or rotation of cell parts. The existence of a true large-scale reversal of asymmetry is confirmed by the way in which such "triplets" eventually become resolved as they lose ciliary rows. They ultimately revert to the singlet state by "suppression" of two of the three oral meridians (pathways 5 and 6). Usually, the ensuing singlets are normal, as depicted in Fig. 11D. But Nelsen (*personal communication*) has repeatedly observed singlets with abnormal oral apparatuses and CVP sets located to the left of the oral apparatus rather than the usual right (Fig. 11E). In these cases, we may presume that the supernumerary reversed oral meridian is the only one to persist, and can then express its reversed organization fully through a reversed direction of CVP determination as well as through oral anomalies (in the "triplets," CVPs are normally absent in the "short arc"). Such "reversed singlets" were sometimes seen undergoing oral replacement; they are presumably in the same morphogenetic dead-end as are stentors with reversed contrast-zones.

Fauré-Fremiet (6) suggested that the appearance of the third oral apparatus

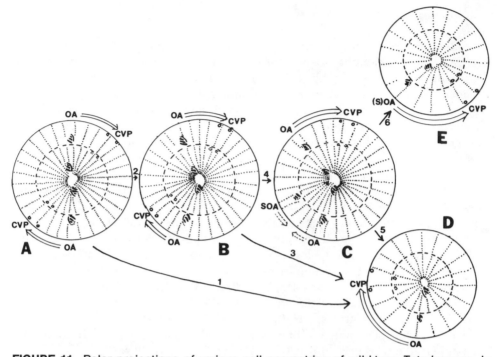

FIGURE 11. Polar projections of various cell geometries of wild-type *Tetrahymena*. In each of these diagrams, the anterior pole is shown in the center of the diagram, and the fission zone is indicated by a dashed circle at the cell equator. The ciliary rows, old and new oral apparatuses, and old and new contractile vacuole pores are shown. The circumferential positions of the meridians with oral apparatuses (OA) and of the contractile-vacuole-pore sets (CVP) are labelled outside of the cell-maps, with arrows showing the presumed direction of CVP determination. (A) A "symmetrical" parabiotic doublet cell. (B) An "asymmetrical" parabiotic doublet. (C) An "asymmetrical" parabiotic doublet with a third meridian with abnormal, supernumerary, oral apparatuses (SOA). (D) A normal singlet cell. Only one normal OA remains, with a CVP set to its right. (E) A reversed singlet cell, with one abnormal OA, presumably derived from the SOA of a triplet, with a CVP set to its left.

indicates some kind of "trouble de la morphogenèse." What kind of "trouble?" Here I will draw on an interpretation by Kumazawa (28) of a very similar pattern reversal in the ciliate *Blepharisma* and extend it by analogy to recent demonstrations of reversed-intercalation in insect segments. Wright and Lawrence (61) found in the bug *Oncopeltus* that if one juxtaposes antero-posterior levels within segments that are *less* than half a segment-width apart, the bug merely regenerates the intervening space. If, however, the juxtaposed regions are *more* than half a segment-width apart, then an amazing thing happens: a new segment border with reversed orientation is regenerated between the juxtaposed surfaces. Wright and Lawrence (61) explained this unanticipated result by applying the shortest-intercalation rule of the polar-coordinate model (13) to a linear segmental gradient. The critical assumption is that the extremes of the segmental gradient are in some sense neighboring values, like 12:59 and 1:00 on a clock. Then, if the artificially juxtaposed surfaces are, for example, like 2 and 10 on a 1 . . . 12 linear gradient (in which the 3–4–5–6–7–8–9 part had been cut out leaving 1–2–(gap)–10–11–12), intercalation might not go by the otherwise expected 2–*3–4–5–6–7–8–9*–10 (intercalated values italicized) route, but rather by the shorter opposite route, 2–*1–12–11*–10. A new, reversed, segment border would then appear in the intercalated 1–12 interval.

The Wright–Lawrence adaptation of the polar coordinate model can be applied to the generation of reversed oral structures in the short arc of parabiotic *Tetrahymena* (and *Blepharisma*) doublets, as shown in Fig. 12. Diagram (A) is merely a straightened-out version of the gradient formulation of the organization of parabiotic doublets shown in Fig. 7C. Diagrams (B) and (C) indicate what might happen as the doublet becomes "asymmetrical." The gradient on the "short arc" becomes steeper (B), the unusually steep gradient becomes unstable, and a discontinuity develops (C). If the positional disparity at the discontinuity is sufficient, reversed intercalation (arrows in C) will take place, generating a pattern border of reversed asymmetry (D). This reversed asymmetry will guarantee mirror-image patterning of the oral apparatus that develops at that border, but at the same time will prevent formation of CVPs in the neighborhood because of the absence of the appropriate gradient contours. The system is unstable, however, because the number of ciliary rows is well above the cell's stability center (31); continual loss of rows will lead to the disappearance of capacity to form oral structures along all but one reference meridian. The eventual outcome will be either a viable normal singlet cell (E) or a presumably inviable reversed singlet (F).

Whereas the type of reversed asymmetry described above is transient and, in most parabiotic doublet clones, rather exceptional, stable expression of a mirror-image pattern is observed in all clones homozygous for a mutant allele called *janus* (10). The *janus* pattern has been described (26, 11) and is illustrated schematically in Fig. 13. The cell has two oral meridians. One, the primary oral meridian, is normal in two senses: first, oral development takes place without any failure, and second, the oral apparatuses thus produced are normal or very nearly normal in organization. However, at a rather constant position nearly 180° opposite from the primary oral meridian there is a second and much more unusual oral meridian, along which oral apparatuses may or may not be generated in particular cells (though always recurring in a clonal lineage), and the oral apparatuses that are formed are always abnormal, frequently with a mirror-image arrangement that is particularly obvious during oral development (26). These abnormal oral structures develop along ciliary rows of normal internal polarity and asymmetry with no inversions, again suggesting that the mirror-image reversal did not result from any prior rotation of the antero-posterior axis.

The positions of contractile vacuole pores of *janus* cells are altered in a characteristic manner. All CVPs are located in one cell half, to the right of the primary oral meridian and to the left of the secondary oral meridian. But rather than appearing along two or at most three adjacent ciliary rows as in wild type cells, CVPs may be formed near as many as four ciliary rows in a broad arc within which CVP-bearing ciliary rows may be interrupted by one or more rows without CVPs (Fig. 13). The impression thus often is conveyed of two CVP sets, one determined to the right of the primary oral meridian, the other to the *left* of the secondary oral meridian (26). Interestingly, the center of the frequently-interrupted CVP-arc

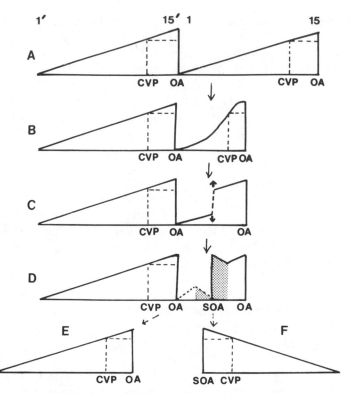

FIGURE 12. A hypothetical scheme for the generation of a reversed third oral meridian in parabiotic doublets. In all of these diagrams, presumed circular gradients are shown in linear form; the right edge of each diagram should be imagined as juxtaposed on the left edge, as in Fig. 7. The locations of the contractile vacuole pores (CVP), original oral apparatuses (OA), and supernumerary oral apparatuses (SOA) are indicated in each diagram. Diagram A represents a "symmetrical" parabiotic doublet as in Figs. 7C and 11A, with positions of selected ciliary rows indicated above the diagram for ease of comparison with Fig. 7C. Diagrams B and C represent "asymmetrical" parabiotic doublets, corresponding to Fig. 11B; the short vertical arrows in diagram C indicate the directions of reverse intercalation in terms of the gradient scheme. Diagram D indicates the gradient-configuration attained at the completion of reverse intercalation (corresponding to Fig. 11C) with the intercalated region stippled. Diagrams E and F show the eventual normal and reversed singlet gradient-configurations, corresponding to Figs. 11D and 11E respectively.

(marked by a solid arrow in Fig. 13) is generally very close to a point midway between the two oral meridians (11; Frankel and Nelsen, *in preparation*).

When *jan⁺/jan⁺* macronuclei are synchronously replaced by new *jan/jan* macronuclei, typical secondary oral structures begin to appear starting 3 cell generations after the genic substitution. As the secondary oral meridian becomes manifest, the number of ciliary rows does not change, and there is no sign of any geometrical distortion (11; Frankel and Nelsen, *in preparation*). The arc of ciliary rows within which CVPs are expressed gradually broadens in both directions, without any major change in the position of the center of the arc. The CVP-arc begins to widen even before oral apparatuses appear at the secondary oral meridian.

These observations clearly indicate that *janus* cells must be regarded not as aberrant parabiotic doublets, but rather as singlet cells in which the dorsal region is transformed into a reversed ventral region. A multicellular analog would be the segment polarity mutants of *Drosophila* (42), in which each segment is transformed into a pattern in which approximately one-half of the segment is replaced by a mirror-image duplicate of the other half. In the *Tetrahymena* case, the dorsal (aboral) half-pattern is missing and replaced by a duplicated but reversed ventral (oral) half-pattern.

Interpretation of this situation is not totally straightforward. A simple, unitary explanation would be to assume that the *janus* gene product (or, more likely, absence of the *jan*+ gene product) wipes out dorsal positional values, and the remaining ventral ones then reverse-intercalate to generate the mirror-image pattern. This, however, does not give a straightforward interpretation of the remarkable CVP pattern of *janus* cells, with variable width yet constant center of the CVP-arc, located halfway between the two oral meridians. This feature appears better explained by postulating the constant presence of a reference meridian near the dorsal midline, with joint determination of CVP position by the patent ventral meridian and the cryptic dorsal one; on this view, developed in detail by Frankel and Nelsen (11), the *janus* allele would alter the expression(s) of an already existing reference border, rather than bringing a new one into existence as is postulated by the reverse-intercalation interpretation.

There are ways of testing these alternatives. It should be noted, however, that different as they are in detail there is a common attitude expressed by both hypotheses: that a mutant allele such as *janus* is not creating something wholly new and unique, but rather is modifying the expression of a condition or potential that already exists in wild-type cells. In my opinion, a conjunction of analysis of symmetry-reversals brought about by altered gene action and of those that result from various geometrical anomalies in wild-type cells offers the best hope for understanding large-scale patterns of ciliates.

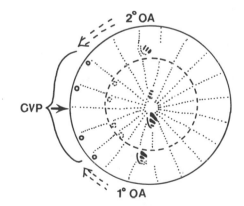

FIGURE 13. A polar projection of a *janus Tetrahymena* that is expressing oral structures along its secondary (2°) oral meridian. Illustrative conventions are as in Fig. 11. The short solid arrow points toward the center of the "CVP arc," while the broken arrows indicate the presumed directions of the influence of the oral meridians in CVP determination.

Conclusions

The major points made in this review can be summarized briefly as follows: There are at least two hierarchically superimposed systems of pattern control, a local system specifically associated with the geometry of ciliary units and a large-scale system that is at least partially independent of that geometry. The pre-existing organization of both of these systems can be propagated longitudinally during growth of a ciliate clone; a by-product of this capacity for propagation is the potential for non-genic clonal inheritance of pre-existing organizational differences. The most probable reference-meridian for the large-scale patterning system is that along which oral apparatuses develop. This meridian is marked in some ciliates by a major pattern discontinuity. Various cellular structures are positioned relationally with respect to this reference meridian. The large-scale positional systems expressed on the cell surface are somehow integrated with a global assessment of the number and locations of ciliary structures. While the local pattern-control system can be inverted but probably cannot be mirror-imaged, the large-scale system can undergo mirror-image reversal by switching either one of two putative orthogonally organized coordinates. Such reversals can either be provoked microsurgically or generated within geometrically undisturbed cells, in the latter cases either by certain abnormal pattern juxtapositions or through the action of a mutated gene.

What do the findings on ciliates have to say to a developmental biologist whose primary interest is in patterning of multicellular organisms? I would suggest three things, in increasing order of specificity and potential explanatory power. First, they indicate that properties describable in terms of fields, gradients, and positional values and signals are just as evident in unicellular organisms as they are in the more familiar multicellular systems; subdivision into many cells is not essential for generating these properties. Second, pattern discontinuities exist within cells as well as between cells, as is evident both in similarities between ciliate division and some types of invertebrate segmentation (11) and in the crucial role of intracellular juxtapositions in organizing developing axes both in ciliates such as *Stentor* and in amphibian eggs (9). Third, similar apparently paradoxical mirror-image patterns can be generated in insects and in ciliates, both by appropriate microsurgery and by the action of certain mutated genes. In both organisms, the paradoxes are resolved if it is assumed that an underlying condition of continuity must be satisfied. This might indicate that it is worthwhile to search for common mechanisms that can operate in both epimorphic and morphallactic regulatory systems in both multicellular and unicellular organisms.

Acknowledgments

I am grateful to Drs. Gary W. Grimes, Pang Yan-bin, and Maria Jerka–Dziadosz for supplying unpublished information concerning development of hypotrich doublets, and to Dr. E. Marlo Nelsen for sharing his current findings on pattern reversals in *Tetrahymena*. I also thank all of these researchers as well as Drs. Julita Bakowska, Anne W.K. Frankel, Robert L. Hammersmith, Stephen Ng, Vance Tartar and Norman E. Williams and Mr. Timothy

Lansing for their constructive criticisms of the manuscript. Recent research in this laboratory included in this review has been supported by grants HD-08485 from the National Institutes of Health (to JF) and PCM 80-21879 from the National Science Foundation (to EMN).

General References

AUFDERHEIDE, K.J., FRANKEL, J., NELSEN, E.M.: Formation and positioning of surface-related structures in *Protozoa*. *Microbiol. Rev.* **44**: 252–302 (1980).

DE TERRA, N.: Some regulatory interactions between cell structures at the supramolecular level. *Biol. Rev.* **53**: 427–463 (1978).

References

1. Allen, R.D.: The morphogenesis of basal bodies and accessory structures of the cortex of the ciliated protozoan *Tetrahymena pyriformis*. *J. Cell Biol.* **40**: 716–733 (1969).

2. Aufderheide, K.J., Frankel, J., Williams, N.E.: Formation and positioning of surface-related structures in Protozoa. *Microbiol. Rev.* **44**: 252–302 (1980).

3. Bakowska, J., Frankel, J., Nelsen, E.M.: Regulation of the pattern of basal bodies within the oral apparatus of *Tetrahymena thermophila*. *J. Embryol. Exp. Morph.* **69**: 83–105 (1982).

4. Beisson, J., Sonneborn, T.M.: Cytoplasmic inheritance of the organization of the cell cortex in *Paramecium aurelia*. *Proc. Nat. Acad. Sci. U.S.* **53**: 275–282 (1965).

5. de Terra, N.: Cortical control of macronuclear positioning in the ciliate *Stentor*. *J. Exp. Zool.* **216**: 367–376 (1981).

6. Fauré-Fremiet, E.: Doublets homopolaires et régulation morphogénétique chez le Cilié *Leucophrys patula*. *Arch. Anat. Microscop. Morphol. Exp.* **37**: 183–203 (1948).

7. Frankel, J.: Positional information in unicellular organisms. *J. Theor. Biol.* **47**: 439–481 (1974).

8. Frankel, J.: Propagation of cortical differences in *Tetrahymena*. *Genetics.* **94**: 607–623 (1980).

9. Frankel, J.: Global patterning in single cells. *J. Theor. Biol.*: **99**: 119–134 (1982).

10. Frankel, J., Jenkins, L.M.: A mutant of *Tetrahymena thermophila* with a partial mirror-image duplication of cell surface pattern. II. Nature of genic control. *J. Embryol. Exp. Morph.* **49**: 203–227 (1979).

11. Frankel, J., Nelsen, E.M.: Discontinuities and overlaps in patterning within single cells. *Phil. Trans. R. Soc. Lond. B* **295**: 525–538 (1981).

12. Frankel, J., Nelsen, E.M., Martel, E.: Development of the ciliature of *Tetrahymena thermophila*. II. Spatial subdivision prior to cytokinesis. *Devel. Biol.* **88**: 39–54 (1981).

13. French, V., Bryant, P.J., Bryant, S.V.: Pattern regulation in epimorphic fields. *Science* **193**: 969–981 (1976).

14. Grimes, G.W.: Morphological discontinuity of kinetosomes during the life cycle of *Oxytricha fallax*. *J. Cell Biol.* **57**: 229–232 (1973).

15. Grimes, G.W.: An analysis of the determinative difference between singlets and doublets of *Oxytricha fallax*. *Genet. Res. Cambr.* **21**: 57–66 (1973).

16. Grimes, G.W.: Laser microbeam induction of incomplete doublets of *Oxytricha fallax*. *Genet. Res. Cambr.* **27**: 213–226 (1976).

17. Grimes, G.W.: Pattern determination in hypotrich ciliates. *Am. Zool.* **22**: 35–46 (1982).

18. Grimes, G.W., Adler, J.A.: Regeneration of ciliary pattern in longitudinal fragments of the hypotrichous ciliate, *Stylonychia. J. Exp. Zool.* **204**: 57–80 (1978).

19. Grimes, G.W., Hammersmith, R.L.: Analysis of the effects of encystment and excystment on incomplete doublets of *Oxytricha fallax. J. Embryol. Exp. Morph.* **59**: 19–26 (1980).

20. Grimes, G.W., L'Hernault, S.W.: Cytogeometrical determination of ciliary pattern formation in the hypotrich ciliate *Stylonychia mytilus. Devel. Biol.* **70**: 372–395 (1979).

21. Grimes, G.W., McKenna, M.E., Goldsmith-Spoegler, C.M., Knaupp, E.A.: Patterning and assembly of ciliature are independent processes in hypotrich ciliates. *Science* **209**: 281–283 (1980).

22. Jerka-Dziadosz, M.: An analysis of the formation of ciliary primordia in the hypotrichous ciliate *Urostyla weissei*. II. Results from ultraviolet microbeam irradiation. *J. Exp. Zool.* **179**: 81–88 (1972).

23. Jerka-Dziadosz, M. Cortical Development in *Urostyla* II. The role of positional information and preformed structures in formation of cortical pattern. *Acta Protozool.* **12**: 239–274 (1974).

24. Jerka-Dziadosz, M.: The origin or mirror-image symmetry doublets in the hypotrich ciliate *Paraurostyla weissei. Wilheim Roux' Arch.* **192**: 179–188 (1983).

25. Jerka-Dziadosz, M., Frankel, J.: An analysis of the formation of ciliary primordia in the hypotrich ciliate *Urostyla weissei. J. Protozool.* **16**: 612–637 (1969).

26. Jerka-Dziadosz, M., Frankel, J.: A mutant of *Tetrahymena thermophila* with a partial mirror image duplication of cell surface pattern. I. Analysis of the phenotype. *J. Embryol. Exp. Morph.* **49**: 167–202 (1979).

27. Kaczanowski, A.: Gradients of proliferation of ciliary basal bodies and the determination of the position of the oral primordium in *Tetrahymena. J. Exp. Zool.* **204**: 417–430 (1978).

28. Kumazawa, H.: Homopolar grafting in *Blepharisma japonicum. J. Exp. Zool.* **207**: 1–16 (1979).

29. Lynn, D.: The organization and evolution of microtubular organelles in ciliated protozoa. *Biol. Rev.* **56**: 243–292 (1981).

30. Nanney, D.L.: Cortical integration in *Tetrahymena*: an exercise in cytogeometry. *J. Exp. Zool.* **161**: 307–317 (1966).

31. Nanney, D.L.: Corticotypic transmission in *Tetrahymena. Genetics* **54**: 955–968 (1966).

32. Nanney, D.L.: Cortical slippage in *Tetrahymena. J. Exp. Zool.*, **166**: 163–169 (1967).

33. Nanney, D.L.: The constancy of cortical units in *Tetrahymena* with varying numbers of ciliary rows. *J. Exp. Zool.* **178**: 177–181 (1971).

34. Nanney, D.L.: Patterns of basal body addition in ciliary rows of *Tetrahymena. J. Cell. Biol.* **65**: 503–512 (1975).

35. Nelsen, E.M.: Transformation in *Tetrahymena thermophila*: development of an inducible phenotype. *Devel. Biol.* **66**: 17–31 (1978).

36. Nelsen, E.M., Frankel, J.: Regulation of corticotype through kinety insertion in *Tetrahymena. J. Exp. Zool.* **210**: 277–288 (1979).

37. Ng, S.F.: Analysis of contractile vacuole pore morphogenesis in *Tetrahymena pyriformis* by 180°-rotation of ciliary meridians. *J. Cell Sci.* **25**: 233–246 (1977).

38. Ng, S.F.: The precise site of origin of the contractile vacuole pore in *Tetrahymena* and its morphogenetic implications. *Acta Protozool.* **18**: 305–312 (1979).

39. Ng, S.F.: Origin and inheritance of an extra band of longitudinal microtubules in the *Tetrahymena* cortex. *Protistologica* **15**: 5–15 (1979).

40. Ng, S.F., Frankel, J.: 180°-rotation of ciliary rows and its morphogenetic implications in *Tetrahymena pyriformis*. *Proc. Nat. Acad. Sci. U.S.*, **74**: 1115–1119 (1977).

41. Ng, S.F., Williams, R.J.: An ultrastructural investigation of 180°-rotated ciliary meridians in *Tetrahymena pyriformis*. *J. Protozool.* **24**: 257–263 (1977).

42. Nüsslein-Volhard, C., Wieschaus, E.: Mutations affecting segment number and polarity in *Drosophilia*. *Nature* **287**: 795–801 (1980).

43. Sonneborn, T.M.: Genetic studies in *Stenostomum incaudatum*. II. The effects of lead acetate on the hereditary constitution. *J. Exp. Zool.* **57**: 409–439 (1930).

44. Sonneborn, T.M.: Does preformed cell structure play an essential role in cell heredity? In: *The Nature of Biological Diversity*. Allen, J.M. (ed), pp. 165–221. New York: McGraw-Hill (1963).

45. Sonneborn, T.M.: The differentiation of cells. *Proc. Nat. Acad. Sci. U.S.* **51**: 915–929 (1964).

46. Sonneborn, T.M.: Gene action in development. *Proc. Roy Soc. Lond. B* **176**: 347–366 (1970).

47. Sonneborn, T.M.: Positional information and nearest neighbor interactions in relation to spatial patterns in ciliates. *Ann Biol.* **14**: 565–584 (1975).

48. Stumpf, H.: Über die Lagebestimmung der Kutikularzonen innerhalb eines Segmentes von *Galleria mellonella*. *Devel. Biol.* **16**: 144–167 (1967).

49. Tartar, V.: Pattern and substance in *Stentor*. In: *Cellular Mechanisms of Differentiation and Growth*. Rudnick, D. (ed.), pp. 73–100. Princeton: Princeton University Press (1956).

50. Tartar, V.: Reconstitution of minced *Stentor coeruleus*. *J. Exp. Zool.* **144**: 187–207 (1960).

51. Tartar, V.: *The Biology of Stentor*. London: Pergamon Press (1961).

52. Tartar, V.: Stentors in dilemmas. *Z. Allg. Mikrobiol.* **6**: 125–134 (1966).

53. Tchang Tso-run, Pang Yan-bin: Conditions for the artificial induction of monster jumelles of *Stylonychia mytilus* which are capable of reproduction. *Scientia Sinica* **14**: 1332–1338 (1965).

54. Tchang Tso-run, Pang Yan-bin: The "ciliary pattern": of jumelle *Stylonychia* and its genetic behavior. *Scientia Sinica* **24**: 122–129 (1981).

55. Tchang Tso-run, Pang Yan-bin: Formation of jumelles and dorsal connected doublets in hypotrich ciliates. *Kexue Tongbao* (Foreign Language Edition) **27**: 1214–1217 (1982).

56. Tchang Tso-run, Shi Xin-bai, Pang Yan-bin: An induced monster ciliate transmitted through three hundred and more generations. *Scientia Sinica* **13**: 850–853 (1964).

57. Uhlig, G.: Entwicklungsphysiologische Untersuchungen zur Morphogenese von *Stentor coeruleus* Ehrbg. *Arch. Protistenk.* **105**: 1–109 (1960).

58. Webster, G.: Morphogenesis and pattern formation in *Hydra*. *Biol. Rev. Cambr.* **46**: 1–64 (1971).

59. Weissmann, A.: Transmissible variations arise from modifications of a hereditary sub-

stance. In: *Source Book in Animal Biology*. Hall, T.S. (ed.), pp. 640–646. New York: McGraw-Hill (based on a translation of "Über die Vererbung," originally published in Jena, 1883) (1951).

60. Wolpert, L.: Positional information and the spatial pattern of cellular differentiation. *J. Theoret. Biol.* **25**: 1–47 (1969).

61. Wright, D. and Lawrence, P.E.: Regeneration of the segment boundary in *Oncopeltus*. *Devel. Biol.* **85**: 317–327 (1981).

Questions for Discussion with the Editors

1. *What bearing, if any, do you believe your discussion of the uncoupling in hypotrichs of large-scale patterns from local patterns might have on pattern specification in the cells and tissues of multicellular (e.g., vertebrate) organisms?*

The aspect of tissue cells that corresponds most closely to the local patterning system of ciliates is the organization of microtubule arrays around centrioles. While large-scale patterns of ciliates have been presented in this review as possible equivalents of the morphogenetic fields of multicellular systems, it is quite possible that such patterns might also exist within individual cells of multicellular organisms. There is evidence for extrinsic control of centriolar orientation in a number of cell types; for example, in neuroblastoma cells, the numerous centrioles are dispersed and organize separate microtubule arrays in interphase and prophase, aggregate into a bipolar system of two centriolar aggregates in later stages of mitosis, and are gathered together into a single compound microtubule organizing center located at the base of the presumptive axon prior to neuronal differentiation (M. Kirschner, in *Developmental Order: Its Origin and Regulation*, S. Subtelny and P.B. Green, eds., pp. 117–132). The important point to emphasize is not any known similarity between the large-scale patterning systems of ciliates and of cells such as differentiating neurons; it is rather that beyond the problem of how cytoskeletal elements are organized in relation to centrioles there lurks the deeper question of how the centrioles themselves are organized in relation to each other and to the overall geometry of the cell.

2. *Would you care to speculate on what types of components might substitute for diffusible morphogens in the analogy you propose between Wolpert's positional information formulation (in which you substitute assembly for gene expression) and your combined views for organelle pattern specification?*

Both site and mechanism of establishment of positional information in ciliates are unknown. Two candidates for the site are the plasma membrane and the fibrogranular layer directly underneath that membrane ("epiplasm"). The mechanism might be diffusion of mobile elements within the plasma membrane, or propagation of structural configurations within the membrane or epiplasm. The apparent reversed intercalation of pattern borders suggests to me that observed gradients might be secondary expressions of different and more continuous underlying spatial parameters; hence even the visualization of a chemical gradient within the membrane or epiplasm would not necessarily reveal the form of the underlying spatial control.

Although the mechanism of local assembly in response to positional cues is unknown, it poses no insuperable theoretical difficulty. Assembly of viral proteins into viral structures is totally dependent upon the appearance of correct structural "cues" within infected bacterial cells. One can imagine that cell-surface positional information, whatever its nature, might regulate the location of comparable "assembly cues" for cell structures. I have no idea, however, of the specific nature of these cues.

CHAPTER 8

Pattern Formation During Embryogenesis of the Multicellular Organism *Volvox*

Manfred Sumper

THE CENTRAL PROBLEM of embryonic development is how genetic information can be translated in a reliable manner to create spatial patterns of cellular differentiation. Development of an embryo requires that cells know where they are in relation to one another (21). Increasing experimental evidence indicates that the cell surface is involved in this social behavior of cells (3, 5). The classical experiment performed by Roux in 1888 (10) perhaps was the first hint for an essential control function of cell to cell contacts: one blastomere in an amphibian 2-cell embryo was killed by treatment with a hot needle. If the killed blastomere remained in contact with the other blastomere only a half-side embryo developed. In sharp contrast, as found later (9), removal of the killed blastomere reprogrammed the surviving blastomere to develop into a complete embryo.

Any promising experimental approach to understanding this intercellular communication system depends on the availability of a simple model organism. By definition, embryonic development of an organism with only two different types of cells would represent the ideal model. Fortunately, such an organism does exist in the plant kingdom: The multicellular green alga *Volvox*. The asexual organism consists of only two cell types: somatic and reproductive. In the case of *V. carteri* about 2000–4000 somatic cells are located as a single layer on the surface of a hollow sphere (spheroid). Eight to sixteen reproductive cells (gonidia) are positioned in a highly regular pattern within one hemisphere of the spheroid.

A new individual is formed from each reproductive cell in a series of cleavages. In this developing asexual embryo, differentiation into somatic and reproductive cells is seen at the division from 32 to 64 cells. At this stage, 16 out of the 32 cells undergo unequal cleavage, forming a small somatic and a large reproductive initial (Fig. 1). While cell division ceases in the reproductive initials, the remaining

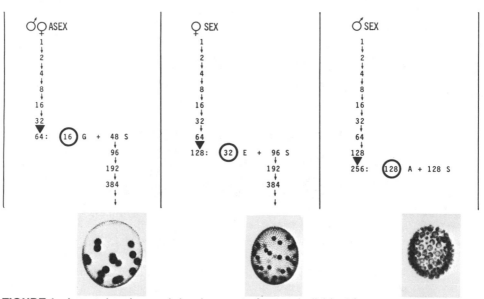

FIGURE 1. Asexual and sexual development of a new individual from a reproductive cell in *Volvox carteri*. Somatic and reproductive cells are differentiated during the 32-64 (asexual), 64-128 (female) or the final division (male). G gonidia (reproductive cells); S somatic cells; E egg cells; A androgonidia (sperm-producing cells).

cells continue their divisions, and finally differentiate into somatic cells. At the termination of cell divisions, the embryo consists of thousands of small somatic cells and 16 large reproductive cells (8, 12, 13). Obviously, some sort of a counting mechanism exists in the developing embryo telling a cell that the embryo is in the 2-, 4- . . . or 32-cell stage. This counting mechanism exhibits a certain degree of flexibility: Under less than optimal growth conditions, the differentiating cell cleavage may occur at the division from 16 to 32 cells. In this case, 8 out of the 16 cells undergo unequal cleavage, thus producing only 8 reproductive initials. On the other hand, a delay of the differentiating cleavage is observed under the influence of a specific glycoprotein, the sexual hormone (14). For instance, in the strain HK 10 of *Volvox carteri* the action of this hormone delays the differentiating cell cleavage until the division from 64 to 128 cells. At this stage, again only half of the embryonic cells undergo unequal cleavage and form 32 reproductive initials. These reproductive cells differentiate to egg cells producing a sexual (female) spheroid with its typical appearance (Fig. 1).

In a male strain, the differentiating cleavage is even more delayed under the influence of the sexual hormone. Differentiation into sperm-producing cells (androgonidia) and somatic cells occurs at the final division, which in male embryos is encountered at the 128- or 256-cell stage.

Besides this cell-counting problem, the embryogenesis of Volvox presents the problem of pattern formation in a fascinating simplicity. The spatial arrangement of the reproductive initials within the developing embryo is exactly controlled. During the whole process of embryogenesis, the reproductive initials remain in the same relative positions where they were formed during early embryogenesis, i.e., the relative positions remain exactly the same in the mature spheroid. The fidelity of this pattern formation in Volvox embryogenesis is documented in Fig. 2, comparing photographs of spheroids containing 8, 12, and 16 reproductive cells, respectively. Eight gonidia are always arranged in two parallel rectangles, turned by 45° against each other. Twelve gonidia are similarily arranged in three parallel rectangles. Sixteen gonidia are arranged in two parallel rectangles (8 gonidia) with the residual 8 gonidia positioned on a zigzag line in between the rectangles.

In the following section a biochemical model is described that is able to predict correctly the characteristics of *Volvox* embryogenesis with respect to pattern formation. In particular, the model offers an explanation to the following issues:

1. What is the counting mechanism which tells a cell that the embryo is in the 2-, 4-, 8- or 32-cell stage?
2. Why do only 16 cells out of the 32-cell embryo undergo the differentiating cell cleavage?
3. What is the control mechanism defining the relative positions of the gonidial initials within the developing embryo?

In addition, experiments performed to test the predictions of this model are described and it is shown that a sulfated cell surface glycoprotein meets the properties postulated by the model.

A Biochemical Model Explaining Pattern Formation During Embryogenesis

Since the publication of this model in 1979 (16), we collected a number of experimental data in favor of these ideas. In turn, these experimental data now allow for a more detailed formulation of the originally proposed model. In its more specified version the model is based on the following three assumptions:

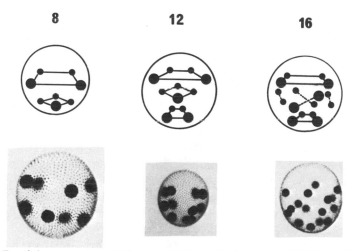

FIGURE 2. Spatial arrangement of reproductive cells in asexual *Volvox* spheroids containing 8, 12, and 16 reproductive cells, respectively.

1. The embryonic cells synthesize a family of structurally related cell surface glycoproteins (denoted in the following as SG 1, SG 2, . . .) which are able to mediate cell to cell contacts. (Contact formation should be established either directly by dimerization of these surface molecules or via receptor molecules.)
2. The synthesis of these glycoproteins is controlled in a hierarchical manner: A new member of the glycoprotein family (e.g., SG 2) is only synthesized if the former member (SG 1) is quantitatively engaged in contact formation.
3. The presence (or absence) of a given type of cell-to-cell contact on the surface of an embryonic cell triggers its determination and subsequent differentiation.

On the basis of these assumptions, the schematic drawings of Fig. 3 demonstrate how the embryonic cells become sorted into subclasses during early embryogenesis: The uncleaved gonidium produces the first member (SG 1) of the contact-forming glycoprotein family. As soon as septum formation of the first division is initiated, a certain aliquot of the SG 1 molecules becomes trapped within the cell-to-cell contact region, simply by random lateral diffusion within the membrane and subsequent contact (complex-) formation. (Alternatively, SG 1 could be stored within the cell and incorporated specifically at the site of the next cleavage furrow.) In the second division, two more cell-to-cell contacts are formed, again consuming a corresponding aliquot of the free (i.e., uncomplexed) SG 1 molecules. Since it is reasonable to assume that the operative cell-to-cell contact area of the first cleavage is not divided by the cleavage planes of the second division, the cells of the 4-cell embryo will be arranged as shown in Fig. 3. Redrawing of the four cells in a more symmetrical configuration results in an arrangement typical of that seen by light microscopic observation: two cells touch one another while the other two have no contact area in common. Light microscopic observation reveals that the 8-cell embryo is shaped like a cup: The four cells at the posterior end of the embryo remain in close contact, while the four anterior cells do not touch each other and form a pore. The model explains this characteristic configuration: the four additional contact sites established during the third division are shown in the schematic drawing of Fig. 3. Folding up this two-dimensional representation of the 8-cell embryo to a hollow sphere necessarily creates a pore, because no contacts have been established between the four anterior cells.

Since the number of cell contacts increases exponentially during embryogenesis, the pool of the free SG 1-molecules becomes exhausted at a sharply defined stage of division. Thereby, the membrane areas not involved in contact formation become cleared from all free SG 1. According to assumption 2 engagement of all SG 1 into cell-to-cell contact formation triggers the production of the second member (SG 2) of this cell surface glycoprotein family. If, for instance, this transition point is reached at the stage of the 8-cell embryo, then the embryonic cells become sorted out into two subclasses by the next cleavage: 8 cells of the 16-cell embryo are equipped with two types of contacts containing either SG 1 or SG 2, whereas the remaining 8 cells completely lack contacts containing SG 1. If, according to assumption (3) of the model, SG 1 suppresses the differentiating cell cleavage

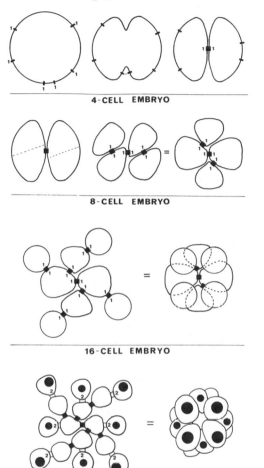

FIGURE 3. Embryonic cells become sorted out into subclasses during early embryogenesis by the formation of cell-to-cell contacts. For details see text; black contacts contain surface glycoprotein 1, white contacts contain surface glycoprotein 2.

whereas SG 2 does not, the characteristics of early *Volvox* embryogenesis can be explained:

1. At the division of the 16-cell embryo, only 8 embryonic cells can undergo the differentiating cell cleavage forming 8 large gonidial initials.
2. The cell lineage and consequently the relative positions of those embryonic cells performing the unequal cleavage can easily be predicted: Cells lacking the SG 1 contacts (and thus producing gonidial initials) are marked with a black dot in the two-dimensional representation of the 16-cell embryo in Fig. 3. By transforming this two-dimensional pattern into a hollow sphere, it follows that the 8 gonidial initials will be arranged in two parallel rectan-

TABLE 1.   Early embryogenesis produces subclasses of cells with respect to their cell-to-cell contacts:

STAGE IN EMBRYOGENESIS	NUMBER OF CELLS IN EACH SUBCLASS
2-Cell	2 Cells with one SG 1 contact
4-Cell	2 Cells with two SG 1 contacts
	2 Cells with one SG 1 contact
8-Cell	2 Cells with three SG 1 contacts
	2 Cells with two SG 1 contacts
	4 Cells with one SG 1 contact
16-Cell	2 Cells with three SG 1 contacts and one SG 2 contact
	2 Cells with two SG 1 contacts and one SG 2 contact
	4 Cells with one SG 1 contact and one SG 2 contact
	8 Cells with *no* SG 1 contact and one SG 2 contact
	↳ *signals unequal cleavage*
32-Cell	2 Cells with three SG 1 contacts, one SG 2 and one SG 3 contact
	2 Cells with two SG 1 contacts, one SG 2 and one SG 3 contact
	4 Cells with one SG 1 contact, one SG 2 and one SG 3 contact
	8 Gonidial initials
	16 Cells with one SG 3 contact

gles, turned by 45° against each other. One rectangle will be located near the equator and while the other rectangle will be located in the anterior half of the sphere, the posterior half of the embryo will contain only somatic cells. This is exactly the pattern found in those *Volvox* spheroids containing only 8 gonidia.

However, an important aspect of the late *Volvox* embryogenesis remains to be discussed: the differentiating (unequal) cell cleavage is observed only at one stage of division (e.g., in the 16-cell embryo). In all subsequent stages of division, only equal cleavages are involved. This means that the suppression of unequal cleavages must be re-established during the whole period of late embryogenesis. In terms of the model, this requirement could easily be met if the established SG 2 contacts immediately trigger synthesis of the next member SG 3 of the surface glycoprotein family. SG 3 like SG 1 has to suppress unequal cell cleavage, in order to guarantee the developmental program of *Volvox*. In this program, no further member of the SG family appears to be required, so that SG 3 should be the only type of these contact forming glycoproteins involved during late embryogenesis (i.e., complexed SG 3 no longer triggers synthesis of the member SG 4).

Table 1 summarizes, in terms of the model, the developmental program for *Volvox* embryos producing 8 gonidia. Development of embryos containing 8 gonidia occurs under less than optimal growth conditions. Under favorable conditions, 16 gonidia are formed by unequal cleavages at the division of the 32-cell stage. R. Starr (13) has carefully compared the patterns of cell lineages in embryos showing differentiation at the two different times in embryogenesis: he clearly demonstrated that the delay in differentiation until the division of the 32-cell stage has involved only an equational division of the same cells which would have been in-

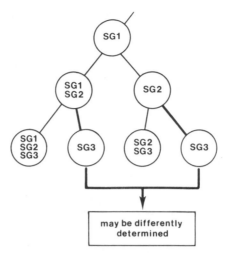

FIGURE 4. Cells having only SG 3 contacts are derived from different stem cells, possibly causing a nonequivalence of these daughter cells.

volved in differentiation, had it occurred at the 16-cell stage.[1] This fact indicates that in any case the process of sorting out two subclasses of embryonic cells is terminated at the 16-cell stage.

In the example outlined so far, the model was adapted to the situation found in Volvox embryogenesis. It is obvious that the processes assumed by the model can create much more complex patterns depending on (1) the available amount of each member of the surface glycoprotein family with determines the number of contacts equipped with that component, and (2) the total number of different members of the glycoprotein family being involved. In addition, it is reasonable to assume that in more complex pattern formation processes two daughter cells having, for instance, only SG 3 contacts may still be differently programmed depending on whether they are derived from a SG 2 containing or a SG 1 + SG 2 containing stem cell (Fig. 4). Clearly by this mechanism, an exponentially growing number of differently programmed embryonic cells would arise. Before discussing some more aspects of the model, I would like to summarize the experimental data supporting these ideas.

Experimental Results

In 1979 we started a search for a cell-surface component with properties meeting the requirements for being involved in a control function as forseen by the model. These initial experiments (18) ended up with the characterization of a highly sulphated surface glycoprotein with an apparent molecular weight of 185 k daltons ("SSG 185"). SSG 185 shows at least three of the properties postulated by the

---

[1]Starr's original conclusions were recently questioned [Green, K.J., Kirk, D.L.: *J. Cell Biol.* **91**: 743–755 (1981)] A re-examination of living embryos by R. Starr showed that this criticism has no basis. Meanwhile Green and Kirk have withdrawn their interpretations concerning cell lineages in *Volvox* (*J. Cell Biol.*, in press).

model: (1) SSG 185 is located on the cell surface; (2) its production rate is prominent only during the short period of embryogenesis (19) and most important; (3) SSG 185 represents a family of different but related substances. The different members of this family are synthesized in a defined order which perfectly matches the developmental program. This point is documented by the experimental results summarized in Figs. 5–7: Short-pulse labeling experiments with $^{35}SO_4^{2-}$ performed over the whole period of embryogenesis of the asexually developing strain HK 10 reveals that $^{35}SO_4^{2-}$ = incorporation into SSG 185 drops to a minimum at the 16-cell stage, i.e., immediately before the differentiating cell cleavage. Beyond the 32-cell stage, $^{35}SO_4^{2-}$-incorporation again increases to a high level. In addition, the apparent molecular weight of SSG 185 systematically drifts towards a lower value until

STAGES IN THE CLEAVAGE

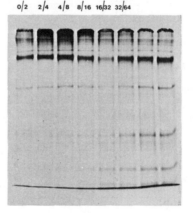

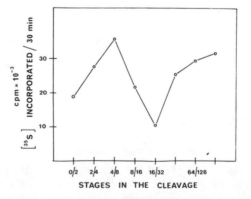

STAGES IN THE CLEAVAGE

FIGURE 5. $^{35}SO_4^{2-}$ incorporation into SSG 185 during embryogenesis of asexual spheroids (strain HK10). After the onset of gonidial cleavage in a highly synchronized population a pulse labeling experiment (30 min) was performed at each stage of division, i.e., every 60 min. (Upper) Total membrane fractions from embryos at the indicated cleavage stages were applied to SDS polyacrylamide (6%) gels and visualized by fluorography. (Lower) SSG 185 bands were cut out of the gels and the incorporated radioactivity was measured in a scintillation counter. Reproduced from (18).

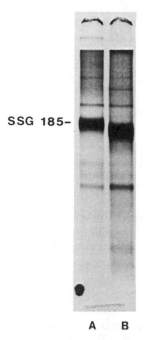

SSG 185–

A   B

FIGURE 6. Comparison of SSG 185 synthesized by 2-cell embroys (lane A) and 64-cell embryos (lane B) on a SDS polyacrylamide (5%) gel.

STAGES IN THE CLEAVAGE

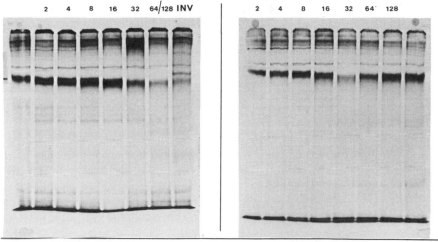

male embryos   female embryos

FIGURE 7. $^{35}SO_4^{2-}$-incorporation into SSG 185 during embryogenesis of sexual spheroids. Stages of cleavage are indicated at the top. Inv., inversion stage. (Left) Sexually induced male embryos. (Right) Sexually induced female embryos. Pulse labeling as in Fig. 5. Total membrane fractions were applied to SDS polyacrylamide (6%) gels and visualized by fluorography. Reproduced from (18).

the 16-cell stage is reached and then remains constant during late embryogenesis. This drift in the molecular weight is more clearly demonstrated in the better resolving gel system shown in Fig. 6, comparing SSG 185 produced during the 2-cell stage with that produced during late embryogenesis.

Under the influence of the sexual hormone, the differentiating cell cleavage is shifted towards later stages: in female strains, the unequal cleavage occurs when the 64-cell embryo divides. Thereafter equal cell cleavages are continued. In male strains, the differentiating cell cleavage is the last one which usually occurs at the stage of the 128-cell or 256-cell embryo. For these reasons it was very interesting to see whether these altered developmental programs lead to (or, in the sense of the model are "caused by") corresponding changes in the SSG 185 production. It is obvious from the results shown in Fig. 7 that this is indeed the case: In male embryos the minimum value of both the apparent molecular weight and the $^{35}SO_4^{2-}$ incorporation is only reached towards the end of embryogenesis. The quite different SSG 185 production pattern of sexually induced females again perfectly reflects the corresponding developmental program: Minimum $^{35}SO_4^{2-}$ incorporation is seen immediately before the differentiating cell cleavage at a time where in addition the drift in the molecular weight of SSG 185 stops. Beyond this stage, the apparent molecular weight remains constant until the end of the embryonic divisions.

The results described so far make the SSG 185 molecules a likely candidate for being involved in the control of embryogenesis. For this reason, we are currently studying the chemical nature of the stage-specific modifications in the SSG 185 molecule. Our preliminary results indicate that the modification does not involve the protein portion of SSG 185, but resides in a highly sulfated saccharide structure containing the neutral sugars mannose and arabinose. Comparing this structural element isolated from SSG 185 synthesized during early and late embryogenesis, respectively, by ion exchange chromatography reveals marked differences in charge density. This suggests that it is the sulfation pattern which is modified during embryogenesis, i.e., the shift in molecular weight could be attributed to the posttranslational enzymatic actions of glycosyltransferases and/or sulfokinases (20).

Model and Experiments

The experimental results outlined in this chapter strongly support the following interpretation: SSG 185 molecules represent a family of at least three different species 1, 2, and 3, with decreasing apparent molecular weights. Production of the different species takes place in a sequential order as schematically shown in Fig. 8. There is experimental evidence that species 2 is less highly sulfated than species 1 and 3 for the following reasons: $^{14}C-CO_2$ pulse labeling patterns of SSG 185 molecules over the whole period of embryogenesis show a fairly constant incorporation, contrasting the sharp minimum incorporation of $^{35}SO_4^{2-}$ immediately before the unequal cleavage. If species 2 fulfills the function of SG 2 in our model (namely being the contact-forming glycoprotein at the stage of division preceding the differentiating one), then both the drift in molecular weight and the minimum of $^{35}SO_4^{2-}$ incorporation are explained. Beyond the differentiating cleavage neither a drift in molecular weight nor a further minimum in $^{35}SO_4^{2-}$ incorporation should be found because species 3 is the only species synthesized during late embryogenesis.

Figure 8 proposes a scheme to explain how the sequential production of species 1, 2, and 3 of SSG 185 molecules could be achieved: The uncleaved reproductive cell synthesizes SG 1. As soon as the first cleavage furrow is formed, SG 1 molecules becomes deposited in the cell-to-cell contact area establishing cell-to-cell contacts by complex formation. The complexed as well as the free form of SG 1 inhibits its own production, whereas only the complexed form activates synthesis of SG 2 (e.g., via a regulatory subunit of a glycosyltransferase or sulfokinase, see above). Thus SG 1 synthesis will be replaced by SG 2 production. By the same mechanism, synthesis of SG 3 is switched on subsequently. If complexed SG 3 does not inhibit its own synthesis, this species remains the only species during late embryogenesis. Several hours after the end of *Volvox* embryogenesis the embryonic cells become separated from each other by extensive synthesis of extracellular sheath material. This spatial separation most likely also breaks embryonic cell-to-cell contacts. As a consequence, inhibition of SG 1 synthesis no longer persists, so that the synthetic machinery becomes reprogrammed to produce SG 1 for the next life cycle. This re-initiation of synthesis of the high molecular weight form is indeed observed experimentally at about the time of cell separation as demonstrated in Fig. 9, showing $^{35}SO_4^{2-}$ pulse labeling patterns obtained at different times of the total life cycle.

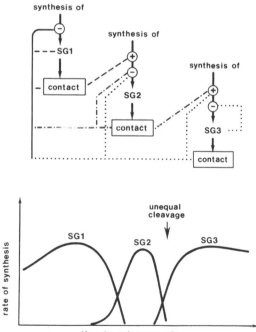

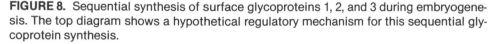

FIGURE 8. Sequential synthesis of surface glycoproteins 1, 2, and 3 during embryogenesis. The top diagram shows a hypothetical regulatory mechanism for this sequential glycoprotein synthesis.

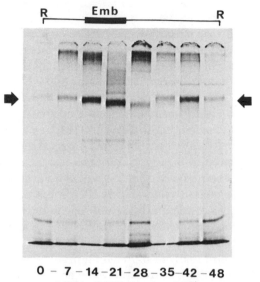

FIGURE 9. Time-dependent changes in the apparent molecular weight of the sulfated surface glycoprotein (SSG 185, arrow) during the life cycle of *Volvox* (HK10 asexual). $^{35}SO_4^{2-}$-pulsed SSG 185 components were compared on SDS polyacrylamide (6%) gels at the times indicated. The developmental program is shown on top: R, release; Emb, embryogenesis.

Concluding Remarks

The experimental results described in the preceding section suggest that SSG 185 is a likely candidate to be involved in the determinative mechanism according to this model. Alternatively, the observed correlation could be a consequence of the determinative process. However, since the observed modifications of SSG 185 precede the onset of the differentiating cell cleavage, this latter alternative is less probable. One crucial property postulated by the model remains to be established for the SSG 185 molecules, namely their capacity to be deposited within the cell-to-cell contact area. Immunofluorescent techniques should resolve this issue:

The model offers a biochemical mechanism for a reliable determination of embryonic cells at a defined stage of division. The proposed mechanism offers another important advantage; there is no requirement for an *a priori* polarity in the uncleaved reproductive cell. Instead, the mechanism not only produces subclasses of cells in defined numbers and positions but also produces polarity within the embryo as demonstrated in Fig 3. For instance, the embryonic cells located at the posterior pole of the embryo have the highest number of SG 1 containing contacts. Towards the anterior pole of the embryo, this number declines to zero, thus establishing a gradient with respect to SG 1 molecules within the extracellular matrix. Gradient formation based on developing cell-to-cell contacts appears much more attractive than a mechanism based on diffusion of low molecular weight substances as postulated by the classical gradient theories. In particular, this mechanism would explain the relative stability of early embryogenesis against mechanical or physical disturbances, a fact known since the beginnings of experimental embryology (reviewed in 11) and hardly to be explained by diffusion-based models.

Outlining the ideas of the model as in Fig. 3, the assumption was made that each newly formed cleavage plane does not divide the previously established contact. In other words, a given contact is preserved only in one of two daughter cells. This requirement could easily be fulfilled if the operative cell-to-cell interactions are mediated by small portions of the cell surface rather than the entire contact area. Experimental evidence from the mating reaction in *Chlamydomonas* makes this a realistic assumption (4). It should be stressed, however, that this assumption is not essential for the model, If, for instance, the operative cell-to-cell contact area is much more extended, the contact zones would no longer become distributed in an all or none fashion among daughter cells. Nevertheless, the extracellular matrix space becomes differentiated during cell cleavages with respect to the spatial distribution of the SG 1, 2, . . . molecules defined in the model, as depicted in the drawings of Fig. 10. While the extracellular matrix becomes a more and more chemically differentiated network, the embryonic cells become embedded in differently composed microenvironments, which in turn would determine the differentiation state. In the example of Fig. 10 adapted to Volvox embryogenesis (i.e., assuming that the supply of SG 1 is consumed in the third division), again exactly those cells of the 16-cell embryo become cleared from contacts to SG 1 molecules, which are known to undergo the unequal cleavage in the next division.

In recent years, a great amount of work primarily with animal sulfated polysaccharides—heparin, chondroitin sulfates and dermatan sulfates—has suggested important physiological functions for these substances (17). Dietrich et al. (1, 2) proposed that sulfated saccharides had a role in cell recognition and adhesiveness in animal cells. Sulfated polysaccharides were also implicated in the embryogenesis of sea urchins; it is assumed that synthesis of sulfated polysaccharides is an indis-

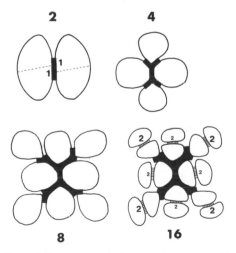

FIGURE 10. Differentiation of the extracellular matrix space with respect to the spatial distribution of SG1-(black) and SG 2-(striped) contacts during early embryogenesis.

pensable step for postgastrular development (6, 7, 15). Taken together, these experimental observations make it likely that in animal systems sulfated polysaccharides are involved in an embryonic control function as well as it is postulated here for *Volvox*.

General References

KOCHERT, G.: Differentiation of Reproductive Cells in *Volvox carteri*. *J. Protozool.* **15**:(3) 438–452 (1968).

STARR, R.C.: Control of differentiation in *Volvox*. *Dev. Biol. Suppl.* **4**: 59-100. Twenty-ninth Symposium Soc. Dev. Biol. (1970).

References

1. Dietrich, C.P., Càssaro, C.M.F.: Distribution of sulfated mucopolysaccharides in invertebrates. *J. Biol. Chem.* **252**: 2254–61 (1977).

2. Dietrich, C.P., Sampaio, L.O., Toledo, O.M.S., Càssaro, C.M.F.: Characteristic distribution of sulfated mucopolysaccharides in different tissues and their respective mitochondria. *Biochem. Biophys. Res. Commun.* **75**: 329–336 (1977).

3. Frazier, W., Glaser, L.: Surface components and cell recognition. *Annu. Rev. Biochem.* **48**: 491–523 (1979).

4. Goodenough, U.W.: Adair, S.W.: Membrane adhesions between *Chlamydomonas* gametes and their role in cell-cell interactions. in: *The cell surface: mediator of developmental processes*. Subtelny, S. (ed.), pp. 101–112. New York: Academic Press (1980).

5. Harrison, F.L., Chesterton, C.J.: Factors mediating cell-cell recognition and adhesion. *FEBS Lett.* **122**: 157–165 (1980).

6. Kinoshita, S.: Heparin as a possible initiator of genomic RNA synthesis in early development of sea urchin embryos. *Exp. Cell Res.* **64**: 403-411 (1971).

7. Kinoshita, S., Yoshii, K.: The role of proteoglycan synthesis in the development of sea urchins. *Exp. Cell Res.* **124**: 361–369 (1979).

8. Kochert, G.: Differentiation of Reproductive Cells in *Volvox carteri*. *J. Protozool.* **15**:(3) 438–452 (1968).

9. McClendon, J.F.: The development of isolated blastomeres of the frog's egg. *Am. J. Anat.* **10**: 425–430 (1910).

10. Roux, W.: Über die künstliche Hervorbringung halber Embryonen durch Zerstörung einer der beiden ersten Furchungszellen, sowie über die Nachentwicklung der fehlenden Körperhälfte. *Virchows Arch.* **114**:(Ges. Abh. II) 419–521 (1888).

11. Spemann, H.: Experimentelle Beiträge zu einer Theorie der Entwicklung. Berlin: J. Springer-Verlag (1936). Nachdruck 1968.

12. Starr, R.C.: Cellular Differentiation in *Volvox*. *Proc. Nat. Acad. Sci. USA* **49**: 1082–88 (1968).

13. Starr, R.C.: Control of differentiation in *Volvox*. *Dev. Biol. Suppl.* **4**: 59–100 (1970).

14. Starr, R.C., Jaenicke, L.: Purification and characterization of the hormone initiating sexual morphogenesis in *Volvox carteri* f. *nagariensis* Iyengar. *Proc. Nat. Acad. Sci. USA* **71**: 1050–54 (1974).

15. Sugiyama, K.: Occurrence of mucopolysaccharides in the early development of the sea urchin embryo and its role in gastrulation. *Dev. Growth Differ.* **14**: 63–73 (1972).

16. Sumper, M.: Control of differentiation in *Volvox carteri*: A model explaining pattern formation during embryogenesis. *FEBS Lett.* **107**: 241–246 (1979).

17. Toole, B.P.: Morphogenetic role of glycosaminoglycans in brain and other tissues. In: Neuronal Recognition. Barondes, S.H. (ed.), pp. 275–329. New York: Plenum Press (1976).

18. Wenzl, S., Sumper, M.: Sulfation of a cell surface glycoprotein correlates with the developmental program during embryogenesis of *Volvox carteri*. *Proc. Nat. Acad. Sci. USA* **78**: 3716–20 (1981).

19. Wenzl, S., Sumper, M.: The occurence of different sulphated cell surface glycoproteins correlates with defined developmental events in *Volvox*. *FEBS Lett.*, *in press* (1982).

20. Willadsen, P., Sumper, M.: Sulphation of a cell surface glycoprotein from *Volvox carteri*: Evidence for a membrane-bound sulfokinase working with PAPS. *FEBS Lett.* **139**: 113–116 (1982).

21. Wolpert, L.: Positional information and the spatial pattern of cellular differentiation. *J. Theoret. Biol.* **25**: 1–47 (1969).

Questions for Discussion with the Editors

1. *In many regards, early* Volvox *development resembles early animal embryogenesis. To what extent might pre-existing internal cytoplasmic components such as the germinal cytoplasm of animal cells play a role in the determination of* Volvox *reproductive cells?*

The idea of a germ plasm localized in specific areas in the developing gonidium and acting in the determination of *Volvox* reproductive cells was put forward by G. Kochert.[1] To test this possibility, he used a classical approach: Unilateral irradiation of an uncleaved gonidium with ultraviolet light in the hope of destroying a cytoplasmically localized morphogenetic substance. In these experiments, the uv-treated cells indeed developed to organisms lacking one reproductive cell at the site of irradiation. These results led to the conclusion of a pre-existing pattern of a cytoplasmic morphogen. However, this conclusion is highly likely to be an over-interpretation. This type of experiment does not at all exclude the additional destruction of a variety of uv-labile substances (e.g., RNAs) that homogenously distributed within the cytoplasm. If one or more of such substances (e.g., simply an enzyme engaged in a biosynthetic reaction) are required in some biochemical reaction during development of a reproductive cell, it is a trivial outcome that a blastomere becoming equipped with irradiated cytoplasm is no longer able to differentiate to a reproductive cell.

2. *Would you briefly explain further the reasons for de-emphasizing the role "diffusion of low molecular weight substances" might play in establishing cell determination patterns?*

Gradient patterns developing by diffusion of low molecular substances result from diffusion constants and from spatial parameters of the egg or embryo. It follows that mechanical deformations of the egg (embryo) or physical influences like temperature gradients

[1]Kochert, G.: Sexual hormones and cell differentiation in *Volvox carteri*. In: Eucaryotic Microbes as Model Developmental Systems. O'Day, D.H., Horgen, P.A., (eds.), pp. 235–251, New York: Marcel Dekker, Inc., 1977.

superimposed along the egg axes should in turn interfere with development. However, experimental experience does not confirm these predictions. To the contrary, early embryogenesis is highly insensitive even to extreme mechanical distortions[2] or to superimposed temperature gradients.[3] Such experimental results fail to confirm the idea of diffusion-based gradients. Gradient formation by covalent modulations of the extra-cellular matrix space as discussed in this chapter clearly overcomes these difficulties.

[2]Vogel, O.: Development of Complete Embryos in Drastically Deformed Leafhopper Eggs. *Wilhelm Roux's Archives.* **191**: 134–136 (1982).

[3]Huxley, J.S.: The modification of development by means of temperature gradients. *Wilhelm Roux's Archives.* **112**: 480–516 (1927).

CHAPTER 9

Patterning in Hydra

Patricia Macauley Bode and Hans R. Bode

THE AIM OF ANY overall study in pattern formation is to be able to elucidate the patterning processes underlying the development of a structure such as a limb or an eye, or even a complete organism. This would include discerning the primary and secondary patterning processes, their sequence of occurence, as well as their interactions. Three aspects of the adult hydra make it attractive for study. It is in a dynamic growth state in which patterning goes on continuously, it has an extensive capacity for regeneration, and it is relatively simple. By observing regulation of the tissue after a variety of experimental manipulations, the patterning processes governing its development are gradually being delineated. It may be possible to learn how a few general pattern formation mechanisms can be utilized to form and maintain the pattern in an entire organism.

Hydra is a freshwater coelenterate consisting of a body tube with head and foot at the ends (Fig. 1). The wall of the body is made up of two concentric epithelia separated by a basement membrane. These layers surround the gastric cavity which extends throughout the length of the body and into the tentacles. The head has two distinct regions, a conical or domed hypostome where the mouth opens and a tentacle zone from which five to seven tentacles normally emerge. The foot or basal disc is a round patch of cells which secrete sticky material enabling this structure to serve as a holdfast. Though the animal is a two-dimensional cylindrical shell, its radial symmetry reduces much of the patterning to changes in one dimension.

Development in hydra consists mainly of head formation and foot formation from tissue of the body column. This development is occurring continuously in the adult, and in that sense the animal behaves as a steady-state embryo. When well fed it is in a dynamic state of growth, pattern formation, morphogenesis, and differentiation. All the cells of the body epithelia are proliferating, and as the tissue expands it is displaced along the column towards the head or foot (7). Upon reach-

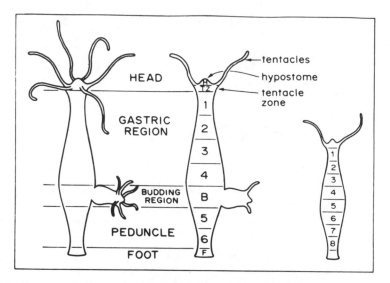

FIGURE 1. Representation of hydra. At left a budding adult. In the middle and at right, schemes for designating parts of the hydra for transplantation and regeneration experiments devised by Wolpert (33) and MacWilliams (15), respectively.

ing the extremities cells differentiate and take part in forming the appropriate structures. Eventually the cells are sloughed from the tips of tentacles, hypostome, and basal disc. Hence, there is the continuing process of nascent noncommitted tissue being patterned to form the head and the foot, and size is maintained by balancing tissue growth with loss.

As a young animal grows larger, cells are also funneled off into a third specialized structure, the bud. Bud pattern formation is initiated approximately two-thirds of the way down the body column, the cell sheet evaginates, and the new axis incorporates more and more tissue as it elongates. After two days, a head begins to form at the distal end, followed a day later by basal disc formation at the point of connection with the parent. Subsequently the bud detaches, resulting in a new individual about one-quarter the adult size.

The embryonic quality of the body tissue and the mechanisms continually operating to pattern head and foot structures are undoubtedly responsible for hydra's most distinctive capability, complete regeneration. If the head and foot are removed from an animal, it will regenerate them both and at the appropriate ends. Any fragment of body tissue down to one-fiftieth of an adult in size can reform into a whole animal having almost normal proportions (3). Even dissociated cells centrifuged into aggregates can reform the tissue layers, regenerate heads and feet, and will eventually separate into complete animals (11).

A characteristic of both extremities is that they have the qualities of an organizer in the same sense that the dorsal lip of an amphibian gastrula is capable of inducing a second embryo when transplanted to another gastrula. This ability of some tissues to control the fate of other tissue has been known in hydra since the early part of the century (5). Pieces of head or basal disc can be transplanted into a host body column and the body tissue will be induced to form a head or basal disc respectively, and part of a secondary body axis. In other words, the two organizers between them have the ability to regulate the development of the entire animal. Therefore, the main aim of pattern formation studies in hydra is to learn how the pattern for the head and the foot can be generated or regenerated.

The second aim of these studies is to understand why the pattern formation processes are initiated in certain positions and not in others. For this the developmental gradients of activation and inhibition play an important role. It appears to be the interactions between these gradients which affect the probability of a pattern formation event occurring at a particular site.

We will describe the hydra pattern formation system in two parts, reflecting these two major topics. In the first part, the evidence for the developmental gradients, their properties, and indications of their roles will be presented. The second part will consider the properties of the pattern formation processes as they occur during regeneration, and what constraints this information puts on models proposed to explain them. The discussion will include the problem of proportion regulation and the secondary patterning of the tentacles.

The Developmental Gradients

Two pairs of gradients have been described, one each for the head and foot. The pair for the head consists of a gradient of head formation capacity, or simply head activation, and a gradient of inhibition of head formation, or head inhibition. Both gradients are maximal in the head, decreasing down the body column. Foot activation and foot inhibition run in the opposite direction with the maxima in the foot. To the extent that it has been tested, the properties of the head and foot gradients are very similar. Only the head pair will be described in detail, since they have been studied more extensively.

The phenomena of activation and inhibition have been known for a long time, but the elucidation of the gradients arose from the original work of Wolpert, Webster and their colleagues (30, 33). This was extended and refined by MacWilliams (8, 15, 16, 17) and most of the detailed properties of the gradients presented here are based on his work.

The Inhibition Gradient

PROPERTIES

Most of the head inhibition emanates from the head which has been demonstrated in a variety of experiments. For example, when a 3-region is transplanted to the middle of the body column of an intact host (Fig. 2), it will induce the formation of a second head with a low frequency of 20%. If the host head is removed at the time of transplantation, the frequency of head induction rises dramatically to 87% (15). The head-removal effect appears to be specific and cannot be mimicked by foot removal or other injury. This result is consistent with the head as the source

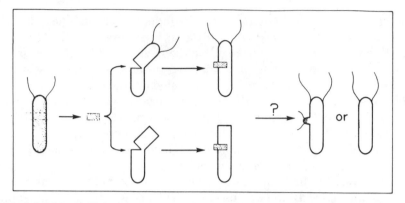

FIGURE 2. Inhibition of head formation by the host head. This format will be used for figures throughout this chapter. Dotted lines indicate where a piece of body column is excised or where an extremity is removed. A small rectangle inserted into the body column indicates where the piece was transplanted into the host. The absence of the head indicates a decapitated host. The possible outcomes follow the question mark: either a head forms or the implant is absorbed.

of an inhibition which suppresses the formation of secondary heads in the body column.

The effectiveness of head inhibition varies down the body column in a graded manner (30, 15). This can be demonstrated by transplanting the 1-region into intact hosts at three positions, one-quarter, one-half, and three-quarters of the way from the head (Fig. 3). The fraction of transplants that formed a head increased with increasing distance from the host head. In other words, the ability of the existing head to inhibit the formation of an additional head was greater when the transplant was close than when it was situated further away (Fig. 4).

When the same experiment was repeated at a higher temperature, 24°C in-

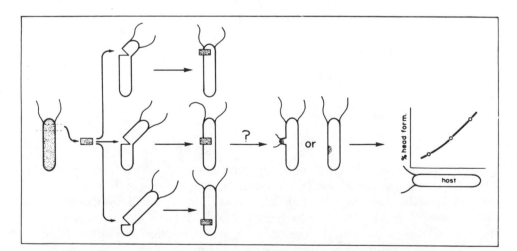

FIGURE 3. Transplantation experiment demonstrating the head inhibition gradient.

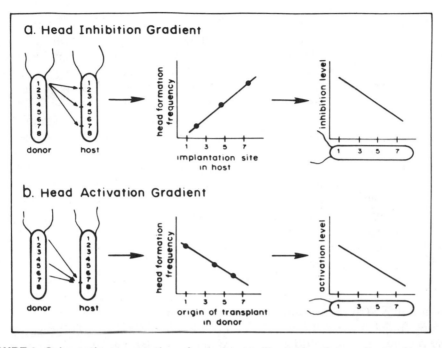

FIGURE 4. Schematic presentation of experiments illustrating the gradients of head inhibition and head activation.

stead of 20°C, the frequency of head formation rose significantly at the one-half and three-quarter positions (15). Inhibition is apparently temperature-dependent, and its range of effectiveness becomes less when the temperature is increased. The data is consistent with a linear gradient of inhibition whose slope is about twice as great at 24° than at 20°.

Head inhibition is apparently a very labile property. Cutting off the head leads to a sharp increase in the fraction of transplants forming heads anywhere along the body column, suggesting that inhibition must disappear rapidly when the source is removed. The kinetics of decay were followed after decapitation, and inhibition was found to reach a minimum at 6 hr (15). By assuming that the breakdown was exponential, the median decay time was determined to be approximately 2 hr.

Another property demonstrated most clearly by Wilby and Webster (31) is that inhibition can be transported in either direction along the body column. The head was removed from the apical end of the animal and grafted onto the basal end. Inhibition emanating from the basal head was able to travel to the apical cut site and prevent head regeneration from occurring there. Hence, the transmission of inhibition is not polarized, and can proceed just as effectively up the body column as down.

The inhibition effects might be attributed to a variety of mechanisms, such as electrical conduction, propagation of a bio-chemical wavefront or mechanical forces. However, it seems plausible that inhibition is due to a substance which is continuously produced in the head and diffuses down through the tissue of the body column. A steady-state gradient would result if the diffusing substance either broke down throughout the body column, or was degraded in a "sink" located at the basal end. The reduced range of the inhibition gradient at the higher temperature is consistent with an overall decay of the substance which occurs at a temperature-dependent rate. The presence of a terminal sink for inhibition is unlikely, since no decrease in secondary head formation (signifying an increase in inhibition level) was detected when the basal portion of the animal was cut away (15).

Assuming a single source at one end and breakdown in all the tissue, MacWilliams (15) showed that his kinetic data was compatible with a diffusing inhibitor molecule. With computer simulations, he calculated a diffusion constant of 2×10^{-6} cm^2/sec, which is about ten times slower than the rate at which small ions diffuse through water. This is similar to the value (0.8×10^{-6}) suggested by Crick (9) for a theoretical morphogen diffusing in cytoplasm. These are only consistency arguments, however, and there is no direct evidence that the inhibition properties are due to a diffusible substance.

In addition to the labile inhibition found in the body column, there is a second minor component to head inhibition which is much more stable (15). After the head is removed and the inhibition has dropped to the lowest level, a higher head-forming frequency is still found in the lower part of the body. This suggests that there is a gradient of inhibition still remaining. To detemine whether this was produced by the regenerating head or was a property of the body tissue, animals were prepared with body columns consisting of either apical halves (Fig. 5A) or basal halves (Fig. 5B). The heads were removed from these animals 3, 6, and 12 hr later. Transplants were tested in each at the midpoint, so if there were any inhibition coming from the regenerating tip, it would be equivalent in both types of animals. The frequency of head formation was significantly lower in animals having apical body tissue. Since the only difference between the animals was the source of body tissue, this suggested that inhibition is also produced in the body and at higher levels in the apical portion. This inhibition must be stable because the effect did not

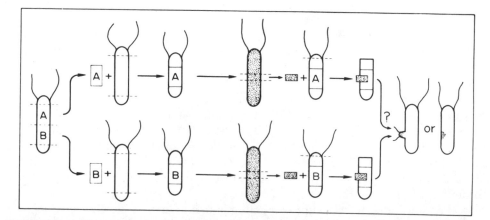

FIGURE 5. Transplantation experiment demonstrating the minor component of the head inhibition gradient produced in the body.

change with time after the animals were assembled. MacWilliams proposed that inhibition-producing "sources" exist in the body as well as the head, and they occur as a gradient.

ISOLATION OF INHIBITORY FACTORS

Factors have been isolated from hydra that have reversible inhibitory effects on morphogenetic processes at concentrations of less than 10^{-8}M when added to the hydra medium. Berking (1) has partially purified a factor which will inhibit the initiation of bud formation, and at slightly higher concentrations will inhibit head or foot regeneration. Schaller (26) reported two inhibitory factors, one of which specifically inhibits head regeneration while the other inhibits foot regeneration. All three factors have molecular weights of less than 500. Berking's inhibitor and the head inhibitor are distributed in a graded fashion in the animal with maxima in the head, while the foot inhibitor is located mainly in the basal portion. It is believed that they exist in a sequestered form from which they can be released, since they are biologically active at concentrations a thousandfold less than found in the animal. The relationship of these three factors to one another and to the measured properties of inhibition remains unclear.

THE ROLE OF THE INHIBITION GRADIENT

Because head inhibition and foot inhibition suppress pattern formation at some sites and allow them in others, they play a part in locating pattern formation events. Thus these two inhibition gradients provide the most likely explanation for why buds, with their developing head and feet, always form at a predictable axial level in the body column.

A variety of experiments indicate that the budding zone is the area where the tissue has escaped the reach of inhibition emanating from head and foot, allowing bud patterning to occur. Removal of head or foot leads to an increase in budding. Reduction of the distance from head to budding zone by removing part of the gastric region will decrease budding. If the animals are reduced in size by starvation, they cease budding, but will resume upon decapitation. Grafting of additional body column tissue into starved animals will also cause budding to begin again. Thus, removing the extremities and thereby the sources of inhibition, or altering the reach of the inhibition gradients by adding or subtracting tissue, all affect the probability of budding being initiated.

Not only do the inhibition gradients control the occurrence and location of buds, but also this same mechanism appears to influence the adult animal size. Since the tissue is growing continuously, size control requires continuous removal of tissue from the animal. In adults 85 % of this loss is into buds (7). The manner in which the inhibition gradients could plausibly control size can be illustrated by considering a freshly detached bud (Fig. 6). Since this young animal does not immediately begin to form new buds, we might assume that the head and foot gradients overlap to some extent. Continued growth of the tissue causes an increase in size reducing the overlap of the gradients. This results in a ring of tissue with lower

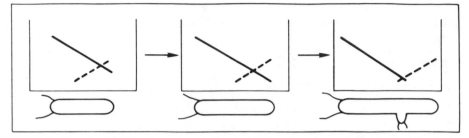

FIGURE 6. Head inhibition (——) and foot inhibition (---) gradients in freshly dropped buds, young adults, and budding adults.

inhibition levels where a bud can form, and the area becomes the budding zone. Once the number of buds reaches a steady-state level appropriate to the food intake, all extra mass is funneled into them and the adult no longer increases in size.

There is some more direct evidence suggesting that the ranges of the inhibition gradients control the size of the animal. At a higher temperature, the inhibition gradient was shown to be steeper and shorter. If the extent of the inhibition gradient controls animal size, an increase in temperature should result in a decrease in size. This is what was found. Hydras are normally cultured at 20°C. When they were grown at a higher temperature (26°C), the animals were much smaller and at a lower temperature (15°C), the animals were much larger. In both cases the absolute distances from head to bud to foot were altered. Thus temperature shifts, which change the extent of inhibition, correspondingly change bud location and animal size.

Another approach, exploited extensively by Sugiyama and colleagues (29), has been to isolate and study mutants with altered morphologies. There are four mutants whose head inhibition gradients were measured by transplantation experiments and found to differ from the wild-type. This allows a direct comparison of the slope and range of inhibition with bud placement and animal size.

Three of the mutants have higher than normal levels of inhibition emanating from the head, but the slopes of the gradients are very different resulting in animals of dramatically different size (Fig. 7). L4 (*H. magnipapillata*) is an extremely long animal that buds at a low rate (29). The inhibition gradient has a much larger absolute extent than in the wild-type. Significant amounts of head inhibition are still found near the basal end of the animal, which is consistent with the budding deficiency. Reg-16 (*H. magnipapillata*) is not as large as L4, but the distance from the head to the budding zone is more than twice that in the wild-type (14). The inhibition levels are very high near the apical end, but slope down to low levels further on in the animal. Thus budding is normal, only displaced, and the size is increased correspondingly. The third mutant, the aberrant (*H. attenuata*), is actually shorter than the wild-type (23). Even though the inhibition starts out higher, the slope of the gradient is very steep, so that levels which permit budding occur closer to the head than in the normal animal.

The last mutant, mh-1 (*H. magnipapillata*), has a considerably reduced inhibition gradient in both initial level and extent (27). However, the animal is only slightly smaller than the wild-type, in fact is similar in size to the aberrant. Buds

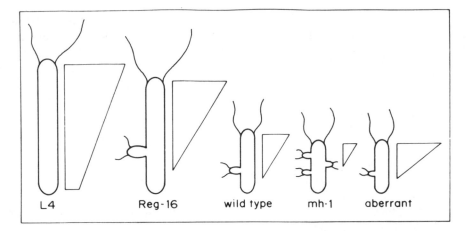

FIGURE 7. Comparison of the head inhibition gradient of several mutants. The first three are mutants of *H. magnipapillata* and the fourth of *H. attenuata*. Wild-type indicates the normal animal of either strain. Triangles represent the head inhibition gradient schematically.

form closer to the head, but not as close as might be expected. Instead, supernumerary heads with short body axes form in the gastric region. For reasons not yet understood, the basal portion of the body is not completed in this area, and the secondary axes do not detach as buds. Since the tissue remains on the animal, its length is not reduced significantly.

In sum, there are several types of data that are all consistent with the distance between head and budding zone, and therefore with the size of the adult animal, being dependent on the range of the inhibition gradient. Animals with extended gradients are larger and those with attenuated gradients are smaller. Second, the level of inhibition affects the chance of heads or buds arising in a given area. If there is too much, it may prevent bud formation where it normally would occur. If too low, it may be insufficient to suppress spontaneous head formation in an inappropriate place.

There is an additional correlation which was noted in temperature shift experiments, feeding regime experiments (22), as well as in comparisons of several mutants and strains of *H. magnipapillata* (29). Big animals form big buds, and small animals form small buds. The size of the bud that pinches off is related to the position at which it forms. This strongly suggests that inhibition is not only involved in bud placement, and thereby adult size, but also in regulating the proportions of a developing animal.

The Activation Gradient

PROPERTIES

Head formation potential or head activation is the capacity of a piece of tissue to induce the formation of a head and secondary axis when transplanted to the body column of a host animal. Head tissue, but not tentacles, has the highest activation level and can induce head formation with a high frequency anywhere in the body, even with the host head present (5). Body tissue also possesses activation, but at lower levels than the head, and the value decreases in a graded fashion down the body column.

The following experiment, which has been carried out in many variations, demonstrates this point (Fig. 8). Pieces of tissue from different axial levels in the donor were transplanted to the middle of the host body columns (16). Regions 1, 2, and 3 were grafted into hosts whose heads were intact (Fig. 8A), while decapitated hosts were used for regions 3, 4, 5, and 6 (Fig. 8B). The head formation frequencies are lower in the intact hosts because the head is transmitting inhibition to the body column. Whether the host head was present or not, the implants showed decreasing ability to form heads as their position of origin in the donor was further down the column. Thus, head activation is distributed as a gradient throughout most of the body column with a maximum in the head (Fig. 4). The two types of hosts were necessary because transplants to either host alone would result in an artificial plateau at minimal (near 0%) or maximal (near 100%) levels of activation.

The activation gradient differs in several ways from the major component of the inhibition gradient emitted by the head. The gradient levels are not immediately changed by head removal, and therefore activation in the body is not a short-lived signal whose source is the head, despite the high level demonstrated there. This implies first that activation in the head has a very limited range of influence as

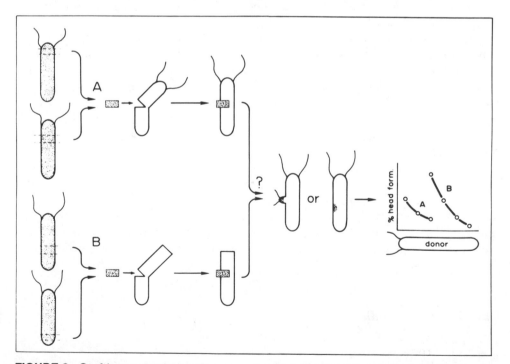

FIGURE 8. Grafting procedure using normal and decapitated hosts to demonstrate the head activation gradient.

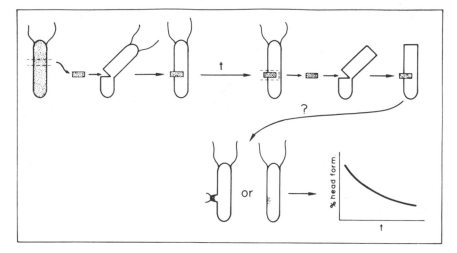

FIGURE 9. Transplantation experiment demonstrating the decay of head activation.

contrasted to inhibition in the head, whose effect on the body is rapid and far reaching. The second implication is that activation in the body column must be a property of the tissue itself. In this respect it is similar to the minor component of the inhibition gradient.

The activation of the body is much less labile than inhibition as measured by the kinetics of its decay (Fig. 9) (16). Donor body tissue was transplanted to a region of slightly lower activation in a host body column. The host head was left intact to prevent the implant from forming a head in most cases. Periodically thereafter, the implant was tested for its ability to induce heads, to determine if its activation level had changed. Activation was found to decrease very slowly to the level of the surrounding tissue, with a median decay time of 30–36 hr. This contrasts sharply with the 2-hr median decay time found for inhibition from the head. Not only is activation a local tissue property, but it is a relatively stable one.

THE ROLE OF THE ACTIVATION GRADIENT

The role of the inhibition gradient is to suppress patterning events within its range, which in addition to maintaining the integrity of the animal from structures arising spontaneously, has important implications for size control. Analogously, the activation gradient plays its role in locating patterning events by stimulating them to occur. It is the relative levels of activation and inhibition that determine whether or not patterning will be initiated. The higher the activation level of the tissue, the greater the chance of overcoming inhibition so that head formation can take place.

The foot activation gradient, which runs in the opposite direction, has similar properties to the head activation gradient. The two gradients together provide an explanation for the most well-known phenomenon of hydra regeneration, namely its polarity. Whenever a piece of the body column is excised, a head regenerates from the more apical end while a foot forms at the more basal end. The apical end will have the relatively highest point of the head activation gradient, which will facilitate head patterning preferentially at that site. The same type of argument can be made for the foot and the foot activation gradient at the basal end.

If there is an extended area of "highest" activation, supernumary structures result (3). This is shown most graphically by comparing the regeneration of two excised strips of tissue, whose orientation in the body column was very different (Fig. 10). The first was taken from the longitudinal axis and included a range of activation values. This piece regenerated reliably into normal miniature hydra. If instead a strip of the same size and shape was taken from the circumferential direction, where activation values were very similar throughout, the incidence of extra structures rose dramatically. Body tentacles, second heads, and extra basal discs formed frequently, in extreme cases obliterating the basic body plan. The longitudinal strip had a small area at the highest activation levels, and thus single end structures resulted. The circumferential strip, with its large area of highest activation levels, could initiate patterning at multiple sites.

It is possible to reverse the polarity of the tissue, but this is a slow process. Wilby and Webster (32) excised sections of body column and grafted a foot to the apical end and a head to the basal end (Fig. 11). At various times thereafter, the grafted heads and feet were removed and the pieces allowed to regenerate. Most regenerated with the original polarity after one day, but after two days the majority regenerated with a head at the original basal end and a foot at the original apical end. If only a basal head was grafted, polarity reversal took four days instead of two. Hence, the reformation of the activation gradient occurs gradually, and like the slow decay of activation levels, indicates the stability of the activation property.

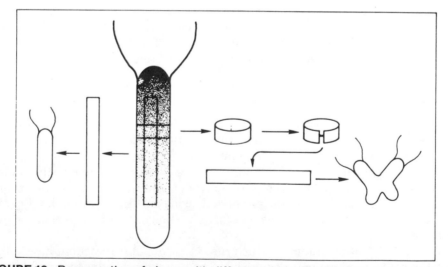

FIGURE 10. Regeneration of pieces with different width to length ratios. Head activation gradient represented as a linearly decreasing density of dots along the body column. The excised pieces are displayed as rectangular planes which roll up and develop into cylindrical animals.

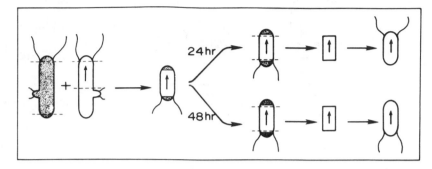

FIGURE 11. Experimental design for demonstrating polarity reversal.

BASIS FOR THE ACTIVATION GRADIENT

The basis for the activation gradient or polarity is currently unknown, but some information is available to narrow down the possibilities. It has been suggested that tissue polarity could derive from an intrinsic cell polarity, in such mechanisms as directed morphogen transport or electrical conduction (30). The polarity of the tissue would be maintained in the same manner as a series of magnets would align their north and south poles appropriately. Then, during regeneration a head would form at the north pole, for example, and a foot at the south pole. Two experiments indicate this is unlikely.

In one (11), the body column was cut into a series of rings (Fig. 12). The rings were flipped over and then grafted back together again. Thus, each piece retained its original axial position, but had an inverted polarity. If polarity were simply a vector property, head regeneration should then occur at the basal end. Instead, it occurred at the original apical end, as expected if polarity were due to a scalar property.

In a second experiment, the original polarity of the tissue was wholly destroyed by dissociating in into a cell suspension (11). Aggregates obtained by centrifuging the cells together can regenerate to form complete animals. Serial aggregates were made of cells derived from the head and from the body in the combinations: head-body-head or body-head-body (Fig. 13). Although body cell

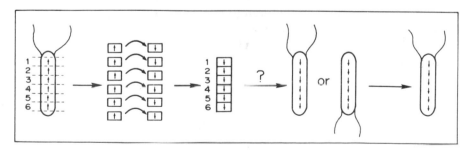

FIGURE 12. Experimental design demonstrating the tissue polarity is not due to an intrinsic cell polarity.

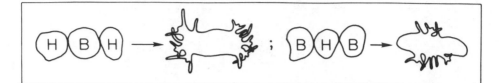

FIGURE 13. Regeneration of head structures in triplet aggregates. The outcroppings represent single tentacles or complete heads (hypostome plus tentacles).

aggregates alone will form heads, in these triplet aggregates head structures appeared mainly where head cells were. A similar experiment was performed using body columns, minus heads and feet. These were cut into apical and basal halves which were dissociated separately. Triplet aggregates composed of apical-half cells and basal-half cells regenerated head structures preferentially from the apical cells, perhaps seeded by undissociated clumps, should have resulted in head formation at any position with equal probability.

The hypothesis which best fits the data at present is that activation is due to a substance produced by "sources" (10, 19). These sources would consist of the machinery of synthesis, and perhaps stored quantities of the material which could be released. Small steady-state amounts of the free substance would be responsible for the biological activity observed. The sources would be distributed down the body column with decreasing density, in the case of head activation, to form the gradient. Formation or destruction of sources of a substance would presumably take longer than synthesis or breakdown of the substance itself, which would account for the stability of the activation property. A "source" mechanism was also suggested as the basis for the minor component of the inhibition gradient which is produced in the body.

Isolation of an "activator"

Schaller (25) has isolated and sequenced a peptide (⟨pGlu-Pro-Gly-Gly-Ser-Lys-Val-Ile-Phe) which specifically stimulates the rates of head regeneration and bud initiation, and raises the number of regenerated tentacles when added to the hydra medium. It has no effect on the rate of foot regeneration. The head activator is distributed in gradient fashion down the body column, with the highest concentration in the head. There is some evidence that the bulk of the factor is stored in neurosecretory granules, and only minute amounts are free and active (24). Exactly how it fits into the pattern formation scheme is not known, since there is no proof that the activator is responsible for the property of activation.

Summary

The role of the developmental gradients is to determine where a head or foot can form (activation gradients) and where an extremity cannot form (inhibition gradients). Their interactions affect the initiation of patterning during regeneration, and the maintenance of pattern during growth. In the following section, the characteristics and dynamics of the patterning processes underlying head and foot formation during regeneration will be discussed.

Pattern Regulation During Regeneration

When the head and/or foot are removed from a hydra, they will regenerate within two to three days. If instead a piece of tissue is excised from the body column, it will first heal into a sphere or short cylinder, and as the head and basal disc regenerate, will gradually elongate to the appropriate form. The entire process is considered to be pattern regulation.

The pattern formation portion of regulation can be divided into a series of events. At the beginning, processes occur which decide whether patterning will be initiated. It is at this stage that the gradients influence where the head or foot will form. In the next few hours, a rapid increase takes place in the activation level at the apical end. This continues until the area attains the head formation capacity of the differentiated head and is "determined." Subsequently, the size of the area is expanded until the presumptive head covers a sufficient fraction of the total tissue to form a proportioned animal. Concurrently, the inhibition gradient returns, and finally the activation gradient below the head area begins to rise slowly to appropriate levels. At some point, a secondary patterning process locates the tentacles.

Initiation of Regeneration

A hydra can be bisected at almost any level of the body column with the same result (Fig. 14). A head will regenerate at the apical end of the lower portion, and a foot will regenerate at the basal end of the upper portion. Since most of the tissue is capable of forming either head or foot, how is the appropriate process initiated? The simplest view has been that removal of the apical end results in a fall in head inhibition to a low enough level to allow head patterning to begin. The foot patterning process would be kept in check by inhibition emanating from the foot. While elimination of the extremity and thus its inhibitory effect is absolutely necessary for regeneration to occur, there is evidence that it may not be sufficient to stimulate the patterning process every time.

For example, Newman (20) has shown that the way in which the head is removed affects the course of regeneration. He ligated hydra just below the tentacles so that the head pinched off leaving only a small hole to damage the integrity of the surface. When compared to controls whose heads had been cut off, regeneration proceeded more slowly. After three days, 10% had not regenerated at all, and 22% had formed only a central tentacle, but no head. All the controls had regenerated, and the incidence of median tentacles was only 8%. Hence, removal of the head with a minimum of injury results in the retardation of development, suggesting that the cut surface plays a role. It has long been noted that regeneration usually takes place at cut ends or junctions of grafts, and that injury has a stimulatory effect.

Two kinds of mechanisms for injury effects have been described recently, one

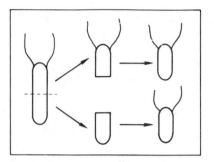

FIGURE 14. Experiment demonstrating that a head or a foot can regenerate at the same axial location.

implicating activation and the other inhibition. MacWilliams (16) has evidence that injuring the tissue will cause a transient increase in activation. Region-3 was transplanted into the middle of a host body column, with the host head left intact to prevent expression. Twenty-four hours later the host head was removed and the animals were injured in different ways. Cuts were made in, above, below, and all around the transplant. The transplants that received the injury directly in the transplant tissue formed heads in 78% of the cases. All other cuts resulted in head formation at frequencies that differed only slightly from the uninjured control of frequency of 26%. The stability of the injury effect was tested and the stimulus was found to disappear within 12 hr.

Because the injury effect was very local, it was interpreted as an elevation of activation. Were it due to a reduction of inhibition, injury outside the transplant should have been just as effective. Inhibition effects are long range; head removal affects transplants rapidly at all axial levels. The activation boost could derive from leakage of an activating substance from sources in or near dying cells. This would add to the local gradient level and the total activation effect would depend on the concentration of sources in that area.

The injury effect has also been measured as a decrease in inhibition, and in this case involves the bound head inhibitor reported by Schaller (26). After decapitation, Kemner and Schaller (13, 14) found that a substance with head inhibitory properties was released into the medium during the first hour. Measurements on the apical third of the body suggested that the loss was mainly from that area and amounted to half of the material normally present in bound form. The release was shown to be specific to the type of regeneration taking place. The same region was tested for head inhibitor content when it was regenerating a head and when it was regenerating a foot. Only during head regeneration were large amounts of head inhibitor lost.

Reg-16, the mutant described earlier, is regeneration deficient, and its reduced ability to regenerate heads is best explained by a suppression of initiation events. The mutant can form buds normally indicating that the patterning process is probably not affected. When the animals were decapitated, regeneration occurred in a fashion similar to Newman's ligated animals, but the insufficiencies were more exaggerated (28). More than half the animals had either no regenerated structures or medial tentacles. Those with heads had fewer tentacles initially which very slowly reached normal numbers.

Achermann and Sugiyama have shown that the mutant is characterized by high levels of inhibition which were found to decay very slowly following decapitation. Significant reduction occurred only after 12 hr. Thus, initiation of regeneration in reg-16 might be impaired after decapitation by the high head inhibition remaining in the body. By 18 hr when inhibition was dropped to a low level, initiation would be less likely, since the injury stimuli would no longer be effective. The wound would have closed, preventing bound inhibitor release, and the activation boost would have decayed. Thus, the prelude to successful regeneration appears to involve reaching a favorable activator-inhibitor ratio at the appropriate site. Once the main inhibition source is removed, initiation is aided by leakage at an injured edge, and the evidence thus far suggests that a free activator spike and a bound inhibitor release occur as a result. Whether these effects are combinative or have a causal relationship is not known.

Head Determination

If the initiation processes have favored activation, the patterning process begins. Activation rises rapidly in the regenerating apical tip until the area attains the organizing capacity of the head, i.e., it will induce head formation at any level in a host body column with a high frequency, in the presence of the host head. At this point the regenerating tissue is said to be "determined" (33).

The kinetics of this activation increase have been measured by transplanting the regenerating tip periodically after decapitation (Fig. 15). (16). There was an initial lag of 2 hr before any significant increase in head formation frequency was detected. Thereafter, the fraction forming heads climbed steadily for 5 hr and then

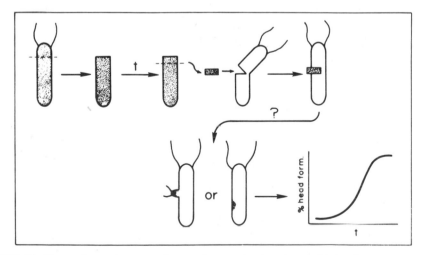

FIGURE 15. Transplantation experiment demonstrating the rapid rise in activation at the regenerating apical tip.

leveled off at a high value typical of differentiated head tissue. According to an analysis by MacWilliams, the data are consistent with a gradual increase in activation as opposed to an all-or-nothing event occurring at different times in individual animals.

To understand the mechanism involved, it was critical to know if the "activation" which increases during regeneration of the head was equivalent to the intrinsic activation gradient. Although the details differed, most models suggested that the highest value of the gradient signaled "make head," and thus the changes measured during regeneration reflected a rapid reestablishment of the gradient endpoint. On the other hand, one model proposed by Gierer and Meinhardt (10, 19) postulated a patterning process that did not involve replacing the missing part of the gradient. The activation gradient would result from a graded concentration of "sources." Each source produces an activating substance at a constant low level. The activator concentration responsible for the biological property of "activation" or head formation potential depends then upon the source concentration at a given position. The activation increase during regeneration would result from a rise in the level of free activator above the steady-state value. The sources would be stimulated locally to produce (or release) large quantities of the substance to pattern the presumptive head area.

In an important series of experiments, MacWilliams decisively demonstrated that there are two components to activation, as distinguished by their labilities (16). Using the experimental design illustrated in Fig. 9, he compared the rates of activation decay of two pieces of tissue that started with similar activation levels. One piece was the 1-region of the body column, whose high activation was due to the activation gradient level. The other was an apical tip from a lower level that had regenerated until the activation had risen to that of a 1-region. The decay rate in the regenerating tissue was found to be much faster, with a median time of 12 hr as compared to more than 30 hr for the nonregenerating tissue.

Because of the clear difference in decay times between the two activations, there is little doubt that they are based on different mechanisms. The lability of the regeneration-provoked activation is consistent with a free activating substance as its basis. A substance would presumably be less stable than the sources which produce it.

A Simple Reaction-Diffusion Model for Hydra

Since of the models presently available (2, 12, 20, 30, 34), the Gierer-Meinhardt reaction-diffusion model (10, 19) best explains the activation rise during regeneration, a brief descriptive version will be presented here. The model centers around two substances, an activator and an inhibitor. The activator stimulates its own production autocatalytically, and it has a short diffusion range. The activator also stimulates production of the inhibitor, which has a long diffusion range. The inhibitor, in turn, blocks activator production. The regeneration polarity of the tissue is maintained by a gradient of activator sources whose concentration is highest apically (Fig. 16a). (The gradient is not necessary for the mechanism. It was added to explain the phenomena in hydra). After the head is removed (Fig. 16b), inhibi-

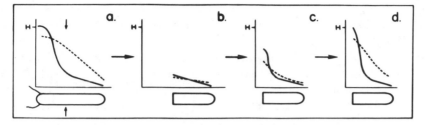

FIGURE 16. Head regeneration as described by a simple reaction diffusion model. Components in the intact animal: activation (———) which is the sum of the basic head activation gradient and the high level of activation in the head due to autocatalysis; head inhibition gradient emanating from head (· · · ·). H: threshold activation level for "head determination." Arrows indicate where animal was decapitated; a represents the gradients in the intact animal. The progressive changes that occur in the gradients after decapitation are shown in b, c, and d.

tion falls to a low enough level to allow activation to increase autocatalytically. Inhibitor produced in conjunction with the activation process diffuses away (Fig. 16c). Activation continues until the inhibitor builds up sufficiently to limit it (Fig. 16d). A stable pattern results when production, diffusion, and decay of these substances reach a steady state. The patterned tissue will then have an activated area committed to head formation, and an inhibited area which remains body tissue.

While the model in this form provides a plausible mechanism for the increase in activation, it is not fully consistent with other data and requires some modification.

Localization of Activation and the Slow Return of the Gradients

The expectation from earlier versions of most models, including that of Gierer and Meinhardt, was that the activation rise during regeneration was initiated throughout the body once head inhibition was removed. The competition to reach the level resulting in head determination would be won by the apical end, which started with an advantage due to the activation gradient. All models assumed that once the head was determined, the inhibition was also fully restored and functioned to limit the size of the presumptive head area by preventing further activation elsewhere. Both of these assumptions were shown to be incorrect.

Recent kinetics experiments have shown that the head inhibition is insignificant long after the head has already been determined (15). The inductive ability of transplants in decapitated hosts were measured periodically to follow the restoration of the host head inhibition. The level of inhibition in the body reached a low point at 6 hr, close to the time when the regenerating apical area becomes committed to head formation. Thereafter, inhibition gradually increased, reaching half strength at 18 hr and almost full strength at 24 hr. The slow return of inhibition was not due to slow equilibration of inhibitor transmission. When the regenerating tip was replaced with a fully differentiated head, inhibition was effective almost immediately, and at the level of the 24-hr regenerating head. Apparently the newly determined head does not produce inhibitor in large quantities, as originally assumed, and many hours of regeneration are necessary before the inhibition gradient is fully restored.

On the other hand, the activation rise during early regeneration was found to be strictly localized, confined to a small area at the apical tip. Wolpert (33) demonstrated this by examining activation changes of the 3-region placed in three different developmental situations. He measured the time required for the head-inducing ability of the 3-region to reach that of a 1-region. When the 3-region was the regenerating tip (34B56F), the increase occurred in a few hours. If the region was immediately below the regenerating tip (234B56F), increase in activation did not begin until after one day, and then the level rose slowly. Even by two days, it had not reached the level of a 1-region which one might expect from its location just below the presumptive head area. If instead a head was grafted onto the 3-region (H34B56F), activation increase began right away. However, it proceeded at the same slow rate as in the 3-region below a regenerating head.

The latter two cases represent changes in the activation gradient. They demonstrate that this component of activation is not adjusted until late in the regeneration process, after the presumptive head has reached some stage of differentiation. Thus the rapid activation rise in occurring only in the apical tip as illustrated by the regenerating 34B56F animal. Since inhibition has not returned in measurable quantities during this period, some other mechanism is needed to confine this activation to a small area.

Head and Foot Proportion Regulation

Since there is evidence of subsequent changes occurring in the regenerating animal, it is likely that "head determination" does not mean that the patterning of the head is complete. A better interpretation is that a threshold activation level has been reached, and the processes leading to head formation are now irreversibly set in motion. One such process has to do with the allotment of only part of the tissue to the head region. Is there a size-sensing mechanism acting to determine the extent of the tissue which is to become the presumptive head and thereby limits pattern formation to a specific area at the apical end?

To gain insight into possible proportioning mechanisms, it was first necessary to know how precisely the various structures regulated with respect to the size of the whole tissue during regeneration. Although small animals appeared to have smaller structures, it had never been shown quantitatively whether this was the case. The approach was to excise pieces of the body column, whose size ranged down to one-hundredth of an adult animal (3). The number of epithelial cells in the head, body column, and foot of the regenerates served as a measure of the fraction of tissue devoted to each structure.

Whole animals developed from pieces as small as one-fortieth of an adult. Tissue pieces, one-twentieth of an adult and larger, produced regenerates that showed good head-body proportioning, although it was not perfect. There was a slight decrease in head fraction from 33% in the smallest animal to 21% in the largest, but this is of a minor order considering the twenty-fold size difference between them.

The basal disc remained in proportion throughout the size range, constituting about 8% of the total number of cells. Clearly, proportion regulation must be included in any model which attempts to explain pattern formation in hydra.

A Proportion-Regulating Reaction-Diffusion Model

The simple reaction-diffusion model presented earlier does not proportion regulate well, certainly not over the size range necessary for hydra. Further, the model predicts that head activation takes place over a considerable extent of the body column, and inhibitor levels rise almost concurrently with activation. Both are inaccurate, since head activation is very local and inhibition does not return to the normal levels for eighteen hours after head determination. Proportion-regulating versions of the Gierer-Meinhardt model have been devised (10, 3), and MacWilliams (16, 17) has recently presented one which satisfies most of the hydra results in a quantitative manner. It limits the head activation area appropriately and prevents activation from occurring elsewhere, while explaining the late return of inhibition.

This model has the same elements as the simple model: autocatalytic production of activator, inhibitor production dependent on activator concentration, and a gradient of sources for the activator. In addition, there is a gradient of sources for inhibitor similar to that of the activator sources. They produce inhibitor at a steady-state rate and the resulting free inhibitor is proportional to the source density of the region. Instead of being distributed in a gradient throughout the body (or an excised piece), the level is almost uniform due to the large diffusion range of the inhibitor. In the normal animal, however, this is swamped by the large quantities which issue from the head and form a gradient.

When the head end is cut away, inhibition drops (Fig. 17b). Activation begins in a few cells at the high point in the activation gradient, boosted by a spike of activator resulting from injury effects. Inhibitor from the body source gradient serves to prevent head activation from occurring at other sites, even if wounded. Assuming the process saturates, a maximum activation level will be reached in the initial cells (Fig. 17d). Then by diffusion, activator will spread to neighboring cells causing expansion of the activated area by further autocatalysis (Fig. 17e). In the early stages when the activated area is small, the correspondingly small amount of inhibitor produced diffuses away allowing the process to continue. As activation proceeds, the inhibitor level gradually builds up throughout the tissue until its concentration near the activating area is sufficient to stop it (Fig. 17f).

The final amount of inhibitor would be directly proportional to the activated cells which produced it and inversely proportional to the total number of cells where it diffuses and breaks down. The resultant activated area would be limited to the same fraction of the tissue regardless of the size of the piece. Thus, the inhibitor acts as a size sensor and operates by a negative feedback loop to regulate the number of cells incorporated into the head.

Like earlier models, the inhibitor coming from the newly patterned head functions to limit the head size. This model differs in that the prevention of activation at locations other than the extreme apical end is taken over by the inhibition

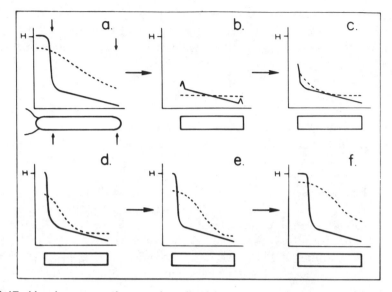

FIGURE 17. Head regeneration as described by a proportion-regulating reaction-diffusion model. The components are the same as in Fig. 15. In b the inhibition levels are due to the basic inhibition gradient that remains after the inhibition from the head was decayed. The small spikes represent transient activator peaks created by injury.

produced in the body. According to computer simulations, the initial rise of activation to saturation level in a small number of cells would be rapid. This could account for the "head determination" which is observed at such an early stage. Further, the simulations show that the spreading of activation and the consequent buildup of inhibitor would proceed slowly, at a rate consistent with the late return of inhibition measured.

The model has only addressed the patterning of the head. However, the basal disc shows similar regeneration properties including early "determination" and better proportioning. The model in this form might be even more satisfactory to explain regulation of the foot. Because it is such a simple structure, namely a disc of differentiated cells, it is easy to imagine that it arose from a circular area of activated cells.

The More Complicated Patterning of the Head

The head, on the other hand, is a complex structure. It consists of a hypostome, a tentacle zone, and tentacles whose numbers vary. When the head parts were examined separately, their proportioning also proved to be more complicated (4). In regenerating animals, the gradually decreasing head fraction with increasing animal size was found to be due to a rather limited proportioning capacity of the hypostome and tentacle zone (Fig. 18). These two regions changed only a factor of five as the total tissue changed a factor of twenty, but they remained proportional to one another.

Interestingly, the total tentacle cells showed a different behavior and remained proportional to the total cells for sizes down to and including one-tenth of

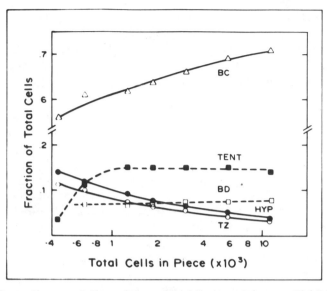

FIGURE 18. Proportion regulation of the several body parts over a 20-fold size range of regenerating pieces. The body parts are the body column (BC:△——△), tentacles (Ten: ■ ————■); basil disc (BD:□ ————□); hypostome (Hyp:•——•) and tentacle zone (TZ: ○——○).

an adult (Fig. 18). However, as the tissue became scarcer and approached the lower size limit, regenerates began to lose the tentacle portion of the pattern. Tentacles were reduced in size, some to a few cells, or were missing altogether although space was available in the tentacle zone.

The finding that hypostome and tentacle zone remain in proportion to one another but not to the rest of the tissue precludes any simple scheme for allotting the proper area to the head. It suggests that the tentacle zone might be patterned with the hypostome as part of the same process, and separate from the tentacle tissue which does proportion regulate. The additional finding that tentacle tissue may fall out of the pattern at very small tissue areas suggests it is lowest in the head hierarchy, i.e., the patterning of the head proper takes precedence over tentacle formation when tissue is scarce. Loss of tentacle tissue when the hypostome and tentacle zone have taken up most of the head tissue also suggests that there might be interactions between their patterning processes.

Because the head has both a nonproportioned and a proportioned part, the possibility exists that there are two primary reaction-diffusion processes. The first would be a simple one to pattern the hypostome and tentacle zone, which could result in the rapid formation of an organizing region. The second and subsequent reaction would delimit the tissue to be incorporated into the tentacles by means of the full proportion-regulating mechanism.

Regulating the Number of Tentacles

The data and models presented so far have dealt with the patterning processes that block out areas for head, body, and basal disc. The result is assumed to be cylindrical tissue shell that has at one pole an activated area destined to become head, at the opposite pole an activated area to become basal disc, and an inhibited area in between to form the body column. The tentacles have been considered only in terms of the total amount of tissue comprising them. This section discusses the secondary patterning process that occurs in the tentacle zone and results in a certain number of tentacles forming.

The tentacles are distributed as a one-dimensional periodic pattern in a ring around the head. Periodic patterns have been grouped in two general categories, spacing patterns and number-regulating patterns. Some classic examples of spacing patterns which have been described are phyllotaxy, bristles in insects, and heterocytes in filamentous algae. In all such cases, there is an average distance between pattern elements, and thus the total number of elements depends on the size of the tissue. The patterning process is thought to involve a zone of inhibition around an activated center which prevents another from near by. When the distance is large enough between centers, an additional one can be added since inhibition there falls to a low level. This may occur when a field is being patterned *de novo*, as well as at a growth zone or during intercalary growth. The spacing is somewhat random, and two structures may form close together if they arise at the same time.

Number regulation is invoked when the number of structures is maintained as the tissue size is altered, and thus the change is reflected in the size of the structures. Digits on a limb or somite number have been studied in this category. Here the processes are much more complicated, and models have been suggested using orienting gradients to initiate a sequence of proportion regulating events.

The tentacles have been something of an enigma because they exhibit elements of both types of patterning. The most common observation is that the average tentacle number of buds is almost the same as the tentacle number of their parents in steady-state cultures. Only a small percentage of buds form additional tentacles as they grow to adulthood, which involves a fourfold increase in mass. Instead, the tentacles also grow larger, as does the size of the head. Thus there is an increase in the size of the repeating structures with increase of overall size, and the number of structures remains almost constant, a characteristic of number regulation.

However, there is a variety of evidence that indicates tentacle number changes with size. For example, in the proportion regulation studies, the average number of tentacles in regenerates increased linearly from three to seven as the total cells increased exponentially by a factor of twenty. The critical dimension which should correlate with the number of tentacles is the circumference of the head. While there is a linear relationship between the number of tentacles and the circumference (21), they are not directly proportional to one another as expected for a spacing pattern. The average distance between structure centers changes with size. Still the fact remains that more tentacles will form when the circumference is larger. This is demonstrated quite dramatically when decapitated animals are strung onto fishline of varying diameters to stretch the tissue in the circumferential dimension. Increased tentacle numbers up to fifteen were obtained, as compared to the normal six.

Two experimental results suggest that a "regulatory spacing pattern" might be more appropriate to describe the tentacles. First, the number of cells separating the tentacles from one another is a constant, averaging around six, regardless of the size of the animal. Second, the size of the individual tentacles changes with the size of the animal, and when added together the total tentacle tissue is proportional to the amount of the body tissue. The tentacle pattern then has two components, the spacing of tentacle organizing centers around a presumptive tentacle zone, and a proportion regulation process which assigns a given amount of tissue to be used for tentacles. A third result amplifies the impression of number regulation when small changes are made in the size of the system. The tentacle zone does not proportion regulate well, and thus its area does not change as much as the rest of the tissue. This means the circumference changes very little.

This leads to a tentative description for the secondary patterning of the tentacles. The tentacle zone is set up below the hypostome as a band of competent tissue. The number of tentacles depends on the circumference of the tentacle zone and would pattern as activated centers at an approximately constant distance from one another typical of spacing patterns. The amount of tissue drawn into the tentacles during morphogenesis would depend on the total allotted during proportion regulation. Once the tentacles are formed, the tentacle zone is broken up into segments, whose width is about six cells, and whose height depends on the size of the tentacles.

The tentacles are thus a spacing pattern, since the distance between repeating structures is maintained. But unlike the typical spacing patterns which add same-size elements when the total area is increased, its structures increase in size also. Therefore, the tentacle patterning has something in common with number-regulating systems, in that it must be linked to a proportioning mechanism.

Conclusion

According to our current understanding, patterning in hydra consists of several parts. There are the inhibition and activation gradients that locate the head, foot or bud, and there are the patterning processes that account for the actual formation of the foot and of the parts of the head. Though most of the processes have probably been identified, it is doubtful that all the mechanisms are understood yet. The complexity of the head and its mixed proportioning capabilities suggest that further mechanisms must be uncovered to explain it fully. They may consist of a mixture of those already postulated, or other secondary events still unknown. Budding is also not fully understood. Though a separate patterning system for bud formation is possible, more than likely a coordination of head and foot patterning is involved.

An understanding of patterning will consist not only of the number and characteristics of the several processes, but it will include knowledge of their interactions. For example, patterning of the head clearly involves a sequence of events. Are the processes underlying these events independent of one another, or do they form a dependent sequence, one triggering the next? Also, it is not known how the activation gradient is set up or adjusted. Could it be a consequence of the inhibition emanating from the (presumptive) head, as Wolpert suggested? (34) There is some circumstantial evidence to support this view.

Interactions may occur between the head and the foot during development.

Formation of the extremities have been considered independent events, probably because most studies dealt with only one of them at a time. However, experiments in which the development of both head and foot occurred simultaneously suggested possible relationships. If head and foot tissue are transplanted together into the same site in a host, no secondary axes of either kind form (20). This brings up the possibility that the two processes extinguish one another. Because of evidence of this kind, some models have postulated that the head and foot mechanisms are in some way the reverse of each other (2, 20, 30) Second, the rate of foot regeneration is faster when a fully differentiated head is present than when it has been removed and is also regenerating. And finally, the amount of tissue allotted to foot formation in small regenerating pieces is far greater in the absence than in the presence of a head, regenerating or complete.

Thus, the challenge of a complete formal description of pattern formation in hydra will be to continue to carefully define the several processes and then unravel their interactions.

Acknowledgments

We thank Harry MacWilliams and Scott Fraser for the thought-provoking discussions, and Marianne Bronner-Fraser, John Dunne, Shelly Heimfeld, Lorette Javois, Lynne Littlefield and Marcia Yaross for their helpful comments on the manuscript. We are grateful to Tsutomu Sugiyama and Sepp Achermann for making their unpublished results available to us. We wish to acknowledge the support of the National Institutes of Health (GM29130) for the work on proportion regulation and the tentacle patterning.

References

1. Berking, S.: Bud formation in hydra: Inhibition by an endogenous morphogen. *Wilhelm Roux' Arch.* **181**: 215–225 (1974).
2. Berking, S.: Analysis of head and foot formation in *Hydra* by means of an endogenous inhibitor. *Wilhelm Roux' Arch.* **186**: 189–210 (1979).
3. Bode, P., Bode, H.: Formation of pattern in regenerating tissue pieces of *Hydra attenuata*, I. Head-body proportion regulation. *Dev. Bio.* **78**: 484–496 (1980).
4. Bode, P., Bode, H.: Proportioning a hydra. *Am. Zool.* **22**: 7–15 (1982).
5. Browne, E. N.: The production of new hydranths in hydra by the insertion of small grafts. *J. Exp. Zool.* **7**: 1–37 (1909).
6. Campbell, R.: Tissue dynamics of steady-state growth in *Hyrda littoralis*. I. Patterns of cell division. *Dev. Bio.* **15**: 487–502 (1967).
7. Campbell, R.: Tissue dynamics of steady-state growth in *Hydra littoralis*. II. Patterns of tissue movement. *J. Morph.* **121**: 19–28 (1967).
8. Cohen, J. E., MacWilliams, H. K.: The control of foot formation in transplantation experiments with *Hydra viridis*. *J. Theor. Biol.* **50**: 87–105 (1975).
9. Crick, F.: Diffusion in embryogenesis. *Nature* **225**: 420–422 (1970).
10. Gierer, H., Meinhardt, H.: A theory of biological pattern formation. *Kybernetik* **12**: 30–39 (1970).

11. Gierer, A., Berking, S., Bode, H., David, C., Flick, K., Hansmann, G., Schaller, H., Trenkner, E.: Regeneration of hydra from reaggregated cells *Nature New Biol.* **239**: 98–101 (1972).

12. Goodwin, B. C., Cohen, M. H.: A phase-shift model for the spatial and temporal organization of developing systems. *J. Theor. Biol.* **25**: 49–107 (1969).

13. Kemmner, W.: Ein Modell der Kopfregeneration in Hydra. Ph.D. Thesis, Ruprecht-Karls-Universitat, Heidelberg (1982).

14. Kemmner, W., Schaller, H. C.: Analysis of morphogenetic mutants of *Hydra*. IV. Reg-16, a mutant deficient in head regeneration. *Wilhelm Roux' Arch.* **190**: 191–196 (1981).

15. MacWilliams, H. K.: Hydra transplantation phenomena and the mechanism of *Hydra* head regeneration. I. Properties of head inhibition. *Dev. Biol. (in press)* (1983).

16. MacWilliams, H. K.: Hydra transplantation phenomena and the mechanism of *Hydra* head regeneration. II. Properties of head activation. *Dev. Biol. (in press)* (1983).

17. MacWilliams, H. K.: Numerical simulations of *Hydra* head regeneration using a proportion-regulating version of the Gierer-Meinhardt model. *J. Theor. Biol. (in press)* (1983).

18. MacWilliams, H. K., Kafatos, F. C., Bossert, W. H.: The feedback inhibition of basal disc regeneration in *Hydra* has a continuously variable intensity. *Dev. Biol.* **23**: 380–398 (1970).

19. Meinhardt, H., Gierer, A.: Applications of a theory of biological pattern formation based on lateral inhibition. *J. Cell Sci.* **15**: 321–346 (1974).

20. Newman, S. A.: The interaction of the organizing regions in hydra and its possible relation to the role of the cut end in regeneration. *J. Embryol. Exp. Morph.* **31**: 541–555 (1974).

21. O'Hern, J., Lenhoff, H. M.: Relationships between size of hypostome and number of tentacles in hydra. *J. Exp. Zool.* **221**: 1–7 (1982).

22. Otto, J., Campbell, R.: Tissue economics of hydra. Regulation of cell cycle, animal size and development by controlled feeding rates. *J. Cell Sci.* **28**: 117–132 (1977).

23. Rubin, D., Bode, H.: The aberrant, a morphological mutant of *Hydra attenuata*, has altered inhibition properties. *Dev. Biol.* **89**: 316–331 (1982).

24. Schaller, H. C., Gierer, A.: Distribution of the head-activating substance in hydra and its localization in membraneous particles in nerve cells. *J. Embryol. Exp. Morph.* **29**: 39–52 (1973).

25. Schaller, H. C., Bodenmuller, H.: Isolation and amino acid sequence of a morphogenetic peptide from hydra. *Proc. Nat. Acad. Sci. USA* **78**: 7000–7004 (1981).

26. Schaller, H. C., Schmidt, T., Grimmelikhuijzen, C. J. P.: Separation and specificity of action of four morphogens from hydra. *Wilhelm Roux' Arch.* **186**: 139–149 (1979).

27. Sugiyama, T.: Roles of head-activation and head-inhibition potentials in pattern formation of hydra: Analysis of a multi-headed mutant strain. *Am. Zool.* **22**: 27–34 (1982).

28. Sugiyama, T., Fujisawa, T.: Genetic analysis of developmental mechanisms in hydra. III. Characterization of a regeneration-deficient strain. *J. Embryol. Exp. Morph.* **42**: 65–77 (1977).

29. Sugiyama, T., Fujisawa, T.: Genetic analysis of developmental mechanisms in hydra. VII. Statistical analyses of developmental-morphological characters and cellular compositions. *Develop., Growth Diff.* **21**: 361–375 (1979).

30. Webster, G.: Morphogenesis and pattern formation in hydroids. *Biol. Rev.* **46**: 1–46 (1971).

31. Wilby, O. K., Webster, G.: Studies on the transmission of hypostome inhibition in hydra. *J. Embryol. Exp. Morph.* **24**: 583–593 (1970).

32. Wilby, O. K., Webster, G.: Experimental studies on axial polarity in hydra. *J. Embryol. Exp. Morph.* **24**: 595–613 (1970).

33. Wolpert, L.: Positional information and pattern formation. *Curr. Top. Dev. Biol.* **6**: 183–224 (1971).

34. Wolpert, L., Hicklin, J., Hornbruch, A.: Positional information and pattern regulation in regeneration of hydra. *Symp. Soc. Exp. Biol.* **25**: 391–415 (1971).

Questions for Discussion with the Editors

1. *Inhibitor gradients appear to lead to a supression of the Hydra body's ability to respond to an implant of, for example, head regions. Other contributions to this volume would interpret that observation in terms of the inhibitor regulating the* competence *of the responding tissue to react to an inductive signal. Is there any reason why you refrained from discussing that phenomenon in terms of the concept of* developmental competence?

Frankly, it never occurred to us to use the term competence. Nor has anyone else used it in discussing hydra. The term is rather vague and has only one relatively well-defined characteristic that is not applicable to hydra. Competence has a temporal aspect to it: a tissue is said to be competent at one stage in development, but not at others. In hydra the tissue of the body column is always competent to form, for example, a head. Whether a head forms or not depends on the interplay of the inductive signal, head activation, and its antagonist, head inhibition. Instead of regulating the competence of the tissue, inhibition is considered to directly control the strength of the inductive signal.

2. *The data on the isolation and characterization of various morphogens you discussed has, quite frankly, not been taken as seriously as it perhaps should be by researchers who work outside of the realm of* Hydra *patterning. Why?*

Most discussions of morphogens in hydra centers on Schaller's factor, the *head activator*. There have been three primary reasons as to why it has not been considered seriously by many people. The first is that the observed effects may have been due to a nonspecific metabolite. The publication of the amino acid sequence of the factor coupled with the fact that it is effective at less than $10^{-12}M$ should reduce these doubts considerably.

The other two reasons have to do with what the head activator might be expected to do. To many people, the initially described effect of the head activator was unimpressive. Because it caused a regenerating animal to form only an extra tentacle or two, the result was dismissed as insignificant. Others had the opposite concern: how can a single morphogen be responsible for something as complex as head formation?

These reactions reflect a general uneasiness with the concept of a morphogen, and they pinpoint a problem. What does one expect a morphogen to do? Should it be able to induce a whole second axis as the dorsal lip did in Spemann's experiments, or as a piece of hydra hypostome implanted into the body column will do? Only if the mechanism were very simple would this be a plausible result. Should the processes be more complex, the expected results may not be obvious. For example, if the mechanism consists of sequential or interlocking processes, a morphogen for one process may have only a minor overall effect, such as making an extra tentacle. Or the result may be counter-intuitive. According to a Gierer-Meinhardt reaction-diffusion model, exposing tissue to an activator may cause inhibition of head formation due to the dynamics of the process.

Still another possibility is that raising the level of only one of a series of morphogens may not affect the number and kind of structures formed so much as the rate at which they are formed. This is another possible outcome of tampering with the activation level in the Gierer-Meinhardt model. Schaller's factor has significant effects on the rates of head regeneration and bud initiation. The increased tentacle number due to the head activator may also be a rate effect. Not all tentacles appear at the same time during head formation so that the number formed at the time of assay may not be the final number. Analogously, Berning's inhibitor retards budding and the regeneration of head and foot.

In sum, we suspect people will take a morphogen seriously only when its function has been more precisely defined and detailed assays devised.

SECTION III

Insects

Embryonic Pattern Formation in Insects: Basic Concepts and Their Experimental Foundations

Klaus Sander

PATTERN FORMATION, although not under that name, was on the minds of the pioneers of experimental embryology when they started experimenting on amphibian and sea urchin embryos nearly a century ago. Experiments on insect embryos were first published 30 years later although descriptive insect embryology was a well developed discipline before the turn of the century. The main cause for this delay was probably the type of cleavage encountered in insects: insect egg cells do not divide in large daughter cells or blastomeres that can easily be separated from each other or killed off singly, as was done in sea urchins and amphibians by Hans Driesch and Wilhelm Roux (32). This handicap and the late start of insect eggs in the field of experimental embryology are the reason that embryonic patterning in insects is rarely mentioned in introductory texts.

If insect embryos ever gain full recognition in the field of biological pattern formation, this may be due to developmental genetics (eg., chapters 11 and 12) rather than to experimental results. Nonetheless, experiments on insect eggs over the years have yielded substantial insights into several steps of embryonic pattern formation. This chapter explains some general concepts of biological patterning that evolved from, or were applied to, results from experimentation on insect eggs. Most of these concepts emerged either in extension of, or in opposition to, former concepts. Therefore our approach will be chronological as well as conceptual. The experiments illustrating the different concepts were selected mainly on their didactic value whenever a choice existed. In discussing the data, we distinguish between "negative" and "positive" evidence. In the former, a conclusion is drawn from *failure* of (in our context) additional pattern elements to appear after experimentation, while in the latter some pattern elements *are generated* as a consequence of experimentation. Conclusions based on positive evidence are preferable since a failure (negative evidence) may have many causes.

None of the concepts presented here is more than a generalization at best, reflecting selected aspects of a series of largely unknown and certainly complex events. The main justification for discussing such concepts is that complex problems like pattern formation are best approached by putting forward informed guesses that are open to experimental testing and thus will guide research until they are modified or rendered untenable by the results of research. Five generalizations emerge from the data presented here.

1. Embryonic patterning *mechanisms differ* considerably between lower and higher insects although the pattern produced is no doubt homologous (discussed also in 24).
2. In higher insects, the longitudinal pattern is strongly influenced by *determinants* localized (or active) near the poles of the egg cell; most pattern elements appear to be specified by some kind of *interaction* between those polar regions which may occur even before the egg cell is divided into daughter cells. The *patterning determinants* thus *differ* from the *germ line determinants* (pole cell determinants) that induce a single cell type (6).
3. Transverse patterning follows *different rules* and may occur later than longitudinal patterning.
4. The signals informing cells about their developmental fates (*instruction*) may differ from those which render a cell incapable of switching to other developmental pathways (*commitment*). This assumption is made in order to reconcile seemingly contradictory sets of data on embryonic "determination" in insects (21).
5. The individual elements of the body pattern, the segments of the germ band, represent functionally isolated cell populations or *compartments* on which further development is based.

Two books can be considered classics in the field of insect embryology: the descriptive text by Johannsen and Butt (8), and the volumes edited by Counce and Waddington (1) which cover both descriptive and experimental aspects. The reference list of this chapter contains many review articles, and for earlier data reference is made to these reviews rather than to the original literature.

Some Traits of Insect Embryogenesis

Embryonic pattern formation in insects occurs between the onset of oogenesis and the end of early embryogenesis, when the basic construction of the larval body becomes visible as a clearcut and fairly stereotyped spatial pattern called the germ band (see below). Oogenesis produces a rather elaborate egg consisting of very resistant egg covers and a large yolky egg cell. Part of the process of pattern formation clearly occurs during oogenesis, as revealed by the visible axial polarization and bilateral symmetry of most insect eggs. Yet this patterned organization of the egg must not be taken to indicate a rigid internal organization or an invisible predetermined pattern in the newly laid egg cell; on the contrary, the germ band pat-

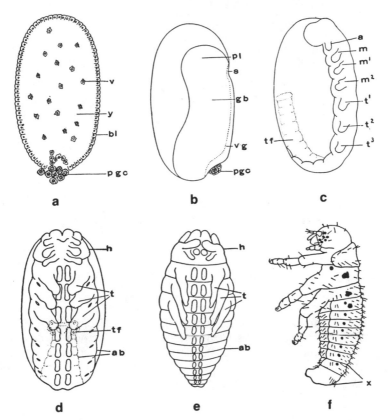

FIGURE 1. An outline of embryogenesis in the Colorado beetle *Leptinotarsa decemlineata*. The egg is about 1.75 mm long and takes 5 days at 25°C until hatching. (a) Section through egg in transition from cleavage to the blastoderm stage. Some cleavage energids (v) are still within the yolk system (y) and some earlier cleavage nuclei were enclosed in the pole cells tied off from the posterior egg pole as germ line or primary germ cells (pgc). The large majority of nuclei have moved close to the egg cell surface and are about to be separated from one another by infoldings of the plasmalemma which delineate the future blastoderm cells (bl). Age 24 hr. (b) The blastoderm has given rise to the future outer embryonic cover (at the left) and the germ anlage (gb) on the right hand (= ventral) side. The head lobes or procephalic lobes (pl) mark the anterior and the pole cells the posterior end of the germ anlage. The prospective foregut or stomodeum (s) and the ventral groove forming the future inner layer (vg) are about to invaginate. The groove separates right and left half of the ectoderm. Age 36 hr. (c) The early segmented germ band is much longer than the germ anlage. Marked are the abdomen (tf) and the appendage buds: a antenna, m, m1, m2 mouth parts, t1–t3 thoracic legs. Age 48 hr. (d) Ventral view of later germ band stage. The abdomen (a) with the tail fold (tf) extends on the dorsal egg side. The head (h) and the thoracic legs (t) are visible. The neurogenic regions of both ectodermal halves have met and fused over the ventral groove and form paired ganglion rudiments (double row of oval outlines). To the side of these rudiments is the pleural region carrying leg buds and slit-like tracheal pits. The rim outside these pits will grow laterally around the yolk system to form the dorsal parts of the larval integument. Age 60 hr. (e) Embryo viewed from ventral after the flanks of the germ band have met dorsally ("dorsal closure"). Age 72 hr. (f) Hatchling larva showing the pattern of landmarks in the cuticle; X marks end of abdomen. From Hegner (3).

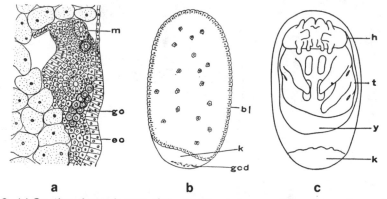

a   b   c

FIGURE 2. (a) Section through part of the beetle abdomen of the stage shown in Fig. 1d. Note the globular germ line cells (gc) in the mesodermal layer covered by the ectoderm (ec) and pervaded by Malpighian tubules (m). At the left are yolk cells derived from the yolk system. (b) Coagulation of the posterior pole region with a hot needle before the onset of cleavage; section through egg at the stage corresponding to Fig. 1a. Note the remnants of the pole disc or germ cell determinants (gcd) in the coagulated cytoplasm (k), and the incipient blastoderm lacking pole cells. (c) Stage as in Fig. 1e obtained after coagulation of posterior cytoplasm before the onset of cleavage. From Hegner (3). Note that hardly the anterior half of the germ band pattern has formed although the coagulated part (k) was restricted to the vicinity of the posterior pole, leaving intact much more than the anterior egg half. The failure of this large anterior egg part to form its normal share of the germ band pattern is reminiscent of the gap phenomenon (see Fig. 6a).

tern in most species studied can be modified drastically by experiments performed after egg deposition. Except for the maturation plasm, there are few if any strictly localized visible components in the egg cytoplasm. The most notable among these are found near the posterior egg pole. Here the ooplasm may contain a blob or disc of somewhat granular material called the oosome that stains well with basic dyes and is rich in RNA (6, 7, 35). This material is part of a developmental mechanism causing determination and early segregation of pole cells or germ line cells at the posterior egg pole (Figs. 1a, 2a, 3). Another localized component is the distinct aggregates of hereditary microbial symbionts that are transmitted through the egg stages in some insects; one instance is the "symbiont mass" in the posterior pole of the leaf hopper egg which is useful as a marker for the location of some important component(s) of the patterning system (see Fig. 7) while the symbionts themselves seem to be without function in embryogenesis (21).

Early embryogenesis starts with egg activation and fertilization (25). The zygote nucleus divides repeatedly and its descendants spread through the egg cell, surrounded by a halo or "island" of yolkfree cytoplasm; nucleus and cytoplasmic island are together referred to as *cleavage energids*. When there are hundreds or thousands of cleavage energids, they approach the surface of the as yet undivided egg cell and individually become encased by infoldings of the (extending) plasmalemma of the egg cell (Fig. 1a). This "superficial" cleavage results in an outer cell layer, the blastoderm, which encloses the yolky remnants of the egg cell called the yolk system.

Part of the blastoderm gives rise to germ anlage and germ band while the rest will form the serosa or outer embryonic cover (Fig. 1b–d). The germ anlage marks the ventral egg side and represents the ventral face of the embryo. It may cover only a small part or nearly the entire surface of the yolk system. In the former case, called the short-germ type by Krause (14), the germ anlage turns into the germ band by proliferative growth, and the majority of body segments form one after the other in antero-posterior succession. In the long-germ type, as the other extreme is called, the germ anlage has the shape and size of the germ band right from the beginning, and the transition from one to the other is essentially an *in situ* subdivision of the germ anlage into a series of segments each of which is represented already in the fate map of the blastoderm stage (28). In both short- and long-germ development, complex morphogenetic movements may occur on the way from the germ anlage to the fully segmented germ band which we neglect here (22).

It is obvious that pattern formation must proceed along partly different lines in short- and long-germ development, and this may be true even for different species subsumed under the same type. Despite the paramount importance of such differences for actual research on pattern formation, we cannot consider the details here. Suffice it to classify the forms mentioned in this chapter: the stone cricket and some beetles have a typical short-germ egg, while some other beetles, the chironomids, and the fruit fly represent the long-germ type; damsel fly, leafhopper, and some further beetles are of intermediate status, with proliferative segmentation restricted mainly to the posterior body half. Finally, silk moth eggs, like all lepidopteran eggs, are hard to classify in this scheme at all.

The germ band, which represents the pattern under consideration in this chapter, is a two-layered structure (ectoderm and mesoderm) expressing an essentially two-dimensional pattern (Fig. 1d). In one dimension of space, this pattern consists of a longitudinal series of pattern elements beginning anteriorly with the head lobes, followed by the series of body segments, and terminating posteriorly with an endpiece called the *telson*. The body segments are grouped in gnathal (3), thoracic (3), and abdominal segments (8–11), most of which can be identified at the germ band or egg larva stages by criteria other than position. In the second dimension of space, the pattern is bilaterally symmetrical and consists of (from medial to lateral) the neurogenic or sternal region, pleural region (which may carry limb buds), and the prospective tergal region. This two-dimensional pattern is transformed into the three-dimensional larval body at the end of the germ band stage when the lateral rims of the germ band grow laterally and extend around the yolk system (Fig. 1d–e) until they meet and fuse in the dorsal midline. The larval body at hatching is covered by a cuticle which in many species carries a pattern of landmarks (segment borders, ripples, bristles, hairs, etc.); these are useful for the identification of individual segments or parts thereof (Fig. 1f).

The Mosaic Hypothesis and the Germ Cell Paradigm

In a very readable summary of his earlier work on beetle eggs, Robert Hegner (4) concluded that "the insect egg at the time of maturation is a mosaic of differentiated cytoplasmic areas predetermined to develop into definite parts of the embryo." This view was formed on the basis of results from two types of experiment: (1) After coagulation of different parts of the egg with a hot needle, the surviving portion seems to produce the same parts of the embryonic pattern which it would have produced in the untreated control egg (Fig. 2c). (2) Stratifying the egg's contents by centrifugation results in the germ band forming at the abnormal position imposed on the periplasm in the centrifuge. As we shall see below, both results can be explained differently.

A concept of special and long-lasting impact came from coagulating the posterior pole region before or after pole cell formation (3). In either case, the individuals developing from such eggs were sterile because their gonads lacked germ cells. This indicated that the pole cells are germ line cells that cannot be substituted for by any other cells in the embryo. More important in our context, the results from cautery before pole cell formation suggested that pole cells can form only from the egg region containing the polar disk granules which Hegner therefore called *germ cell determinants* (Fig. 2b). He tried to provide positive evidence for the role of these granules by shifting them to other parts of the egg cell in the centrifuge.

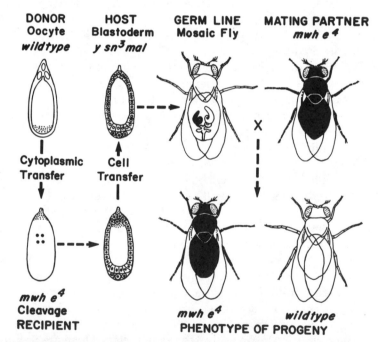

FIGURE 3. Induction of pole cells in the fruit fly *Drosophila melanogaster* by transfer of cytoplasm from the oosome region (stippled) to the anterior pole region of a recipient egg. The recipient was genetically marked as shown in the figure. The pole cells induced ectopically in the recipient egg were removed and injected among the pole cells of a host carrying different marker genes. After metamorphosis, the host flies were mated with flies of the (recessive) recipient genotype (shown in black). Progeny homozygous for the recipient's marker genes then proved that the globular cells induced in the recipient by injected donor pole plasm were capable of functioning as germ line cells. From Illmensee (7).

Much to his disappointment, however, he could not accurately locate the germ cells in embryos developing from centrifuged eggs.

It took more than half a century before the transplantation experiments of Karl Illmensee (6, 7) proved positively that cytoplasm taken from the posterior egg pole can induce pole cell formation in other regions of the fruit fly egg (Fig. 3). But right from. the beginning, Hegner's thoughts on germ cell determinants were extended to other egg regions, thus serving as a paradigm for somatic pattern formation directed by a mosaic of localized determinants. This concept is untenable, as we shall see; for whatever Hegner's "mosaic of differentiated cytoplasmic areas" might consist of, it certainly is not a mosaic of specific determinants for individual somatic cell-types, nor for the different segments or other elements of the body pattern, excepting perhaps the most anterior and the most posterior structures.

Embryonic "Regulation" in Lower Insects

Experiments published by Friedrich Seidel and his student Gerhard Krause provided striking proof that embryonic pattern formation, at least in lower insects, could not be explained on the basis of a detailed mosaic of determinants prelocalized in the egg cell. Cauterizing the posterior part of a damsel fly egg, Seidel (26) obtained two embryos instead of one—a result clearly incompatible with the mo-

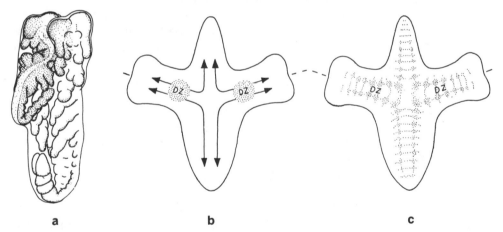

FIGURE 4. Cruciform duplication (*duplicitas cruciata*) in the stone cricket *Tachycines asynamorus* (Orthoptera). (a) Specimen in the germ band stage. One half of each head (top) is marked by stippling, the other left blank. The stippled halves join to form the stippled throax and abdomen at the left while the white halves continue as the white germ band at the right. The heads turn their dorsal sides towards each other, the posterior germ bands their ventral sides. This alignment is imposed on the parts mainly by the common amnion (thin outer contour); if the heads were bent down ventrally by 90°, and the abdomens bent up dorsally, the crosslike configuration shown at the right would result. (b) Model of "determinative streaming" originating in the differentiation center (DZ); note splitting of left and right halves of the streams (arrows) in the midline. Pointed arms of cross signify tails, broad arms are heads. (c) The same blastema envisioned as an oscillating system which generates pattern by resonance phenomena. From Krause (13).

saic hypothesis. The "regulative" capacity of the stone cricket egg was even more striking. After inflicting various cuts on the early germ anlage with a glass needle, Krause (12) obtained all sorts of partial duplications and triplications, and also complete twins. A particular type of result was the *duplicitas cruciata* (Fig. 4a). In vertebrates and annelids, this patterning anomaly can be explained as resulting from two opponent archenteron roofs or budding embryonic rudiments, respectively, which were growing towards each other and split up on meeting; the left body half of one partner would then fuse with the right half of the other (and vice versa), and each fused pair continued growing as a uniform structure (13). In the stone cricket, however, this simple explanation cannot hold. Here, the duplicitas cruciata must result from complex inductive or field effects as sketched in Fig. 4b and c. Crosswise duplications are not known from long germ-type eggs and thus may reflect specific traits of short germ patterning.

Morphogenetic Centers and Pattern Formation

Since the data from eggs of lower insects clearly contradict the mosaic hypothesis, embryonic pattern formation in these forms required other explanations. Combining evidence from species belonging to several insect groups, Seidel suggested that pattern formation ensues from a region near the middle of the prospective germ anlage, called the *differentiation center* (27). In many insects, this region can be recognized because morphogenetic steps like gastrulation and segmentation occur there first, and then spread anteriorly and posteriorly; in terms of the germ band pattern, the center seems linked to the anterior thoracic region. Seidel assumed that this "leading" region is also a region that *instructs* adjacent regions as to the parts of the pattern they should form. He was not able to provide positive evidence for this function, e.g., by transplanting material from the center and demonstrating its influence on patterning in new and indifferent surroundings, as had been done shortly before with the amphibian organization center by Spemann and Mangold (32). The negative experimental evidence brought forward in favor of this function (27), on the other hand, can possibly be explained on a different assumption, namely that the capacity of cells to take over the function of disabled neighbors *is lost first* in this region, and thereafter in progressively more anterior and more posterior regions. The center would then not be the place from where patterning instructions (e.g., orders for forming the different segments) emerge and spread into the rest of the germ anlage, but rather the place where the capacity of cells to deviate from their normal developmental pathways is first lost. This is not implausible, in view of the fact that morphological differentiation occurs there first; such visible specialization, according to commonly held views, should be preceded or accompanied by impairment of the capacity to switch to nonrelated developmental functions, and this impairment (commitment) should therefore take the same spatio-temporal course as visible differentiation. An explanation on this basis was provided in Fig. 13 of (21).

Earlier, Seidel (27) did another experiment that proved seminal, although again, his interpretation may need to be modified or extended in the light of more recent data. If a small posterior portion of the damsel fly egg is eliminated before a cleavage nucleus arrives there, the blastoderm cells all over the egg will turn into serosa cells, although most of them would have become part of the germ anlage in the untreated egg. Seidel called this region *Bildungszentrum (formative center)*, since in its absence the germ anlage fails to form. The term was translated later on as *activation center* which is unfortunate because the region has nothing to do with egg activation as commonly understood (25). Seidel's interpretation was that the immigrating cleavage energid triggers in the polar ooplasm a change of physiological state which propagates anteriorly until it reaches and "activates" the differentiation center (Fig. 5); thereupon the center can attract blastoderm cells, and a germ anlage ensues. In the light of more recent data, what is spreading from the posterior pole region might not (only?) be a general prerequisite required for a germ anlage to form instead of serosa, but (also?) factors or a gradient essential for patterning within the germ anlage, as will be discussed in the next section.

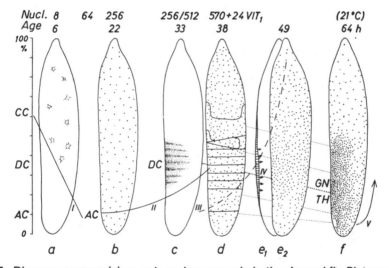

Figure 5. Diagram summarizing early embryogenesis in the damsel-fly *Platycnemis pennipes* (Odonata) and evidence for a differentiation center in pattern formation. Scale at the left serves to determine positions in the egg; at the top, the number of nuclei is shown. Ameboid outlines in (a) represent cleavage energids which spread from a cleavage center (CC). Dots in (b) and (d—f) represent nuclei. Sloping dotted lines between d) and f) delimit prospective territories (GN = gnathocephalon, TH = thorax). Arrow V indicates morphogenetic movements of the germ anlage blastema. II and III symbolize functional and structural changes spreading from the activation center (AC) after interaction with a cleavage energid which migrated there (curve I); label IV denotes transformation of blastoderm into germ anlage, beginning in the differentiation center. (d) shows fate map established with UV lesions at late blastoderm stages. Dots in (c) indicate sites where corresponding defects can be induced in the early blastoderm. After Seidel, from (14). Note that the early locations (c) are clustering closer together than the corresponding anlagen in the late blastoderm (d). Seidel took this to indicate a spreading of segmental instructions from the differentiation center. However, the change might as well be due to "regulative" capacities being lost first in that region; see Fig. 13 in (21).

Posterior Determinants and Morphogenetic Gradients

Planning to test the relevance of Seidel's ideas for eggs from another insect group, the writer performed large series of ligation experiments on leafhopper eggs, varying both the site and stage of ligation. The outcome was compatible neither with the mosaic hypothesis nor with a differentiation center comprising less than two-thirds of the entire germ anlage (or one-third of the entire egg) (Fig. 6a); moreover, elimination of the posterior pole region did not suppress germ anlage formation altogether as in the damsel fly. On the other hand, the pole region was found to exert a profound effect on pattern formation (Fig. 6b). The later it was tied off, the more complete was the segment pattern formed by the surviving anterior part of the egg. This graded temporal change was paralleled by an effect graded in space: the larger the posterior region eliminated at a given stage, the more segments were lacking in the surviving anterior part. When more than 30–40 % of the egg were removed (up to and beyond line V in Fig. 6b, c), the effect was like removing the activation center of the damsel fly egg at an early stage; that is, only serosa was

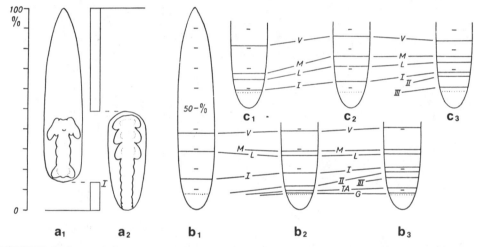

FIGURE 6. Constriction experiments on eggs of the leafhopper *Euscelis plebejus* (Homoptera). (a) Diagrams showing minimum fragment size required for specification of a prothorax in anterior (a1) or posterior fragments (a2); note that this segment, which is the bottom segment at the left but the top segment at the right, forms in quite different locations under different experimental conditions. Columns represent parts of the egg removed. (b and c) Stage dependence of minimum sizes for anterior fragments capable of forming various body segments as the most posterior pattern elements (see lettering). These diagrams reflect conditions prevailing in eggs of "good" (b) and "poor" (c) capacity for pattern formation in anterior fragments. b1, c1 = ligation before cleavage, b2, c2 = ligation during cleavage, b3, c3 = ligation at an early germ anlage stage. The capital letters indicate the most posterior segment to be expected with ligation at the level shown (e.g., prothorax in Fig. a1). Note the addition of posterior segments to the pattern after prolonged contact with the posterior egg region. Labels indicate: V, procephalon; M, mandibular/maxillary segment; L, labial segment; I–III thoracic segments; TA anterior abdominal segments; G, posterior abdominal segments. From (18).

formed although some blastoderm cells (or their descendants) should have contributed to the germ anlage. This suggests that the posterior pole region plays a comparable role in both species. The difference in the results from early elimination arises because the system is more "immature" at oviposition in the damsel fly than in the leafhopper, reacting therefore in the damsel fly by an initial all-or-nothing switch instead of the graded response observed in the leafhopper. Additional findings can be recruited for supporting this view (see Fig. 12 in 21).

The data shown in Fig. 6b convey the impression that prolonged contact with the posterior pole region somehow pushes the "borders" for individual segments anteriorly. That is, it improves the patterning capacity of the surviving anterior part. Positive evidence for this role of the posterior pole region was provided by shifting material from there anteriorly and enclosing it in an anterior egg fragment by ligature (19). In two-thirds of all cases, such combined fragments produced the complete pattern, while control fragments not supplied with pole material were barely capable of forming the anterior head structures (Fig. 7). To explain how this pole material could so profoundly influence patterning, not only in its vicinity, but also over quite some distance, e.g., in the thorax, the author developed a gradient interpretation (Fig. 8) on the basis of the data shown in Fig. 6b, c. In this hypothesis, a gradient of some unidentified egg component(s), sloping from the posterior region, steepens during early development, and this shifts anteriorly the sites where the component(s) in question would reach any given scalar value. The scalar value of the graded property in a given place might then specify which segment should form there. However, other data suggested the introduction of a second gradient of opposite slope, and the assumption was then made that the local relation between anterior and posterior "strength" (= scalar value) served as the parameter directing differentiation (19). Which molecules or structures represent the graded prop-

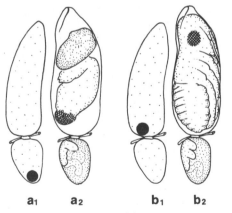

a₁　　**a₂**　　　**b₁**　　**b₂**

FIGURE 7. Influence of posterior pole material (represented by black disc) on pattern formation in the leafhopper *Euscelis plebejus* (Homoptera). (a) After constriction of the egg during cleavage (a1) the anterior fragment (top) forms an oversized procephalon with fused eyes at the posterior end (a2). When supplied with pole material (b1), the fragment as a rule will form the complete pattern, i.e., a viable larva (b2). The posterior fragments (bottom) can produce approximately the posterior half of the pattern, which here is shown at the germ band stage. From (21).

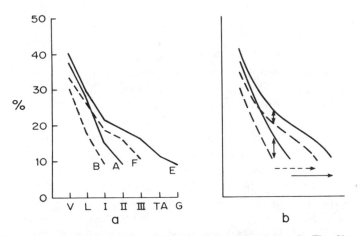

FIGURE 8. Graphical transformation of the *Euscelis* data shown in Fig. 6b, c into gradient representations; the vertical orientation of the egg axis (ordinate) has been retained from Fig. 6. Each pattern element is assigned a scalar value (abscissa, for abbreviations see Fig. 6), and the locations on the ordinate (see Fig. 6) for these pattern elements are plotted against the respective scalar values. A and B, "gradients" before cleavage (see Fig. 6 b1, c1), E and F, same at the early germ anlage stage (see Fig. 6 b3, c3). The right panel shows smoothed "gradients" with arrows at the bottom indicating the increase in scalar value near the posterior pole during early development. Double-headed arrows mark locations of a given scalar value at the early and at the late stage (in eggs with good and poor capacity for anterior pattern formation, upper and lower arrowhead, respectively). Turn the figure by 90° (left side towards the bottom) in order to visualize the increase with age in steepness of the gradients. From (19).

erties remains open in either case, as does the question whether the gradient(s) are located in the cortical region of the egg cell and the blastoderm cells derived from it, or (also) in the central portion and the yolk system.

One deduction from the gradient interpretation was that the pole material should build up a gradient of inverted polarity if confined to the anterior end of a posterior fragment. Inverted patterns compatible with this deduction were indeed obtained in most cases after this experiment. Besides completely reversed partial germ bands, "double abdomen" patterns combining original and reversed polarity were found (21) (Fig. 9). These "longitudinal mirror duplications" were explained by assuming that part of the effective pole material had been left behind at its original location so that a double-sloped gradient resulted. By delayed ligaturing, which would allow the pole material to act both anteriorly and posteriorly from its new location, even longitudinal triplications of the posterior body half were obtained (Fig. 10f).

These duplications and triplications cannot be explained satisfactorily with the quantitative assumptions of the original two-gradient model (31 and 0. Vogel, personal communication). Moreover, some evidence and theoretical considerations suggest that the gradient(s) may arise already during oogenesis. The experiments described above would thus not interfere with the mechanisms that establish the gradient, but rather with mechanisms which maintain the (labile) initial gradient(s) until the blastoderm or germ anlage stages.

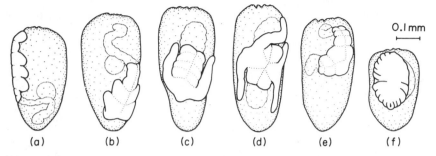

FIGURE 9. Effects of transposing material from the posterior pole to the anterior end of a posterior fragment in the leaf hopper (*Euscelis*. (a) control fragment (pole material left in place) forms partial germ band with abdomen immerged in the yolk system posteriorly; (b and c) partial germ bands of reversed longitudinal polarity; (d and e) germ bands of the longitudinal mirror image type, immerged abdomens anteriorly and posteriorly. Note oblique plane of symmetry in (e). (f) Late Stage double abdomen. From (21).

These assumptions are part of Hans Meinhardt's dynamic gradient interpretation which by computer simulation was shown to be compatible with a large body of experimental data from leafhopper and other insect eggs (see Chapter 3). Yet this compatibility does not exclude other possible mechanisms, and some findings can better be explained by assuming local interaction and intercalation as a patterning principle (23). This holds for double abdomens with oblique planes of mirroring symmetry, as found both in the leafhopper (Fig. 9) and in a beetle (29). A complete reinterpretation of the leafhopper results, on the basis of local interaction

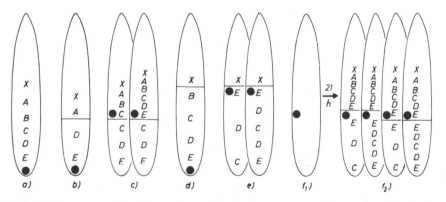

FIGURE 10. Summary of patterns formed in leafhopper eggs. Transverse line indicates level of ligation, black dot marks position of pole material. Letters approximate the blastodermal fate map under normal (a) and various experimental conditions (b–f). X extraembryonic blastoderm, A procephalon, B gnathocephalon, C thorax, D anterior abdomen, E posterior abdomen. Overlapping eggs illustrate different results from the same type of experiment. (b and d) ligations causing a gap in the pattern. (c and e), ligation after transposition of pole material; pattern In the right egg of (c) is shown in Fig. 8b, patterns in (e) are shown in Fig. 9b–e. (f) Delayed ligature results in longitudinal mirror duplications or triplications, with pattern E-D-C extending both anteriorly and posteriorly from the location of the pole material. Experiments b to f1 were performed during cleavage, ligature in f2 was delayed until an early germ anlage stage. From (21).

rather than global gradients, has been attempted by Otto Vogel (31 and *unpublished*).

The Gap Phenomenon: Proof of Antero-Posterior Interaction in Patterning

If both fragments of the leafhopper egg continue developing after early ligation, the partial germ bands they produce do not add up to the complete pattern: some segments are formed neither in the anterior nor in the posterior fragment, and thus a gap remains somewhere in the middle of the germ band pattern. This phenomenon was first observed in the leafhopper (Fig. 10b), but later found to occur in most insect eggs so far cut in two fragments during early development (9, 21, 23, 29). The gap is not due to local destruction of determinants (as the mosaic egg hypothesis might suggest), but to the prevention of further interaction between the parts separated. This follows from the fact that an additional operation, namely transposition of posterior pole material, in the leafhopper not only prevents the gap, but actually increases the patterning capacity of the system above the normal level (longitudinal duplications and triplications, Figs. 7, 10c, f). The gap can be explained by the gradient concepts outlined earlier in this chapter (16, 19), but might as well occur with other types of interactive patterning.

Anterior Determinants and Bipolar Pattern Specification

Details of the gap phenomenon suggested that interaction between posterior and anterior egg regions in the leafhopper (and most other species) is not one-way but mutual, and this lead to the assumption of the second (anterior) gradient mentioned above (19). Striking evidence for an anterior influence, in this case exerted by the anterior pole region, on embryonic patterning came from experiments on eggs of chironomid midges. Hideo Yajima (33, 34) showed that centrifugation or local uv-irradiation can inflict global changes which lead to "double abdomen" or "double cephalon" patterns instead of a normal larva (Fig. 11). He concluded that the periplasm of the midge egg contains two formative localities, one in each end of the egg, which carry a tendency for head or abdominal development, respectively, and by interaction determine the thoracic structures.

The analysis of ooplasmic determinants in Yajima's "anterior formative locality" was carried down to the molecular level by Klaus Kalthoff and his collaborators in an impressive series of experiments on double abdomen induction in chironomids (9). A recent experiment demonstrating the effects of local uv-irradiation near the anterior egg pole of a midge egg is shown in Fig. 12. Action spectra for uv induction of double abdomens indicated that the uv targets, i.e., the structures or molecules in the anterior cytoplasm whose inactivation triggers formation of a tail instead of a head, contain both a protein and a nucleic acid component. The latter component permits photoreversion of double abdomen induction: the percentage of abnormal embryos decreases drastically when uv-irradiation is followed by irradiation with longer wavelengths under conditions known to support repair of uv-damage to nucleic acids. Thereafter it was shown by Kandler-Singer and Kalthoff

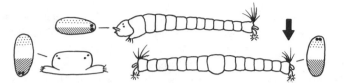

FIGURE 11. Effects of centrifugation on pattern formation in *Chironomus dorsalis* (Diptera). Top: normal larva, obtained after longitudinal stratification of egg contents. Bottom: double cephalon and double abdomen resulting from transverse stratification. Arrow marks direction of centrifugal force, large dots in stratified eggs represent pole cells. In other chironomids, the type of double monster is not strictly linked to the orientation of the egg in the centrifuge. After Yajima, from (21).

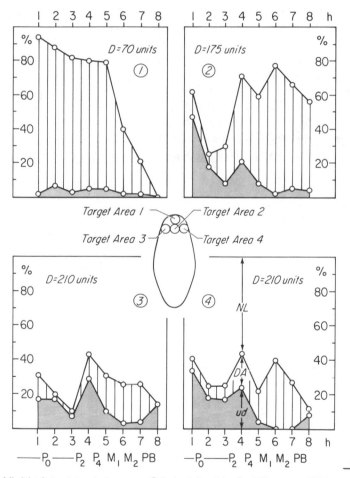

FIGURE 12. Yield of double abdomens (DA, hatched) in *Smittia* spec. (Chironomidae, Diptera) after uv irradiation of the target areas indicated in the inset. NL means normal larvae, ud undifferentiated eggs; letters at bottom indicate stages from early cleavage to early blastoderm. D is uv dose in relative units. Note that double abdomen yield is highest with the most anterior target area (No. 1), although the dose used there was much lower than in areas 2–4. In the latter, the yield increases with age while it decreases in area 1; this may indicate that the "anterior determinants" of Smittia slightly shift positions during early development. From Kalthoff (9). Copyright Alan R. Liss, Inc. (1983).

(10) that RNase S injected into the anterior pole region induces double abdomen formation while the two polypeptides which constitute RNase S fail to do so if injected singly. Taken together these results mean that in midge eggs the decision for "anterior" development depends on ribonucleoprotein particles located in the cytoplasm of the anterior egg region. Recent results indicate that this decision is made in competition or cooperation between anterior and posterior determinants rather than by a mere threshold effect of the anterior determinants. How the pattern develops once this decision has been made in each egg pole remains open as yet; formal possibilities were discussed by several authors (9, 23, 34).

Before leaving the problem of longitudinal patterning, let us reconsider Hegner's claim for a mosaic status of the beetle egg. Extensive investigations in recent years have shown that the "gap phenomenon" is very prominent in beetles (21, 30); indeed, some results from Hegner's coagulation experiments are indicative of the gap phenomenon (Fig. 2c). Moreover, the double abdomen aberration can be induced in beetles, too (29, 30). The formation of a germ anlage in the wrong place, another of Hegner's key results, is easily achieved in a number of definitely nonmosaic insect eggs (Fig. 10c, f). Thus it seems that Hegner's findings were insufficient as a basis for his conclusions: the (invisibly) different cytoplasmic areas in the insect egg are not "predetermined to develop into definite parts of the embryo," but rather are distinguished in the sense that they harbor different essential components of the global mechanism that specifies the basic body pattern.

Integrated Two-Dimensional Patterning or Linear Patterning in Two Dimensions of Space?

As explained earlier, the basic body pattern is two-dimensional at the germ band stage; it displays a longitudinal and a transverse aspect. So far we have considered only the longitudinal aspect, that is, the series of different body segments lined up in the germ band. Is this self-imposed restriction to a single dimension of space justified by experimental data? This question is urgent in view of the elegant model of Kauffman et al. (11; and see Chapter 4) which achieves integrated patterning in two dimensions of space.[1] However, several findings speak against patterning by such a mechanism in insect embryogenesis, and implicitly justify our approach of studying pattern formation as a linear event. One recent finding is that adjacent longitudinal strips of blastoderm may produce patterns of opposite axial polarity and discrepant regional character, e.g., a tail end forming in one strip next to head parts in the adjacent strip (29, 23). Another argument comes from experiments altering the geometrical proportions of the patterning system, to which Kauffman-type models are very sensitive. The geometry of the patterning system is altered for instance in the ligaturing experiments described previously, and even more so when egg or germ anlage are separated lengthwise (see below). The patterns formed after such experiments differ strongly from those predicted from integrated two-dimensional patterning by Kauffman's mechanism.

[1]Those looking for a tangible analogy to this model should consider Chladni's figures on musical instruments, see e.g., Hutchins, C.M., *Sci. Am.* **245**: 127–135 (1981), and Rossing, Th.D., *Sci. Am.* **247**: 147–152 (1982).

Parallel Twins: Transverse Pattern
Regulation in Embryonic Blastemas

The transverse aspect of embryonic pattern formation becomes apparent in the subdivision of the germ anlage blastema into the left and right half, with the various structures or organ rudiments arranged from medial to lateral in mirror image symmetry on either side of the midline (Fig. 1b–e). The mosaic egg hypothesis would postulate a corresponding bilaterally symmetrical pattern of prelocalized determinants but here, too, the experimental data are conflicting with this view. When the germ anlage of the stone cricket is cut in the median plane, both halves will form bilaterally symmetrical germ bands instead of left and right half, respectively, of a single germ band. The way by which this "regulation" is achieved in "half-blastemas" varies depending on stage (Fig. 13). After early separation, each half "symmetrizes" and gains a new midline. In a half-blastema isolated at a later stage, the original midline is retained and the lacking half is replaced by changing the developmental fate of the adjacent prospective amnion (12). This type of experiment was carried one step further in the leafhopper where the entire egg can be separated lengthwise with an appropriate gadget (20). The astonishing result was that not only left and right egg halves can each produce a bilaterally organized germ band, but even a dorsal half can do this although in normal development it would yield only the most lateral parts of the germ anlage (Fig. 14). Finally, symmetrization of both halves of the germ anlage was achieved in vitro in a holometabolous insect, the silkmoth, by Krause and Krause (21). Also, Miya and Kobayashi induced a beetle egg by cold shock to produce up to four bilaterally symmetrical embryos side by side (21).

These data indicate that bilateralization and transverse patterning can occur fairly late in development, and are systemic properties of the embryonic blastema rather than of the egg cell, despite the bilaterally symmetrical shape of the latter and its shell. Moreover, in the leafhopper the reaction of egg cell or blastema to

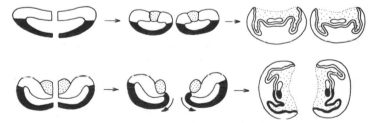

FIGURE 13. Transverse pattern regulation in the stone cricket *Tachycines asynamorus* (Orthoptera). Shown are transverse sections through the amniotic cavity at different stages. Top row: Longitudinal cutting of early germ anlage and amnion (black) is followed by resymmetrization of the halves of the germ anlage before gastrulation, as shown by the mesoderm anlagen (stippled) and by the relative positions of the twin germ bands (at the right). Bottom row: A similar cut during gastrulation is followed by "assimilatory induction" (arrows) in the amnion which heals to the germ anlage blastema at the cut; the twin germ bands in this case take up a back-to-back position. After Krause, from (21).

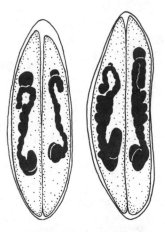

FIGURE 14. Bilaterally symmetrical twin germ bands in eggs of the leafhopper *Euscelis plebejus* (Homoptera). The eggs were pinched lengthwise during late cleavage with a blunted razor blade. Left egg divided in lateral halves, right egg separated in ventral and dorsal half. Due to secondary movements, the head lobes are seen close to the posterior pole (bottom) and the abdomens point anteriorly (top). Drawn after photographs in (20).

lengthwise splitting differs profoundly from the reaction to transverse disrupture (by ligation, etc.). After lengthwise separation, both fragments produce much more than their normal share of the body pattern ("positive regulation"), and the capacity to do so *decreases* quite late in development. After transverse separation, on the other hand, both parts (or at least one of them) produce less than their normal share of the pattern ["negative regulation" leading to the gap phenomenon, (20)], and their pattern-forming capacity *increases* with age at separation until at some late stage they form those parts of the pattern which they also would have formed in normal development. Lateral halves isolated at this "mosaic" stage, however, are still capable of positive regulation. We take these differences to mean that the germ band pattern is established by linear patterning mechanisms acting successively and differently along two axes of space set at right angles, rather than by simultaneous two-dimensional patterning.

Clonal Restriction and the Compartment Hypothesis

So far we have treated the egg cell and the germ anlage blastema as a single patterning unit. However, the germ band consists of many segments that look like variants of one and the same repeat unit, particularly when they develop intrasegmental patterns during subsequent stages. This raises the question of whether and when during development the segments become functionally distinct patterning units. An answer comes from experiments that reveal the developmental fates of the descendants of individual cells marked at different stages. Peter Lawrence (15) irradiated early embryos of a bug with x-rays. This treatment caused hereditary changes in individual cells which become manifest as pigmentation anomalies in descendant cells if these happen to contribute to the integument of the larva. The distribution of the descendants of a single altered cell, which by definition represent a clone, showed a striking dependence on the stage of irradiation (Fig. 15). Following irradiation before the blastoderm stage, patches of descendant cells were scattered over several segments in the larva, and some patches extended across segment borders. Irradiation at or after the blastoderm stage led to the confinement of all descendants to a single segment (or even a ventral or dorsal half-segment), and patches were seen to extend along the borders of that particular segment but never across it (clonal restriction). The conclusion is that at (or briefly after) the blastoderm stage the progenitor cells for a given segment lose the capacity to contribute to adjacent segments. A cell population defined by clonal restriction was called a compartment by Antonio Garcia-Bellido (15), and this seems to be a functional unit for subsequent patterning. This view is based on genetical evidence from the fruit fly which moreover indicates a progressive subdivision of segmental compartments into smaller compartments (see Chapters 12 and 13).

Mesodermal Patterning by Induction from the Ectoderm

The ectodermal pattern is externally visible and therefore attracts most attention. However, it is paralleled by a patterned internal organization, and we may ask the

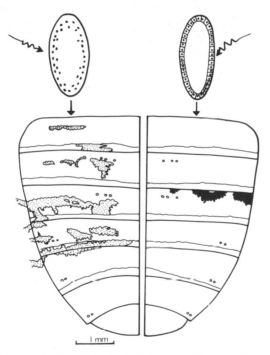

FIGURE 15. Compartments in the ventral abdomen of larvae of the milkweed bug *Oncopeltus fasciatus* (Heteroptera). Irradiation of early embryonic stages with X-rays causes hereditary changes in individual cells which become visible as pigmentation anomalies in the epidermal descendants of these cells. A clone induced during cleavage (left half) may contribute to several larval segments and to the dorsal side while a clone induced at the blastoderm stage (right half) is restricted by segment (= compartment) borders. From Lawrence (15). Copyright MIT, 1981.

question how this arises. This question is of particular interest because in amphibians nearly all externally visible features are somehow dependent on influences from archenteron and mesoderm, that is, from internal structures. The inverse seems true for insects. Mutant embryos that locally or entirely lack the mesodermal structures show nonetheless recognizable segmental patterns in the cuticle derived from the ectoderm (17, see Chapter 11). Experiments performed on germ anlage and later stages in some insects revealed that the internal pattern is "imprinted" upon the inner layer from the patterned ectoderm. The first relevant set of data came from the work of Eberhard Bock (12). He eliminated small portions of the prospective germ layers before these layers were stacked onto each other by the process of gastrulation. Gaps made in the ectoderm were accompanied by failure of the adjacent mesodermal structures to form. More convincingly, elimination of the prospective lateral mesoderm caused the prospective medial mesoderm cells to move laterally where they differentiated according to their new positions, and not according to their prospective (medial) fates.

Cell Determination in the Fruit Fly

Like most biological terms, cell determination is vaguely defined and inconsistently used. It may indicate a progressive process or a state. If meant to indicate a process, determination is thought to cause a cell to enter a specific pathway of development, and/or to render it incapable of switching (under the conditions of an experiment) to another developmental pathway. Seemingly contradictory results from insect embryos suggests that these may be two separate steps for which the writer earlier used the terms instruction and commitment (21). A "naive" cell might consecutively receive instructions or orders for entering several developmental pathways (without obeying these orders), but then comes the signal for commitment and the cell is bound to follow the last instruction received, and to differentiate accordingly by using the relevant parts of its genetic outfit (for instance those required in the metathorax). Experiments on fruit fly embryos have yielded some information on the timing of determination (sensu commitment) which we will discuss because the timing might decide to which pattern element a cell will ultimately contribute.

Treating fruit fly embryos with a heat pulse or ether vapor, Karl Henke et al. (5) and Hans Gloor (2) found that (some) progenitor cells for the halteres (organs on the third thoracic segment of the fly) were apparently switched to the pathway leading to wing development (typical of the second thoracic segment). They thereby "phenocopied" homeotic mutants of the *tetraptera* or *bithorax* type (21). Interestingly, this effect can be achieved only during a narrow "time window" around the blastoderm stages (5). This means that the cells by that time have been (or are being) instructed about their segmental affiliation, and that thereafter they are committed to one single segment-specific pathway, unless deprived of the regulatory genes required for this restriction (21).

By another approach, Karl Illmensee showed that the determined or committed state is a property of the whole cell rather than its nucleus (6, 7). He injected genetically marked blastoderm cells or nuclei from stages up to gastrulation into host eggs. Anterior lateral blastoderm cells transplanted to the posterior pole of blastoderm hosts failed to execute the functions appropriate to their new surroundings and instead formed structures related to their original positions. On the other hand, when genetically marked nuclei from gastrula cells were injected into unfertilized eggs, these eggs occasionally gave rise to flies carrying the marker, thus showing that (some?) nuclei are still unrestricted in their developmental potential at blastoderm or even gastrula stages although the cells from which these nuclei came were probably restricted.

Outlook

Insect eggs provide a variety of technical obstacles not encountered in the more frequently studied developmental systems. Yet they also offer unique possibilities which have not yet been fully exploited. It is these possibilities which can be expected to advance our understanding of pattern formation in insect embryos and perhaps on a more general level. For instance, the wealth of detail in the patterns formed (compare Fig. 1c–f with a hydra!) could provide much valuable information if properly utilized after experiments altering the pattern. The biggest advantage, however, of insects is the well-known genetics of some species, mainly the fruit fly and the silk moth. Many mutants known in these species are by themselves marvelous tools for research on pattern formation (see Chapter 11) yet some should prove even more fruitful in connection with appropriate experiments on oogenesis and early embryonic stages. We thus may look forward to new insights into pattern formation ranging from the molecular level to the formal rules for complex molecular and supramolecular interactions which guide patterning systems, based on work with insect embryos.

Acknowledgments

I am deeply indebted to the Deutsche Forschungsgemeinschaft for supporting my research work, to my colleagues in Austin, Cambridge, Freiburg, Geneva, Mito, and Würzburg who provided data and figures, and to Mrs. M. Scherer for invaluable help with the manuscript.

General References

Counce, S.J., Waddington, C.H. (eds.): *Developmental systems: insects*. 2 Vol. London: Academic Press (1972).

Sander, K.: Specification of the basic body pattern in insect embryogenesis, *Adv. Insect Physiol.* 12:125–238 (1976).

References

1. Counce, S.J., Waddington, C.H. (eds.): *Developmental systems: insects*. 2 Vol. London: Academic Press (1972).

2. Gloor, H.: Phänokopie-Versuche mit Äther an *Drosophila, Rev. Suisse Zool.* 54:637–712 (1947).

3. Hegner, R.W.: Experiments with Chrysomelid beetles. III. The effects of killing parts of the eggs of *Leptinotarsa decemlineata. Biol. Bull.* 20:237–251 (1911).

4. Hegner, R.W.: The genesis of the organization of the insect egg, *Am. Naturalist* 51:641–661 and 705–718 (1917).

5. Henke, K., v.Fink, E., Ma, S.-Y.: Über sensible Perioden für die Auslösung von Hitzemodifikationen bei *Drosophila* und die Beziehungen zwischen Modifikationen und Mutationen. *Zeitschr.indukt. Vererbungslehre* 79: 267–316 (1941).

6. Illmensee, K.: Nuclear and cytoplasmic transplantation in *Drosophila.* In: *Insect development,* Lawrence, P.A. (ed.), pp. 76–96. Oxford: Blackwell (1976).

7. Illmensee, K.: *Drosophila* chimeras and the problem of determination. In: Genetic mosaics and cell differentiation Gehring, W.J. (ed.), pp. 51–69. Berlin: Springer Verlag (1978).

8. Johannsen, O.A., Butt, F.H.: Embryology of insects and myriapods. New York: McGraw-Hill (1941).

9. Kalthoff, K. Cytoplasmic determinants in dipteran eggs. In: Time, Space and Pattern in Embryonic Development. New York: Alan R. Liss (1983).

10. Kandler-Singer, I., Kalthoff, K.: RNase sensitivity of an anterior morphogenetic determinant in an insect egg (*Smittia* spec., Chironomidae, Diptera). *Proc. Nat. Acad. Sci. USA* 73:3739–3743 (1976).

11. Kaufman, St.A., Shymko, R.M., Trabert, K.: Control of sequential compartment formation in *Drosophila, Science* 199:259–270 (1978).

12. Krause, G.: Induktionssysteme in der Embryonalentwicklung von Insekten. *Ergeb. Biol.,* 20:169–198 (1958).

13. Krause, G.: Die Entwicklungsphysiologie kreuzweise verdoppelter Embryonen. *Embryologia,* 6:355–386 (1962).

14. Krause, G., Sander, K.: Ooplasmic reaction systems in insect embryogenesis. *Adv. Morphogen* 2:159–303 (1962).

15. Lawrence, P.A.: The cellular basis of segmentation in insects, *Cell* 26:3–10 (1981).

16. Meinhardt, H.: A model of pattern formation in insect embryogenesis. *J.Cell Sci.* 23:117–139 (1979).

17. Nüsslein-Volhard, Ch.: Maternal effect mutations that alter the spatial coordinates of the embryo of *Drosophila melanogaster.* In: Determinants of spatial organization, Subtelny, St. Konigsberg, I.R., (eds.), pp. 185–211. Academic Press (1979).

18. Sander, K.: Analyse des ooplasmatischen Reaktionssystems von *Euscelis plebejus* FALL. (Cicadina) durch Isolieren und Kombinieren von Keimteilen. I *Wilhelm Roux' Arch.* 151:430–497 (1959).

19. Sander, K.: Analyse des ooplasmatischen Reaktionssystems von *Euscelis plebejus* FALL. (Cicadina) durch Isolieren und Kombinieren von Keimteilen II., *Wilhelm Roux'Arch.* 151:660–707 (1960).

20. Sander, K.: Pattern formation in longitudinal halves of leafhopper eggs (Homoptera) and some remarks on the definition of "embryonic regulation". *Wilhelm Roux' Arch.* 167:336–352 (1971).

21. Sander, K.: Specification of the basic body pattern in insect embryogenesis. *Adv.Insect Physiol.* 12:125–238 (1976a).

22. Sander, K.: Morphogenetic movements in insect embryogenesis. In: Insect development, Lawrence, P.A. (ed.), pp. 35–52. Oxford: Blackwell (1976 b).

23. Sander, K.: Pattern generation and pattern conservation in insect ontogenesis—problems, data and models. In: Progress in developmental biology, Sauer, H.W. (ed.). *Fortschritte Zool.*, **26**:101–119. Stuttgart, Gustav Fischer Verlag (1981).

24. Sander, K.: The evolution of patterning mechanisms: gleanings from insect embryogenesis and spermatogenesis. In: *Development and evolution*, Goodwin, B.P. (ed.), pp. 137–159. Cambridge: Cambridge University Press (1983a).

25. Sander, K.: Biology of fertilization in insects. In: Biology of fertilization, Metz, C.H. Monroy, A. (eds.) Academic Press (1983b; in press).

26. Seidel, Fr.: Die Determinierung der Keimanlage bei Insekten III, *Biol. Zentral*bl. **49**:577–607 (1929).

27. Seidel, Fr.: Entwicklungsphysiologische Zentren im Eisystem der Insekten. *Zool. Anz. Suppl.* **24**:121–142 (1961).

28. Schubiger, G., Newman, Jr. S.M.: Determination in *Drosophila* embryos. *Am. Zool.* **22**:47–55 (1982).

29. van der Meer, J.M.: Region specific cell differentiation during early insect development. *Ph.D. thesis*, University of Nijmegen (1978).

30. van der Meer, J.M., Kemmner, W., Miyamoto, D.M.: Mitotic waves and embryonic pattern formation: no correlation in *Callosobruchus* (Coleoptera), *Wilhelm Roux's Arch.* **191**:355–365 (1982).

31. Vogel, O.: Experimental test fails to confirm gradient interpretation of embryonic patterning in leafhopper eggs. *Dev.Biol.* **90**:160–164 (1982).

32. Willier, B.J., Oppenheimer, J.M. (eds.): Foundations of experimental embryology, Englewood Cliffs, N.J.: Prentice-Hall, Inc. (1964).

33. Yajima, H.: Studies on embryonic determination of the harlequin-fly, *Chironomus dorsalis*. I. Effects of centrifugation and of its combination with constriction and puncturing. *J. Embryol. exp. Morph.* **8**:198–215 (1960).

34. Yajima, H.: Studies on embryonic determination of the harlequin fly, *Chironomus dorsalis*. II. Effects of partial irradiation of the egg by ultra-violet light. *J. Embryol. Exp. Morphol.* **12**:89–100 (1964).

35. Zissler, D., Sander, K.: The cytoplasmic architecture of the insect egg cell. In: *Insect ultrastructure* King, R.C. Akai, H. (eds.). Vol. 1, pp. 189–221. New York: Plenum Press (1982).

Questions for Discussion with the Editors

1. *Several examples of insect pattern formation you described are best explained, in formal terms, by morphogen gradients models. What are the near-term prospects for actually obtaining unequivocal experimental proof for the existence of such gradients?*

Prospects for finding gradients are not bad (there are indeed insect eggs which show a gradient of basophilic components), prospects for proving unequivocally that a given gradient is involved in patterning are poor, and chances that a single parameter functions as *the* gradient specifying pattern are next to nil in my in my view [because of the interactions to be expected in any system as complex as a living cell or sheet of cells (24)]. The ideal proof would be to expose "naive" cells to different levels of the parameter in question, and get the pattern elements formed which are to be expected for the respective level (along the lines of Toivonen and Saxen, but much greater power of resolution would be required). I do not expect such an experiment to work in insects. But even if it does, it proves only that things in the complete system *could* work that way; remember the effects of some most unlikely "inductors" in amphibians! However, with luck some mutants could provide cues as to the molecules really involved—but only in connection with appropriate experiments!

2. *Which of the insect systems described would yield the most useful information on pattern formation if it were pursued at the genetic level with the same vigor as* Drosophila *has been studied?*

Lack of vigor is not the main impediment for genetic studies in other insects; the drawback is much longer life cycles—and perhaps the inability of other species to survive with the most impossible genetic defects. I am told that Morgan or someone in his lab tried to replace *Drosophila* by *Coelopa*, and failed because he did not get viable mutants. The hymenopterans, especially *Mormoniella* (*Nasonia*), are rather similar on principle to *Drosophila* as far as embryogenesis goes; so despite their amenability to genetic work I would not expect new principles to arise from it. The most novel insights were to be gained from short germ eggs like lower beetles or crickets, because in these patterning must be largely independent of the yolk system and ooplasmic determinants. The intrablastemic reactions which gradually and successively establish the pattern in this type of development could be studied using mutants, but the technical odds are against such an approach. I once had a cricket strain which produced germ bands consisting only of procephalon and telson, but I could not maintain it.

3. *Insect embryologists generally regard pattern formation as the result of activities associated with specific cells, cell lineages, or groups of cells. Do subcellular phenomena, such as those associated with cilia row organization in protozoa, contribute to insect pattern specification?*

A ciliate-like cortical organization (both structural and functional) in the oocyte or egg cell seems improbable to me. However, properties in embryonic blastemas, for instance, polarity, might be encoded in the cytoskeleton of the individual cell, and might be signalled from cell to cell by surface properties influenced by the cytoskeleton. This seems to be the case in the larval hypoderm (according to findings of K. Nübler-Jung in our department) but might apply to any stage from the blastoderm onward.

CHAPTER 11

Genetic Analysis of Dorsal-Ventral Embryonic Pattern in *Drosophila*

Kathryn V. Anderson and Christiane Nüsslein-Volhard

THE SPATIAL PATTERN OF embryonic development is ultimately directed, like all other biological processes, by the coordinated activity of genes. The goal of developmental genetics is to identify and characterize genes specifically required in particular developmental processes. The genetic dissection of a process into its component parts can be applied just as profitably to the understanding of embryonic pattern formation in higher animals as it has been in the analysis of bacterial metabolic regulation.

Although genes direct pattern formation in all higher organisms, *Drosophila* is particularly well-suited to the genetic analysis of embryonic pattern formation, both because of the relative ease of genetic manipulation and because of the existence of well-defined spatial landmarks in the developed larva. Genetic studies are facilitated by *Drosophila*'s relatively small genome of about 5000 genes, its rapid generation time of two weeks, and the special genetic markers and chromosomes developed during the 70 years of research in *Drosphila* genetics. The larval pattern is characterized by readily identifiable cuticular structures that are formed at well-defined, reproducible positions of the larva (Fig. 1). Mutants that specifically alter spatial pattern can be recognized by changes in the organization of the cuticular markers in the differentiated larva.

Mutants affecting embryonic development can be divided into two functional classes. The first class is the group of zygotic lethal mutations, where the mutant phenotype is the result of the genetic constitution of the embryo itself. The second class of mutants is the set of maternal effect mutations, in which the mutant embryonic phenotype is independent of the genetic makeup of the embryo, and instead is determined by the genotype of its mother. Maternal effect mutations affect

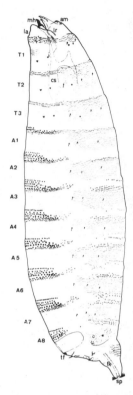

FIGURE 1. Cuticular landmarks of the first instar *Drosophila* larva. The cuticle of the larva provides a large number of useful markers defining both the anterior-posterior and dorsal-ventral patterns. In the anterior-posterior axis (anterior up), the sequence of three thoracic and eight abdominal segments is characteristic. In the dorsal-ventral axis (ventral left), the most prominent difference is in the presence of the thick, short ventral setae arranged in segmental bands in the ventral cuticle, with fine dorsal hairs on the dorsal side. mh: mouth hooks; la: labial segment; lr: labrum or median tooth; am: antennal and maxillary sense organs; cs: cephalopharyngeal skeleton; tf: tuft; fk: filzkörper; sp: spiracles.

the substances synthesized during oogenesis and stored in the egg cytoplasm for use during embryogenesis, while zygotic mutations alter substances normally synthesized during embryonic development under the direction of the nuclei in the embryo. In general, the maternal effect mutations that have been studied exert their influence throughout the embryo, while zygotic mutations tend to have more restricted, local effects.

Maternal effect mutations that alter the spatial determination of embryonic cells fall into two classes: those that affect anterior-posterior fate and those that affect dorsal-ventral fate (15). Both classes of mutations alter the fate of every cell of the early embryo, but with respect to only one of the two dimensions. In *bicaudal* embryos, for example, the fate of cells is shifted in the anterior-posterior axis, while the dorsal-ventral pattern is unaffected. In *dorsal* embryos the anterior-posterior polarity is retained, while the normal dorsal-ventral pattern is radically altered. These mutations reveal that there are two independent, perpendicular systems of information both of which operate over the field of the entire embryo.

These mutant phenotypes make it clear that embryonic determination is not the result of a mosaic of localized cytoplasmic determinants, but rather of an epigenetic, graded system of positional information. Unfortunately, the concept of positional information has remained rather abstract. The detailed analysis of mutant phenotypes and of the gene activities involved in normal pattern formation should make the concepts of the establishment and interpretation of positional information more concrete.

The studies described here illustrate how the analysis of mutants that alter spatial coordinates along one of the two primary embryonic axes, the dorsal-ventral axis, can illuminate the mechanisms involved in establishing positional information. Careful genetic and embryological analysis of mutant phenotypes makes it possible to make specific inferences about normal gene function. Systematic identification of all the genes important in the establishment of dorsal-ventral pattern defines the number of all of the components specifically required in this process. In addition, the mutations define the particular biochemical processes that are used in establishing pattern, and provide an access to a molecular understanding of pattern formation.

Dorsal-Ventral Pattern and the Dorsal Phenotype

The body plan of the *Drosophila* larva can be described in terms of an anterior-posterior series of repeating segments and a dorsal-ventral pattern perpendicular to the anterior-posterior pattern. The major germ layer decisions are subdivisions of the dorsal-ventral pattern: the separation of the mesoderm from ectoderm, neural ectoderm from epidermal ectoderm, and of ventral from dorsal epidermal ectoderm occur in the dorsal-ventral axis.

This essentially two-dimensional body plan can, in the case of *Drosophila*, be followed back in development to the cellular blastoderm stage. Cellular blastoderm is at 3.5 hr after fertilization, the first stage at which true cells are formed. For a description of early *Drosophila* embryogenesis, see Poulson, (21). The blastoderm is a single-layered sheet of cells covering the surface of the embryo. At cellular blastoderm, cells assume at least some positional identity: anterior or posterior epidermal cells removed from the blastoderm differentiate according to their original location (3, 7).

Fate-mapping studies performed at cellular blastoderm reveal the one-to-one correspondence between position in the blastoderm and final fate of the cell. Poulson (21) traced the movements of cells from the single-layered sheet of cells at blastoderm through morphogenesis histologically, and was able to define which cells of the blastoderm give rise to the various cell types of the larva. Killing of cells in specific regions of the blastoderm results in the loss of specific larval and adult structures. The cells that are lost are not compensated for by their neighbors, and

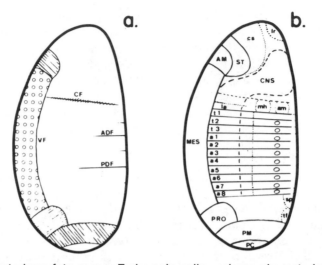

FIGURE 2. Blastoderm fate maps. Embryonic cells undergo characteristic morphogenetic movements and differentiate into particular structures according to their position in the single layer of blastoderm cells. In the fate maps anterior is up, and dorsal is to the right. (a) Invaginations and folds at specific positions of the blastoderm are characteristic of the morphogenetic movements of gastrulation. The ventral furrow (VF), a longitudinal strip encompassing the 20% of the cells at the ventral midline, invaginates in the first movement of gastrulation. The cepahlic fold (CF) is a lateral transverse fold. On the dorsal side of the embryo is a series of parallel folds, with anterior dorsal fold (ADF) and the posterior dorsal fold (PDF) being particularly prominent. (b) The differentiated structures of the larva have been traced back to their blastoderm anlage by both histological (21) and cell ablation (16) techniques. With respect to the dorsal-ventral axis, there are a number of useful structures which can be used as characteristic landmarks. The 20% of the cells near the ventral midline of the blastoderm give rise to mesoderm (MES). The anlage for the ventral nerve cord and the ventral cuticle overlap in the region 20–40% of the egg circumference away from the ventral midline. (The nervous system anlage is delimited by the broken line; the ventral cuticle anlage ends at the dotted line.) The dorsal larval cuticle derives from the region 60–95% away from the ventral midline. Within the anlage of the dorsal cuticle are the anlage of the tracheal pits (circles), which derive from 15% off the dorsal midline. The dorsal-most 5% of the blastoderm gives rise to the extra-embryonic membranes, amnion and serosa. In the head and telson (tail), there are a number of specialized structures that map at defined dorsal-ventral positions. The labrum (lr) (median tooth) derives from the dorsal midline. The supraesophageal ganglion of the central nervous system is also derived, in part, from an extreme dorsal position. More laterally, at approximately 15% off the dorsal midline, lie the anlage of the antennal and maxillary sense organs (am). The major parts of the head skeleton (cs), including the mouth hooks (mh) derive from more lateral positions. In the telson, both the spiracle (sp) and the tuft (tf), a clump of setae (tf), derive from at or near the dorsal midline. At approximately the same dorsal-ventral position as the tracheal pits is the origin of the Kilzkörper, the fluorescent and highly refractile internal specialization of the posterior openings. AM: anterior midgut; ST: stomodeum; PRO: proctodeum; PM: posterior midgut; PC: pole cells.

the specific defects that result have been used to establish larval fate maps that have some minor modifications of the blastoderm fate map of Poulson (12, 27).

From fate-mapping studies, it is possible to map cuticular structures of the normal *Drosophila* larva to specific dorsal-ventral positions in the blastoderm (12) (Fig. 2b). There is a progression of anlagen for different cell types in the dorsal-ventral axis. The mesodermal anlage lies most ventrally. More laterally is a region that gives rise to both ventral epidermis and the ventral nerve cord. The anlage for the dorsal epidermis lies in the dorsal 40% of the blastoderm. Numerous specializations in the head and the telson also derive from defined dorsal-ventral positions of the blastoderm.

Embryos produced by females homozygous for the maternal effect mutation *dorsal*[1] are completely dorsalized. The cells of the blastoderm live and undergo differentiation, but they all behave like the cells normally located in the dorsal positions of the blastoderm. The differentiated embryo, sitting in its egg case at the time when a normal larva would hatch, consists of a long hollow tube of dorsal cuticle. The labrum is present anteriorly and the spiracles posteriorly, but all more ventral pattern elements, including tracheae, Filzkörper, head sense organs, ventral cuticle, ventral nervous system and mesoderm, are lacking (Figs. 3c, 4d). The only internal structures are neurons in the region of the brain, which normally derive from a dorsal position of the blastoderm fate map.

Females homozygous for a weaker mutation at the same locus, dl^2, produce some more ventral structures. Filzkörper are always seen in the telson, and antennal and maxillary sense organs in the head, and occasionally narrow bands of ventral setae are produced (Figs. 3e, 4c). The weakest *dorsal* phenotype is observed in the offspring of females hemizygous for *dorsal*[+], that is with only one copy of the normal dorsal gene. At high temperature and in particular genetic backgrounds, dl^1/ + females produce embryos with the dorsal "dominant" phenotype: the embryos make Filzkörper, head sense organs, bands of ventral setae of nearly the normal width, but lack normal mesoderm, and are characteristically twisted within the egg case (Fig. 3d).

The *dorsal* allelic series, from strong phenotypes in which the embryo is totally dorsalized, with the smooth progression through weaker phenotypes, involves the gradual addition of more ventral structures in the same order in which they appear in the embryonic fate map (Fig. 2). This progression reveals that the *dorsal* gene is involved in the establishment of positional information along a dorsal-ventral continuum. The strong *dorsal* phenotype (dl^1/dl^1), considered in isolation, is such a dramatic departure from the normal larva that it is difficult to recognize it as a mutant affecting a basic embryonic axis. The weaker alleles at the same locus, however, make it apparent that the activity of the dorsal gene is an essential component of dorsal-ventral positional information.

The direct demonstration that in mutant *dorsal* embryos the position-dependent fate of the cells in the blastoderm is shifted was accomplished by fate map studies of dorsal dominant embryos (16). In normal embryos, cell ablation at the ventral midline produces no cuticular damage, since the cells at the ventral midline normally give rise to only mesoderm. In dl^D embryos cell ablation at the ventral midline produces defects in ventral setae, indicating that the cells at the ventral

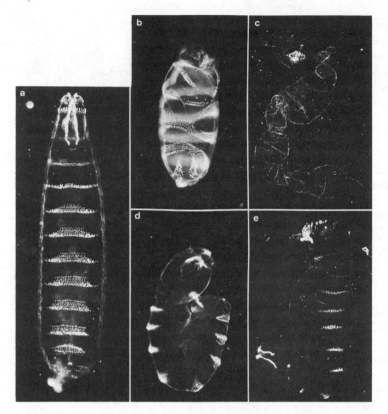

FIGURE 3. Dark field views of the cuticle produced by embryos with altered dorsal-ventral pattern. (a) The wild-type larva. The ventral setae are organized segmentally in bands on only the ventral surface (toward the reader) of the larva. (b) The ventralized embryo produced by females heterozygous for the mutation *Toll*. Ventral setae are produced in continuous rings around the entire dorsal-ventral circumference. (c) A *dorsal* embryo, with no ventral setae. The cuticle is made up entirely of dorsal cuticle with its characteristic fine dorsal hairs. (d) A *dorsal*-dominant embryo produced by a $d1^1/+$ female at 29°. Ventral setae bands are of nearly normal width, and the head skeleton is nearly normal. The embryo has assumed the characteristic tail-up position. (e) A $d1^2$ embryo, showing narrowed setae bands and abnormal head skeleton. The Filzkörper, the refractile posterior tracheal openings are readily seen.

midline no longer give rise to mesoderm, but instead contribute to the ventral cuticle. Similarly, lateral irradiation of normal blastoderm embryos leads to damage in ventral setae belts, but dorsal cuticular irregularities in dl^D. Thus in dl^D the fate of cells in all parts of the blastoderm is shifted such that they behave as if they derived from a more dorsal position than their actual position in the blastoderm.

The *Dorsal* Phenotype: Gastrulation

An additional demonstration that the *dorsal* phenotype is the result of a reprogramming of all blastoderm cells comes from observations of gastrulating *dorsal*

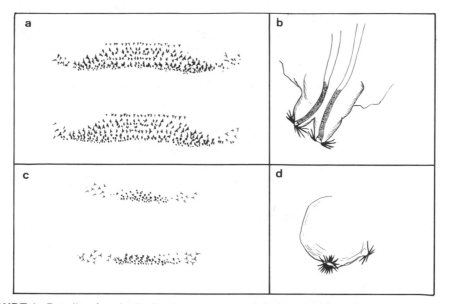

FIGURE 4. Details of cuticular landmarks. (a) Ventral setae bands of the normal larva. (b) Filzkörper and spiracles of the normal larva. The spiracles are rings of fine hairs on the outside of the tracheal openings, and the Filzkörper are the refractile specializations on the interior of the tracheal opening. (c) The narrowed ventral setae bands seen in *dorsal* [2]. (d) Spiracles, without Filzkörper or tracheae, seen in *dorsal* [1].

embryos. The normal movements of gastrulation, which begin immediately after blastoderm cells are completed, are highly asymmetrical in the dorsal-ventral axis (Fig. 5a). The first event of gastrulation is the invagination of the 20% of the cells on the ventral midline in the ventral furrow, to form the presumptive mesoderm. At the same time, the cephalic fold appears as a transverse fold on the lateral sides of the embryo. Immediately after the completion of the mesodermal invagination, the cells at the posterior of the embryo begin to move dorsally and anteriorly in the movement known a germ band extension, in which the presumptive ectoderm and underlying mesoderm become stretched out to almost twice the length of the embryo. Germ band extension appears to be a combination of stretching out of the cells in the germ band and pulling by the cells on the dorsal side, as they are thrown up into a series of dorsal folds.

Eggs laid by females homozygous for *dorsal* are indistinguishable from wild-type during early cleavage, nuclear migration and the early stages of cell formation. The first detectable difference between *dorsal* and wild-type embryos is seen at the time of the completion of the blastoderm cells. Normally the ventral cells are completed approximately 5 min before the dorsal blastoderm cells. In *dorsal*, however, the ventral and dorsal cells are completed at the same time, the first indication of the lack of dorsal-ventral polarity. In general, gastrulation in *dorsal* is characterized by the lack of dorsal-ventral asymmetry. There is no invagination of the ventral furrow. Instead of the dorsal-ward movement of germ band extension, the entire blastoderm forms a series of dorso-ventrally symmetrical transverse folds (Fig. 5c). These transverse folds can be considered, at least in part, homologous to

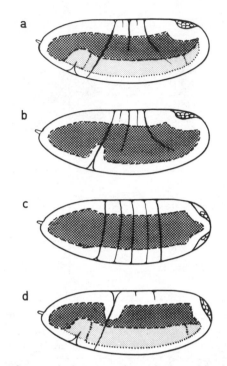

FIGURE 5. The pattern of gastrulation seen in embryos with altered dorsal-ventral pattern (anterior left, dorsal up). (a) Wild-type gastrulation, approximately 10 min. after the onset of gastrulation. The ventral furrow (light grey) has invaginated along the ventral midline. The head fold, the prominent transverse fold one third of the way from the anterior pole, lies predominantly laterally. Dorsal folds lie dorsally and laterally. Germ band extension has begun, as indicated by the pocket of pole cells on the dorsal side of the embryo. (b) Gastrulation in *dorsal* ² a weak allele at the *dorsal* locus. The ventral furrow is absent. The cephalic fold is now prominent ventrally. The dorsal folds appear normal, and the pole cells have been pushed to the dorsal side. (c) Gastrulation in the extreme *dorsal* phenotype, *d1*¹. All dorsalventral polarity is lacking. There is no ventral furrow, and no true head fold. The dorsal folds extend around the circumference of the egg. The pole cells remain posteriorly at their site of origin. (d) Gastrulation in the ventralized embryo produced by *Toll* + females. The ventral furrow is deeper than in wild-type. The cephalic fold is prominant on the dorsal side of the embryo, indicating the absence of dorsal pattern elements. The dorsal folds are greatly reduced. Germ band extension does not occur.

the dorsal folds seen in wild type embryos that seem to pull the germ band dorsally, but in *dorsal* these folds are produced at all dorsal-ventral positions.

    As in the case of the differentiated structures, weaker alleles at the *dorsal* locus also show the gradual appearance of more ventral characteristics in the gastrulation pattern (Fig. 5b). In *dl*² the transverse folds seen in *dl*¹ do not extend to the ventral side of the embryo and dorsal-ward extension of the presumptive ectoderm occurs, but no ventral furrow, and therefore no mesoderm, is produced. The position of the cephalic fold is particularly helpful in recognizing the shift of dorsal-ventral positional information in weak alleles of *dorsal* such as *dl*²: when ventral position values are missing, the normally lateral cephalic fold is shifted to the ventral side of the gastrulating embryo. The *dl*ᴰ embryos appear nearly wild-type during gastrulation, with the exception that no ventral furrow is formed.

Thus as soon as gastrulation, only 3½ hr after fertilization, dorsal-ventral positional information has already been interpreted and is expressed in the pattern of morphogenetic movements. The effect on morphogenesis is so early in development and so dramatic that the phenotype could be interpreted as primarily a blockage of the normal changes in cell shape and position of gastrulation. That this is not, in fact, the case can be inferred from the phenotypes of mutants which affect the movements of gastrulation, but do not cause the loss of particular pattern elements in the dorsal-ventral axis. For example, in the zygotic lethal mutation *twisted gastrulation* (29), germ band extension does not occur, but all germ layers are produced and the spatial pattern of the developed lethal larva is fundamentally normal.

Allelic Series and Amorphic Phenotypes

The careful analysis of mutant phenotypes can provide considerable insight into the function of the wild-type gene product. In the analysis of mutant phenotypes, it is important to keep in mind that different mutations can cause different sorts of alterations in gene activity, and therefore different phenotypes (14). The most common mutations are those that lead to complete or partial lack of gene activity. Mutation can also result in altered gene activity, and in these cases the function of the wild-type gene can not be readily inferred from the mutant phenotype. In order to infer what the function of wild-type gene is from mutant phenotypes, it is therefore essential to obtain multiple mutations in the same gene and to identify the phenotype associated with lack of function at that locus.

As described above, it is the weak alleles that provide the clearest demonstration that the *dorsal* gene product is part of a system of graded positional information. It is therefore critical to establish that the allelic series represents the decreasing function of the *dorsal* gene. The *dl*ᴰ phenotype is clearly the result of partial lack of function, since females heterozygous for a deficiency for *dorsal* over the wild-type gene can produce embryos of the *dl*ᴰ phenotype. Two weak alleles have been identified, *dl*² and *dl*⁵, which allow the production of Filzkörper and occasional narrow bands of ventral setae. These phenotypes must also represent partial lack of function, because the mutant phenotype is exaggerated over a deficiency for the *dorsal* gene.

Maternal effect mutations are particularly interesting because they do not affect viability of the fly or male fertility. In order to conclude, however, that the wild-type product of the mutant gene has a very specific role in directing embryogenesis, it is critical to demonstrate that the phenotype associated with the total absence of gene function, the amorphic phenotype, has no effect on viability. It is particularly difficult to demonstrate that a maternal effect phenotype represents the complete lack of gene activity. Cytologically visible deficiencies for that locus currently provide the only clear proof that the gene function must be totally absent. However since virtually all visible deficiencies are homozygous lethal, the ar-

gument that a maternal effect phenotype is the amorphic phenotype must be somewhat indirect.

In the case of *dorsal*, the amorphic phenotype appears to be the phenotype
produced by dl^1/dl^1 females. The phenotype of dl^1 over a deficiency for *dorsal* is not
stronger than the dl^1 homozygous phenotype. In a series of seven alleles picked on
the basis of the dominant *dorsal* phenotype, none shows a stronger phenotype over
a deficiency than the dl^1/dl^1 homozygous phenotype. In addition, in a screen for lethal mutations uncovered by a deficiency including the *dorsal* gene where an average of 3.5 lethal mutations per complementation group were identified, no lethal
alleles of *dorsal* were found (25). Therefore the *dorsal* gene codes for a substance
needed to establish embryonic pattern that is not essential for any other function in
the fly. This contrasts the prediction of Wolpert (31) that the molecules important
in specifying positional information would be used over and over again in the life
cycle, and indicates instead that at least some aspects of the system establishing embryonic dorsal-ventral pattern are unique.

Identification of Genes Controlling Dorsal-Ventral Polarity

Using the phenotypes of the *dorsal* allelic series as a guide, it is possible to identify
mutants in other genes which also cause dorsalization of the embryo by examination of the cuticle phenotype of other mutant embryos. By this criterion, the first
chromosomal maternal effect mutation, *gastrulation defective* (*gd*), also appears to
be an essential gene in the establishment of dorsal-ventral polarity. Alleles of *gastrulation defective* were isolated independently in two laboratories in systematic
searches for female sterile mutations on the first chromosome [the fs573 mutation
of Gans et al. (5) and the 14–743 complementation group of Mohler, (13)]. The
phenotype of strong alleles of *gd*, for example, 14–743, is indistinguishable from
dl^1, both in the pattern of gastrulation and in that the cuticle secreted by mutant
embryos contains only dorsal pattern elements. Weaker alleles at the *gd* locus show
the same progressive addition of more ventral pattern elements seen in the *dorsal*
allelic series.

In a screen for zygotic lethal mutations on the second chromosome, a maternal
effect mutation on the third chromosome was fortuitously isolated. The phenotype
of strong alleles at this locus, the *easter* locus, is again indistinguishable from dl^1,
and weaker alleles show the progressive addition of ventral pattern elements.

Mutants in two other maternal effect loci on the third chromosome affecting
the pattern of gastrulation were identified by Rice (22) in a screen for female sterile
mutations. Rice described the pattern of gastrulation in mutants mat(3)2 and
mat(3)4 as consisting of a series of transverse folds, with no ventral furrow formation and no germ band extension. Subsequent analysis of these mutants and more
recently isolated alleles at these two loci has shown that these two genes are also
dorsalizing. Like *dorsal*, both the pattern of gastrulation and of the final cuticle indicate that the mutant embryos are completely dorsalized.

The *Drosophila* embryo develops inside an egg shell which itself has a definite
dorsal-ventral polarity. The egg shell, the chorion, is more curved on the ventral
than on the dorsal side and make two appendages on the anterior dorsal surface of

the egg. In the mutant *fs(1)K10*, homozygous females produce eggs in which the egg shell is less asymmetric in the dorsal-ventral axis, and the appendages are fused and lie ventrally rather than dorsally. In the rare cases where embryos develop inside these abnormal egg shells, the embryos are dorsalized (28). A third chromosomal female sterile locus, *spindle*, produces eggs which appear to be ventralized in shape. The eggs are convex at all dorsal-ventral positions, and there are no dorsal appendages. These eggs are, unfortunately, never fertilized, so it is not possible to determine whether the pattern of embryonic developement would undergo a corresponding ventralization. The identification of these two genes reveals that the dorsal-ventral polarity of the egg case is also under specific genetic control, and the *K10* phenotype suggests that there exists a stage when egg-shape polarity is coupled with embryonic polarity.

With the fortuitous identification of six genes which mutate to give a pure maternal effect altering dorsal-ventral embryonic pattern, it becomes important to know exactly how many genes are involved in the establishment of dorsal-ventral pattern. A systematic screen for a particular type of mutation involves testing large numbers of mutagenized chromosomes, until every gene of interest is represented by three to four mutant alleles. Such saturation mutagenesis experiments, for zygotic mutations affecting embryonic pattern for the three major chromosomes of *Drosophila* have been completed (18, 8, 29). Saturation mutagenesis experiments for maternal effect mutations on the second and third chromosomes are in progress (24, 9).

In the screens for zygotic mutations giving a visible embryonic lethal phenotype, 120 loci were identified affecting embryonic pattern. Of these loci, two were identified whose primary effect is an obvious alteration of dorsal-ventral pattern. These two loci, *twist* and *snail*, have a cuticle phenotype indistinguishable from the *dorsal*-dominant phenotype: homozygous larvae have slightly narrowed ventral setae belts, and are twisted inside the egg case. They are also indistinguishable from *dorsal*-dominant during gastrulation, in that they fail to make a ventral furrow, but perform the movements of dorsal extension approximately normally.

It is striking that these zygotic mutants causing abnormal dorso-ventral pattern have such relatively minor aberrations when compared to the maternal effect mutations. Since in the saturation mutagenesis experiments no zygotic lethal genes were identified which exert a more dramatic effect on dorsal-ventral pattern, it is safe to conclude that such zygotic lethal genes do not exist. In the absence of zygotic gene expression, only the ventral-most pattern elements are lost, while the remaining dorsal-ventral pattern is complete. This implies that on the whole, the elements of dorsal-ventral polarity are established during oogenesis including both the system which establishes a graded distribution of positional information and the system of reception that measures the concentration of the graded substance, since the lack of either the graded substance or its receptor would affect the entire dorsal-ventral pattern.

The saturation screens for maternal effect mutations have not identified any second chromosomal loci except *dorsal* that affect dorsal-ventral pattern. On the third chromosome, in contrast, there are at least five complementation groups in addition to *easter*, *mat(3)2* and *mat(3)4*, which mutate to phenotypes indistinguishable from *dorsal*[1]. Analysis of these newly identified genes including the identification of the amorphic phenotype and characterization of weak alleles should establish whether these genes can be phenotypically distinguished from *dorsal* and what stage of the process of dorsal-ventral pattern formation they affect.

Dominant mutations are relatively rare, and reveal either unusual genes whose function is dosage-sensitive (one intact copy of the gene is insufficient) or rare mutations in which the regulation of normally recessive genes is altered. Dominant mutations affecting dorsal-ventral pattern can not be identified in systematic searches for recessive mutations, because they are lethal or sterile at least one generation before the tested generation. An examination of the apparently dominant female-sterile lines in the search for zygotic lethal mutations did allow the recovery of a dominant maternal effect mutation, *Toll*, which causes ventralization of the embryonic pattern (30). The ventralized embryos, unlike totally dorsalized embryos, retain a dorsal-ventral asymmetry but the ventral pattern elements appear to expand in the fate map at the expense of dorsal pattern elements. At gastrulation, the ventral furrow is deeper than in wild-type, and the normally lateral cephalic furrow is found at the dorsal side (Fig. 5d). The cuticle of the differentiated *Toll* embryo has ventral setae all around the dorsal-ventral circumference of the embryo, and no dorsal hairs are seen (Fig. 3b).

The genes identified as affecting dorsal-ventral polarity on the basis of their embryonic phenotypes are absolutely required for the correct establishment of dorsal-ventral embryonic pattern. In the case of maternal effects, the only essential activity of these genes during the entire life cycle is in the production of embryonic polarity. These genes need not, however, be the only genes whose function is required in the establishment of polarity. It is possible to imagine that there are genes important in this process which are required for other functions during the life cycle, and would therefore mutate to lethality. It is not currently possible to assess the numbers or importance of such lethal genes, but it should eventually be possible to identify such genes by virtue of their functional interaction with other genes, or by making homozygous clones for such genes in the germ line of otherwise heterozygous flies. The very fact that there exists a set of nonlethal maternal effect genes required for the establishment of dorsal-ventral polarity suggests, however, that the unique problem of embryonic determination has required a unique solution. Although these maternal effect genes may act on normal, mundane substrates, it should be possible to study the essential aspects of the system by studying these genes whose phenotypes indicate a particular importance in embryonic pattern formation.

Distinguishing Between the Functions of the Genes Controlling Dorsal-Ventral Pattern

Using the approaches just described, it should be possible to identify all, or nearly all, of the genes whose specific function is required to specify dorsal-ventral embryonic pattern. This comprehensiveness is a necessary prerequisite for understanding how the system works, but does not define the function of the individual genes. The mutations affecting dorsal-ventral polarity fall into three groups on the basis of their amorphic phenotypes and apparent time of action (Table 1). The genes in

TABLE 1. Genes controlling embryonic dorsal-ventral pattern

	Genetic Location	Cytological Location in Polytene Chromosomes	Numbers of Alleles Identified	Strong Phenotype
Maternal effect loci affecting egg shape				
K10[28]	1–0.5	2B17–3A3	4	Dorsalization of egg and embryo
Spindle	3–	—	3	Ventralized egg shell; unfertilized
Maternal effect loci affecting embryonic pattern; no effect on egg shape				
Dorsal	2–52.9	36C	11	Dorsalized embryos
Easter	3–57	88C–F	13	Dorsalized embryos
Gastrulation defective	1–37	11A1-7	8	Dorsalized embryos
Mat (3) 2	3–17[22]	—	8	Dorsalized embryos
Mat (3) 4	3–51[22]	—	3	Dorsalized embryos
Toll	3–91	—	3	Dominant; partially Ventralized embryos
Zygotic lethal loci affecting-embryonic pattern				
Twist	2–100	59B6–8;D4–5	4	Partially dorsalized embryos
Snail	2–51	35C3–D1;35D4–7	4	Partially dorsalized embryos

the first group (*K10* and *spindle*) are pure maternal effects and affect the polarity of the egg shell, and therefore must act relatively early during oogenesis. Mutations in the genes in the second group (*dorsal, gastrulation defective, easter, mat(3)2, mat(3)4,* and other third chromosomal loci) cause the total dorsalization of the embryonic pattern as a maternal effect, without affecting the polarity of the egg shell. The third group of mutants (*twist* and *snail*) are zygotic in their function, causing only partial dorsalization of the pattern, and are apparently the last in the temporal series of gene functions.

It is striking that within the group of recessive maternal effect mutations resulting in complete embryonic dorsalization without affecting egg shape, the strong phenotypes are indistinguishable. This implies that these genes are responsible for components of a single, integrated process. This process is most probably the actual determination and interpretation of the pattern of dorsal-ventral positional information. The earlier maternal effects may be responsible for orienting the dorsal-ventral axis, and the zygotic mutations may stabilize or enhance positional information, but the *dorsal*-like maternal effect genes probably are primarily responsible for the primary process of dorsal-ventral determination. It is hardly surprising that a number of gene functions are necessary for this process. In order to understand the actual mechanisms by which dorsal-ventral polarity is established and interpreted, it is essential to define differences between these gene functions in all possible ways.

Some phenomenological differences between these gene functions can be inferred from the allelic series of the different genes. The weak alleles of *dorsal, gastrulation defective*, and *easter*, although very similar, are distinguishable. For instance, while weak alleles of *dorsal* appear uniformly dorsalized, weak alleles of *gastrulation defective* are more dorsalized to different extents in the posterior compared to the anterior region of the embryo; that is, the ventral setae bands become progressively narrower posteriorly. The frequency of recovery of weak alleles is also probably a reflection of underlying differences between genes or gene products. Only two of eight EMS-induced *dorsal* alleles allow the production of Filzkörper or more ventral structures, while five of eight *gastrulation defective* alleles produce Filzkörper and more ventral structures.

Another technique for distinguishing between gene functions is to compare temperature-sensitive periods, if temperature-sensitive alleles are available. Temperature-sensitive periods have been determined for both *dorsal* (23) and *gastrulation defective* (10). In both cases, the temperature-sensitive period begins several hours before egg deposition and ends after fertilization. The temperature-sensitive period of *gd*, however, ends at 1.5 hr after fertilization, the time at which the nuclei arrive in the superficial cytoplasm of the egg, while the temperature-sensitive phenotype of *dorsal* can be influenced by temperature until the beginning of cellularization of the blastoderm, 2.5 hr after fertilization.

It should be possible to draw additional functional distinctions between gene functions by analysis of double mutants. For example, the dominant ventralizing mutation *Toll* produces different phenotypes in combination with the different recessive dorsalizing mutations. Females of the genotype dl^1/dl^1; *Toll*/ + produce dorsalized embryos that are indistinguishable from *dorsal* alone, while gd^{14-743}/gd^{14-743}; *Toll*/ + females produce embryos with a new synthetic phenotype that is neither dorsalized nor ventralized. The *gd-Toll* phenotype can be considered as lateralized, in that both the most ventral and most dorsal pattern elements are lacking. At gastrulation, these embryos do not make a ventral furrow and the normally lateral cephalic furrow is seen both at the dorsal and at the ventral side of the embryo. The differentiated *gd-Toll* lack dorsal cuticle, and instead form ventral setae at all dorsal-ventral positions.

These observations reveal functional distinctions between the *gd* and the *dorsal* gene activities. In the absence of gd^+ activity, it is possible to produce some ventral structures. The $dorsal^+$ gene activity seems to be required for ventral structures, even in a *Toll* background. Further genetic combinations should make distinctions between the other dorsalizing loci possible.

Rescue of Maternal Effect Mutations by Injection of Wild-Type Cytoplasm

A mutation defines a specific biochemical lesion. In principal, it should be possible to add back the missing substance and repair the mutant phenotype. In multicellular organisms, this is impractical in most cases, unless the animal can digest the missing molecule in its normal diet. The fertilized zygote is, however, a special case during the life cycle, when the missing molecules injected into the cytoplasm should be able to reach all parts of the future animal. Maternal effect mutations therefore provide a favorable case for attempts at rescuing the mutant phenotype by injection of the missing wild-type product.

The first demonstration that is possible to rescue a maternal effect mutation by injection of wild-type cytoplasm was the rescue of the *o* gene maternal effect mutation of the axolotl *ambystoma* (2). Females homozygous for the *o* mutation produce embryos which die at gastrulation. The same embryos, when injected with cytoplasm from normal eggs, frequently gastrulate normally and occasionally develop into larva.

Two maternal effect mutations in *Drosophila*, *deep orange* and *rudimentary*, have been demonstrated to be rescuable by injection of wild-type cytoplasm (6, 20). Both these mutations apparently affect nucleotide metabolism, lead to nonspecific embryonic cell death, and are both rescuable by fertilization with a wild-type sperm. The rescue of the mutant phenotype by wild-type sperm implies that the activity of the gene in the zygote during embryonic development is sufficient to allow normal embryogenesis. This class of maternal-effect, paternal-rescuable mutations is particularly likely, therefore, to be rescuable by injection into early embryos. Embryos produced by *deep orange* females, when mated to mutant males, normally die during late gastrulation. Injection of wild-type cytoplasm permits a large percentage of injected embryos to proceed to late embryonic stages where muscle movements are visible. The pyrimidine biosynthetic pathway deficient mutant *rudimentary* causes, as a maternal effect, death at late embryonic stages, 13–14 hr after fertilization. Injection of wild-type cytoplasm allows hatching of the larva in a reasonable percentage of cases, and occasional development into adults. The injection of purified pyrimidines into mutant *rudimentary* embryos leads to rescue at a higher frequency than that of wild-type cytoplasm.

The *dorsal* mutation is a pure maternal effect: the phenotype is unmodified by fertilization by wild-type sperm. It is therefore perhaps somewhat unexpected that the *dorsal* phenotype can be partially rescued by the injection of wild-type cytoplasm at precellular stages (23). The injection of a small volume of wild-type cytoplasm into *dorsal*[1] embryos results in the appearance of more ventral cuticular structures in the differentiated, but still lethal, larva. The most reliable marker for rescue is the Filzkörper, the highly refractile, fluorescent specializations at the posterior tracheal openings. Filzkörper, which normally derive from a position in the blastoderm fate map approximately 15% off the dorsal midline, are never seen in uninjected *dorsal* embryos, but are found in more than 70% of *dorsal* embryos injected with wild-type cytoplasm (Table 2). This partial rescue can be seen not only in the differentiated cuticle, but also in the pattern of gastrulation. The injected *dl*[1] embryos extend the ventral ectodermal cells towards the dorsal side in a manner analogous to *dl*[2], although they fail to form a ventral furrow (Table 2).

The dorsalizing mutant *easter* is also rescuable by the injection of wild-type cytoplasm. As in *dorsal*, a high percentage of embryos injected with wild-type cytoplasm at precellular stages differentiate Filzkörper, while Filzkörper are never seen in uninjected embryos (1) (Table 2).

TABLE 2. Partial rescue of dorsalized mutant embryos by injection of wild-type cytoplasm.

| | GASTRULATION ASSAYS [% RESCUED (N)] | | CUTICLE ASSAYS [% RESCUED (N)] | |
RECIPIENT EMBRYOS	D/V Asymmetry	"Germ Band" Extension	Filzkörper	Ventral Setae
Dorsal (dl¹/dl¹ mothers)	79% (79)	49% (97)	77% (74)	0% (74)
Easter (ea¹/ea¹ mothers)	70% (47)	29% (47)	53% (149)	5% (148)

Cytoplasm from the ventral side of cleavage stage wild-type (Oregon R) embryos was injected into a ventral, posterior position in the mutant embryos at the syncytial blastoderm stage. The volume of cytoplasm transferred corresponds to about 2% of the total egg volume.

Rescue can be assayed both at gastrulation and in the cuticle produced by developed embryos. Rescue at gastrulation can be seen by the appearance of dorsal-ventral asymmetry at gastrulation: the transverse folds shift to the dorsal side of the embryo and the cephalic fold appears on the ventral side of the embryo. A smaller percentage of injected embryos show extension of the "germ band" to the dorsal side, analogous to that seen in dl^2 (Fig. 5b). Rescue can also be assayed as the percentage of differentiating embryos producing Filzkörper or ventral setae. These structures are never produced by uninjected embryos or by mutant embryos injected with cytoplasm from other embryos mutant at the same locus; for example, cytoplasm from dorsal embryos never shows rescue of dorsal by any of these assays.

The fact that the *dorsal* and *easter* phenotypes can be partially rescued by cytoplasmic injection reveals several aspects of the process of establishing dorsal-ventral polarity. Since the injections are performed after fertilization, positional information is not irreversibly fixed during oogenesis. This observation agrees well with the observation that the temperature-sensitive period of *dorsal* extends until 2.5 hr after fertilization. In both *easter* and *dorsal*, the rescuing principle must be a diffusible substance, or it could not be transferred from one embryo to the other. The fact that the phenotypes are only partially rescued is not surprising, since the volume of cytoplasm transferred is less than 2% of the total egg volume. It is likely that the injection of more concentrated cytoplasmic extracts would lead to a greater extent of rescue.

The rescuability of these two mutations, both of which produce a similar phenotype, makes it possible to compare the qualities of the rescuing substances. That two different substances are involved is clearly demonstrated by the fact that cytoplasm from *dorsal* embryos can rescue the *easter* phenotype and vice versa. Thus the *dorsal* and *easter* mutations cause the absence of two independently regulated substances, both of which are absolutely required in the same process. The rescuing principles also have different physical, and perhaps functional, properties. The rescue of *dorsal* is a local phenomenon: production of Filzkörper is seen only when the wild-type cytoplasm is injected posteriorly. The rescue of *easter*, in contrast, seems to be a more spreading phenomenon: wild-type cytoplasm injected anteriorly also leads to the production of Filzkörper. From this it appears that the *easter* rescuing substance has a wider range of effective diffusibility than the *dorsal* rescuing substance. Such simple observations on the quality of rescue may help to distinguish between the activities of wild-type gene products.

This biological assay provides a means of identifying and biochemically characterizing the rescuing principles. Homogenates of wild-type cytoplasm can also partially rescue the *dorsal* phenotype. It should, therefore, be possible to perform at least a preliminary purification of the *dorsal* rescuing principle, assaying purified fractions by cytoplasmic injections.

The Process of Establishing Dorsal-Ventral Pattern

The mutations affecting the dorsal-ventral pattern of the *Drosophila* embryo that have been isolated strongly support theories of graded distribution of positional information rather than theories of localized determinants. The allelic series of *dorsal*, together with the fate-mapping studies of *dorsal* dominant conclusively show that there is a dorsal-ventral continuum of positional information which is shifted in *dorsal*.

All of the mutants affecting dorsal-ventral pattern can be described as changing the distribution of a graded system of positional information, as illustrated in Fig. 6. Perhaps the most remarkable aspect of the system that has been thus far revealed is that mutations in the maternal effect loci *easter* and *gastrulation defective*, and probably the other less-well characterized third chromosomal loci, produce the same strong phenotype and have a very similar allelic series to *dorsal*. In the absence of any one of these gene functions, the pattern collapses to a ground state of dorsalization. The hypomorphic alleles at each locus, which represent partial lack of function, all cause partial dorsalization of the embryonic pattern. This similarity of phenotypes makes it impossible to single out one element and hypothesize, for example, that the normal *dorsal* gene product is a morphogen with a higher concentration on the ventral than on the dorsal side of the embryo. The involvement of multiple components, all of which are essential to prevent dorsalization of the embryo, suggests that the gene products are interacting components of a coordinated, self-regulating system of positional information.

The phenomenon of a group of genes involved in the same process which mutate to similar, if not identical, phenotypes, is not unique to the system of dorsal-ventral polarity. There are at least six loci which mutate to produce neuralization of the embryo (11). Nine loci have been identified which produce deletions in every other embryonic segment, and seven loci produce mirror-image duplications within segments (17, 19). Several loci have been identified which produce Polycomblike homeotic transformations (26, 4). These systems differ in the ease with which it is possible to distinguish the mutant phenotypes of related genes: of these systems the genes affecting dorsal-ventral pattern appear to be the most difficult to distinguish on the basis of phenotype. In all of these fundamental developmental processes, though, the major current problem is to distinguish between gene functions in order to better understand the system.

Virtually all the information necessary for the establishment of over all embryonic dorsal-ventral pattern is maternally derived, since zygotic mutations affect only a portion of the system. This maternally provided system includes the production of the morphogen, the means of distributing the morphogen in a concentration gradient, and the receptor system which measures the morphogen concentration. Loss of any part of this system results in total dorsalization of the embryo.

The maternally derived system can be influenced after fertilization, since the temperature-sensitive periods of both *gd* and *dl* extend after fertilization, and *ea*

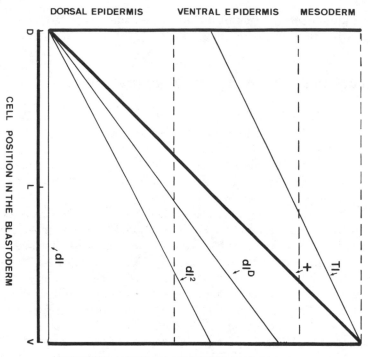

DORSAL EPIDERMIS   VENTRAL EPIDERMIS   MESODERM

CELL POSITION IN THE BLASTODERM

POSITIONAL VALUE (ARBITRARY UNITS)

FIGURE 6. The phenotypes described in terms of changes in graded positional value. All of the phenotypes described here can be formally characterized as shifting the positional values of the cells in the blastoderm embryo. In wild-type ($+$), there is a continuous increase in positional value from dorsal to ventral. In strongly dorsalized mutant embryos (d1), all cells behave according to the dorsal ground state. Weaker dorsalizing mutations (dl^2 and dl^D) produce embryos which have some dorsal-ventral polarity, but the highest ventral positional values are not obtained. The ventralized *Toll* embryos lack the lowest, dorsal, position values. The horizontal lines indicate the threshold levels for dorsal epidermis, ventral epidermis and mesoderm. They are chosen such that the intersections with the positional value lines, projected onto the abscissa indicate the positions of the respective anlagen on the blastoderm along the dorso-ventral axis.

and *dl* can be partially rescued by injection of wild-type cytoplasm during early embryonic development. The system remains flexible until the time of cell formation. As soon as cells are formed, they behave differentially, according to their dorsal-ventral position. There is less than an hour between the end of the time when it is possible to influence the mutant phenotypes by temperature-shift or cytoplasmic injection and the time when dorsal-ventral positional information is expressed in the morphogenetic movements of gastrulation. Interpretation of positional information and expression of position-dependent differences are closely coupled in time, and perhaps also in mechanism.

Further genetic analysis of the establishment of dorsal-ventral pattern should establish some hierarchy of gene functions. The ultimate understanding of how the individual components differ and what their functions are will depend on the biochemical characterization of the gene products, made possible by the genetic identification of the molecules that play a unique role in the establishment of embryonic pattern.

General References

Nüsslein-Volhard, C.: Maternal effect mutations that alter the spatial coordinates of the embryo of *Drosophila melanogaster*. *Symp. Soc. Dev. Biol.* **37**: 185–211 (1979).

Poulson, D.F.: Histogenesis, organogenesis and differentiation in the embryo of *Drosophila melanogaster* Meigen. In: *Biology of Drosophila*, Demerec, M. (ed.) New York: Hafner, pp. 168–274 (1950).

Wright, T.R.F.: The genetics of embryogenesis in *Drosophila*. *Adv. Gen.* **15**: 261–395 (1970).

References

1. Anderson, K.V. Nüsslein-Volhard, C.: Cytoplasmic transfer partially rescues three maternal effects mutations affecting embryonic pattern formation. In *Drosophila.*, Manuscript *in preparation.*

2. Briggs, R., Cassens, G.: Accumulation in the oocyte nucleus of a gene product essential for embryonic development beyond gastrulation. *Proc. Nat. Acad. Sci. USA.* **55**: 1103–1109 (1966).

3. Chan, L.-N., Gehring, W.: Determination of blastoderm cells in *Drosophilia melanogaster*. *Proc. Nat. Acad. Sci. USA.* **48**: 2217–2221 (1971).

4. Duncan, I.M.: Polycomblike: A gene that appears to be required for the normal expression of the bithorax and antennapedia complexes of *Drosophila melanogaster*. *Genetics* **102**: 49–70 (1982).

5. Gans, M., Audit, C., Masson, M.: Isolation and characterization of sex-linked female sterile mutants in *Drosophila melanogaster*. *Genetics* **81**: 683–704 (1975).

6. Garen, A., Gehring, W.: Repair of lethal developmental defect in *deep orange* embryos of *Drosophila* by injection of normal egg cytoplasm. *Proc. Natl. Acad. Sci. USA.* **69**: 2982–2985 (1972).

7. Illmensee, K.: Nuclear and cytoplasmic transplantation in *Drosophila*. *RES Symposium.* **8**: 76–96. (1976).

8. Jürgens, G., Wieschaus, E., Nüsslein-Volhard, C., Kluding, M.: Lethal mutations affecting the larval cuticle pattern in *Drosophila melanogaster*. II. Identification of third chromosomal loci. Manuscript *in preparation.*

9. Jürgens, G., Nüsslein-Volhard, C.: Manuscript *in preparation.*

10. Konard, K., Mahowald, A.P.: *personal communication.*

11. Lehmann, R., Dietrich, U., Jiménez, F., Campos-Ortega, J.A.: Mutations of early neurogenesis in *Drosophila*. *Wilhelm Roux' Arch.* **190**: 226–229 (1981).

12. Lohs-Schardin, M., Cremer, C., Nüsslein-Volhard, C.: A fate map for the larval epidermis of *Drosophila melanogaster*: localized cuticle defects following irradiation of the blastoderm with an ultraviolet laster microbeam. *Dev. Biol.* **73**: 239–255 (1979).

13. Mohler, J.D.: Developmental genetics of the Drosophila egg. I. Identification of 59 sex-linked cistrons with maternal effects on embryonic development. *Genetics.* **85**: 259–272 (1977).

14. Muller, M.J.: Further studies on the nature and causes of gene mutations. *Proc. Sixth Int. Cong. Genet.* 213–255 (1932).

15. Nüsslein-Volhard, C.: Maternal effect mutations that alter the spatial coordinates of the embryo of *Drosophila melanogaster*. *Symp. Soc. Dev. Biol.* **37** 195–211 (1979).

16. Nüsslein-Volhard, C., Lohs-Schardin, M., Sander, K., Cremer, C.: A dorso-ventral shift of embryonic primordia in a new maternal effect mutant of *Drosophila*. *Nature.* **283**: 474–476 (1980).

17. Nüsslein-Volhard, C., Wieschaus, E.: Mutations affecting segment number and polarity in *Drosophila*. *Nature.* **287**: 795–801 (1980).

18. Nüsslein-Volhard, C., Wieschaus, E., Kluding, M.: Lethal mutations affecting the larval cuticle pattern in *Drosophilia melanogaster*. I. Identification of 49 loci on the second chromosome. Manuscript *in preparation*.

19. Nüsslein-Volhard, C., Wieschaus, E., Jürgens, G.: Segmentierung in *Drosophila*: Eine genetische Analyse. *Verh. Deutsch. Zool. Gesell. in press* (1982).

20. Okada, M., Kleinman, I.A., Schneiderman, H.A.: Repair of a genetically-caused defect in oogenesis in *Drosophila melanogaster* by transplantation of cytoplasm from wild-type eggs and by injection of pyrimidine nucleotides. *Dev. Biol.* **37**: 55–62 (1974).

21. Poulson, D.F.: Histogenesis, organogenesis and differentiation in the embryo of *Drosophila melanogaster* Meigen. In: *Biology of Drosophila*, Demerec, M. (ed.). New York: Hafner, 169–274 (1950).

22. Rice, T.: Isolation and characterization of maternal effect mutants: an approach to the study of early determination in *Drosophila melanogaster* Ph.D. thesis. Yale University. (1972).

23. Santamaria, P., Nüsslein-Volhard, C.: Partial rescue of *dorsal*, a maternal effect mutation controlling dorso-ventral polarity in the *Drosophila* embryo, by injection of wild type cytoplasm. Manuscript *in preparation*.

24. Schüpbach, T., Wieschaus, E.: Manuscript *in preparation*.

25. Steward, R.: *personal communication.*

26. Struhl, G.: A gene product required for correct initiation of segment determination in *Drosophila*. *Nature.* **293**: 36–41 (1981).

27. Underwood, E.M., Turner, F.R., Mahowald, A.P.: Analysis of cell movements and fate mapping during early embryogenesis in *Drosophila melanogaster*. *Dev. Biol.* **74**: 286–301.

28. Wieschaus, E.: fs(1)K10, A female sterile mutation altering the pattern of both the egg coverings and the resultant embryo in *Drosophila*. In: *Cell Lineage, Stem Cells and Cell Determination*, Le Douarin, N. (ed.) Amsterdam: Elsevier-North Holland, pp. 291–302 (1979).

29. Wieschaus, E.: Lethal mutations affecting the larval cuticle pattern in *Drosophila melanogaster*. III. Identification of first chromosomal loci. Manuscript *in preparation*.

30. Wieschaus, E.: *personal communication.*

31. Wolpert, L.: Positional information and the spatial pattern of cellular differentiation. *J. Theoret. Biol.* **25**: 1–47 (1969).

Questions for Discussion with the Editors

1. *How broadly do you define the term "positional information" when describing the dorsal allelic series? For example, do you mean to go so far as to equate positional information with the putative gradient/receptor system you mentioned?*

The phenotypes of the *dorsal* allelic series (and of the allelic series of the other maternal effect loci producing dorsalized embryos without affecting egg shape) reveal that positional information in the dorsal-ventral axis is a continuum, and that positional value is proportional to the amount of wild-type gene activity. We believe that the most direct way in which to visualize such a system of positional information is in terms of the graded distribution of a morphogenetic substance in the dorsal-ventral axis and the subsequent interpretation of local morphogen concentration by specific receptors. This model is almost certainly an oversimplification, but hopefully will provide a useful framework for generating specific hypotheses about the functions of the individual components coded for by the maternal effect genes.

2. *To what extent do you expect that recombinant DNA technology (e.g., isolation and sequencing of specific gene complexes) could—in the near future—contribute to understanding pattern specification in systems such as those you described?*

In *Drosophila*, the genetic identification of specific loci necessary for pattern specification makes the isolation of the corresponding DNA sequences by recombinant DNA techniques straightforward, if still laborious. However, since the maternal effect genes described here are transcribed during oogenesis, but exert their phenotypic effect after fertilization, characterization of the molecular structure of the genes themselves will probably not reveal anything directly about the process of pattern specification. The usefulness of purified genes will, instead, be as a means to isolate purified gene products. While rescuing substances can be partially purified from homogenates of embryonic cytoplasm, complete purification will probably only be feasible using purified DNAs to purify transcripts and then using isolated transcripts to synthesize protein products *in vitro*.

CHAPTER 12

Cell Interactions Controlling the Formation of Bristle Patterns in *Drosophila**

Lewis I. Held, Jr. and Peter J. Bryant

Bristle Patterns in *Drosophila*

ONE OF THE MAIN REASONS for using *Drosophila* in the study of pattern formation is the availability of numerous genetic mutations which affect patterns in potentially informative ways, and the ease with which the *Drosophila* genome can be manipulated to make genetic combinations for use in gene dosage, gene interaction, and genetic mosaic analysis. Another reason, not often appreciated, is the intricacy and reproducibility of the patterns themselves, which make it possible to recognize subtle alterations caused by genetic mutations or developmental manipulations. The elements making up these intricate patterns include cell hairs (trichomes), and various cuticular sense organs such as sensilla trichoidea and sensilla campaniformia, but the most conspicuous pattern element is the bristle. Bristles are sometimes categorized into macrochaetes and microchaetes based on size, but in fact there is a continuum of sizes and the basic structure of these elements is quite uniform. Bristles are usually fluted and pointed, but in some cases smooth and/or blunt. Surrounding the base of each bristle is a cup-shaped socket, which distinguishes the bristle from a cell hair. Most of the bristles are innervated, the cell body of the sensory neuron lying just below the socket (Fig. 1A).

On the surface of a *D. melanogaster* adult there are approximately 5000 bristles, organized in a variety of patterns (Fig. 2). In the compound eye, the ommatidia (facets) form a hexagonal array, and the interommatidial bristles are placed at

*We dedicate this article to the memory of Professor Curt Stern, who pioneered the use of *Drosophila* bristle patterns in developmental genetics research.

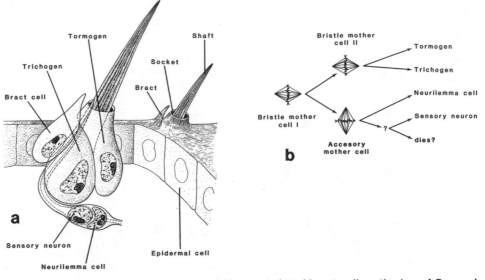

FIGURE 1. (a) A typical bristle organ and its associated bract cell on the leg of *Drosophila melanogaster*. Courtesy of Dr. Heinrich Walt. (b) Cell lineage of the bristle organ of *Oncopeltus*. From Lawrence (31).

the anterior end of each horizontal interface between ommatidia. Elsewhere, the most common bristle arrangement is the row, which can be either parallel or perpendicular to the direction in which the bristles point. Rows of bristles are found around the rim of the eye, on the thorax and legs, along the anterior margins of the wings and the posterior margins of the abdominal tergites, and on the genitalia. Bristles not aligned in rows include the large bristles on the head and thorax, which occupy fixed sites, and the small bristles on the abdominal tergites, which are evenly spaced but apparently randomly arranged.

Figure 2 shows examples of the kinds of pattern whose development we are attempting to understand. These patterns develop from relatively homogeneous groups of cells called imaginal discs and histoblast nests, which are found in the developing larva and which differentiate during the pupal stage. They are products of a single cell layer, so that they can be considered as two-dimensional patterns for the purpose of this analysis. The question we are attempting to answer is: what mechanisms establish the fate of particular cells within the epithelium so that they ultimately differentiate as bristle organs rather than as cell hairs or other structures?

The Role of Cell Lineage

From time to time, attempts have been made to invoke the pattern of cell lineage within the imaginal disc as somehow establishing cell fates within the developing pattern. It could be imagined, for example, that a "quantal mitosis" (29) at some developmental stage could give rise to nonequivalent daughter cells, each of which

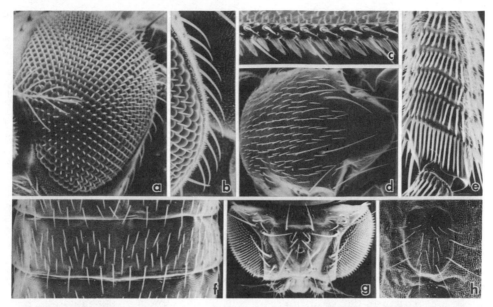

FIGURE 2. Examples of bristle patterns in *D. melanogaster*. (a) hexagonal lattice of facets in the eye with interommatidial bristles at the anterior end of each horizontal interface between ommatidia; (b) row of bristles on the posterior rim of the eye; (c) anterior margin of the wing; (d) dorsal thorax; (e) distal end of the tibia of a first leg; (f) third abdominal tergite; (g) top of the head; (h) third abdominal sternite.

would then be destined to produce descendants with a common fate such as to produce all the bristles of a row or all the cells of an ommatidium. In theory, a succession of such quantal mitoses could generate patterns of great complexity and reproducibility, provided the divisions themselves were accurately controlled in both time and space.

Differentiative Divisions

Evidence from cell lineage studies indicates that the quantal mitosis model as described above can be applied only to the development of individual bristles and sensilla, and not to patterns of such structures. The sequence of divisions has not been carefully studied in *Drosophila*, but a variety of patterns of differentiative divisions has been described for the bristles, scales, hairs and sensilla of other insects. For example, in the development of bristles in the bug *Oncopeltus* (31) the "bristle mother cell I" first swells and then divides horizontally, the mitotic spindle being in the plane of the epidermis (Fig. 1B). One of the two daughter cells (the "bristle mother cell II") later divides again horizontally to produce the trichogen and tormogen cells, which produce the bristle shaft and socket respectively. The second daughter cell from the first division (the "accessory mother cell") divides vertically to produce the neurilemma or sheath cell and a cell which apparently goes through one more division after which one daughter cell dies and the other becomes the sensory neuron. In the case of chemosensilla, additional divisions occur in the accessory cell lineage so that four sensory neurons are associated with the final sensillum.

In *Drosophila* Lees and Waddington (34) were able to distinguish bristle-forming cells from the surrounding epidermis at 15 hours after puparium formation by their size, and they found many pairs of such cells indicating that at least one of the differentiative divisions had occurred by this time. The two cells of each pair differed in position; that lying nearest the apical surface of the epithelium became the tormogen while that lying below it became the trichogen. Lees and Waddington (34) did not study the origin of the sensory neuron and neurilemma cells.

It is difficult, of course, to establish cell lineage patterns from single observations of each specimen, and continuous observations of the differentiative divisions have not yet been reported in any insect. Experimental studies of the divisions have been quite limited, and, in one case, difficult to reconcile with the conventional interpretation. Poodry (37) studied the effects of radiation on the development of specific bristles of *Drosophila*. He found that the precursors for each bristle became radioinsensitive at a characteristic time, which was interpreted as the time when cell division ceased in the precursor cells for that structure. According to this interpretation, structures derived from sister cells should show a loss of radiosensitivity at the same time. Since the tormogen cells became radioinsensitive as much as two hours later than the corresponding trichogen cells, Poodry suggested that the trichogen and tormogen may not, in fact, be sister cells but may be "first cousins."

The lineage model for insect cuticular structures was apparently quite attractive to Bernard (3) who concluded from histological studies of several insects that each ommatidium is derived from a single initial cell which goes through a specific series of divisions to produce the different cell types of the differentiated structure. However, the scheme does not apply to *Drosophila*, at least as far as the pigment-producing cells (14 out of the 22 cells of the ommatidium) are concerned. Ready *et al.* (39) used x-ray-induced mitotic recombination to produce clones of unpigmented cells in an otherwise red-pigmented eye. The results showed that the ommatidium is not derived from a single initial cell, and that the pigment-producing cells of the ommatidium bear no consistent clonal relationship to one another (Fig. 3). Any two cells of the ommatidial pattern could be in the same clone in some ommatidia, but in different clones in other ommatidia. In a more extensive analysis, Hofbauer and Campos-Ortega (28) showed that photoreceptor cells 1, 6, and 7 of the ommatidium were more likely to be clonally related to one another than to the other cells, but this seems to result from the fact that these three cells are produced later in the developmental sequence than are the remainder of the pigment cells, rather than from an obligatory lineage pattern.

Although bristles and sensilla have been shown to have a clonal basis, lineage does not appear to be important in the arrangement of these elements into larger scale patterns; clones of marked cells can occupy partly overlapping territories on different animals, showing that clonal relationships, in general, are variable (5). Moreover, the cell lineage pattern within a developing imaginal disc can be manipulated so that one marked clone grows to several times the normal size, while other, relatively slow-growing clones are reduced or eliminated by an unknown mechanism termed "cell competition" (36). Lineage patterns are also modified as a

FIGURE 3. A clone of marked cells in the retina of *Drosophila*. Black indicates unpigmented mutant cells arising via mitotic recombination induced by x-rays in the late first instar. Dotted circles indicate interommatidial bristles (unscored). Anterior is to the left. From Ready et al. (39).

result of x-ray-induced cell death, followed by compensatory overgrowth of the surviving clones of cells in radiation-induced mitotic recombination experiments (24). Neither of these types of lineage-pattern modification has any detectable effect on the final pattern, again suggesting that lineage is not a determinant of cell fate within the pattern.

Compartmentalization

One way in which lineage does appear to play a role in the development of imaginal disc patterns is in the early subdivision of cell populations into polyclonal compartments. These compartments are revealed as lineage restrictions that divide the imaginal disc into anterior and posterior (A/P) or dorsal and ventral (D/V) halves at specific developmental stages (see 17). At the time the compartmentalization events occur, each compartment consists of a group of cells that are not necessarily clonally related; it is not initiated as a single cell as would be the case in a strict lineage mechanism. Presumably the cells that initiate each compartment are related simply by their positions in the disc, rather than by their lineage. Once a compartment is established, even unusually fast-growing clones do not grow beyond its boundaries. Furthermore, at least in the case of the anterior and posterior compartments, each compartment coincides with a group of cells that is under the influence of a distinct control gene. The function of the *bithorax*[+] gene is apparently limited to the anterior compartment within the haltere, and the function of *post-bithorax*[+] is limited to the posterior compartment (17). *Engrailed*[+] seems to be involved in maintaining the developmental program in the posterior compartment, since posterior clones that are mutant for this gene can cross the compartment boundary into the anterior compartment, and posterior *engrailed* cells often produce cuticular structures characteristic of the anterior compartment (32).

The role of compartmentalization in pattern formation in imaginal discs is not fully understood. One suggestion is that the disc may be subdivided, not only by the initial A/P and D/V compartmentalizations, but by subsequent compartmentalization events that would sequentially partition it into further smaller and smaller compartments (17). If these compartmentalization events were associated with specific restrictions on the developmental fate of the cell groups, then this model would be, in effect, a polyclonal analogue of a strict lineage model, in which developmental fates would be progressively restricted in groups of cells rather than in individual dividing cells. However, there appears to be no good evidence for any compartmentalization events beyond A/P and D/V. Therefore, it seems difficult to avoid the conclusion that the formation of spatial patterns in imaginal discs must involve interactions between cells based either on their positions with respect to one another [i.e., based on their "positional information" (59)], or on their prospective fates, without regard to their exact lineage relationships. In the remainder of this article, we discuss the variety of such interactions that have been identified.

Interactions Between Cells Based on Position

Pattern Regulation in Imaginal Disc Fragments

The position-dependent cell interactions controlling pattern formation in imaginal discs have been investigated by three experimental procedures (see 4 for a detailed review). In the first procedure, a number of different specific fragments of the disc are implanted into mature larvae, where they undergo metamorphosis with the host. The transplanted fragments produce distinct sections of the pattern, permitting the establishment of a detailed fate map of the disc. The fact that such fate maps can be made shows that the presumptive pattern of differentiation has already been established in the imaginal disc before differentiation occurs. Furthermore, when discs are taken from younger larvae and implanted into hosts which are ready to metamorphose, they produce partial patterns which suggest that the pattern is built up in a specific sequence during growth of the tissue.

In the second experimental procedure, specific disc fragments are given the opportunity for growth and pattern regulation before metamorphosis. This is done by culturing the fragments in adult female hosts, where they grow by cell multiplication but do not differentiate until transferred to larval hosts for metamorphosis. It has been shown for several imaginal discs that some fragments can regenerate missing parts during adult culture, but that other fragments, in contrast, undergo pattern duplication. The results of a large series of adult culture experiments on the wing disc are summarized in Fig. 4. In most cases, the disc was cut into three pieces by two parallel cuts and the resulting three fragments cultured separately. In many of these series, the central fragment was able to regenerate the missing parts of the disc in both directions, whereas the two edgepieces underwent pattern duplication during culture. This revealed the principle that when one fragment of an imaginal disc shows regeneration during culture, the complementary piece will usually show duplication.

These three-segment experiments were carried out with cuts at several levels in orthogonal and diagonal directions, with similar results. However, in two series

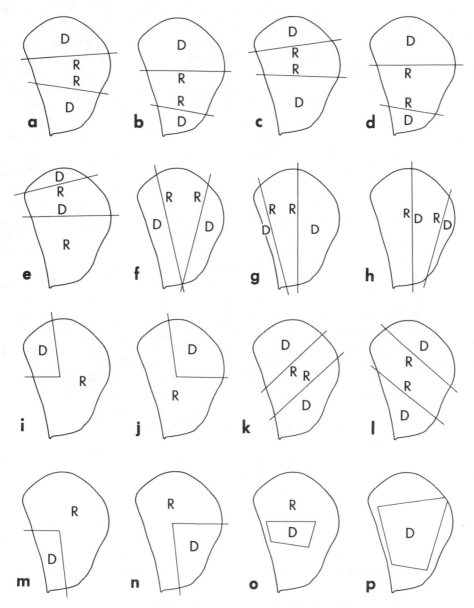

FIGURE 4. Summary of the observed regulative behavior of various fragments of the imaginal wing disc. R, regeneration; D, duplication. From Bryant (4).

where one of the edge pieces was large, a different result was obtained (Fig. 4E, H). In these cases, the large edge piece could regenerate the remainder of the disc, but the central piece underwent duplication and at the same time regenerated the small edge piece. We assume that regeneration and duplication occur by growth from opposite edges of these central pieces, thus preserving the rule of regeneration and duplication from complementary fragments.

An interpretation of regeneration and duplication in imaginal discs was suggested by the results of mixing experiments. In the controls for this experiment, two identical pieces of the imaginal wing disc, which would duplicate when cultured intact, were mechanically mixed together so that they formed a coherent mass, and then cultured *in vivo* for seven days. When the resulting implant was transferred to host larvae for metamorphosis, the final implant contained no structures other than those that would have been produced by similar fragments subjected to immediate metamorphosis. However, when two genetically marked fragments taken from opposite ends of the disc were mixed and cultured, each one was stimulated to grow and to regenerate structures that were never made in the control implants. This was interpreted as an example of intercalary regeneration occurring between normally nonadjacent parts of the pattern that are brought into contact during the tissue mixing procedure. Position-specific interactions of this type also occur when one of the two components consists of haltere cells and the other of wing cells (2) showing that these interactions depend on the relative positions of cells in the pattern, rather than on the specific pattern elements they are destined to produce. Regeneration of wing disc fragments can also be stimulated by mixing them with other discs, and although these interactions are weaker than between haltere and wing (58, 7) and have not been shown to be position-specific, they do nevertheless suggest that the same positional information system may be used in all imaginal discs. A similar conclusion has been reached from the finding that both leg and antenna cells can differentiate in a position-specific way in the antennae of *Antennapedia*, which consist of a mixture of leg and antenna cells (38).

Intercalary regeneration also appears to account for the pattern regulation which occurs in many intact cultured fragments of imaginal discs. Such fragments initially have a free edge produced by the cutting procedure, but within one or two days this is eliminated by wound healing between different parts of the edge (41, 42). Disc fragments often show intense local incorporation of tritiated thymidine, suggestive of local cell proliferation, near the position of the cut, with a maximum at about three days after the beginning of the culture period (12).

Clonal analysis of duplicating leg-disc fragments (1) shows that, early in the culture period, the majority of dividing cells are localized in the region close to the wound. During the later stages of growth, clonal analysis indicated that almost all of the cell proliferation was occurring in the growing duplicate, with hardly any divisions taking place in the original fragment. This is a clear demonstration that pattern regulation in these tissues is epimorphic, that is, occurring during proliferative growth, in contrast to morphallactic regulation which occurs by cell reprogramming without the need for proliferation. Furthermore, when disc fragments were cultured in sugar-fed flies, in which cell proliferation is inhibited, regeneration was blocked, indicating that cell proliferation is not only associated with regeneration, but is required for it (45).

The Polar Coordinate Model

The remarkable complementarity of behavior of imaginal disc fragments in undergoing either regeneration or duplication has been explained by the polar coordinate model (13), which was devised to account for pattern regulation in cockroach legs and amphibian limb regeneration blastemas, as well as imaginal discs. In fact, some aspects of the model are best illustrated by the cockroach experiments, so some of these will be discussed before the applicability of the model to *Drosophila* imaginal discs is explained. The model begins by assuming that positional information is specified by a system of polar coordinates, in which the radial coordinate specifies the proximo-distal axis (peripheral-central in an imaginal disc) and the circumferential coordinate specifies circumferential positions in the field. It is then assumed that when cells are confronted (as a result of grafting or wound healing) with other cells that are not their normal neighbors, the discontinuity in either the radial or the circumferential coordinate of positional information stimulates local cell proliferation, and the cells produced by such proliferation adopt the intervening positional values in whichever coordinate is affected, giving intercalary regeneration. In the case of the circumferential coordinate, it is assumed (based on evidence from cockroach legs) that intercalation reconstructs the shorter of the two sets of positional values that would resolve the discontinuity. It is this "shortest intercalation rule" that allows the model to account for the regeneration and duplication of complementary disc fragments; if one of the two fragments contains more than half of the circumference it will regenerate, whereas the complementary fragment would then necessarily contain less than half of the circumference and would duplicate (Fig. 5). Such complementary regulative behavior has been demonstrated for the wing, haltere, leg, eye-antenna, and genital discs. In many cases and most clearly in the wing disc, it is the larger fragments which regenerate and the smaller fragments which duplicate. However, the leg disc provides examples of large duplicating fragments and small regenerating ones, which have been explained by postulating unequal spacing of positional values around the circumference of this disc (13).

Intercalation in the proximo-distal axis has been well documented in the cockroach leg, but it occurs only within the limits of each leg segment. For example, when a distal level of the tibia was grafted onto a proximal level of the femur, intercalation occurred until the compound segment (a mosaic of tibia and femur cells) reached about the length of a single normal segment (see 13). The distal femur, femuro-tibial joint, and proximal tibia, which are the intervening pattern elements between graft and host boundaries, were not produced in these experiments, so the leg itself was one segment shorter than usual. From results such as these it has been concluded that proximo-distal intercalation is controlled by a gradient of positional information which is repeated in each leg segment. When intercalation occurs it restores the continuity of a single segmental gradient of positional information, but the overall leg pattern can be deficient or otherwise abnormal.

In addition to the segmentally repeating gradient of positional information, the cockroach leg must also be controlled by a positional information system spanning the entire length of the leg, since legs from which the distal segments are amputated can regenerate the missing segments. This can also occur from the originally proximal face of a section of the leg that is removed and reattached by its original distal face ("reversed regeneration"). Hence, from all levels there seems to be a tendency to regenerate distally, a rule called "distal transformation" (see 13).

When the distal end of an amputated cockroach leg is converted surgically into a symmetrical double dorsal or symmetrical double ventral pattern, in some

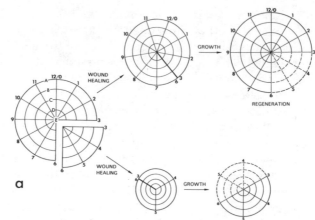

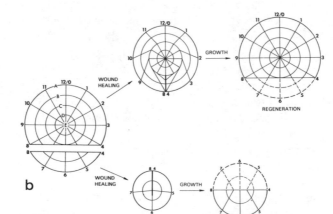

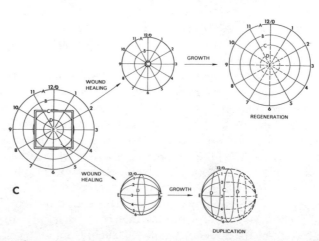

FIGURE 5. Interaction rules of the polar-coordinate model applied to the results obtained by culturing fragments of imaginal wing discs. (a) Three-quarter and one-quarter sectors. In the larger and smaller pieces alike the cut edges heal together. This brings position 3 and position 6 together in both instances. The shortest intercalation rule governs, and the

missing values, 4 and 5, are intercalated (dashed lines). For the one-quarter sector this means duplication; for the three-quarter sector it means regeneration. (b) Bisection of the disc. Both fragments heal by contraction along the cut edges, bringing positions 8 and 4 together. Intercalation again gives rise to a regenerate and a duplicate. (c) Central fragment and peripheral fragment of the wing disc. A complete circular sequence of positional values is present at the cut edge of each contracted fragment. Distal transformation results in duplication for the central fragment and regeneration for the peripheral fragment. From French et al. (13).

---

cases subsequent growth produces a variable, symmetrical and usually distally incomplete regenerate, but in other cases there is no regeneration. Such results have been interpreted by assuming that the extent and character of the distal regenerate is controlled by interactions between cells from different circumferential positions that come together during healing at the amputation surface (8). In the case of a double dorsal, double ventral, or any other symmetrical amputation surface, one type of healing (Fig. 6A, mode 1) can bring together identical positions from the two symmetrical halves. This is assumed to be a stable configuration since there are no positional information discontinuities to stimulate growth. However, another possible mode of wound healing brings together cells differing in circumferential positional information at the amputation surface (Fig. 6A, modes 2 and 3), just as any type of healing on the amputation surface of a normal, asymmetrical leg leads to the confrontation of circumferentially different positions (Fig. 6B). Under these conditions, it is assumed that growth is stimulated and that circumferential intercalation occurs, and that the newly produced cells are prevented from adopting positional values identical to those of their pre-existing neighbors, but are instead forced to adopt more distal positional values. In the case of a normal, asymmetrical amputation surface this process will continue until all the more distal structures are replaced, whereas with a symmetrical double-half stump the process will generate more distal double partial circumferences which contain less of the circumference with each cycle of intercalation. Eventually the double partial circumferences are reduced to nothing, giving a variable, tapered and distally incomplete symmetrical regenerate.

Most of the cockroach-leg experiments just quoted have their counterparts in the work on *Drosophila* imaginal discs (see 8 for review). When the central portion of either the leg or the wing disc is removed, the remaining peripheral part can regenerate it; this corresponds to normal distal regeneration following amputation of the distal part of the cockroach leg. When the central part of the wing disc is cultured, it produces an additional copy of the distal structures, corresponding to reversed regeneration (Fig. 4O, P). The central part of the leg disc is a very small fragment, which tends not to grow during culture.

Many proximal segments of leg or wing imaginal discs duplicate during culture, and the duplicated fragments then correspond to surgically produced double partial circumferences in the cockroach. As expected, these fragments do show a variable amount of distal regeneration, and those fragments with larger fractions of the circumference tend to produce complete distal regenerates more frequently than those with smaller fractions.

Mixing experiments have been carried out using proximal and distal fragments

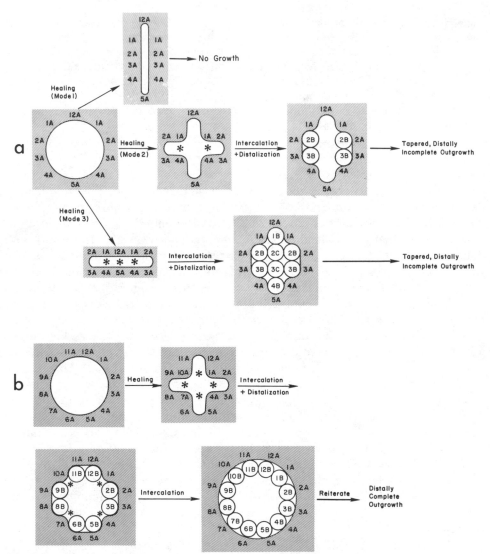

FIGURE 6. Model of distal outgrowth from (a) symmetrical and (b) asymmetrical wound surfaces. In both cases the tissue remaining after removal of B, C, D, and E levels of the pattern is shaded and the wound edge at level A is outlined by the circle. (a) A symmetrical wound surface. This diagram could represent a surgically produced double-half limb in the cockroach, or a duplicated proximal segment of an imaginal disc. With a symmetrical starting configuration, the outcome depends on the kind of wound healing that occurs. Mode 1 healing gives no positional value confrontations to stimulate intercalation and thus no distal outgrowth occurs. With mode 2 and mode 3, circumferential intercalation (*) would produce cells with positional values identical to those of preexisting adjacent cells, and so the new cells are forced to the next most distal level B. In mode 3 additional cells are produced and these adopt the C level. In both cases the final diagram shows the final product with no positional value discontinuities, which represents a symmetrical and distally incomplete outgrowth. Mode 3 gives a more complete outgrowth than mode 2. (b) A normal, asymmetrical wound surface. This diagram could represent an imaginal disc with the center removed or the stump of a cockroach leg after amputation

of the terminal parts. Circumferential intercalation between different parts of level A gives rise to cells which adopt level B. Subsequent intercalation completes the B level, and reiteration of the whole process generates the remaining distal levels. In outgrowth from complete circumferences, the process is essentially independent of variations in the directions of wound healing. From Bryant et al. (8).

---

from either wing (25) or leg (49) discs. In both cases, the proximal fragment regenerated distally (as it would have done in isolation) but the distal fragment failed to regenerate proximally. This has been interpreted as suggesting that these discs have segmentally repeating gradients like those proposed for the cockroach leg, and that proximo-distal intercalary regeneration may have occurred within the limits of one segment but that this amount of intercalation would not have been detectable given the limitations of the markers available.

In general, the polar coordinate model has been very successful in rationalizing the enormous body of experimental results on imaginal disc pattern regulation, and in drawing attention to the local cell interactions which appear to control growth and changes of pattern. The phenomenon of intercalation may also be involved in growth control and the elaboration of patterns of positional information in normal development, but this is less accessible to experimentation due to the small size of the developing cell populations. However, the genetic methods so well developed in *Drosophila* provide some new approaches to the analysis of development, which will be discussed in the next two sections.

Pattern Abnormalities Caused by Cell-Death Mutations

One of the great advantages of *Drosophila* in developmental biology is the availability of mutant strains that have been accumulated by geneticists over the last 74 years. Many of these cause local cell degeneration in imaginal discs, and the regulative responses of the remaining viable cells can be studied just as in the surgical experiments. The following kinds of abnormalities have been studied in this way:

PATTERN DEFICIENCIES

When pattern elements are missing from the adult fly, the cause often turns out to be cell degeneration occurring during larval or pupal development. Cell death has been demonstrated in the wing discs of the wing-reduction mutants *vestigial*, *scalloped*, *apterous*, *cut*, and *Beadex* (14, 15, 30), in the eye discs of the eye-reduction mutants *Bar*, *eyeless*, and *l(1)ts726* (15, 44, 9) and in the leg discs of heat-treated *l(1)ts504* and *l(1)ts726* larvae, which give rise to adults with leg deficiencies (44, 46, 9). In the mutant *decapentaplegic* (47) where many of the adult appendages are distally deficient, there is extensive cell death in the central (presumptively distal) regions of the corresponding imaginal discs (P. Bryant and S. Nguyen, *unpublished results*). The specificity of such mutants can be quite astonishing, as in *retinal degeneration* B^{KS222} where photoreceptor cells 1–6 degenerate while cells 7–8 are left reasonably intact in each ommatidium (23).

PATTERN DUPLICATIONS

In the imaginal discs of some mutants, cell death appears to be sufficiently extensive that the remaining viable fragment is not large enough to regenerate the missing parts, but instead produces a mirror-image duplication of the remaining partial pattern, as in the surgical experiments. This sequence of events seems to be the cause of pattern duplication in the wing disc of *vestigial, scalloped* (30) and *apterous* (M. Stevens, *personal communication*), in the eye-antenna disc of heat-treated *l(1)ts726* (44, 9), and in the leg discs of heat-treated *l(1)ts726* and other temperature-sensitive cell-lethal mutations affecting the leg (44, 46, 9). In *l(1)ts726* and *l(1)ts504* it has been shown that, in genetic mosaics consisting partly of wild-type and partly of mutant tissue, the wild-type tissue can participate in forming the duplicate pattern (44, 46), supporting the idea that pattern duplication is an indirect effect of mutant action elsewhere in the imaginal disc. Furthermore, clonal analysis with *l(1)ts726* has shown that cell proliferation parameters are different between the original and the duplicate patterns (22); in the latter, proliferation appears to occur later in development, after the heat pulse used to cause the duplication. This is of course, consistent with the idea that pattern duplication is stimulated as a response to the cell death which occurs during or shortly after the heat pulse.

PATTERN TRIPLICATIONS

The mutation *l(1)ts726* also gives rise to another type of abnormality when a two-day pulse of 29°C is initiated at about the beginning of the third larval instar (21). This seems to be a response to a restricted, internal zone of cell death within the leg disc. Such a zone, if it does not include the center or edge of the disc, must be bounded by a perimeter which is symmetrical as far as circumferential positional information is concerned. This apparently leads to the same type of symmetrical, tapering and distally incomplete outgrowth that is produced by surgically constructed symmetrical amputation surfaces in the cockroach leg. Since the outgrowth contains two copies of pattern elements that are also produced by the remaining viable parts of the leg disc, the result is a pattern triplication. Details of the cell death patterns, and the results of clonal analysis of the outgrowths, are consistent with this interpretation (J. Girton, *personal communication*).

Mutations Affecting Cell Interactions in Imaginal Discs

Mutations at several different loci in *Drosophila melanogaster* appear to lead to alterations of the cell interactions that control growth and pattern formation, so that imaginal disc growth continues far beyond the normal limits. In some of these mutants the loss of growth control is associated with loss of specialized junctions and decreased cell contacts.

In the lethal mutant *lethal (2) giant discs* the larval period in the homozygote is prolonged for up to nine days, and during this time the imaginal discs grow to several times their normal size (6). The imaginal discs retain the ability to metamorphose in wild-type hosts although only about 20% of the normal number of bristles and sense organs are differentiated. In the case of the third leg disc, a second set of concentric folds develops during the extended larval period, and after metamorphosis a duplicate leg, in mirror image with the primary leg, is differentiated. During the extended larval period, extra cell proliferation is also seen in the imaginal rings of the salivary gland, foregut and hindgut, and in the brain and testis.

When imaginal discs or imaginal rings from *l(2)gd* are transplanted into wild-type adult hosts, they continue to proliferate at a high rate whereas the corresponding wild-type tissues cease growth at characteristic cell population sizes (P. Bryant and P. Levinson, *unpublished results*). Thus the growth-control defect is intrinsic to the imaginal primordia, rather than being a result of the extended larval period.

In other lethal mutants, an extreme disruption of cell interactions is associated with loss of growth control as well as loss of the ability to differentiate cuticular patterns (20). In *lethal (2) giant larvae*, the mutant imaginal discs show loss of the monolayered epithelial structure seen in the wild type, the cells often entirely lacking specialized junctions and contacts with other cells. The discs are often enlarged, sometimes fused with one another, and they show rapid growth when transplanted into wild-type hosts.

From all of the experiments on pattern regulation in wild-type imaginal discs, as well as the genetic experiments, a rather consistent picture emerges in which both pattern formation and growth are controlled by interactions between the disc cells and their neighbors. While disc-extrinsic factors such as nutrients and hormones can presumably control the rate at which growth occurs, the control over where it occurs in the disc and over the final cell population which is achieved seems to be a function of cell interactions within the disc itself.

Interactions Between Cells Based on Presumptive Fate

An important question concerning the notion of positional information is the scale of graininess in the coordinate system: Are discrete positional values assigned to groups of cells, single cells, or even parts of single cells? The finer the grain of the system the greater must be the coding capacity of the genome in order to match positional values with appropriate states of cellular differentiation. If bristle patterns in *D. melanogaster* were constructed entirely via a positional-information mechanism, the intricacy and reproducibility of most of these patterns would require an accuracy approaching one positional value per cell. Whether this is feasible or not, it is impossible to say until we know more about the positional information mechanism. However, the results of a variety of experiments point to the existence of "fine-tuning" mechanisms, in which cells already specified to become bristles interact with other cells to control their fate, or in which already specified bristle cells adjust their physical positions in order to achieve the final pattern. In the remainder of this article, we discuss the evidence for such mechanisms.

Spacing Mechanisms

For bristles such as the head and thoracic macrochaetes, whose positions are constant, it is easy to imagine that the bristle sites are specified by a positional-infor-

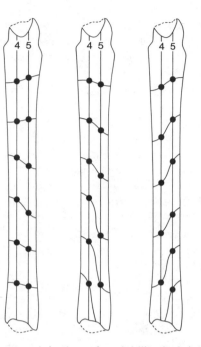

FIGURE 7. Illustration of an explanation of variability in bristle number and position based on positional information. Camera lucida drawings of three second-leg basitarsi, seen from the dorsal side (proximal end at top), with the positions of the bristle sockets in rows 4 and 5 represented by circles. The lines intersecting the bristle positions are contour lines within a hypothetical coordinate system of positional information which is assumed to specify the positions of the bristles.

mation system. However, for bristles whose positions vary from fly to fly, difficulties arise. For example, consider two rows of bristles on the second-leg basitarsus: rows 4 and 5 (Fig. 7). Neither the number nor the positions of the bristles in these rows are constant from fly to fly, and the number and positions of the bristles in each row vary independently of those of the other row. If bristles were specified at specific positional-information contours, then such fluctuations would imply that the contours are greatly distorted in some flies, as shown in Fig. 7. Simple inaccuracy in reading positional information would not account for this behavior, since regular bristle spacing is maintained in spite of the variability in bristle number. An alternative explanation, not based on positional information, would be to suppose that bristle positions within each row are specified by some sort of spacing mechanism designating bristle sites at regular intervals. Slight variations in the length of the interval would generate variable numbers of bristles occupying variable positions, and fluctuations in one row would have no effect on the pattern in other rows. The same spacing mechanism could generate all rows in *D. melanogaster* which have evenly spaced bristles, and most rows are of this type.

The second-leg basitarsus provides a convenient bristle pattern with which to investigate mechanisms of patterning (Fig. 8). On this segment there are, on average, about 80 bristles, most of which are organized in eight longitudinal rows. Within each row, the bristles are uniformly spaced at a characteristic interval. Every bristle in the longitudinal rows is associated with a minute hairlike structure called a "bract," just proximal to the bristle socket. (See Fig. 1A; bracts are also found associated with bristles on other distal leg segments as well as on the proximal anterior wing margin.) There are a few other bracted bristles at the proximal end of the second-leg basitarsus (between rows 1 and 2 and between rows 7 and 8), and there are typically five bractless bristles at specific locations between the rows. Thus the pattern includes both evenly spaced bristles, whose numbers and positions vary, and uniquely identifiable bristles (the bractless bristles) which occupy characteristic sites.

A spacing mechanism could theoretically measure each bristle interval as a

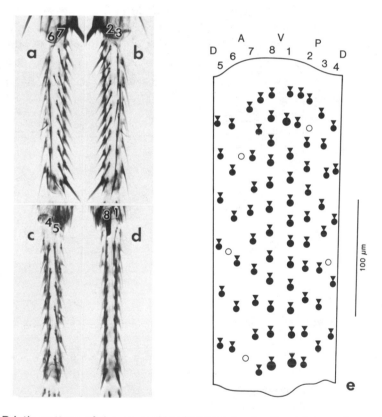

FIGURE 8. Bristle pattern of the second-leg basitarsus of a *D. melanogaster* male. The segment is shown from the anterior (a), posterior (b), dorsal (c), and ventral (d), and its bristle pattern is diagrammed (e). The proximal end in each case is at the top. In *e*, bracted bristles are symbolized by filled circles, bractless bristles by open circles, and bracts by triangles. D, dorsal; A, anterior; V, ventral; P, posterior. Numbers refer to longitudinal rows of bristles. Diameters of circles reflect actual diameters of bristle sockets. The segment is about 68 cells long and 23 cells around.

fixed distance or as a fixed number of cells. If each interval were measured as a physical distance, then it should be possible to change the number of bristles in each row by altering the length of the segment. If, on the other hand, each interval were measured as a constant number of cells, then the number of bristles in each row could be modified by altering the number of cells along the segment, and bristle-interval could be modified by altering cell diameter. The dimensions and cellular parameters of the second-leg basitarsus are altered in flies that have been starved during the larval period, in triploid flies, and in the mutants *dachs* and *four-jointed* (54, 26). Examination of bristle number and bristle interval in these various types of flies (Fig. 9) reveals a constant proportionality between bristle interval and cell diameter, and between the number of bristles in each row and the number of cells along the segment. These dependent relationships are consistent with the hypothesis that the bristle intervals in each row are measured as a constant number of cells. That is, the results are explicable in terms of a spacing mechanism which counts cells, instructing a cell every "x" cells along each row to become a bristle cell, where "x" is an integer that varies from row to row.

In order to account for these results in terms of positional information, it would be necessary to suppose that the number of positional values along the basi-

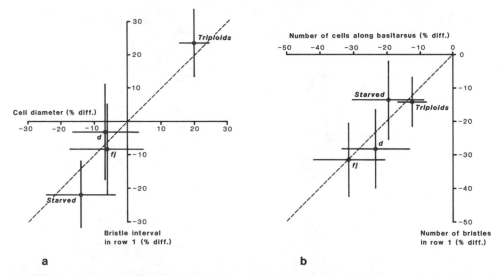

FIGURE 9. Alterations in the bristle pattern of the second-leg basitarsus as a function of the size and number of cells on the segment, in starved flies, triploids, and the mutants *dachs* (*d*) and four-jointed (*fj*). Changes are expressed as percent increase or decrease relative to well-fed, wild-type flies. (a) Percent change in the average interval (absolute distance measured in microns) between neighboring bristles in row 1 versus percent change in the average cell diameter on the segment. (b) Percent change in the average number of bristles in row 1 versus percent change in the average number of cells along the basitarsus. Error bars indicate ± 1 standard deviation (*N* = 10 for each point). Dashed lines have a slope of 1. (Although only data for row 1 are shown, all rows give similar plots since neither the relative numbers of bristles nor the relative bristle intervals of the eight rows are appreciably altered in any of these types of flies.) Data from Held (26).

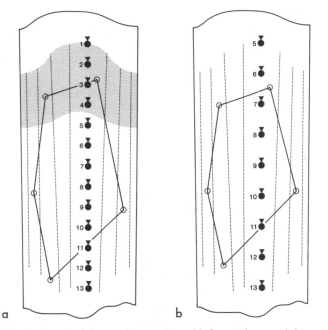

FIGURE 10. Illustration of the failure of a positional-information model to explain starvation-associated alterations in the bristle pattern of the second-leg basitarsus. These diagrams depict the average number of bristles in a representative row, row 1, and the average proximo-distal positions (measured as percent segment length from the distal end) of the five bractless bristles on: (a) 10 large basitarsi (average length $= 354\mu$m) from well-fed wild-type males, and (b) 10 small basitarsi (average length $= 198\mu$m) from wild-type males starved during the larval period. Solid lines connect bractless bristle positions. In (a) the bristles in row 1 are numbered consecutively from the proximal end. The shaded area includes 28.4% of the length of each bristle row; this is the fraction of proximo-distal positional values that would have to be eliminated (assuming that bristle positions are specified by a positional information mechanism) in order to reduce the average number of bristles in all eight rows from 76.4 for (a) to 57.0 for (b). The removal of such a large fraction of the pattern area from either the proximal end [as shown in (a)], or the distal end, or partly from both ends of the pattern area should cause some bractless bristles to be eliminated and others to be drastically relocated. In fact, however, all five bractless bristles are present on the smaller basitarsi, and their locations differ from those of the larger basitarsi by no more than 6% segment length. Data from Held (26).

tarsus is proportional to the number of cells along its length. In the case of the basitarsi of starved flies, therefore, there should be fewer positional values along the segment since there are fewer bristles in each row. Continuing this line of reasoning, one would expect that a reduction in the range of positional values along the basitarsus would also cause a relocation of the five bractless bristles. In fact, however, the expected relocations are not observed (Fig. 10). Instead, the bractless bristles occupy positions which can be described as a fixed percentage of segment length from either one of the segment boundaries. This leads us to suggest that the overall pattern framework of the segments including the positions of bristle rows and of bractless bristles, may be specified by a positional informational system, but that the positions of bristles within rows may be established by a spacing mechanism.

Inhibitory Fields

Given that each bristle organ is formed by the clonal descendants of a single bristle mother cell, whatever mechanism is responsible for bristle patterning must be precise enough to single out individual cells within the epidermis. A study by Stern (48) suggests that this process of bristle-cell specification takes place in two steps. Stern examined genetically mosaic flies composed partly of tissue hemizygous for the recessive X-chromosomal mutation *achaete* and partly of heterozygous (phenotypically nonmutant) tissue. *Achaete* causes a specific macrochaete on the thorax (in addition to other bristles) to be missing. In some of the mosaics, this macrochaete was missing from its normal site owing to the presence there of *achaete* tissue, and a new macrochaete was formed by nearby heterozygous cells at a location different from the normal position of the bristle (Fig. 11). Stern's interpretation of this result was that the specification of the normal macrochaete site takes place as follows: (1) a group of cells is designated, each of which is competent to form the macrochaete (this step could be accomplished by a positional-information system of only modest accuracy), (2) a single cell within the group begins to undergo macrochaete differentiation and simultaneously inhibits the remaining cells from doing so. Usually the cell that becomes the macrochaete cell would be located at a certain position within the group, but when this site is occupied by a cell unable to commence macrochaete differentiation (e.g., when the cell is *achaete*) then another cell within the group would assume this fate. Presumably this mechanism would be responsible for specifying the positions of all of the macrochaetes on the head and thorax and all other bristles whose positions are normally constant.

Richelle and Ghysen (43) have proposed a model that explains how the crude designation of a group of cells could automatically lead to the reproducible pin-

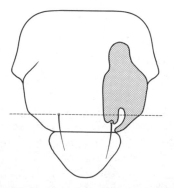

FIGURE 11. Diagram of the thorax of a mosaic fly composed partly of *achaete* tissue (shaded area) and the remainder of heterozygous (phenotypically nonmutant) tissue. The dashed line indicates the antero-posterior location of the posterior dorsocentral macrochaete in wild-type flies. Redrawn from Stern (48).

pointing of a particular cell within the group (step 2 of Stern's mechanism). In their model, this "fine-tuning" step is accomplished by having each competent cell produce a bristle-inducing substance that diffuses away within the epithelium. The concentration of this substance would, of course, be highest in the center of the group of competent cells, and the first cell whose concentration of this substance exceeds a certain threshold—usually the cell in the center—would commence differentiation and inhibit the remaining cells from undergoing the same differentiation.

Inhibitory fields could conceivably be used for another purpose besides ensuring precision in the designation of single cells—namely, for bristle spacing. A bristle-spacing model based upon inhibitory fields was first proposed by Wigglesworth (57) for the bug *Rhodnius*, and subsequent versions of this model have been devised for various bristle patterns in *Drosophila*. For example, a row of evenly spaced bristle cells can be specified within a file of competent cells if (1) the cells acquire the tendency to become bristle cells sequentially from one end of the file to the other, and (2) each cell that becomes a bristle cell generates around itself an inhibitory field within which no other cell can become a bristle cell (11). The row of uniformly spaced bristles created in this way would have its bristle interval equal to the radius of the inhibitory field. Since the size of an inhibitory field could depend upon the size of the bristle cell, a dependence of bristle interval upon cell size (Fig. 9) is as explicable with this kind of spacing mechanism as it is with a mechanism which measures bristle intervals as a fixed number of cells. Thus inhibitory fields might be involved in the "fine-tuning" of accurately specified bristle sites as well as in the patterning of evenly spaced bristles. They might also be involved in the establishment of more complex patterns, such as in the sternites (Fig. 2H) where there is a linear relationship between bristle size and distance to the nearest neighbor bristle (10).

Sequential Specification

If the positions of the bristles within a row are designated sequentially from one end of the row to the other, then this temporal order might be maintained through the period when the bristle cells undergo their differentiative divisions. In order to check for a spatiotemporal pattern of bristle-cell divisions on the second-leg basitarsus, pupae were exposed to 10,000 rads of gamma rays at various times after pupariation and then allowed to complete development. Their second-leg basitarsi were then examined to determine which bristles had differentiated normally. Ten thousand rads is presumed to prevent further cell divisions (37) so that bristles which differentiate normally can be assumed to have completed their differentiative divisions prior to the time of irradiation. A general distal to proximal sequence in the appearance of bristles was observed within each of the longitudinal bristle rows on the basitarsus (Fig. 12; L. Held, *unpublished results*). However, there is an abrupt transition from basitarsi possessing about ¼ of the normal number of bristles (16 h) to nearly unaffected basitarsi (17 h). The abruptness of this transition argues against the idea that bristle sites are specified sequentially from one end of each row to the other, since in that event (assuming a uniform rate of specification

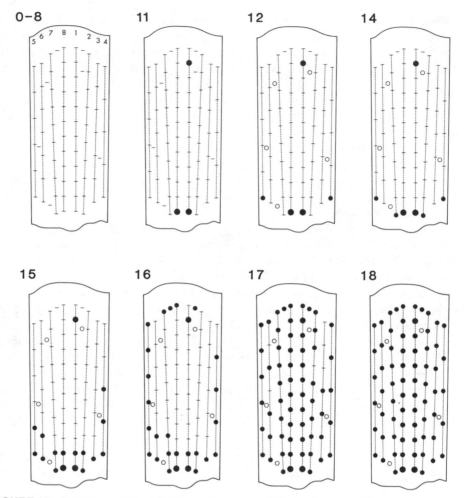

FIGURE 12. Positions of the bristles on the second-leg basitarsus which differentiate after irradiation of whole pupae with 10,000 rads of gamma rays at various times after pupariation. Bold face numbers indicate the age (in hours after pupariation) when the pupae were irradiated. Each filled circle indicates a bristle which differentiated (shaft and socket both present) in 50% or more of the basitarsi examined. Basitarsi were from five males for each age group.

along each row) a steadily moving wavefront of insensitivity would have been expected.

Another prediction of the idea that bristle intervals are measured sequentially from one end of each row is that errors in bristle positioning (due to variation in interval length within a row) should accumulate. Thus one might expect the first bristle site to be more constant than the last one. A study of the variability of the positions of the bristles in rows 4 and 5 on the second-leg basitarsus has shown that (1) the positions of the two terminal bristles vary to approximately the same extent, and (2) the positions of the terminal bristles are relatively constant, whereas the positions of the remaining bristles vary, depending on the number of bristles in the row (Fig. 13; L. Held and T. Pham, *unpublished observations*). These results argue against a unidirectional specification mechanism, unless the length of each row is subsequently adjusted so that the endpoints come to occupy definite positions. (Sequential specification from the two ends toward the middle is still a possibility.) An example of a theoretical spacing mechanism which does not involve sequential specification is the reaction-diffusion model of Turing (55; see also 35). With this type of model, uniform intervals are created as a result of reactive molecules that have different rates of diffusion, and presumably rows could be generated whose endpoints are fixed.

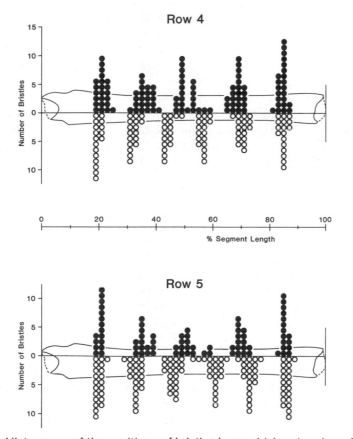

FIGURE 13. Histograms of the positions of bristles in row 4 (above) and row 5 (below) on basitarsi having either five (filled circles above each horizontal axis) or six (open circles below each horizontal axis) bristles in the depicted row. Each circle represents a single bristle. Data from 40 second-leg basitarsi (20 male, 20 female) are shown. Bristle positions (recorded as the position of the center of the bristle socket) are grouped at intervals of 2% segment length. Ordinate: Number of times a bristle occupied a particular 2%-segment-length interval on the abcissa. (Three of the 40 basitarsi had either four or seven bristles in row 5, and row-5 bristle positions for these basitarsi are not shown.)

Inductive Interactions

In some rows of bristles in *D. melanogaster*, the bristles have no intervening spaces (for example, the transverse rows on the tibia of the first leg, Fig. 2E). If all rows were to originate in this manner—as solid files of bristle cells—then such files could easily be generated in a sheet of uncommitted cells by the designation of one bristle cell at the end of each file and the transmission of an inductive signal (which would instruct each uncommitted cell to become a bristle cell) from cell to cell along the file. Nonbristle cells could later be introduced into each file by cell divisions or cell movements, thereby spacing the bristle cells out at regular intervals.

This notion of a sequential induction mechanism in bristle row development is made more plausible by evidence indicating another type of inductive interaction in bristle pattern formation: namely, bract induction. Each bract is made by a single cell (40), which is not a member of the clone of cells that makes the bristle organ (53). When larvae are injected with either Mitomycin C (50) or nitrogen mustard (51) and allowed to complete development, many bristles lack their sockets, and bracts form only next to those bristles which have a socket. Also, in the mutants *Hairless* and *shaven-depilate*, many sockets develop without a shaft, and bracts form only next to those sockets which have a shaft (52). Thus, the development of a bract seems to depend upon the normal development of its accompanying bristle, and the implication is that the bract cell is actually induced by the cells of the bristle organ.

One way of testing the hypothesis that the bristle cells of each row are sequentially induced by one another would be to prevent cells at various locations along a row from becoming bristle cells. According to this hypothesis, the inductive signal would be unable to traverse the inhibited cells, so that no bristles should form downstream from the blockage. A mutation whose phenotype closely approaches the phenotype expected for a mutation that blocks the cellular decision to become a bristle cell is $Df(1)sc^{10-1}$, a deletion at the *scute-achaete* locus. Flies hemizygous for this deletion are missing most of their bristles, especially on the legs. In a sample of 14 $Df(1)sc^{10-1}$ males that were examined, only a single bristle was found on the second-leg basitarsi. Wherever the bristles are missing, there are no blemishes or other indications of any attempt at bristle differentiation. Perhaps the strongest evidence for the early action of mutations at this locus is the fact, discussed above, that *achaete* interferes with bristle development early enough for nearby cells to compensate for the absence of the affected bristle by changing their developmental fate. Second-leg basitarsi from gynandromorph mosaics of $Df(1)sc^{10-1}$ showed numerous examples of gaps within or at both ends of longitudinal rows (Fig. 14; L. Held and N. Ha, *unpublished observations*). This result argues against a sequential induction mechanism, at least for these rows.

A different sort of sequential induction mechanism has been invoked to explain the regular pattern of ommatidia and bristles in the eye. Based upon their finding that ommatidia are not clonal units and their observation that a wave of mitosis sweeps across the eye disc during the third instar, Ready *et al.* (39) have proposed that uncommitted cells in front of the mitotic wave are recruited into a repetitive matrix of already specified cells behind the wave, much like molecules

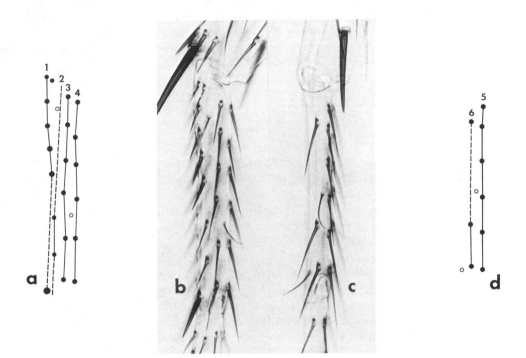

FIGURE 14. Examples of second-leg basitarsi from gynandromorph mosaics of *Df(1)sc* [10-1] which were missing bristles within or at both ends of one or more longitudinal bristle rows. (a, b) Bristle pattern of the posterior side of a mosaic basitarsus which has a gap inside row 1 and at each end of row 2. (c,d) Bristle pattern of the anterior side of a mosaic basitarsus which has a gap inside row 6. (Rows 7 and 8 are completely missing.) Filled circles represent bracted bristles; open circles represent bractless bristles. Assignment of bristles to specific rows is based upon bristle position and bristle morphology. (Bristle shapes vary from row to row.)

are incorporated into the lattice of a growing crystal. Each newly recruited cell, upon entering the matrix, would be induced (by its neighboring committed cells) to adopt a fate consistent with its position in the matrix. Since the interommatidial bristles are separated by intervening pigment cells, these sequential inductions would presumably involve the repetitive induction of bristle cells by nonbristle (pigment) cells and *vice versa*.

Cell Movements

When the bristle cells of a cuticular region are originally specified, their pattern may theoretically differ considerably from the arrangement that they eventually come to assume through either growth (18) or cell movements (56). The most striking rearrangement of a bristle pattern for which there is evidence involves a rotation of the sex comb, a row of thick, blunt bristles on the male foreleg basitarsus. Marked clones which otherwise are straight tend to detour where they pass through the sex comb area, and the detours can be explained by assuming that the row of sex comb cells rotates by approximately 90° relative to the rest of the leg during a late stage of development (53).

On the tergites, individual bristle cells can evidently migrate within the epidermis since tergite clones that are marked with mutations affecting bristles and hairs are often found to have marked bristles outside the marked-hair territory and unmarked bristles inside it (19).

On the legs, marked clones tend to be oriented longitudinally, yet they often include bristles and not their accompanying bracts or *vice versa*, suggesting lateral movements of bract cells relative to bristle cells (5). The clonal affiliations of bracts and bristles indicate that these cell movements have characteristic directions which vary among the bristle rows (27, 33, Fig. 15). Such movements may constitute part of the general scheme of cell rearrangements which is thought to occur during evagination of the leg discs (see 16).

Bristle cell movements may also be involved in straightening bristle rows. Indirect evidence for this notion comes from genetically mosaic legs in which tarsal segments have been prevented from reaching their full lengths by tissue expressing the mutation *four-jointed* (which causes reduced growth). Wild-type portions of such segments possess crowded bristles arranged in zigzag rows, suggesting that bristle rows normally pass through a stage of only crude alignment which, in this instance, has been preserved due to insufficient elongation of the segments (54). Another fact which suggests corrective cell movements in bristle row development is that the bristles in row 1 on the tarsi of the first and second legs can be embraced

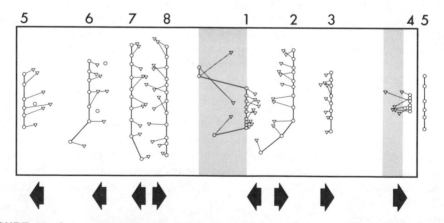

FIGURE 15. Gynandromorph map of the bristles (circles) and bracts (triangles) on the second-leg basitarsus (from 27). Bristles and bracts which are near one another on the map are closely related in their cell ancestry; those far apart are distantly related. Arrows below the map indicate the inferred directions of movements that the prospective bract cells must undergo at some time during development. (Shaded areas indicate the zones where the anterior/posterior compartment boundary passes through the basitarsus.)

by clones belonging to either the anterior or the posterior compartment (27, 33). If, as seems likely, the boundaries between compartments are straight during development, then the cells of this bristle row would have to be initially misaligned. Finally, some evidence comes from experiments in which pupae were subjected to heat shocks, and their second-leg basitarsi were examined for abnormal bristle arrangements (L. Held, *unpublished observations*). When pupae were heat-shocked between 0 and 12 hr after pupariation, the bristles within the longitudinal rows were often misaligned and unevenly spaced, and when the pupae were shocked at 21–24 hr after pupariation, the bristles were aligned but still unevenly spaced (Fig. 16). These results suggest that the patterning of bristle sites within each longitudinal row takes places as follows: (1) when the bristle cells are first specified, they are somewhat scattered and only roughly aligned, (2) at 0–12 hr after pupariation the bristle cells within each row become aligned, and (3) at 21–24 hr after pupariation they become evenly spaced.

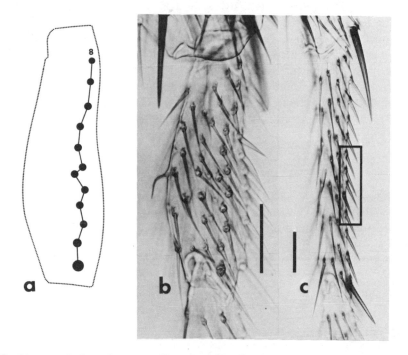

FIGURE 16. Abnormal phenotypes on the second-leg basitarsus caused by heat shocks administered to pupae at various times after pupariation. (a, b) Anterior side of a second-leg basitarsus from a female heat-shocked at 12 hours after pupariation. [Only row 8 is diagrammed in (a)]. The bristles of row 8 are misaligned and unevenly spaced. (c) Posterior side of a second-leg basitarsus from a female heat-shocked at 21 hours after pupariation. The rectangle encloses a section of row 1 in which the bristles are unevenly spaced. (Some slight misalignment is also evident.) Assignment of bristles to specific rows is based upon bristle position and bristle morphology. Note the difference in magnification between (b) and (c). (The scale bar in each case represents 50μm.) Basitarsi from flies heat-shocked at 0–12 hr after pupariation are frequently short and wide.

Conclusions

Although investigators of pattern formation have been motivated by a search for a single basic mechanism, we are now at a stage where it is difficult to avoid the conclusion that several mechanisms are involved in the formation of cuticular patterns in *Drosophila*. At least in some of the imaginal discs, the field of cells becomes subdivided into several compartments which represent, in some sense, units of the control of gene activity. Interactions between cells then seem to set up position-specific differences of "positional information" irrespective of lineage or compartmental affinity of the interacting cells. Even this process may consist of two steps, as in the leg where the overall proximo-distal axis may be specified separately from the proximo-distal patterns within each segment. Positional information, however, seems to assign bristle sites in only an approximate fashion, and inhibitory interactions between bristle cells and non-bristle cells seem to be involved in "fine tuning" the pattern. Still another step, involving cell movement, seems to be involved in aligning bristles into rows and assuring even spacing, and finally there appears to be a specific inductive event which initiates bract development. Whether some of these events are restricted to bristle patterns, it is too early to tell, but in any case the study of these patterns should alert us to the possibility that, in the formation of patterns in general, there may be many steps between the initially homogeneous cell population and the final pattern of differentiation.

Acknowledgments

The authors' investigations are supported by grants HD 06082, HD 16663, AG 01979, CA 09054 and HD 05584 from the National Institutes of Health, DHEW, and grant PCM-8011777 from the National Science Foundation. It is a pleasure to thank the colleagues mentioned in the text for the use of their unpublished results. Critical comments on the manuscript were kindly provided by Danny Brower, Ken Muneoka, David O'Brochta, and Mary Stevens.

General References

BRYANT, P. J.: Pattern formation in imaginal discs. In *The genetics and biology of Drosophila*, Vol. 2c Ashburner, M., Wright, T. R. F. (eds.). New York: Academic Press, pp. 229–335 (1978).

BRYANT, P. J., BRYANT, S. V., FRENCH, V.: Biological regeneration and pattern formation. *Sci. Am.* 237: 66–81 (1977).

LAWRENCE, P. A.: The development of spatial patterns in the integument of insects. In *Developmental systems: insects*, Vol. 2 Counce, S. J., Waddington, C. H. (eds.). New York: Academic Press, pp. 157–209 (1973).

References

1. Abbott, L. C., Karpen, G. H., Schubiger, G.: Compartmental restrictions and blastema formation during pattern regulation in *Drosophila* imaginal leg discs. *Dev. Biol.* **87**: 64–75 (1981).

2. Adler, P. N.: Position-specific interaction between cells of the imaginal wing and haltere discs of *Drosophila melanogaster. Dev. Biol.* **70**: 262–267 (1979).

3. Bernard, F.: Recherches sur la morphogénèse des yeux composés d'arthropodes. Développement, croissance, réduction. *Bull. Biol. Fr. Belg., Suppl.* **23**: 162 pp (1937).

4. Bryant, P. J.: Pattern formation in imaginal discs. In "The *genetics and biology of Drosophila*, Vol. 2c Ashburner, M., Wright, T. R. F. (eds.) New York: Academic Press, pp 229–335 (1978).

5. Bryant, P. J., Schneiderman, H. A.: Cell lineage, growth and determination in the imaginal leg discs of *Drosophila melanogaster. Dev. Biol.* **20**: 263–290 (1969).

6. Bryant, P. J., Schubiger, G.: Giant and duplicated imaginal discs in a new lethal mutant of *Drosophila melanogaster. Dev. Biol.* **24**: 233–263 (1971).

7. Bryant, P. J., Adler, P. N., Duranceau, C., Fain, M. J., Glenn, S., Hsei, B., James, A. A., Littlefield, C. L., Reinhardt, C. A., Strub, S., Schneiderman, H. A.: Regulative interactions between cells from different imaginal discs of *Drosophila melanogaster. Science* **201**: 928–930 (1978).

8. Bryant, S. V., French, V., Bryant, P. J.: Distal regeneration and symmetry. *Science* **212**: 993–1002 (1981).

9. Clark, W. C., Russell, M. A.: The correlation of lysosomal activity and adult phenotype in a cell-lethal mutant of *Drosophila. Dev. Biol.* **57**: 160–173 (1977).

10. Claxton, J. H.: Some quantitative features of *Drosophila* sternite bristle patterns. *Aust. J. Biol. Sci.* **27**: 533–543 (1974).

11. Claxton, J. H.: Developmental origin of even spacing between the microchaetes of *Drosophila melanogaster. Aust. J. Biol. Sci.* **29**: 131–135 (1976).

12. Dale, L. Bownes, M.: Is regeneration in *Drosophila* the result of epimorphic regulation? *Wilhelm Roux' Arch.* **189**: 91–96 (1980).

13. French, V., Bryant, P. J., Bryant, S. V.: Pattern regulation in epimorphic fields. *Science* **193**: 969–981 (1976).

14. Fristrom, D.: Cellular degeneration in wing development of the mutant *vestigial* of *Drosophila melanogaster. J. Cell Biol.* **39**: 488–491 (1968).

15. Fristrom, D.: Cellular degeneration in the production of some mutant phenotypes in *Drosophila melanogaster. Molec. Gen. Genet.* **103**: 363–379 (1969).

16. Fristrom, D., Fristrom, J. W.: The mechanism of evagination of imaginal discs of *Drosophila melanogaster.* I. General considerations. *Dev. Biol.* **43**: 1–23 (1975).

17. Garcia-Bellido, A.: Genetic control of wing disc development in *Drosophila*. In: "Cell Patterning" Ciba Foundation Symposium **29**: 161–182 (1975).

18. Garcia-Bellido, A.: From the gene to the pattern: Chaeta differentiation. In *Cellular controls in differentiation* (eds). Lloyd, C. W., Rees, D. A., pp. 281–301. New York. Academic Press (1981).

19. Garcia Bellido, A., Merriam, J. R.: Clonal parameters of tergite development in *Drosophila. Dev. Biol.* **26**: 264–276 (1971).

20. Gateff, E.: Malignant neoplasms of genetic origin in *Drosophila melanogaster. Science* **200**: 1448–1459 (1978).

21. Girton, J. R.: Pattern triplications produced by a cell-lethal mutation in *Drosophila. Dev. Biol.* **84**: 164–172 (1981).

22. Girton, J. R., Russell, M. A.: A clonal analysis of pattern duplication in a temperature-sensitive cell-lethal mutant of *Drosophila melanogaster. Dev. Biol.* **77**: 1–21 (1980).

23. Harris, W. A., Stark, W. S., Walker, J. A.: Genetic dissection of the photoreceptor system in the compound eye of *Drosophila melanogaster. J. Physiol. Lond.* **256**: 415–439 (1976).

24. Haynie, J. L., Bryant, P. J.: The effects of X-rays on the proliferation dynamics of cells in the imaginal wing disc of *Drosophila melanogaster. Wilhelm Roux' Arch.* **183**: 85–100 (1977).

25. Haynie, J., Schubiger, G.: Absence of distal to proximal intercalary regeneration in imaginal wing discs of *Drosophila melanogaster. Dev. Biol.* **68**: 151–161 (1979).

26. Held, Jr., L. I.: Pattern as a function of cell number and cell size on the second-leg basitarsus of *Drosophila. Wilhelm Roux' Arch.* **187**: 105–127 (1979a).

27. Held., Jr., L. I.: A high-resolution morphogenetic map of the second-leg basitarsus in *Drosophila melanogaster. Wilhelm Roux' Arch.* **187**: 129–150 (1979b).

28. Hofbauer, A., Campos-Ortega, J. A.: Cell clones and pattern formation: genetic eye mosaics in *Drosophila melanogaster. Wilhelm Roux' Arch.* **179**: 275–289 (1976).

29. Holtzer, H., Rubinstein, N., Fellini, S., Yeoh, G., Chi, J., Birnbaum, J., Okayama, M.: Lineages, quantal cell cycles and the generation of cell diversity. *Quart. Rev. Biophys.* **8**: 523–557 (1975).

30. James, A. A., Bryant, P. J.: Mutations causing pattern deficiencies and duplications in the imaginal wing disc of *Drosophila melanogaster. Dev. Biol.* **85**: 39–54 (1981).

31. Lawrence, P. A.: Development and determination of hairs and bristles in the milkweed bug *Oncopeltus fasciatus* (Lygaeidae, Hemiptera). *J. Cell Sci.* **1**: 475–498 (1966).

32. Lawrence, P. A., Morata, G.: Compartments in the wing of *Drosophila*: a study of the *engrailed* gene. *Dev. Biol.* **50**: 321–337 (1976).

33. Lawrence, P. A., Struhl, G., Morata, G.: Bristle patterns and compartment boundaries in the tarsi of *Drosophila. J. Embryol. Exp. Morphol.* **51**: 195–208 (1979).

34. Lees, A. D., Waddington, C. H.: The development of the bristles in normal and some mutant types of *Drosophila melanogaster. Proc. Roy. Soc. Lond. B.* **131**: 87–110 (1942).

35. Meinhardt, H.: Models for the ontogenetic development of higher organisms. *Rev. Physiol. Biochem. Pharmacol.* **80**: 47–104 (1978).

36. Morata, G., Ripoll, P.: *Minutes*: mutants of *Drosophila* autonomously affecting cell division rate. *Dev. Biol.* **42**: 211–221 (1975).

37. Poodry, C. A.: A temporal pattern in the development of sensory bristles in *Drosophila. Wilhelm Roux' Arch.* **178**: 203–213 (1975).

38. Postlethwait, J. H., Schneiderman, H. A.: Pattern formation and determination in the antenna of the homoeotic mutant *Antennapedia* of *Drosophila melanogaster. Dev. Biol.* **25**: 606–640 (1971).

39. Ready, D. F., Hanson, T. E., Benzer, S.: Development of the *Drosophila* retina, a neurocrystalline lattice. *Dev. Biol.* **53**: 217–240 (1976).

40. Reed, C. T., Murphy, C., Fristrom, D.: The ultrastructure of the differentiating pupal leg of *Drosophila melanogaster*. *Wilhelm Roux' Arch.* **178**: 285–302 (1975).

41. Reinhardt, C. A., Hodgkin, N. M., Bryant, P. J.: Wound healing in the imaginal discs of *Drosophila*. I. Scanning electron microscopy of normal and healing wing discs. *Dev. Biol.* **60**: 238–257 (1977).

42. Reinhardt, C. A., Bryant, P. J.: Wound healing in the imaginal discs of *Drosophila*. II. Transmission electron microscopy of normal and healing wing discs. *J. Exp. Zool.* **216**: 45–61 (1981).

43. Richelle, J., Ghysen, A.: Determination of sensory bristles and pattern formation in *Drosophila*. I. A model. *Dev. Biol.* **70**: 418–437 (1979).

44. Russell, M. A.: Pattern formation in the imaginal discs of a temperature-sensitive cell-lethal mutant of *Drosophila melanogaster*. *Dev. Biol.* **40**: 24–39 (1974).

45. Schubiger, G.: Regeneration of *Drosophila melanogaster* male leg disc fragments in sugar fed female hosts. *Experientia* **29**: 631 (1973).

46. Simpson, P., Schneiderman, H. A.: Isolation of temperature sensitive mutations blocking clone development in *Drosophila melanogaster*, and the effects of a temperature sensitive cell lethal mutation on pattern formation in imaginal discs. *Wilhelm Roux' Arch.* **178**: 247–275 (1975).

47. Spencer, F. A., Hoffman, F. M., Gelbart, W. M.: *Decapentaplegic*: a gene complex affecting morphogenesis in *Drosophila melanogaster*. *Cell* **28**: 451–461 (1982).

48. Stern, C.: Genes and developmental patterns. *Proc. 9th Int. Cong. Gen.* **1**: 355–369 (1954).

49. Strub, S.: Leg regeneration in insects: An experimental analysis in *Drosophila* and a new interpretation. *Dev. Biol.* **69**: 31–45 (1979).

50. Tobler, H.: Beeinflussung der Borstendifferenzierung und Musterbildung durch Mitomycin bei *Drosophila melanogaster*. *Experientia* **25**: 213–214 (1969).

51. Tobler, H., Maier, V.: Zur Wirkung von Senfgaslösungen auf die Differenzierung des Borstenorganes und auf die Transdeterminationsfrequenz bei *Drosophila melanogaster*. *Wilhelm Roux' Arch.* **164**: 303–312 (1970).

52. Tobler, H., Rothenbuhler, V., Nothiger, R.: A study of the differentiation of bracts in *Drosophila melanogaster* using two mutations, H^2 and sv^{de}. *Experientia* **29**: 370–371 (1973).

53. Tokunaga, C.: Cell lineage and differentiation on the male foreleg of *Drosophila melanogaster*. *Dev. Biol.* **4**: 489–516 (1962).

54. Tokunaga, C., Gerhart, J. C.: The effect of growth and joint formation on bristle pattern in *D. melanogaster*. *J. Exp. Zool.* **198**: 79–96 (1976).

55. Turing, A. M.: The chemical basis of morphogenesis. *Philos. Trans. R. Soc. London (Biol.)* **237**: 37–72 (1952).

56. Ursprung, H.: The formation of patterns in development. In "Major problems in developmental biology" Locke, M., (ed.) *Symp. Soc. Dev. Biol.* **25**: 177–216. New York: Academic Press (1966).

57. Wigglesworth, V. B.: Local and general factors in the development of 'pattern' in *Rhodnius prolixus* (Hemiptera). *J. Exp. Zool.* **17**: 180–200 (1940).

58. Wilcox, M., Smith, R. J.: Regenerative interactions between *Drosophila* imaginal discs of different types. *Dev. Biol.* **60**: 287–297 (1977).

59. Wolpert, L.: Positional information and the spatial pattern of cellular differentiation. *J. Theoret. Biol.* **25**: 1–47 (1969).

Questions for Discussion with the Editors

1. *Your concluding remarks imply that several mechanisms may function to generate pattern in cuticular structures. Yet as elegant as the imaginal disc system is, most of the models it has generated depend upon data obtained by transplanting discs into hosts. How do you imagine the inherent complexity of a transplantation/host system can be circumvented in future studies so that the several mechanisms which may function in disc development can be distinguished from one another?*

We doubt whether it will be possible to analyze such a complex process using only transplantation experiments. Our present understanding of disc development is based not only on transplantation experiments, but it also draws heavily on clonal analysis, the study of mutant phenotypes and of mosaics between mutant and normal cells. Future studies will also use all of these methods, but of course the supreme advantage of *Drosophila* is that in this organism it should now be possible to use a combination of molecular genetics and developmental analysis to identify some of the actual gene products that are involved in the cell interactions controlling pattern formation.

2. *The next stage in the analysis of cuticular pattern formation could conceivably involve biochemical analyses. Briefly, what is known about the biochemistry of bristle differentiation?*

Very little. Herschel Mitchell and his colleagues have described the protein synthetic patterns associated with the morphogenesis of wing hairs, but we have no information on bristle cells because there is no way of obtaining a pure preparation of bristle cells for study. Again, genetics might come to the rescue, since there is at least one mutation (*shibire*) which apparently converts most of the hair cells on the thorax to bristle cells. This could be used to identify bristle-specific proteins, but we suspect that these would be more relevant to bristle morphogenesis than to pattern formation.

CHAPTER **13**

Morphogenesis and Compartments in *Drosophila*

Jane Karlsson

MANY OF THE MOST INTERESTING QUESTIONS in developmental biology center around the causal relationships between gene expression and morphogenesis. The complexities of development in higher animals are such that it is natural to assume that each process must have its own detailed program encoded in the genome. The obvious dependence of differentiation of the various cell types on differential gene expression, the ease with which morphological variants can be selected for by animal breeders, and the dependence of all processes in prokaryotes on the genome, all lend weight to this view.

It is a view nevertheless that still awaits experimental verification, and there seems to be a growing feeling among developmental and cell biologists that cells can do remarkable things without referring to the genome for instructions at every step.

Both of the current concepts about pattern formation in *Drosophila*, the gradient type and the homeotic gene/compartment type, rely on the genome for "interpretation" of positional cues, and therefore necessarily predict the existence of large amounts of detailed genetic information, as discussed in detail below. So far, no evidence of the existence of this information has been obtained, either from morphological mutants or from molecular genetics, and for this reason one cannot help wondering whether there might be alternative explanations.

There is in fact an old idea about morphogenesis which does not have this "interpretation" step, and makes no predictions about morphological mutants which are not borne out by the evidence. This idea has been out of favor for a while, but it seems now to be regaining acceptance. It is simply that the morphological changes occurring at any one stage are the necessary consequence of those occurring at earlier stages; complexity increases gradually, and no more is done at each stage than is necessary to get to the next [(20), and J. Gerhart, *personal communication*]. It is not difficult to see why this idea has been out of favor; it does not say enough, it does not include a set of rules for getting from one stage to the next, and for very few systems do we have any idea of what such rules might consist. Nevertheless, because this type of idea dispenses with the need for large amounts of detailed genetic information, it seems worthwhile exploring it to find whether it has other difficulties that might lead one back to the interpretation-type of model.

As it happens, the evidence relating to that part of *Drosophila* morphogenesis based on cell division does suggest that the generation of shape by cell division might operate according to simple rules. These and other observations lead one to wonder whether there might be some sort of complicated cellular machinery involving the cytoskeleton, cell junctions, and the cell division apparatus, which is capable of generating specific shapes with great precision, and which operates according to as yet unknown rules [see (31)]. The machinery for making complex shapes without cell division certainly exists; protozoans make patterns just as complicated as those of multicellular animals, and it is difficult to imagine them doing this by means of a gradient/interpretation system because this requires differential gene expression in different cells. It could be that this pattern-generating machinery was extended, many millions of years ago, to include a mechanism for directed cell division and the means for communicating structural information between cells, to make a sort of supracellular skeleton that is responsible for specific multicellular morphologies. The rules governing the operation of this supracellular scaffolding might be in the same general category as those governing assembly of structural proteins.

This is speculation at present, and it is clear that this third concept, although it does not suffer from the interpretation problem, has the same type of drawback as the other two, namely that it is necessary to postulate complicated gadgetry for which there is as yet little concrete evidence. Clearly, it is not possible at present to state definitely which type of system actually operates in *Drosophila*.

A Simple Gradient Model

Drosophila differs from other animals in which pattern formation is studied in that growth is not accompanied by differentiation but precedes it. Presumptive adult tissue grows as "imaginal discs". Each adult appendage plus associated thorax or head are the contribution of one disc; the thoracic segments each have two, one dorsal and one ventral. The abdomen is made from much smaller primordia. Disc tissue can be cultured, although not yet very well *in vitro*; two consecutive hosts must be used, first an adult female, for growth, and then a mature larva, for differentiation. A striking result to come out of this work is that of two complementary disc fragments, one regenerates during culture while the other duplicates, in mirror image, structures already present (4). It is this observation, among others, on which the polar coordinate model is based (see Chapter 14).

The polar coordinate model is not necessarily a gradient model, although a logical extension of the basic idea would be that the two coordinates correspond to two gradients. It does propose that all imaginal discs use the same coordinates, which would mean that they use the same two gradients, and indeed the idea that different organs use the same positional information stems in part from observations on *Drosophila* which relate to homeotic genes.

Homeotic mutants cause one body part to develop as another; most viable ones replace only part of an organ, and replace it with a corresponding part of a different organ. Thus, some alleles of *Antennapedia* replace distal antenna with distal leg, and others replace proximal with proximal. In alleles with variable expression, it can be shown that a given part of the antenna is always replaced by the same part of the leg (25), indicating that the two organs respond to the same positional cues. Hence, the idea that all discs use the same positional information (33), and that this positional information might be in the form of simple monotonic gradients, interpreted differently in each disc.

Supposing then that each disc has two gradients, two being the minimum to specify position uniquely in a two-dimensional sheet of cells. The mature wing disc has about 50,000 cells, and the first question one might want to ask is how many of these need separate instructions. Imaginal disc cells have to first divide, and do so in such a way as to produce the precise shape of the mature disc (see Fig. 1). Second, they must undergo rearrangements at metamorphosis to convert the flat-folded epithelium into a three-dimensional appendage. Third, the cells themselves must change shape; in the wing blade for instance, the cells are highly columnar before metamorphosis, and flattened in the adult, increasing their diameter many times. Other more complicated shape changes occur in hinge and joint regions. Fourth, the cells must deposit cuticle, of a particular type and thickness. Fifth,

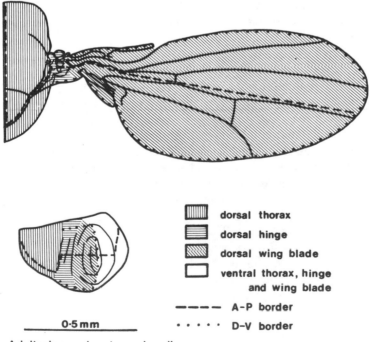

dorsal thorax
dorsal hinge
dorsal wing blade
ventral thorax, hinge
and wing blade

- - - - A-P border
· · · · · D-V border

0·5 mm

FIGURE 1. Adult wing and mature wing disc.

most cells must secrete one or several hairs of various lengths. In addition, some cells must make sense organs, and others must have the information necessary to join up accurately with neighboring discs. This list considers only the epidermal cells; discs also contain presumptive muscle cells, the muscle pattern probably arising as a result of induction by epidermal structures (17).

Let us consider the situation in the wing hinge. This consists of a highly complicated system of interlocking plates and rods; it may well be the most complex nonrepeating pattern of those commonly studied. It would probably not be safe to assume that any neighboring pair of cells, out of thousands, behaves identically with respect to all of the above parameters. By the late third instar, each of the plates and rods knows exactly where to form, and knows this without being told by neighboring structures; this must be so because small fragments of imaginal discs differentiate autonomously. Thus, fragments implanted directly into mature larvae differentiate without (or with very little) growth, and they produce those structures they would have produced *in situ* (3). In fact, very small fragments of the presumptive wing hinge do not differentiate particularly well, perhaps for mechanical reasons (*unpublished observations*), but the same kind of information can be gained from the mirror-image duplications produced by many fragments after culture in adult females. These duplications in the best cases are quite perfect down to the smallest detail, and the mirror plane follows the line of the original cut, wherever it may be (3). This means that no detail of the pattern is produced as a result of induction by another structure, since any part can be missing and still its neighbors differentiate perfectly. Therefore, the gradient/interpretation system must be able to cope with the entire complexity of the pattern on its own.

So, each cell, or at most each very small group of cells out of thousands, must respond to a particular value in the gradient system and interpret that value according to instructions in the genome. This means that hundreds, or even thousands, of different values of the gradients must be "written" separately in the genome, each value accompanied by instructions regarding the parameters outlined above. These instructions would have to be quite detailed, in some cases at least; none of the five parameters can be specified by anything as simple as an on-off switch. Since the wing hinge is only a small part of the fly, this represents a small part of what has to be present in the genome, which would have to contain an extremely large amount of detailed information.

Nevertheless, the existence of such information cannot be ruled out yet, as the function of large amounts of DNA remains unknown. Suppose, for instance, that it were scattered around the genome, then a mutation might only affect a single cell, or very small group of cells, and might well go undetected. But this idea has problems of its own; the occurrence of morphological evolution is not easy to imagine if mutation can only affect a few cells at a time. If a large patch "needed" to change, it might wait a very long time indeed.

Supposing then that this information were clustered in some way in the genome; it might be that each of the genes involved in shape change and cuticle deposition has a list of the positional values requiring its activity "written" next to it. If this were the case, then a small deletion might affect a larger part of the pattern. But then one would expect a class of mutations corresponding to small deletions that neatly removed, or changed, a predictable part of the pattern, and the size of the DNA deletion would be expected to bear some sensible relation to the size of the deleted or changed part. Such a class of mutations is not found (19). There are certain types of mutant which have invariant defects, for instance the segmentation mutants isolated by Nuesslein-Volhard and Wieschaus (21), and the polarity mutants of Gubb and Garcia-Bellido (11). These are unlikely to be mutations in a "list" of positional values since each allele affects the same part of the pattern. Apart from these and the homeotic mutations, the visible mutations of *Drosophila* are much more easily described as grossly disruptive, consistent with the idea that they affect the actual proteins required for differentiation, rather than the details of their deployment.

The basic problem is that any gradient system whose form bears no relation to that of the final pattern constitutes a "code"; codes bear no logical relation to what they mean, and so their meaning, in the form of a complete dictionary, must be written down somewhere. The homeotic gene/compartment type of model has the same problem.

A Homeotic Gene/Compartment Model

The first part of this theory concerns homeotic or "selector" genes. There is a fair number of homeotic mutants scattered around the genome, most of them changing haltere into wing, antenna into leg, or second and third legs into first leg (26). The homeotic genes in the two major clusters, the *bithorax* complex (BX-C) and the *Antennapedia* complex (ANT-C), are organized in such a way that it is difficult to avoid the conclusion that they have something very fundamental to do with spatial organization. The ANT-C seems to deal with the anterior part of the animal while the BX-C deals with the posterior part, and the genes of the BX-C are with one exception arranged in a way which reflects the order of segments in the animal (18).

The second part of the theory involves compartments. These were discovered in 1973 (10); the observation was quite unexpected, and was made with the use of mitotic recombination clones. Such clones can be generated at any desired time in development of *Drosophila* by irradiating larvae heterozygous for one or more cuticular "markers" such as *yellow* or *ebony* which alter cuticle color, or *multiple wing hairs* which adds extra hairs to each cell. Irradiation induces recombination and a heterozygous cell produces two homozygous daughters, one wild-type and one homozygous for the marker.

Mitotic recombination clones had been made before 1973, and it had already been observed that clone shape is not predictable, meaning that cells divide according to their position and not according to their lineage. In 1973, a new technique was used which generates very large clones (the "*Minute* technique"). These large clones revealed certain hitherto unsuspected lines, or compartment borders, which the clones had been unable to grow across (10). One such line runs all the way down the middle of the wing blade, quite straight but not following any landmark such as a vein (see Fig. 1). This line is set up very early; at least one more is set up later on, separating dorsal from ventral wing blade. Most imaginal discs have at least one such line, usually running down the proximo-distal axis of the appendage. It is possible that the process of compartmentalization continues until the last compartments are very small or even single cells. This is not known as the later the clone is made, the smaller it is, even if it contains fast-growing cells, and small clones do not define borders well.

The type of pattern-generating mechanism based on these observations arises from certain homeotic mutants which appear to transform compartments one into another instead of segments one into another. The various alleles of *bithorax*, for instance, after which the BX-C is named, cause transformations of haltere into wing only within the anterior compartment, and the two *postbithorax* alleles likewise only affect the posterior compartment. The *engrailed-1* mutant transforms at least some posterior structures into their corresponding anterior ones (16). Supposing that homeotic mutants could be found that transformed other compartments one into another, it would be difficult to avoid the conclusion that these genes are involved in giving developmental instructions to each compartment. These would not be direct instructions for differentiation, since much the same structures are found in both of a pair of compartments. Splitting a population of cells into two, and switching on a gene in one but not the other, and continuing this process until all cells are in different compartments, might be a way of assigning spatial information which could then be interpreted according to instructions in the genome. If compartmentalization stops well short of single cells, a gradient system could be used to assign position within each compartment.

This latter idea was suggested by early work on the *engrailed-1* mutant, whose phenotype seemed to indicate that complete absence of this gene would cause a transformation of the posterior compartment into a second anterior compartment in mirror image, and therefore that the two compartments have identical mirror-imaged gradients of positional information. However, it is now known that absence of this locus gives virtually no transformation of posterior to anterior [D. Gubb, *personal communication*, and (14)], and since no other evidence indicates that different compartments use the same positional information, this idea may be questioned. Indeed, the whole idea that compartments are units of function for homeotic genes may be questioned; if this were the case, it should be relatively easy to obtain mutants which transform a complete compartment while leaving its sister unaffected, and in fact this is very difficult. Most alleles either transform part of a compartment, or one (or more) whole compartment plus part of another.

It is not easy to see how this system of assigning spatial information could work in the way suggested if homeotic genes are not responsible for making the cells of one compartment different from those in another. It will be suggested later that compartments do indeed play an important role in spatial organization, but not because they themselves differ, which as far as differentiation is concerned they do not, but because cells lying along compartment borders are different from cells elsewhere.

The predictions about the genome made by the models discussed above arise because positional information is assigned in a form that bears no relation to that of the final pattern. An obvious way out of this would be a system whose form does resemble the pattern. But there is evidence, from partial replacement of structures in homeotic mutants mentioned earlier, that all discs, or at least some pairs, use the same positional cues. Since final differentiation in each of the pairs is different, how can the form of the cues resemble the pattern?

Similarity of Positional Cues in Different Discs

It is worthwhile looking at all the available evidence regarding how similar do positional cues have to be? The best possible way to answer this would be to make very small mitotic recombination clones of homeotic mutants, and look at the structures produced in each clone in relation to its position. This is not possible since very small clones of homeotic mutants do not differentiate autonomously. The next best method is to study larger clones, or alleles with variable expression; a given structure from one organ should always be replaced by the same structure from the other. This is less satisfactory because of the possibility that the interface between the two tissues lies in preferred positions; it could be that cells of one origin, finding themselves adjacent to cells of another origin which they do not recognize, merely divide until they produce cells which do recognize them. There is evidence from dissociation experiments, discussed below, which suggests that this is sometimes the case. Thus Postlethwait and Schneiderman (25) found very good positional correspondence between leg and antenna structures in their study of the variable allele *Antp-R*, but this cannot be taken as evidence that each cell or small group of cells in one organ has its precise positional counterpart in the other.

Correspondence in the other common partial transformation, haltere into wing, is good proximally but poor distally; the distal tip of the haltere can lie adjacent either to proximal wing (*bx-3* flies) or distal wing (*pbx* flies; *unpublished observations*). A good correspondence is reported for the proboscis-leg-antenna trio (28), but here replacement of whole compartments was studied rather than single structures, and there is another report of rather poor correspondence (12).

Other evidence which bears on this question comes from experiments in which parts of different organs are grafted together, intercalary regeneration being stimulated in certain cases (4). However, such experiments can only show a crude correspondence at best, and more information can be gained from dissociation-reaggregation experiments.

This type of experiment has certain limitations in *Drosophila*; imaginal discs can be dissociated with proteolytic enzymes, and reaggregated in a centrifuge, but disc tissue must be cultured *in vivo*, and the behavior of the cells cannot be continuously monitored. The relative contributions of cell death, cell movement and growth cannot therefore be determined directly. It has been estimated by indirect means that very considerable cell death occurs in such experiments (32), and since dissociation in not complete, the likelihood is that only clumps of cells survive (4). These can grow enormously, and it seems probable, since disc cells do not normally change their fates except by cell division, that the pattern reconstruction which takes place, does so mainly by this means. For instance, if wing discs from two different genotypes (say *yellow; multiple wing hairs* and *ebony*) are dissociated and reaggregated together, large parts of the pattern are reconstructed during culture, and in a high proportion of cases the different structures are made of both *y;mwh* and *e* cells in an integrated mosaic (7). Whether or not the cells or clumps of cells have actually migrated within the pellet, the conclusion remains that $y;mwh$ and e cells from similar parts of the disc stayed together to make the mosaic, and must therefore have recognized each other. Thus, cells from different parts of a single disc are able to recognize each other on the basis of position.

Despite its limitations, then, this method does give information about position-specific recognition, and therefore the behavior of pairs of discs of different origin is of considerable interest. If for instance it were found that all pairs of discs make integrated mosaics with each other, the conclusion that each cell or small group of cells in one disc has its exact positional counterpart in the other would be inescapable. This is however the case only with the different leg discs; the three legs do make integrated mosaics with each other, except in those places (the transverse rows) where morphology is different (7). Wing and leg cells do not make mosaics at all, and appear to be trying to minimize contact (24). Between these two extremes of good mosaics and no mosaics at all, there is a third category, termed by Garcia-Bellido and Lewis (9) "ditypic territories," in which the two types of cell do not interdigitate but do not relinquish contact altogether, and the same type of positional correspondence is seen as in partial homeotic replacement, proximal structures from one disc associating with proximal structures from the other, and distal with distal. This type of association is seen in the leg-antenna pair (8) and in the wing-haltere pair (9). The lack of interdigitation suggests that cells from the different organs only recognize one another at certain positions.

Examining the patterns of association in these experiments, one gets the impression that what matters is similarities in morphology. The three legs, morphologically very similar, make good mosaics with each other, but not where morphology differs. The mesonotum (dorsal mesothorax) and humerus (dorsal prothorax), both made of thick bristled cuticle with a slight curvature, also make mosaics, although here the observations must be considered preliminary as the number of cases was small (30). No other pairs make mosaics, although in the abdomen, where this type of experiment cannot be done as imaginal tissue develops from primordia much smaller than discs, partial homeotic replacement of one segment by another does occur (6); again, these segments are morphologically very similar. Pairs of discs with very different morphology such as wing and leg do not make mosaics or even ditypic territories, and partial homeotic replacement is not observed either. Antenna and leg make ditypic territories; both are segmented, and both develop as a series of concentric rings corresponding to one or more whole segments. Wing and haltere also make ditypic territories, and both are unsegmented and have three rather different parts, a thorax part, a hinge part, and an inflated distal part.

The patterns of association discussed above are often thought to result from the activity of homeotic genes, which are thought to confer surface properties on cells so that discs with different active genes will "sort out" from one another (16). However, this idea does not seem to fit the data as well as the idea that association is based on morphological similarities; for instance, the three legs are all from different segments, and differ by the activity of genes such as those of the BX-C and the ANT-C and *Polycomb*, and yet they make good mosaics with one another *except* in those places where morphology is different, an observation very difficult to explain along these lines. Mesonotum and humerus also make mosaics in spite of being in different segments. Wing and second leg are in the same segment, and no homeotic mutant is known causing a transformation between the two, and yet they will not even make ditypic territories. As Poodry *et al.* (24) suggest, wing wants to make a sheet and leg wants to make a tube, so how can they be expected to cooperate?

The fact remains, though, that in order to explain these results along the lines of morphological similarities, "tubeness" or "sheetness" must be properties of cells, clumps of cells if not single ones, as well as being properties of the tissue as a whole. Cells in different parts of imaginal discs have different shapes (23), and it seems possible that this might be the basis for differential recognition. The edges of any fragment of the epithelial sheet might have a specific configuration which would fit best with the complementary configuration from the same disc, but which might also fit quite well with a fragment from a specific part of another disc. It could be that the curvature of the concentric rings which make leg and antenna segments is reflected in the shapes of cells or cell clumps, proximal antenna cells preferentially associating with proximal leg cells on this basis. Certainly, leg and antenna discs have quite similar diameters [Fig. 2 in (22)].

It remains to be seen whether cells can recognize one another on the basis of their shape. It is clear though that the dissociation data discussed above cannot be explained in terms of identical gradients of positional information alone, or all pairs of discs would make good mosaics with each other; additional assumptions about morphological similarities and differences must be made. Thus a "morphological" explanation makes no predictions that are not also made by conventional explanations.

Growth

Let us turn now to the question of growth. This is particularly interesting in *Drosophila* for several reasons: first, because the shapes of imaginal discs are built up almost exclusively by cell division, other possible factors such as cell death, migration, or cell shape change playing a minor role if any (22). Second, regeneration is very successful in *Drosophila*, regenerated structures being in many cases indistinguishable from nonregenerated ones, and it therefore seems reasonable to suppose that regeneration uses the same mechanisms as does normal growth, and that studying regeneration could tell us how patterned growth is achieved. Third, the deployment of single cells and their progeny during growth can be followed in *Drosophila* by means of mitotic recombination clones.

Let us consider for a moment exactly what is involved in making a mature wing disc. At hatching, this disc consists of about 40 cells and has a simple spindle shape. By the late third instar, this has become a flattened sac composed of a peripodial layer of very flat cells and an epithelial layer of columnar cells, which make most of the adult derivatives. This epithelial layer has a pattern of folds which at metamorphosis unfold to make wing blade and hinge. The mature disc thus has a fairly simple but quite predictable shape (see Fig. 1).

Supposing that cell division were controlled according to the simplest possible gradient model, one might expect that the instructions given to cells at various positions would define parameters such as the timing or the orientation of cell division. If cells at all positions had precise instructions about timing and orientation, one would expect the shapes of mitotic recombination clones to be predictable, because each cell would be doing exactly the same thing as the cell in that position in other animals. This idea can be ruled out at once because the precise shape of clones is not predictable. So if both these parameters are not precisely controlled, perhaps one of them is; it cannot however be timing, because it is possible to generate clones of cells which divide faster than their neighbors, and even if such clones grow to many times their usual size, the wing is nevertheless normal in shape. Cells thus appear to be permitted to divide more or less when they like. Therefore we are left with the orientation; is it possible that the orientation of each cell division is controlled but not the timing? It is difficult to see how this could be the case; since cell division is happening continuously, if a particular division is to be programmed, in any way at all, its timing must be programmed to a large extent as well. It would work if all cell divisions had the same orientation, but at least in some places, orientation is almost random (2).

One is forced to the surprising conclusion that cell division is not controlled precisely at all. But if this is so, how does the disc acquire such a predictable shape? A re-examination of the data shows a possible way out of this dilemma; there are places at which cell division is controlled very precisely. Compartment borders are straight, and lie in predictable places; cell division that might lead to distortion of these lines apparently does not occur. This raises the possibility that the predictable shape of the disc results because *some* cells have precise instructions, although *most* do not. Indeed, it is easy to see how such a system could work; in the case of the wing blade, since this is flat in the adult, only the outline needs to be precisely specified; cells not lying along the edge need merely maintain a certain density. And there is indeed a compartment border running all the way around the edge of the wing blade.

If this is the case, and certain cells know very precisely what to do while others do not, this behavior should be reflected in regeneration. Cells at compartment borders should be able in some sense to "do more"; disc fragments containing compartment borders should behave differently from those that do not. And this is in fact what is found. Distal regeneration does not occur from proximal wing disc fragments that do not contain the ventral end of the antero-posterior border, although fragments containing it regenerate distally at high frequency (13). Similarly, fragments which normally only duplicate are able to regenerate as well if they contain the dorsal end of this same border. The structures regenerated lie along the border, and when the proximal end of the border is reached, regeneration stops. This occurs at high frequency, while duplicating fragments lacking any part of this border never regenerate any structures at all (15). Similar correlations between compartment borders and regenerative capacity are reported for the leg disc (5, 27).

Thus the clone data and the regeneration data seem to say the same thing, that the precise shape of the mature disc results from precise control along certain lines. At least some of these lines lie in the places one might expect if they were part of a supracellular system of scaffolding; the dorso-ventral border lies along the wing margin, and the antero-posterior border lies along the line separating the anterior rigid part of the wing from the more flexible posterior part (C. Ellington, *personal communication*). One might call these lines "growing girders." The existence of such things should perhaps not come as too much of a surprise. D'Arcy Thompson remarks (29), "It is clear, I think, that we may account for many ordinary biological processes of development or transformation of form by the existence of trammels or lines of constraint, which limit and determine the action of the expansive forces of growth that would otherwise be uniform and symmetrical."

It is noteworthy too that compartment borders from different discs seem to recognize one another; in partial homeotic transformations, the antero-posterior border of the two organs is in register. Thus, a composite organ made of proximal antenna and distal leg would have a continuous antero-posterior border running down the proximo-distal axis (28). This might explain at least some of the good positional correspondence between leg and antenna structures in the mediolateral axis.

If the above analysis of the clone and regeneration data is anywhere near correct, the shape of the disc will be determined by two factors: the position of compartment borders, and the rules governing their elongation. It could be that there is a gradient along each border which cells translate into cell division; this could probably be made to fit quite well with the regeneration data and cannot be ruled out at the moment. Following the idea of shape changes producing further shape changes, though, there may be another way of doing it. Is it possible that the signals that result in cell division are not chemical but mechanical? Could it be that cells produce mechanical changes by dividing, and use these same changes as signals for further division? If the borders elongate, the cells in the interior might be stretched, and if stretching leads to cell division, expansion of the interior might in turn stretch the borders. Positional information might thus be to some extent self-generating.

The idea that the morphology of the mature wing disc is determined by compartment borders does seem to be quite well supported by the data, but what of the morphology of the adult wing? If the idea that shape changes produce further shape changes is to be a satisfactory explanation, then the pattern of cell division must lay the foundations for what happens at metamorphosis, and this means that cells in the middle of compartments as well as those at the borders must know what to do. Compartment border cells may be responsible for the overall shape of the wing and for some details of the final pattern, such as the bristles and hairs along the wing margin, but there are complex structures at some distance from any border, and here other mechanisms must be at work.

In some cases, it seems possible that the pattern of final differentiation is laid down to some extent in the shapes which the cells have been made to adopt as a result of the pattern of cell division. It could be that the activities of compartment borders during growth sets up patterns of mechanical stress that results in cells undergoing specific deformations in specific places. These deformations might be registered by the cells in the configuration of their cytoskeletons, and these configurations might undergo specific transformations at metamorphosis. For instance, the distal part of the wing hinge contracts enormously at metamorphosis, and before metamorphosis, the cells in this region appear somewhat elongated in the direction in which they are to contract (1). The rule here might be: "the more elongated you are in a particular direction, the more you must contract (or reshuffle) in that direction at metamorphosis." In the wing blade, the rule might be "the more columnar you are now, the more you must flatten later." The wing hinge is the most difficult part to imagine differentiating according to such simple rules, and it is here that one feels most tempted back to gradient models, in spite of their problems. The deformations resulting from the folding pattern might conceivably lay down at least parts of the outlines of the various plates; as metamorphosis proceeds and the outlines appear, their appearance might cause further deformations within the plates which could add the final details of shape.

If the pattern of stress forces is indeed responsible for differentiation, it might explain why duplications in the wing hinge region are so perfect; a mirror-image duplication might be the only configuration, apart from the normal one, in which the forces are exactly balanced.

Suppose that final differentiation could in time be fitted into this scheme, where does that leave homeotic genes? It is clear that if morphogenesis is indeed governed by machinery involving the cytoskeleton and the cell division apparatus, some aspect or other of this machinery must be exposed to mutation and hence to evolution. There must be some way in which the amount of cell division and cell shape change occurring at specific places and/or times can be altered by means of mutation. This could be what homeotic genes do, if they are indeed responsible for the shape differences between different organs.

One of the most striking things about partial homeotic transformations is that the interface between the two tissues is a sharp one; cells with intermediate shapes are not usually found. This seems to suggest that differences in cell shape are somehow quantal, that there is a finite number of shapes cells can adopt, or shape changes they can undergo from given starting conditions. The idea of one shape change leading to another might be modified to read "a specific shape change leads to one of a small number of further shape changes," and perhaps homeotic genes are involved in decisions regarding which of this small number is actually realised. Certainly it will be very interesting finding out exactly what these genes do.

We are left, in this alternative view of morphogenesis, with several very interesting and difficult problems; exactly what homeotic genes do, how mechanical changes are translated in some circumstances into cell division and in others into differentiation, how regeneration rebuilds the pattern of cell shape distortion, what controls elongation of compartment borders, how cells recognize one another on the basis of morphology, and doubtless many others. These are problems many will consider just as difficult as the interpretation problem, if not more so. Perhaps there will be some who believe it worthwhile trying to think in new ways about old questions even if these new ways raise more problems than they solve.

General References

LOVTRUP, S.: *Epigenetics*, Chichester, UK: Wiley (1974).

POODRY, C. A.: Imaginal discs: morphology and development. In *The Genetics and Biology of Drosophila*, (eds.) Ashburner, M., Wright, T. R. F., Vol. 2d, London: Academic Press (1980).

References

1. Brower, D. L., Smith, R. J., Wilcox, M.: Cell shapes on the surface of the *Drosophila* wing imaginal disc. *J. Embryol. Exp. Morph.* **67**: 137–151 (1982).

2. Bryant, P. J.: Cell lineage relationships in the imaginal wing disc of *Drosophila melanogaster*. *Dev. Biol.* **22**: 389–411 (1970).

3. Bryant, P. J.: Pattern formation in the imaginal wing disc of *Drosophila melanogaster*: fate map, regeneration and duplication. *J. Exp. Zoo.* **193**: 49–78 (1975).

4. Bryant, P. J.: Pattern formation in imaginal discs. In *The Genetics and Biology of Drosophila*, (eds.) Ashburner, M., Wright, T. R. F., (eds.) Vol. 2c, London: Academic Press (1978).

5. Bryant, P. J., Girton, J. R.: Genetics of pattern formation. In *Development and Neurobiology of Drosophila*, (eds.) Siddiqi, O., Babu, P., Hall, L. M., Hall, J. C., New York: Plenum (1980).

6. Capdevila, M. P., Garcia-Bellido, A.: Genes involved in the activation of the *bithorax* complex of *Drosophila*. *Wilhelm Roux' Arch.* **190**: 339–350 (1981).

7. Garcia-Bellido, A.: Pattern reconstruction by dissociated imaginal disc cells of *Drosophila melanogaster*. *Dev. Biol.* **14**: 278–306 (1966).

8. Garcia-Bellido, A.: Cell affinities in antennal homeotic mutants of *Drosophila melanogaster*. *Genetics* **59**: 487–499 (1968).

9. Garcia-Bellido, A., Lewis, E. B.: Autonomous cellular differentiation of homeotic *bithorax* mutants of *Drosophila melanogaster*. *Dev. Biol.* **48**: 400–410 (1976).

10. Garcia-Bellido, A., Ripoll, P., Morata, G.: Developmental compartmentalisation in the dorsal mesothoracic disc of *Drosophila*. *Dev. Biol.* **48**: 132–147 (1976).

11. Gubb, D., Garcia-Bellido, A.: A genetic analysis of the determination of cuticular polarity during development in *Drosophila melanogaster*. *J. Embryol. Exp. Morph.* **68**: 37–57 (1982).

12. Kamrat, D., Falk, R.: Positional precision of homeotic transformations. Abstract, 7th European *Drosophila* Research Conference (1981).

13. Karlsson, J.: Distal regeneration in proximal fragments of the wing disc of *Drosophila*. *J. Embryol. Exp. Morph.* **59**: 315–323 (1980).

14. Karlsson, J.: Homeotic genes and the function of compartments in *Drosophila*. *Dev. Biol.* In press (1983).

15. Karlsson, J., Smith, R. J.: Regeneration from duplicating fragments of the *Drosophila* wing disc. *J. Embryol. Exp. Morph.* **66**: 117–126 (1981).

16. Lawrence, P. A., Morata, G.: The compartment hypothesis. *Symp. Roy. Ent. Soc. Lond.* **8**: 132–149 (1976).

17. Lawrence, P. A., Brower, D. L.: Myoblasts from *Drosophila* wing disc can contribute to developing muscles throughout the fly. *Nature* **295**: 55–57 (1982).

18. Lewis, E. B.: A gene complex controlling segmentation in *Drosophila*. *Nature* **276**: 565–570 (1978).

19. Lindsley, D. A., Grell, E. H.: Genetic variation in *Drosophila melanogaster*. Carnegie Inst. Washington, Publ. No. 627 (1968).

20. Lovtrup, S.: *Epigenetics*, Chichester, UK: Wiley (1974).

21. Nuesslein-Volhard, C., Wieschaus, E.: Mutations affecting segment number and polarity in *Drosophila*. *Nature* **287**: 795–801 (1980).

22. Poodry, C. A.: Imaginal discs: morphology and development. In *The Genetics and Biology of Drosophila*, (eds.) Ashburner, M., Wright, T. R. F. Vol. 2d, London: Academic Press (1980).

23. Poodry, C. A., Schneiderman, H. A.: The ultrastructure of the developing leg of *Drosophila melanogaster*. *Wilhelm Roux' Arch.* **166**: 1–44 (1970).

24. Poodry, C. A., Bryant, P. J., Schneiderman, H. A.: The mechanism of pattern reconstruction by dissociated imaginal discs of *Drosophila melanogaster*. *Dev. Biol.* **26**: 464–477 (1971).

25. Postlethwait, J. H., Schneiderman, H. A.: Pattern formation and determination in the antenna of the homeotic mutant *Antennapedia* of *Drosophila melanogaster*. *Dev. Biol.* **25**: 606–640 (1971).

26. Postlethwait, J. H., Schneiderman, H. A.: Developmental genetics of *Drosophila* imaginal discs. *Ann. Rev. Genet.* **7**: 381–433 (1974).

27. Schubiger, G., Schubiger, M.: Distal transformation in *Drosophila* leg imaginal disc fragments. *Dev. Biol.* **67**: 286–295 (1978).

28. Struhl, G.: Anterior and posterior compartments in the proboscis of *Drosophila*. *Dev. Biol.* **84**: 372–385 (1981).

29. Thompson, D'A. W.: *On Growth and Form*, Cambridge, UK: Cambridge University Press (1942).

30. Tobler, H., Schaerer, H. R.: Formation of chimeric pattern and reaggregation of cells in varied combinations of thorax forming imaginal discs of *Drosophila melanogaster*. *Wilhelm Roux' Arch.* **167**: 367–369 (1971).

31. Tucker, J. B.: Cytoskeletal coordination and intercellular signalling during metazoan embryogenesis. *J. Embryol. Exp. Morph.* **65**: 1–25 (1981).

32. Wilcox, M., Smith, R. J.: Compartments and distal outgrowth in the *Drosophila* imaginal wing disc. *Wilhelm Roux' Arch.* **188**: 157–161 (1980).

33. Wolpert, L.: Positional information and pattern formation. *Curr. Top. Dev. Biol.* **6**: 183–224 (1971).

Questions for Discussion with the Editors

1. *According to your view, several* engrailed *alleles do not obey all the rules they should in order to provide strong supporting evidence for the compartment hypothesis. Which genes, if any, come the closest in your opinion to supporting the original compartment idea?*

The original idea was that at compartmentalisation, a previously homogenous population of cells is split into two, and a selector gene is activated in only one of them and causes it to develop differently from the other. Absence of such a gene would result in the two compartments developing identically, as the phenotype of the *engrailed-1* mutant suggested would happen in the absence of the *engrailed* gene. This has turned out not to be the case, however, and no other mutant causes a convincing transformation between sister compartments.

2. *Assuming that you are correct in postulating that shape is specified by the cells in compartment borders, is it accurate to conclude that you believe the major key to understanding compartments will come from focusing attention on the mechanics of cell divi-*

sion along borders, rather than, for example, further genetic analyses of homeotic mutants?

I would imagine that regeneration and homeotic mutants will give results which are complementary in this respect. Cutting up imaginal discs and asking how they respond should give information about the mechanical forces responsible for shaping the disc, and homeotic mutants should give information about shape alternatives and how one alternative is chosen rather than another.

CHAPTER 14

A Model of Insect Limb Regeneration

Vernon French

THE COMPLEX SPATIAL ORGANIZATION of the mature animal develops from the initial conditions provided by the fertilized egg. Throughout the history of embryology, it has frequently been argued that this is only an apparent increase in complexity: the future animal is already fully mapped out in the egg, in the form of a tiny organism (e.g., the homunculus) or its modern equivalent, an exquisitely detailed mosaic of cytoplasmic determinants. This position is logically untenable (the spatial organization of each generation must have always pre-existed within that of the previous one, like an endless sequence of Russian dolls), and it is disproved by the experimental evidence. While cytoplasmic determinants are present in some eggs and may orientate the axes of the embryo, it has been repeatedly shown that detailed patterns of cell fate are gradually elaborated upon during development through interactions between embryonic cells: patterns are *formed*, not preformed.

The interactions of pattern formation are usually studied by altering the spatial arrangements within the developing embryo, depriving cells of their normal neighbors or placing them against cells that are normally far away. In early vertebrate or insect embryos, these experiments result in widespread changes in cell fate, *regulation*, and these results give information on the normal processes whereby cells acquire fates appropriate to their position in the undisturbed embryo (see Chapters 3, 20). The later embryo becomes subdivided into regions having a commitment to form a particular structure (e.g., a vertebrate forelimb, or an insect mesothoracic segment) but within which the precise fate of cells still depends on their position, hence on their interactions with their neighbors.

The most influential recent approach to pattern formation has been that of *positional information* (32), which suggests that spatial patterns of developmental fate arise in two separate steps. The cells interact to generate a *map of positional values* across the tissue, with cells at the edges having special boundary properties. The cells then individually *interpret* their positional value (an extremely demanding process, see Chapter 13), becoming committed and eventually differentiating to form particular structures. A simple form of map is a monotonic gradient perhaps in the concentration of a diffusible substance. Gradients account naturally for the observation that regulation after a graft or excision usually involves averaging, so that inappropriate neighbors change their fates to form structures normally lying between them (see Chapters 3, 20 for a discussion of gradients and other types of map, and of different forms of interpretation).

A one-dimensional pattern such as the (simplified) Hydra can be analyzed in terms of the formation and interpretation of a gradient, or gradients, running down a single body axis (32). In a more complex three-dimensional structure, such as a limb, cellular fate must be decided with reference to a more complicated map. One form of three-dimensional map could consist of three independent, mutually perpendicular gradients, perhaps running along the anatomist's axes: proximal-distal, anterior-posterior and dorsal-ventral. Each cell would have a unique combination of three map co-ordinates and could behave appropriately. Pattern formation in the embryonic chick limb has usually been analyzed in this way (see Chapter 1), suggesting that cells develop according to their position assessed independently, and by rather different mechanisms, along the three axes.

In many developing organs, such as the chick limb, the cells differentiate into recognizable cell types and lose the ability to respond to an absence or change of neighbors. In other animals, including some amphibians and insects, functional limbs and other body parts can regenerate. Dedifferentiation and growth occurs, and new patterns are formed within the new tissue (see Chapter 22). The larval legs of insects (such as cockroaches) are particularly convenient for analysis of pattern formation, since they regenerate very reliably, have a number of cuticular and other morphological markers characteristic of particular positions on the limb, and are suitable for precise grafting operations.

I shall briefly describe the embryonic development of insect limbs, and then discuss a wide range of results from regeneration experiments that suggest that the larval leg epidermis has a particular form of map of positional values, and responds to a disturbance in accordance with two specific rules. I shall then discuss some limitations of this model of regeneration, and finally return to the embryonic limb to ask whether regeneration tells us anything about embryonic development.

The Development of Insect Legs

The segments of the insect embryo are first visible at the germ band stage and, in most insects, the thoracic segments form protruding limb buds that grow and eventually become segmented. The head and abdominal segments also form buds, but these either regress or eventually develop into specialized appendages such as mouthparts, antennae, and cerci. In the late embryo, the leg epidermis secretes cuticle-bearing, characteristics bristles, spines, and claws, and the limb mesoderm differentiates into a precise pattern of muscles. The larva hatches with functional legs (Fig. 1A) that grow and secrete a new cuticle before each larval moult. In hemimetabolous insects (such as cockroaches and crickets), the larva closely resem-

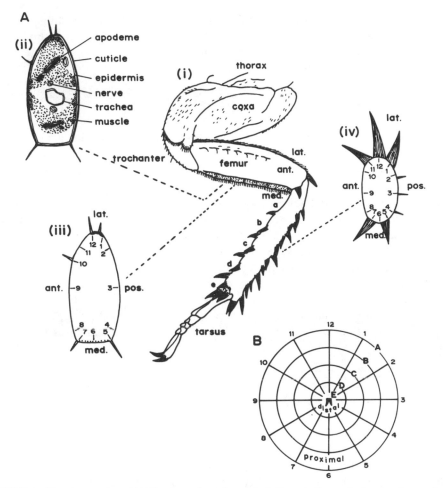

FIGURE 1. The insect leg. (A) The larval cockroach, left metathoracic leg is shown in anterior view (Ai), in transverse section through the femur (Aii), and in schematic transverse sections at the level of femur (Aiii) and tibia (Aiv). Ai shows five proximal-distal levels (a–e) marked down the tibia. Aiii and Aiv show 12 positions (1–12) marked around the leg circumference, as well as the anterior (*ant*), posterior (*post*), medial (*med*), and lateral (*lat*) faces of the leg, and the distribution of rows of bristles, spines, and characteristically pigmented cuticle. Aii shows the surface cuticle and the various internal tissues of the leg. (B) The leg epidermis, as represented by the polar coordinate model. Thoracic structures form at the periphery of the field and distal structures (claws) form in the centre. Cells have positional values characteristic of their level on a radius (A–E) and their position on a circumference (1–12). There is no discontinuity between 12 and 1, so a continuous sequence of values runs around the limb.

bles the adult, and the larval leg gives rise directly to the adult leg. In some holometabolous insects (such as beetles), the larva undergoes great changes in growth and structure in passing to the pupal and adult stages but, again, the larval leg develops into the (rather different) adult leg.

In other holometabola (including *Drosophila*), the prospective adult thoracic cells invaginate in the late embryo to form internal imaginal discs (e.g., the 6 leg discs) which grow but do not differentiate in the larva. In the pupal stage, the mature leg disc evaginates, secretes the cuticle and forms muscles to make the leg and surrounding thorax of the adult fly (see Chapter 12).

Regeneration and the Polar Coordinate Model

In the polar coordinate model (PCM) the epidermis of the insect leg is represented as a two-dimensional field with the distal tip in the center and the proximal structures around the periphery (Fig. 1B). The field thus corresponds to an "end-on" view of the larval leg, or to the fate map of the *Drosophila* imaginal disc. Cells are assumed to have 2 properties characteristic of their position (i.e., positional values): a radial value in a sequence (A–E) running down the leg, and a circular value in another sequence (1–12) running continuously around the circumference of the leg (16, 6). The notation of 12 circular and 5 radial values is arbitrary. There is likely to be a far larger number of values as the map is likely to be much "finer-grained." A cell's positional values govern which cuticular structures it will form, and the interactions between cells with different values can be studied in regeneration following amputation or grafting experiments on the larval leg, or following comparable fragmentation or mixing experiments on imaginal discs (see Chapter 12).

Proximal-Distal Intercalary Regeneration

When the distal level of the tibia of a cockroach larval leg is grafted onto the proximal level of a host tibia, local growth occurs at the junction and the intervening mid-tibia is formed by *intercalary regeneration* (4, 8). The resulting tibia is almost normal in size and pattern (Fig. 2A). If proximal tibia is grafted onto the distal level of the host tibia, a similar intercalary regenerate is formed, but now in reversed orientation so that the resulting leg, although very abnormal in size and pattern, has no positional discontinuities between adjacent cells (Fig. 2B).

A graft between mid-tibia and mid-femur results in healing and no intercalary regeneration (Fig. 2C), but grafting the distal femur onto the proximal level of the host tibia results in the formation of an intercalary regenerate, approximately half a segment in length and largely femur in type (4). Similarly, a graft between distal levels of the coxa and tibia provokes no intercalary regeneration (8), but grafting distal coxa onto proximal tibia produces an intercalary regenerate which is largely coxa in type (Fig. 2D).

In terms of the PCM, these results show that interaction between cells with different radial positional values stimulate local growth and the *intercalation* of intermediate positional values. They also show, however, that the leg does not contain a single radial sequence of values: the sequence A-E is repeated in the coxa, the femur, the tibia, and (as shown by other experiments) the tarsus. Hence Figure 1B is a simplified representation of the positional value map of the leg.

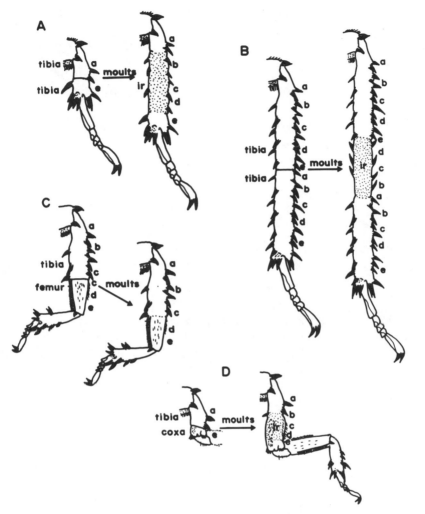

FIGURE 2. Proximal-distal intercalary regeneration. (A, B) Interaction between proximal (a) and distal (e) levels of the cockroach tibia leads to local growth and the formation of an intercalary regenerate (stippled) consisting of the intermediate levels (b, c, d). (C) When the mid (c) levels of host tibia and graft (prothoracic) femur are combined, they heal without intercalary regeneration. (D) A graft between proximal (a) level of the tibia and distal (e) level of the (prothoracic) coxa results in an intercalary regenerate, largely of coxa type. Note that the distal parts of the graft, which were removed, have regenerated.

Circumferential Intercalary Regeneration

When a longitudinal strip of epidermis and cuticle is moved around the circumference of the femur of the cockroach larval leg, and grafted into a different position, normally nonadjacent cells interact along the graft-host junctions. After one moult, the graft and host retain their differentiated structures (e.g., bristles), but

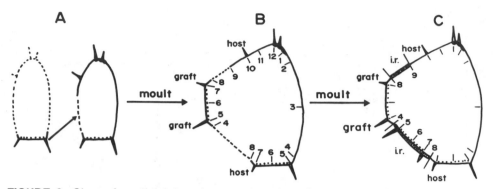

FIGURE 3. Circumferential intercalary regeneration. Schematic sections through the cockroach femur (see Fig. 1Aiii) showing the graft of medial face of left femur to the anterior face of the host left femur (A), the result after one moult (B) with a band of new tissue (dotted) separating graft and host, and the final result after two moults (C). The intercalary regenerate (ir) consists of the shortest section of circumference normally separating graft and host positions, so a new medial face (postitions 5, 6, 7) forms between positions 4 and 8.

are separated by an intercalary regenerate which usually bears few recognizable structures (Fig. 3B). New bristle rows and other markers appear at the next moult and show that the new tissue has formed the section of circumference that normally separates the original graft and host positions (12). The leg now has an abnormal but continuous circumferential pattern that remains stable through growth during succeeding larval instars. If the leg circumference is labeled as in Fig. 1Aiii, then intercalary regeneration reliably occurs by the *shortest route* (so interaction between cells from positions 4 and 8 generates tissue characteristic of positions 5, 6, 7, not 3, 2, 1, 12, 11, 10, 9). This rule is followed after a wide variety of grafts, and the confrontation of opposite positions (e.g., 10 and 4) leads to regeneration of either intervening half-circumference (12).

Distal Regeneration

If a cockroach larval leg is amputated at any level, the haemolymph clots over the wound and the stump epidermis migrates in to heal under the scab. Cell division forms the regenerate, which is folded beneath the old cuticle of the stump. The epidermis secretes a new cuticle differing only slightly in pattern from that of the original leg, and new muscles form in approximately the original pattern. The regenerate freed at the next moult is therefore a fair copy of the structures normally present distal to the amputation site. If the graft is made with reversed proximal-distal orientation, then its free end, which originally faced proximally, still regenerates the *distal* structures.

The PCM originally proposed that distal regeneration occurs only from a complete leg circumference, such as that normally present at an amputation site (16). However, it is now known that incomplete circumferences can form partial distal regenerates (6, and Chapter 22). These results can be understood if cells with dif-

ferent circumferential values come together during wound healing after amputation and this interaction leads to distal outgrowth (Fig. 4). The interaction will
lead to intercalation of cells with intermediate circumferential values, but if these
values already exist on adjacent cells, then the new cells must adopt a *more distal*
radial value.

After amputation through a normal circumference, almost any healing pattern will generate a new circle of cells with more distal values, and repeated rounds
of intercalation and distalization will produce a complete regenerate (Fig. 4A).
The extent of distalization from an abnormal double-half circumference will depend critically on the direction of wound healing. As shown in Fig 4B, regular
healing will produce a tapering regenerate which may be distally incomplete.
Healing directly along the plane of symmetry will provoke no regeneration, while
healing perpendicular to this will confront opposite values and may lead to intercalation of the values of the missing half circumference, forming an outgrowth which
will branch into two complete mirror-image regenerates (6, and Chapter 22).

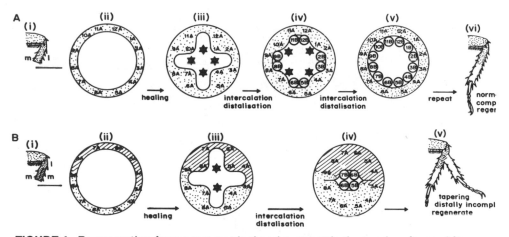

FIGURE 4. Regeneration from asymmetrical and symmetrical wound surfaces. After amputation through a normal tibia (Ai), complete regeneration occurs (Avi). If the medial half
(m) of the left tibia is grafted in place of the lateral half (1) of the right tibia and the leg is
then amputated (Bi), a symmetrical "double-medial" circumference is exposed. This
forms a partial distal regenerate (Bv). Note that a normal regenerate is formed from the
complete circumference at the proximal end of the graft. Aii–Av and Bii–Biv show intercalation and distalization at the amputation site. The proximal tissue of the stump (level A)
is stippled or hatched, and the wound edge is outlined by the inner circle. (A) The normal
circumference will heal to confront cells with different values (iii). Shortest route intercalation (stars) will produce new cells with values identical to adjacent stump cells, so they
adopt *more distal* radial values (iv), becoming 2B, 3B etc. Subsequent intercalation completes the B level (v) and the process is repeated until all distal levels have been formed.
This result is almost independent of healing pattern. (B) The result from the double medial
circumference will depend on the pattern of wound healing. If healing occurs to confront
4A and 4A, 5A and 5A etc., then no regeneration will occur. If the wound edge heals fairly
regularly (as shown), intercalation and distalization will occur at some positions but the
ring of more distal, new cells will have lost values in the line of symmetry: the regenerate
will taper and (in this case) will be distally incomplete.

Supernumerary Regeneration

If the right leg of a larval cockroach or cricket is amputated and the distal part is grafted onto the stump of the left leg, then one transverse axis of the graft is necessarily reversed relative to the stump. After reversal of the anterior-posterior axis, the graft heals and a supernumerary distal regenerate forms from the junction at

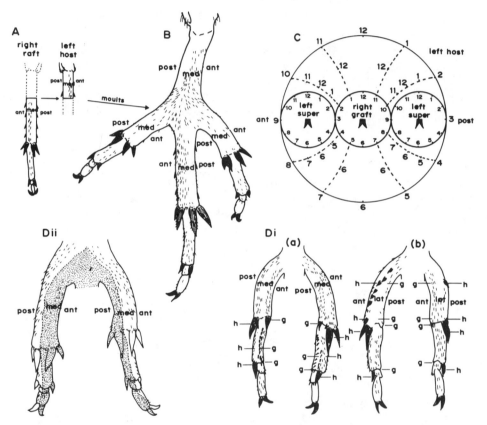

FIGURE 5. Formation of supernumerary regenerates in the leg of the cricket, *Achaeta*. (A) Diagram of graft at proximal tibia level between left and right metathoracic legs. The medial-lateral (med–lat) axes are in agreement but the anterior-posterior (ant–post) axis of the graft is reversed relative to the host. (B) Medial view of resulting leg showing supernumerary regenerates in anterior and posterior positions. (C) PCM interpretation of supernumerary formation. The host circumference is represented by the outer circle, the graft by the inner circle, and between these are the values formed by shortest route intercalation between cells confronted at the junction. Complete circumferences are formed anteriorly and posteriorly and from each a supernumerary (super) will regenerate. (D) When the graft tibia is taken from the prothoracic leg (i) or from another cricket species (ii) the supernumeraries are seen to be reliably half-host/half-graft in structure. D(i) shows medial (a) and lateral (b) views of the two supernumeraries with host (metathoracic) structures (*h*) and graft (prothoracic) structures (*g*). D(ii) shows a medial view of the two supernumeraries with graft (*Teleogryllus*) tissue stippled.

anterior and posterior positions (Fig. 5A). After a medial-lateral axis reversal, supernumeraries form in medial and lateral positions (3, 5, 7). The supernumerary regenerates are of host handedness and orientation.

If the leg is amputated, rotated by 90° and replaced on the stump, it heals and de-rotates back into alignment with the host, showing that cells at the junction can change contacts to minimize positional discrepancies. If the leg is rotated by 180° and replaced, both transverse axes will have been reversed. The graft can de-rotate in either direction, and often either one symmetrical, distally incomplete supernumerary or two complete supernumeraries are formed at the junction (3, 5, 7).

Figure 5B illustrates the PCM interpretation of the formation of supernumerary distal regenerates. After reversal of one transverse axis, intercalary regeneration occurs at the junction between confronted graft and host cells. This generates a complete circumference at each of the positions where opposite positions are confronted. The extra circumferences each form a distal regenerate protruding from the side of the main leg and having the observed handedness and orientation. The complete supernumeraries sometimes formed after the 180° rotation can occur in a variety of positions and orientations, and these can be explained in a similar way, by patterns of intercalation at the junction (16). The incomplete symmetrical supernumeraries may form from a part of the junction which fails to heal as the graft de-rotates.

The Formal Polar Coordinate Model

The PCM proposes that the insect limb epidermis has a two-dimensional map of positional values arranged along its longitudinal and circumferential axes, and that its regenerative behavior results from two rules for cellular interaction (6). The *shortest intercalation rule*: interaction between cells with different positional values provokes local growth, producing cells with the shortest set of intermediate values. The *distalization rule*: if intercalated cells have values identical to those of adjacent pre-existing cells, then the new cells adopt a more distal value.

This provides a simple and plausible interpretation, in terms of only the local interactions between adjacent cells, of a wide range of experimentally-produced and naturally occurring insect (and also crustacean and amphibian) limbs which show regeneration of missing structures, duplication of structures, and the formation of complete or tapering or branching supernumeraries. A crucial feature of the model is intercalation, and I shall now discuss other aspects of this process, some problems and uncertainties about the model and, finally, the way in which regeneration of insect legs may be related to their embryonic development.

Intercalation: Growth and Lineage

Does Intercalary Regeneration Occur by Local Growth?

Morgan (27) suggested two extreme ways in which patterns could be regenerated: morphallaxis where "a part is transformed directly . . . without proliferation at the cut surface," and *epimorphosis* where "a proliferation of material precedes the development of the new part." However, he cautioned that regeneration may involve both processes. Intercalary regeneration in cockroach legs is broadly epimorphic. Observation of the cuticular patterns formed at the first and subsequent moults gives indirect evidence that growth and change of positional values only occurs adjacent to the graft/host junction (Fig. 3), but careful histological studies are needed to establish the extent of the cell division. Is cell division initiated just in those cells contacting abnormal neighbors? Does it then spread back into the graft and host? Does division occur throughout the developing intercalary regenerate or is it restricted to a narrow front? Do regenerative cell divisions occur at the same time in the moult cycle as the divisions producing normal growth from instar to instar?

The results of a histological study of grafted cockroach legs (2) indicate that circumferential intercalation involves a burst of cell division localized in the region where graft and host epidermis have healed together (Fig. 6). These divisions start earlier in the moult cycle than the widespread divisions of growth and the specific differentiation divisions associated with the formation of new bristles.

In *Drosophila*, there is evidence from direct observation and clonal analysis

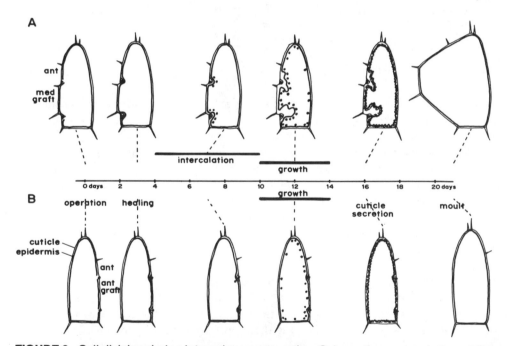

FIGURE 6. Cell division during intercalary regeneration. Schematic representation of distribution of colchicine-arrested mitoses (dots) in the epidermis of the cockroach femur at various times between the graft operation and the first moult. Legs are shown in diagrammatic cross section through the grafted region (see Fig. 1Aiii). (B) is control graft of anterior (ant) cuticle plus epidermis into the anterior face of the right femur, (A) is the experimental graft of medial face (med) into the anterior face of the left leg (see Fig. 3). Thick horizontal bars indicate the duration of the localized cell divisions of intercalation (far more numerous in B), and the widespread divisions of growth and bristle differentiation.

(1) that regeneration of imaginal disc fragments depends on cell division which is localized at the edge of the fragment and produces the new tissue within which new pattern elements are formed (see Chapter 12).

Do the Cut Edges Interact and Both Contribute to the Intercalary Regenerate?

In cockroach leg grafting experiments, graft and host cells taken from different positions are placed together and they heal. To determine whether both graft and host cells are involved in circumferential intercalation, I grafted between different leg segments, since they bear recognizably different cuticular structures (13). Grafting strips of tibia or coxa to a corresponding position on the host femur results in simple healing, while grafting to a noncorresponding position provokes intercalary regeneration (Fig. 7). In accordance with the shortest intercalation rule, the new tissue corresponds to that part of the circumference between the original positions of graft and host cells, and it is composed partly of graft-segment and partly of host-segment structures. These results show that all segments have the same circumferential organization and, assuming that segment character does not change during intercalation, they show that both edges contribute tissue. Hence the nature of the intercalary regenerate is determined by an *interaction* between the two edges: during intercalation, cells from a particular position can form structures lying in either direction around the circumference, depending on the position of origin of the other cut edge. Similar grafts between different segments (Fig. 2C, D) show that both edges contribute during longitudinal regeneration and hence that cells can form more proximal or more distal structures, depending on the level of the other cut edge (4).

If the cells from two different positions in a *Drosophila* imaginal disc are

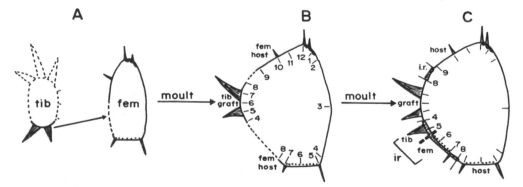

FIGURE 7. Circumferential intercalary regeneration between tibia and femur of the cockroach leg. Schematic cross sections of tibia (tib) and femur (fem) showing the graft of medial face of left tibia into the anterior face of left femur, (B) the result after one moult with new tissue (dotted) separating graft and host, and (C) the final result after two moults, showing intercalary regeneration (ir) by the shortest route between original graft and host positions. Compare with Fig. 3. Along the medial side of the graft, the border between tibial and femur tissue *always* forms between positions 5 and 6 (heavy dashed line).

mixed together, they both contribute to an intercalary regenerate (20). A disc fragment is assumed to regenerate by intercalation after it has folded and healed its cut edges together (20, 16). Healing cannot be directly observed and has to be deduced from the appearance of fragments when dissected out after a period *in vivo*. In some wing disc fragments, healing has been convincingly demonstrated (see Chapter 12), but the position is much less clear for other fragments and in leg disc fragments it seems that regeneration occurs sequentially from *one* of the cut edges (1). Healing and interaction may occur, but it remains a possibility that imaginal disc cells may systematically change their positional values while growing out as a free edge.

Are There Lineage Restrictions on Intercalary Regeneration?

During circumferential intercalation following strip grafts between different segments of the cockroach leg, graft- and host-derived cells regenerate the intermediate structures, but it is not always the case that each forms approximately half of the intercalary regenerate. The tibia and coxa bear few markers that are formed reliably on intercalated tissue, so interpretation depends mainly on the presence or absence of femur structures. In a wide range of different graft combinations, anterior femur cells can form medial structures up to *but not beyond* position 5½ (as in Fig. 7A), or they *can* form the recognizable lateral structures of position 12–1. Reciprocally, posterior femur cells can intercalate up to *but not beyond* position 5½ and *cannot* form the lateral structures. These results suggest that the leg is divided into two lineage *compartments* at positions 5½ and approximately 1 (13). During intercalary regeneration, cells cannot cross the borders to make structures of the other compartment.

If the larval insect leg consists of compartments that are respected during regeneration, and if supernumerary regenerates form by distalization (see Fig. 4) from circumferences generated by intercalation (see Fig. 5), then grafts between distinguishable left and right legs can be used to further define the compartment borders.

In the beetle, *Tenebrio molitor*, the larval pro- and meta-thoracic legs are very similar, but the adult legs differ reliably in the structure of the distal coxa and the tarsus (Fig. 8A). If the larval pro-thoracic leg (leg I) is removed at the based and grafted in normal orientation to the site of the metathoracic leg (leg III), the graft simply heals and forms an adult leg of typical leg I structure (15). If the right leg I is grafted with reversed anterior-posterior orientation to the site of the left leg III, the supernumeraries are *reliably* of dual structure, forming graft (leg I) structures on the graft side and host (leg III) structures on the host side (Fig. 8B). If the graft is made reversing the medial-lateral axis, then supernumeraries arise in medial and lateral positions and they are variable in composition (Fig. 8C).

The larval pro- and meta-thoracic legs of the cricket, *Achaeta domesticus*, also differ in structure of the tibia and tarsus, and grafting between left and right legs produces similar results to those from *Tenebrio*: anterior and posterior supernumeraries are reliably half graft-half host (Fig. 5Di), while medial and lateral supernumeraries are variable in composition (14, 15). In experiments where the markers for graft or host origin consist of presence or absence of cuticular structures, it is not possible to determine the exact position of the border. However, when similar grafts are made between metathoracic legs of *Achaeta* and the darker cricket, *Teleogryllus oceanicus*, a distinct border can usually be traced between light and dark cuticle. After reversal of the anterior-posterior axis, the supernumeraries have longitudinal borders in constant positions, slightly posterior to mid-lateral and mid-medial (Fig. 5Dii). When the graft is oriented to reverse a different transverse axis, supernumeraries may be of dual origin or formed largely (or occasionally completely) from either the graft or host (14, 15). The borders may run irregularly or obliquely down the leg.

These results can be easily interpreted in terms of the PCM. After anterior-posterior axis reversal the circumferences at the base of the supernumeraries will originate by intercalation between graft and host tissue at positions well away from the proposed compartment borders. Both components will intercalate up to the borders and therefore the circumferences (and hence the supernumeraries) will be reliably half-and-half in origin (Fig. 8B). After medial-lateral axis reversal, intercalation will start from tissue which spans the borders and will thus not be subject to lineage restrictions: graft and host will contribute variable amounts to the circumferences, and the supernumeraries will be of variable composition (Fig. 8C).

These results and those from the strip grafts (Fig. 7) suggest that intercalary regeneration in larval insect legs occurs by the shortest route within a continuous circumferential sequence, but that it is unable to cross two positions which divide the leg into anterior and posterior compartments.

Compartments were first found by clonal analysis of the development of *Drosophila* imaginal discs (18). From blastoderm stage the wing or leg disc consists of two adjacent populations of cells which will form precisely defined territories, or compartments, on the surface of the fly. In normal development, marked clones never cross the borders and will not do so even if the rest of the disc is growing much more slowly. The compartment restriction can be broken, however, by dividing cells in the early stages of regeneration of fragments of *Drosophila* wing or leg disc (20, 1) and it is intriguing that apparently this does not occur during intercalation in cockroach, cricket or beetle legs.

Difficulties with the Polar Coordinate Model

The PCM arose from analysis of the results of regeneration experiments on several types of vertebrate and invertebrate limb (16). It has stimulated a great deal of work on different aspects of regeneration (wound healing, cell division, pattern formation) in these and other experimental systems and, in its slightly modified form (6) it still provides a unified interpretation of most of the available data (see also Chapters 12, 22 this volume). There are problems, however, and I shall briefly discuss some of these.

Is Distal Regeneration Driven by Intercalation?

The model proposes that distal regeneration occurs after amputation because healing across the end of the stump provokes intercalation. The presence of adjacent

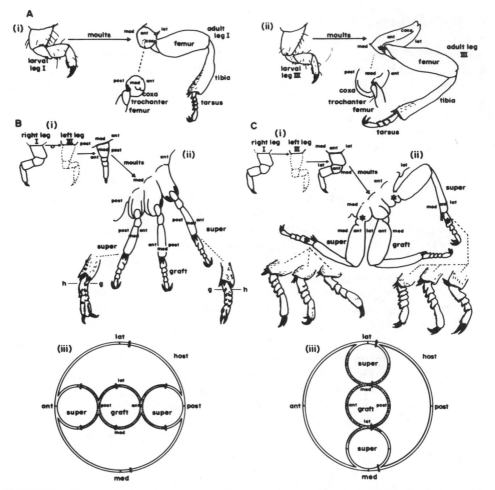

FIGURE 8. Supernumerary leg regeneration in the beetle, *Tenebrio*. (A) The structure of (i) the prothoracic leg (*Leg I*) and (ii) the metathoracic leg (*Leg III*) of the larval and adult beetle. Abbreviations as in Figs. 1 and 5. Legs are shown in anterior view (plus medial views of the coxae of adult legs). Note the difference in structure of the adult coxa/trochanter articulations and of the adult tarsi. (B) The right larval leg I is grafted to the site of the left larval leg III, reversing the anterior-posterior axis (i). Diagrammatic medial view (ii) of the resulting adult leg showing supernumerary regenerates in approximately anterior and posterior positions. Note that the anterior supernumerary is of leg III (host) structure on its anterior side and of leg I (graft) structure on its posterior side, while the posterior supernumerary has graft structures on the anterior and host structures on the posterior side. Medial views of the supernumerary tarsi show clearly their composite structure. Interpretation of the composition of the supernumeraries in (iii). Host, graft, and supernumerary circumferences are shown as in Fig. 5C with the position of the proposed compartment borders marked by heavy lines. Intercalation between graft and host anterior and posterior regions will generate the supernumerary circumferences and will be restricted by the border positions: each supernumerary will be half graft/half host in origin. Graft and graft-derived tissue is stippled. (C) Graft of larval leg I, reversing the medial-lateral axis (i). Diagrammatic anterior view (ii) of the resulting adult leg showing supernumer-

ary regenerates in approximately lateral and medial positions. At the distal coxa and tarsus, the supernumeraries may be of graft or host structure (stars). Anterior views of several supernumerary tarsi demonstrate the variability in their structure. Interpretation of the composition of the supernumeraries in (iii). The intercalation which generates the extra circumferences occurs from regions spanning the borders, so the graft and host will contribute in variable proportions to the supernumeraries. Thus a particular position (e.g., anterior distal coxa) will be of host origin on some supernumeraries and of graft origin on others.

---

stump cells with their unaltered positional values causes the new cells to take up more distal values. In Fig. 4, we have assumed the most simple and extreme form of epimorphosis where cellular positional value changes *only* during division provoked by a discontinuity between adjacent cells, and then only *one* of the daughters adopts a new intermediate value. We have also assumed that new cells forced to distalize will adopt the *next* most distal value. If all these assumptions are valid, then distal regeneration should involve only those cells confronted at the site of healing and should occur by a localized terminal "growth zone" forming successively more distal levels of the limb (Fig. 9A). However, if distalization involves the new cells adopting a *much more* distal value than their pre-existing neighbor, subsequent proximal-distal intercalation will restore continuity to the sequence of values. This process would involve cell divisions in a terminal "growth zone" (at least in early stages of regeneration), plus division throughout the developing regenerate (Fig. 9B).

In a recent histological study of regeneration following amputation (30), the initial cell divisions were found at the distal end of the stump, just away from the site of healing (some divisions may occur *before* healing is completed). Cell division then appeared to spread back into the tissue of the stump (Fig. 9C). Proliferation produces new tissue which will push the stump epidermis back from its once-overlying cuticle, so it is difficult to be sure how much of the stump is involved in generation. However, cell division may extend back into the stump epidermis for a distance of around 50 cell diameters (30).

In terms of the PCM, the observed pattern of cell division suggests that intercalation results in *both* daughter cells changing positional values. In this case, the effect of a positional discontinuity will gradually spread out from a graft/host junction, and spread back into the stump from an amputation (24, 30). This still does not explain why the initial cell divisions do not appear precisely at the positional discontinuity at the site of healing. If change of positional values can occur *without* cell division (i.e., by morphallaxis), then a change may spread back from the site of healing and be followed by any pattern of cell division within the band of altered tissue (25). However, the PCM cannot really be modified in this way, because it proposes that more distal values are generated precisely *because* the adjacent cells have unchanged proximal values. A model which does not link distal regeneration to averaging between different circumferential positions (e.g. 25) loses the ability to explain and predict regeneration patterns from abnormal or extra circumferences (Figs. 4, 5).

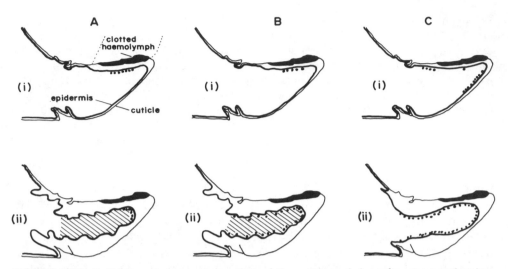

FIGURE 9. Cell division during regeneration of the cockroach leg after amputation between the trochanter and femur. (A, B) Patterns of epidermal cell division expected from two slightly different versions of the PCM. (A) If distal regeneration occurs because intercalation at the healing tip drives *one* of the daughter cells to the *next* distal level (as in Fig. 4), then early mitoses (dots) should occur under the wound (i) and should produce new tissue (hatched) that will displace the old tissue back into the stump. At later stages (ii), new cells will still only be produced at the tip of the regenerate. (B) If the new daughter cell can adopt a *much more* distal level (see text), then the regenerated tissue (hatched) will still only be derived from the tip of the stump, but later mitoses will be occurring throughout the regenerate (ii). If intercalation at the tip changes the level of *both* daughter cells, then division will spread back into the stump tissue. (C) Observed pattern of cell division [adapted from (30)]. After 3 days (i), the epidermis has healed under the wound, and mitoses are seen on either side of this position. By 4 days, mitosis has spread back to the level of the distal coxa (ii).

Repetition Within the Proximal-Distal Axis

The results of grafting experiments between different segments of the cockroach (or stick insect or cricket) leg show that the segments each contain the same proximal-distal sequence of positional values (Fig. 2C, D). Hence the formal PCM (Fig. 1B), with one A–E sequence, is a considerable simplification and does not explain, for example, why distal regeneration does more than complete the segment in which the amputation or graft was made. The segments are clearly different, with cells interpreting their proximal-distal and circumferential values to form different patterns of structures. There must be some relationship between the segment states such that they are generated in a reliable order during distal regeneration. However there is no evidence for a continuous sequence (or gradient) of segment states down the whole limb: intercalation between coxa and tibia does *not* give femur (Fig. 2D). All studies of grafting between segments have produced minority results not predicted by repeating segmental sequence, but these are best interpreted in terms of delayed healing and partially independent regeneration by the graft and host (4, 8).

Regeneration and Compartments

The PCM does not illuminate the relationship between pattern and lineage, nor give any clues about the function of compartments. Compartments were originally defined as lineage territories and have since been assumed to be regions of differently determined cells (17). Some aspects of the phenotype of the *engrailed* mutant of *Drosophila* suggest that the posterior compartment of wing or leg is partially transformed into a mirror-imaged anterior compartment, so it has been suggested (9) that the compartments are equivalent fields of positional information. They may normally be interpreted differently because of the "on" (posterior) or "off" (anterior) state of the *engrailed*+ gene. This view of compartments seems untenable as the imaginal disc and larval leg responds to fragmentation or grafting as if it contains a *single* continuous sequence of positional values around its entire circumference (21). Compartments, however, *are* lineage restrictions observed during normal development and, at least in larval legs, during regeneration. What is their function?

There is some evidence from imaginal discs that compartment borders are important for growth control (see Chapter 13), but in larval legs growth seems to be stimulated locally by pattern discontinuities, regardless of the position of compartment borders (Figs. 2, 3, 8). Borders may be required to initiate distal regeneration in imaginal disc fragments, and may also be involved in circumferential regeneration (Chapter 13). Again, this does not seem to be true of the anterior/posterior compartment border in larval leg regeneration (however, see Chapter 3). This may be a basic difference between regeneration in larval legs and imaginal discs, or it may reflect the difference between a graft combination and an isolated fragment that may or may not heal its cut edges together. For the moment, however, the PCM does not help in resolving these problems.

Is the Map Two-Dimensional?

The insect leg is a complicated, three-dimensional structure (Fig. 1A). It has a surface layer of epidermis and cuticle, internal apodemes (formed by extensive invaginations of epidermis at the leg joints), patterns of muscles (attached to apodemes and regions of surface epidermis), nerves (formed by axon growth out from the segmental ganglia or in from surface receptors) and trachea (formed by invagination of body epidermis). Can regeneration of this pattern really be due just to properties of the surface epidermis?

There is no indication that the internal tissues have any role in the formation of surface epidermal pattern. Predictable abnormal epidermal patterns are produced by epidermal grafting experiments that leave internal tissues almost untouched (e.g., Fig. 3). The distal regenerates formed after amputation, and the supernumeraries formed after grafting (Fig. 5) contain normal patterns of invaginated epidermal apodemes and more-or-less normal muscle patterns. There is considerable evidence that insect muscle patterns depend on epidermal pattern. The formation of muscles in embryonic legs seems to be directed by the developing apodemes around which the myoblasts aggregate and differentiate (10). Similarly, the position of the new muscles differentiated in the beetle pupa depend on the epidermal sites of attachment (31), and the complex pattern of adult muscles formed from the mesoderm cells associated with *Drosophila* imaginal discs probably depends on the properties of the cells of the disc epidermis (23). The muscles of regenerating cockroach legs are probably formed when the myoblasts coming from dedifferentiated stump muscle aggregate around developing apodemes and surface attachment sites (29). Furthermore it is highly probable that the nerves and trachea growing into the regenerate from cut ends in the stump, form patterns depending on the arrangement of muscles and epidermis.

Thus it is likely that, given the permissive hormonal conditions, leg regeneration occurs because of properties of the epidermis, and the patterns of structures formed depends only on the distribution of positional values and the interactions occurring within this cell layer.

Is the Map Polar?

The PCM suggests that the range of results of regeneration experiments indicates that the two-dimensional larval leg or imaginal disc epidermis has a specific map of positional values (organized in continuous sequences along polar axes) and responds to a disturbance in accordance with two rules. It is possible to explain many of these results (e.g., the production of distal and supernumerary regenerates) by

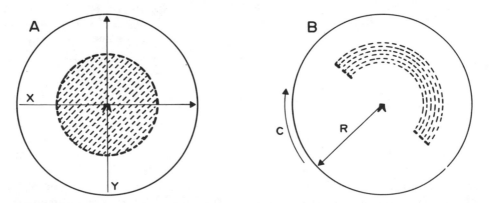

FIGURE 10. Polar and Cartesian maps of the insect leg epidermis. The epidermis is represented as a disc with the distal tip (marked by claws) in the center. (A) If the map is Cartesian, cells have positional values along X and Y axes. When the distal part of the limb is removed, simple averaging along the two axes will replace the missing tissue (dashed lines). (B) If the map is polar, cells have positional values along a radius (R) and around the circumference (C), as in Fig. 1B. Regeneration of the distal tip requires a special rule since the regenerated structures are in so sense "between" the remaining proximal ones. However, after a strip graft that confronts opposite positions (shown as the two separate cut surfaces), averaging along the circumferential axis produces the observed arc of tissue (dashed lines). Averaging along Cartesian axes would form distal structures, so a Cartesian model requires a special rule to accommodate this result.

any form of continuous map and a rule of re-establishing *continuity* (24, Chapter 5). However, without a particular coordinate system and particular rules it is not possible to give definite predictions for many grafting experiments. Since one coordinate system can always be mathematically transformed into another, the pattern of experimental results cannot *prove* that a tissue employs a particular form of map. However different coordinate systems require different additional rules to account for the results (see Fig. 10). The two rules of the PCM generate clear predictions and explain most of the available data in a simple and plausible way.

Leg Development and Leg Regeneration

So far I have discussed the regeneration of the large functional legs of larval insects (and have mentioned similar results from regeneration experiments on mature imaginal discs of *Drosophila*). I have argued that the leg epidermis bears a very detailed map of stable cellular positional values. How was this map formed? Can we, in fact, deduce anything of the embryonic development and normal growth of limbs from their regenerative behavior?

If the imaginal disc primordia of *Drosophila* embryos at the immature discs of young larvae are damaged surgically, then the eventual adult structures have duplications and deletions which are broadly similar to those resulting from the regeneration of fragments of the mature disc (16). Similarly, exposing young larvae of temperature-sensitive cell-lethal mutants to a period at the restrictive temperature produces leg duplications and triplications (19). These can be analyzed in terms of regeneration after the loss of parts of the positional map of the mature disc (see Chapter 3, 12). These results suggest that, even from early embryonic stages, the disc primordium contains the basic outline of its eventual fine-grained map. The cellular interactions which direct disc repair after a disturbance at any stage may be those which control the elaboration of the map and the growth of the disc during normal development.

Very few experiments have been done on the development of embryonic legs of other insects. From descriptive studies, it seems that the leg segments are formed in a proximal-to-distal sequence during early growth of the leg bud, and in some insects, the leg buds will regulate to form complete legs after removal of the distal parts, either before or after they become visibly segmented (11).

A rather different sort of experiment also indicates that the cellular interactions demonstrated in regeneration studies may be involved in embryonic leg development. The locust embryo gradually forms visible thoracic and abdominal segments in an anterior-to-posterior sequence as it grows. If early embryos are given a brief temperature shock, they eventually develop characteristic abnormalities in several segments just posterior to those visible at the time of operation (26). In other words, the temperature treatment is affecting a band of tissue which is in the growing tip of the embryo and is not yet visibly segmented. A common abnormality is the fusion of two adjacent segments, usually only on one side of the body. If very early embryos are temperature-shocked, they frequently develop with a fused prothoracic/mesothoracic or mesothoracic/metathoracic segment, which bears a single leg base. This leg may remain a single structure down to its distal tip; it may

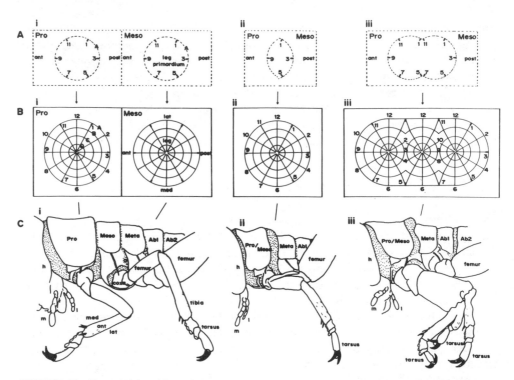

FIGURE 11. Formation of legs in normal and temperature-shocked locust embryos. Side view (C) of first instar larvae developing from a control embryo (i), or from embryos temperature-shocked before segments are visible (ii, iii). Nonsclerotised cuticle is stippled. Drawings show the head (h), with maxillary palp (m) and labial palp (l), and also the prothoracic (Pro), mesothoracic (Meso), metathoracic (Meta), first abdominal (Ab1) and second abdominal (Ab2) segments. Temperature-shocked embryos may develop a fused prothoracic/mesothoracic segment with a single leg base which forms a single leg (Cii) or a leg which branches into two or three (Ciii) distal parts. (Note that the leg in Cii bears an unusual outgrowth, and the middle branch of the leg in Ciii has a fused tibia-tarsus). (*A, B*) Interpretation of segment fusion and leg formation in terms of the PCM. When segments are formed in the normal embryo (Ai), the leg primordium consists of a few cells with proximal positional values (1A, 3A, 5A etc.). If these cells interact by intercalation and distalization (see Fig. 4A), then the primordium will grow and the rest of the polar co-ordinate map will be formed (Bi). Cells will interpret their values, forming a complete, normal leg on each thoracic segment. After early heat shock, the embryo may form a fused prothoracic/mesothoracic segment with a single leg primordium. (Alternatively, the segments may form normally and then fuse secondarily.) If the leg primordium consists only of anterior and posterior edges (Aii), intercalation will nonetheless complete the normal map (Bii). As in the normal primordium, this result is more or less independent of precisely which cells interact. If the primordium consists of large anterior and posterior parts (Aiii), interactions perpendicular to the anterior-posterior plane will insert two sets of inappropriate values across the middle of the primordium, leading to the formation of a complex map containing three centers (Biii)—the resulting leg will branch into three distal tips (Ciii).

diverge to give two distal parts; or it may form *three* distal tips (each bearing a set of claws) lying in the anterior-posterior plane (26), as shown in Fig. IIC.

When proposing the PCM, we suggested (16) that the leg primordium is first established within the developing body segment with just a few of the proximal circumferential positional values. If interaction between the embryonic cells with these values follows the rules deduced from their much later regenerative behavior, then growth will be stimulated within the primordium and the map will be built up as the more distal values and the intervening circumferential values are formed (Fig. 11Ai, Bi).

The temperature shock causes fusion of thoracic body segments not visible at the time of treatment. It probably causes abnormal, partial segments to be formed but, if a pattern of invisible segments is already present, the temperature shock may result in fusion by killing a band of cells including the intersegmental margin. In either case, the fused body segment may develop a single, abnormal leg primordium. If this has only the very anterior values appropriate to one segment plus the very posterior values from the next segment, intercalation will fill in missing values and the leg will grow out as a single, complete structure (Fig. 11Aii, Bii). However, if the primordium contains more than the anterior half plus more than the posterior half of the positional values, then as it grows out interactions within the leg may complete *three* sets of circumferential values, forming a triple structure (Fig. 11Aiii, Biii). Of course, this explanation does not account for cases where a single leg base diverges into *two* distal tips.

If most of the ventral thorax between successive cockroach larval legs is removed, the posterior of one leg base heals to the anterior of the next leg base and an extra, anterior-posterior reversed larval leg is regenerated (16, and Chapter 3). The similarity is striking between this larval leg *regeneration* result and the result of temperature shock disruption of the very early stages of locust leg *formation* especially since locust legs, once segmented, are incapable of regeneration in embryonic or larval life. These sorts of results encourage us to regard the mechanisms of regeneration as being, at least to some extent, those of embryonic development. Insects such as cockroaches and crickets may regenerate larval legs because they retain the abilities for cellular interaction by which the embryonic legs were formed, and which are then lost in other insects such as locusts.

Conclusions and Prospects

Pattern formation in insect development is currently an exciting and productive field of research. Insects exhibit rich, precise patterns and are particularly amenable to the techniques of experimental embryology and of developmental genetics. I have only discussed a small area of insect development and have argued that the regenerative behavior of the insect leg suggests a particular model of the spatial cues which give cells information about their position. This model, the PCM, proposes that the epidermis has a detailed, two-dimensional map of positional values, arranged along polar coordinates down and around the limb. Positional values are stable cellular properties except in new cells produced by division provoked by a discontinuity between the values of neighboring cells. The new cells adopt values intermediate between those of their neighbors, thereby removing the discontinuity. However, if this process intercalates cells with values identical to those of adjacent cells, then the new cells change to a more distal value. These processes will replace the tissue missing after accidental damage or autotomy of the limb, and will produce the bizarre, abnormal structures observed after a variety of grafting experiments.

One function of phenomenological models such as the PCM is to relate a variety of experimental observations and focus attention on particularly important or obscure aspects. The PCM has been derived mainly from analysis of the pattern of cuticular structures formed after amputations and grafts. Clearly, we also need much more direct information on epidermal cell behavior during healing and regeneration. What are the relationships between wounding, the contact of dissimilar cells, cell division and a change of cellular positional values? Is intercalation really tied to cell division which is a very local response to positional discontinuity? The PCM has no particular role for compartments or compartment borders, but these lineage restrictions are prominant features of normal development and of regeneration. Are they important in the control of pattern or growth? The PCM proposes a close link between pattern formation and growth. Cell division is certainly stimulated by pattern discontinuities but what is the relationship between this and the normal growth of the intact leg? In the normal leg, neighboring cells will have slightly different positional values and may be stimulated to divide, intercalating new cells with intermediate values. As long as neighboring cells are sensitive to the degree of discrepancy between their positional values, the leg will continue to grow through larval life (8). There seem to be several problems with this view of growth as "fine-grained intercalation." Why do the cell divisions of intercalation which follow an amputation or graft occur much earlier in the moult cycle than those of normal growth? How can larval locust legs, that are incapable of intercalation after grafting or amputation, still grow reliably through larval life? The PCM proposes that the pattern of the internal tissues is derived secondarily from that of the epidermis. Further studies are needed of the formation of the muscles and other tissues, in embryonic development and during regeneration after a variety of grafts.

The PCM suggests that intercalation and distalization are involved in the development of the limb primordium (see Fig. 11), but how is that primordium first formed as part of the embryonic thoracic segment? One suggestion (16, 13) is that the circumferential positional values of the leg extend proximally from the coxa onto the surrounding thorax so that, apart from intersegmental "buffer zones," all of the ventral part of the thoracic segment has a polar positional map (see Chapter 3). Similarities between regeneration in the *Drosophila* leg and wing discs (16, 6) may even suggest that a similar polar map covers the dorsal thoracic segment. It is difficult to relate this view of the organization of the thoracic segment with what we know of pattern formation in the abdominal segment. Numerous grafting experiments on the dorsal and ventral abdomen suggest that each abdominal segment contains a *linear* anterior-posterior gradient of positional values (22) within which epidermal cells can respond to a discontinuity, possibly by intercalation (28). Since insect thoracic and abdominal segments have presumably evolved from the similar segments of multi-legged ancestors, it is unlikely that they are patterned in completely different ways.

A second function of phenomenological models is to define the cellular properties which are sufficient to explain the observed multicellular behavior. This can give clues about the molecular mechanisms of pattern formation. The PCM involves only local interactions between cells and their nearest neighbors, so it suggests that "positional value" may be a property of the cell surface rather than an intracellular concentration of a diffusible morphogen. Similarly, the first stage of cellular interaction may involve modifications of the cell surface. If pattern formation does occur in this way, at least in regenerating limbs, it will involve molecular mechanisms very different from those suggested by source-diffusion or reaction-diffusion models (see Chapter 3).

The PCM has been somewhat modified (6) in the light of results obtained since it was initially proposed (16). It accounts for most of the available data (although there are some problems, as I have indicated) and at present, it provides the most coherent basis for discussing and experimenting upon the insect limb. It suggests a wide range of experimental approaches to pattern formation in insect epidermis. Further results may support the present model or may lead to further modification. However, if its major assumptions are disproved or its major predictions not upheld, the PCM will go the mortal way of most models.

Acknowledgments

I thank Hilary Anderson, Nigel Holder, Jane Karlsson, Jane Mee, and Dave Wright for many useful discussions. This work is supported by grants from the Science Research Council, the Royal Society and E.M.B.O.

General References

FRENCH, V.: Pattern regulation and regeneration. *Phil. Trans. R. Soc. Lond.* B. 295: 601–617 (1981).

KUNKEL, J.: Cockroach Regeneration. in: *The American Cockroach* Bell, W., Adiyodi, K. (eds.), pp. 425–443. London: Chapman & Hall (1981).

References

1. Abbott, L., Karpen, G., Schubiger, G.: Compartmental restrictions and blastema formation during pattern regulation in *Drosophila* imaginal leg discs. *Dev. Biol.* 87:64–75 (1981).

2. Anderson, H., French, V.: *Manuscript* in preparation.

3. Bohn, H.: Analyse der Regenerationsfähigkeit der Insekten-extremität durch Amputations- und Transplatationsversuche an Larven der afrikanschen Schabe *Leucophaea maderae* Fabr. (Blattaria). II. Mitt. Achsendetermination. *Wilhelm Roux' Arch.* 156:449–503 (1965).

4. Bohn, H.: Interkalare Regeneration und segmentale Gradienten bei den Extremitäten von *Leucophaea*-Larven (Blattaria). I. Femur und Tibia. *Wilhelm Roux' Arch.* **165**:303–340 (1970).

5. Bohn, H.: The origin of the epidermis in the supernumerary regenerates of triple legs in cockroaches (Blattaria). *J. Embryol. Exp. Morph.* **28**:185–208 (1972).

6. Bryant, S., French, V., Bryant, P.: Distal regeneration and symmetry. *Science* **212**:993–1002 (1981).

7. Bullière, D.: Interprétation des régénérats multiples chez les Insectes. *J. Embryol. Exp. Morph.* **23**: 337–357 (1970).

8. Bullière, D.: Utilisation de la régénération intercalaire pour l'étude de la détermination cellulaire au cours de la morphogenèse chez *Blabera craniifer* (Insecte Dictyoptère). *Dev. Biol.* **25**:672–709 (1971).

9. Crick, F., Lawrence, P.: Compartments and polyclones in insect development. *Science* **189**:340–347 (1975).

10. Fournier, B.: Contribution à l'étude expérimentale des relations entre l'ectoderme et le mésoderme au cours du développement embryonnaire de la patte de *Carausius morosus* Br; les ébauches d'apodèmes et la ségrégation des masses musculaires présomptives. *C.R. Acad. Sci. (Paris)* **266**: 1864–1867 (1968).

11. Fournier, B.: Expériences d'amputation faites sur les embryons du Phasme *Carausius morosus* Br; en vue d'apprécier l'aptitude de la patte embryonnaire à reconstituer ses portions manquantes. *C.R. Acad Sci. (Paris)* **269**:2401–2404 (1969).

12. French, V.: Intercalary regeneration around the circumference of the cockroach leg. *J. Embryol. Exp. Morph.* **47**:53–84 (1978).

13. French, V.: Positional information around the segments of the cockroach leg. *J. Embryol. Exp. Morph.* **59**:281–313 (1980).

14. French, V.: Leg regeneration in insects; cell interactions and lineage. *Am. Zool.* **22**:79–90 (1982).

15. French, V.: *in preparation* (1983).

16. French, V., Bryant, P., Bryant, S.: Pattern regulation in epimorphic fields. Science **193**:969–981 (1976).

17. Garcia-Bellido, A.: Genetic control of wing disc development in *Drosophila*. In: *Cell patterning*. Ciba Foundation Symp. 29 (ed.) Lawrence, P., pp. 161–182. Amsterdam: A.S.P. (1975).

18. Garcia-Bellido, A., Ripoll, P., Morata, G.: Developmental compartmentalization of the wing disc of *Drosophila*. *Nature, New Biol.* **245**:251–253 (1973).

19. Girton, J.: Pattern triplications produced by a cell-lethal mutation. In *Drosophila*. *Dev. Biol.* **84**:164–172 (1981).

20. Haynie, J., Bryant, P.: Intercalary regeneration in the imaginal wing disc of *Drosophila melanogaster*. *Nature*, **259**:659–662 (1976).

21. Karlsson, J.: Homeotic genes and the function of compartments in *Drosophila*. *Dev. Biol. in press* (1983).

22. Lawrence, P.: The development of spatial patterns in the integument of insects. In: *Developmental systems: insects*. *(eds.)* Counce, S. Waddington, C., Vol. 2, pp. 157–209. London & New York: Academic Press (1973).

23. Lawrence, P.: Cell lineage of the thoracic muscles of *Drosophila Cell* **29**:493–503 (1982).

24. Lewis, J.: Simpler rules for epimorphic regeneration: the polar co-ordinate model without polar co-ordinates. *J. Theor. Biol.* **88**:371–392 (1981).

25. Maden, M.: The regeneration of positional information in the amphibian limb. *J. Theor. Biol.* **69**:735–753 (1977).

26. Mee, J.: D. Phil. Thesis, Univ. of Edinburgh. *in preparation* (1983.)

27. Morgan, T.: Regeneration. New York: Macmillan (1901).

28. Nübler-Jung, K.: Pattern stability in the insect segment. I. Pattern reconstitution by intercalary regeneration and cell sorting in *Dysdercus intermedius. Wilhelm Roux' Arch.* **183**:17–40 (1977).

29. Steinbach, P., Josephson, R.: Pattern regulation of mesoderm in transplanted and supernumerary limbs of the cockroach. I. The final morphology. *J. Exp. Zool.* **211**:331–341 (1980).

30. Truby, P.: Blastema formation and cell division during cockroach limb regeneration. *J. Embryol. Exp. Morphol. in press* (1983).

31. Williams, G., Caveny, S.: A gradient of morphogenetic information involved in muscle patterning. *J. Embryol. Exp. Morph.* **58**:35–61 (1980).

32. Wolpert, L.: Positional information and pattern formation. *Curr. Top. Dev. Biol.* **6**:183–224 (1971).

Questions for Discussion with the Editors

1. *Clearly, intercalation of cockroach appendage grafts involves cell proliferation and is, therefore, epimorphic. Please speculate on the possible roles cell division plays in leg pattern regulation at (a) the cellular level (does it, for example, generate a population of dedifferentiated cells?); and (b) the molecular level (does it, for example, serve to dilute out pre-existing regulatory molecules?).*

The epidermis of an intercalary regenerate is formed by proliferation of epidermal cells at a positional discontinuity (new muscle is probably formed by more extensive dedifferentiation, division and redifferentiation of the mesoderm cells). The local growth response, and the fact that the resulting continuous sequence of positional values is stable even on a position of local symmetry (see Figs. 2B, 3C) suggest that "positional value" may be a nondiffusible qualitative property of the cell surface, rather than an intracellular concentration of a "diffusible morphogen." If this is the case, we have very little idea at present about how neighboring cells could sense a discontinuity, and how this could result in cell division and adoption of intermediate cell states. Alternatively, "positional value" could be the concentration of a morphogen which, in some way, could only pass between very dissimilar cells and which would provoke division in cells experiencing a change in concentration. There remains, of course, the problem of coding for circumferential positional value; a quantitative (gradient) mechanism would produce a boundary region having special properties.

2. *What are the prospects for enlarging the scope of cockroach appendage regeneration studies to include genetic analyses (e.g., use of cell markers, mutants, etc)?*

Larval insect limbs are very suitable for precise grafting operations but not for developmental genetics. There are no informative morphological mutants available, and the cuticle color mutants are not sufficiently clear and cell-autonomous for use in clonal analysis. In the bug *Oncopeltus*, abdominal segment formation has been analyzed using x-ray-induced clones of abnormal cell pigmentation and this technique is being used to study embryonic leg development. Regeneration of leg tissue is pathetic in *Oncopeltus*, however. The technique is unlikely to work on other insects where the chromosomes are not holokinetic and the chromosome breaks are likely to be cell-lethal.

SECTION IV

Plants

CHAPTER 15

Controls of Cell Patterns in Plants

Tsvi Sachs

CONSIDER THE POSSIBLE DISTRIBUTIONS of two cell types in a two-dimensional tissue formed from one original cell. Of the many possibilities, the fundamental arrangements to be dealt with are strands and spacing patterns (Fig. 1). In both, one structure or cell type forms the matrix for the orderly distribution of the other. These patterns are important not only in themselves but also as components of more complex cases and as possible models for the more general three-dimensional patterns.

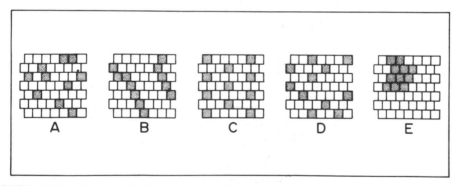

FIGURE 1. Possible distributions of one cell type in a matrix of another. An example of what could be a random distribution is shown in (A), though additional samples would be required for confirmation. A strand arrangement (B) is a central topic of this chapter. The pattern in (C) is precisely spaced, yet the relations between the cells are not the same in different axes. A less orderly spacing pattern (D) involves a minimal distance between the specialized cells but no exact determination of their location. Stomata patterns considered below are of this type. A group arrangement (E) is one of the many patterns not dealt with here.

Environmental influences on development can be dramatic, especially in plants, but they could not exert the detailed control that could specify the relative location of cells. The environment, therefore, can act as a switch determining the expression of different possible patterns and the duration and rate of specific processes. The location of cell types and the relations between them must depend on internal processes; these are interactions between the different parts (spatial controls) and the dependence on past developmental events (determination, or, more generally, temporal controls). The evidence considered here is largely of three kinds: precise observations of mature and developmental stages, studies of regeneration after various wounds, and results of experiments where organs are replaced by substances they are known to produce. The purpose is to find the basic parameters of time, orientation and distance that could characterize the control systems involved.

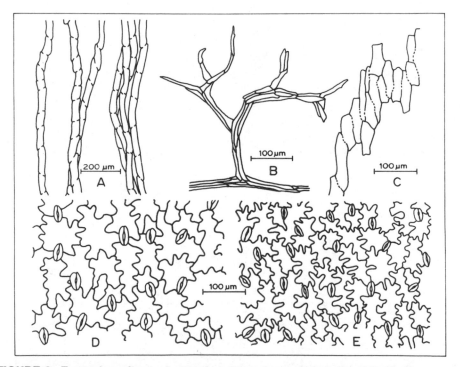

FIGURE 2. Examples of actual strand and spacing patterns. Vessels (A–C) are readily distinguishable from the surrounding tissues and serve in water transport. They consist of files of thick-walled cells, dead when mature, connected by special openings. Projection microscope drawings of cleared turnip storage tissue (A), pea leaf (B) and a regenerative vessel above a wound in a bean stem (C). The distribution of stomata is spaced but not precise (D, E). Each stoma consists of two specialized cells whose shape can change and regulate the size of pores in leaf surfaces. Projection microscope drawings of nail polish copies of leaf surfaces of *Anagallis* (D) and pea (E). For details of the development of these patterns see Figs. 14 and 15.

It is shown that the arrangement of specialized cells in extended files, or vessels, depends on a gradual, long-term feedback between an inductive flow of hormones and the differentiation of cells as preferred channels for its transport. This continued differentiation canalizes the flow to defined strands and accounts for the determinate nature of the processes, for the mutual induction of differentiation by cells along the same file and for the inhibition of similar events in other, neighboring cells, from which the hormone flow is diverted. The second fundamental arrangement, the spacing pattern of stomata, is shown to depend primarily on polarized intracellular developmental events that form the specialized stoma and the neighboring cells together as large units. These can be in direct contact with one another, and no inhibitory interactions between neighboring stomata or other cells need be involved. In both cases, therefore, there is no clear separation between pattern specification and its overt expression. It is the processes of differentiation themselves that are oriented and polarized resulting in the patterned localization of specialized structures.

Both examples considered here come from plants, which have a number of important advantages for the study of patterns and their controls. Though plants include the largest and oldest organisms, they are relatively simple in numbers of cell, tissue, and organ types. They also have remarkable regenerative capacities, revealing the action of developmental controls even in large, nonembryonic structures. Two additional traits are, however, of greatest importance. Plant development, though based on the same general biochemical and ultrastructural principles as that of animals, occurs without any movement of the cells relative to one another. Thus the origin of a cell is clearly related to its location. Finally, embryonic development, including the formation of new organs and new patterns, continues at shoot and root tips throughout the life of the plant. There is no problem in obtaining thousands of genetically uniform seedlings with large nutrient stores. These are more robust, and of more convenient size, than embryos of either plants or animals.

The Formation of Vascular Strands

Inductive Control by Hormone Flow

New vessels, and other vascular tissues, differentiate throughout the life of a plant, supplying new shoot and root tissues as they form. Most plants also add vascular tissues along the parts of the axis whose length does not change (3, 4). This growth in girth permits an adjustment of the size of the functional vascular system to the leaf and root systems it connects. In addition, the vascular system is commonly capable of regeneration by a redifferentiation of other cells to form vascular channels (Fig. 2C and 3A). These facts show that the formation of new vessels is under continuous control, even in relatively mature parts of the axis, and indicate that this control is associated with the development of the shoot and root.

The nature of these controls can be studied by simple experiments of organ removal, regeneration, and grafting (for review see 21). The absence of the young parts of the shoot, especially the developing leaves, reduces the formation of vessels

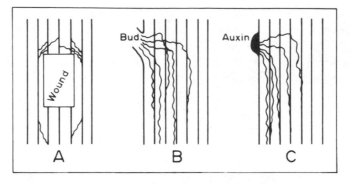

FIGURE 3. Experiments demonstrating controls of vessel formation and orientation. The drawings are diagrammatic with only a few vessels indicated: straight lines show those present before the experiment started and wavy lines show where new vessels form. Vessels differentiate around a wound (A), in connection with bud growth below a removed shoot (B) or, in a similar way, in connection with an exogenous source of auxin (C). Such patterns have been seen in cleared material from various plants, for example bean and *Coleus* stems.

in the stem below (9). This is true of all types of vessel formation: within apices, along the axis, and around wounds and has been observed in a wide variety of species and conditions. It is easy, for experimental purposes, to find systems where vessel formation is completely prevented, rather than reduced, when the growing shoot above the region of observation is removed. It may be concluded that signals from developing shoot tissues are one of the controls of vessel differentiation. This conclusion is supported by new vessel formation that is associated with the development of buds below removed shoots (Fig. 3B). Connecting vessels also appear when a growing shoot is grafted to a cut stem or root (Fig. 4A). Evidence, as yet inconclusive, for the complementary control, of leaf initiation by vascular tissues (13) is not contradictory (21).

Exogenous applications of the plant hormone auxin (indol-3-acetic acid) cause the differentiation of new vessels in a wide variety of systems: within growing shoot apices, along the plant axis, in wounded tissues (Fig. 3C) and in tissue cultures. Auxin, furthermore, is known to be formed in relatively large quantities in growing shoot tissues, especially in developing leaves. Since in causing vessel differentiation auxin replaces organs in which it is formed, it must be at least one component of the signals of natural induction (7, 21). Auxin, therefore, is a known inducing principle or evocator of cell differentiation. The role of auxin, however, is by no means limited to vessel differentiation; it also induces other vascular tissues and has many other effects of plant development, including the promotion and inhibition of cell growth, cell division, and a wide variety of differentiation processes. Though it was first discovered as an inducer of cell growth (27), and this is often considered its "typical" effect, auxin appears to be a general directing signal of developing shoot tissues, and the response it evokes depends not on its chemical nature but rather on the competence, or developmental history, of the receiving tissue (19, 21).

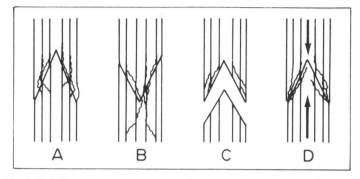

FIGURE 4. Vessel differentiation in grafts. Diagrammatic drawings, showing only very few, representative vessels. The ones formed after the plants were cut are indicated by wavy lines. The difference between the grafts in (A) and (B) shows an inequality in the effects of the shoot and the root. Leaving the shoot and the root separate confirms that the shoot is a source of inductive stimuli for vessel differentiation, (C). An influence of tissue polarity (the original shoot to root orientation, arrows) is seen most clearly when one of the two graft members is inverted (D). The drawings diagram the results seen in cleared material using a variety of plant stems, for example those of bean seedlings.

Vessel differentiation is controlled and oriented not only by the developing shoot but also by the root (Fig. 4A, B). The removal of the roots does not, however, prevent continued vascular differentiation, at least not in short-term experiments, where the dependence of shoot development on the roots is not expressed (Fig. 4C). This suggests that the roots are not active sources of signals for vascular differentiation but rather act as sinks for substances originating in the shoots. If so, vessel differentiation would occur along the flow of signals, one of them auxin, from the developing shoot tissues to the root apices. This concept, of differentiation along channels of flow, is also suggested by a number of observations:

1. Vessel differentiation follows the original shoot-to-root polarity of the tissues in which it is induced. This is seen most dramatically when tissues of opposite polarity are grafted together [Fig. 4D, (27)]. This polarity of differentiation corresponds to the polarity of auxin transport, which is known to depend on determined properties of the transporting tissue and not on the location of sources of auxin and its sinks (5; for polarity in plants see 22, 24). The dependence of vessel differentiation on tissue polarity thus indicates a relation to auxin flow.

2. Vessels induced by a growing bud or by an exogenous source of auxin (Fig. 3B, C) are not restricted to a cupshaped region near the source of induction. Instead, they form long strands connecting this source with the direction of the roots, along the expected channels of auxin flow.

3. Treatments that raise auxin concentration in the tissue without causing flow do not result in vessel differentiation. Thus when competent tissue (Fig. 5A) is surrounded by a source of auxin from all sides (Fig. 5B) it does not differentiate, regardless of the concentration of the applied auxin, unless there is an "outlet" to the rest of the plant, through which flow can occur (Fig. 5C).

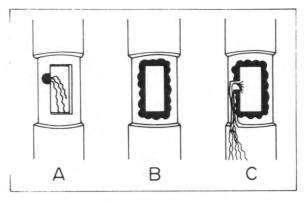

FIGURE 5. Importance of auxin flow for vessel induction in girdled tissues. Competent stem regions were isolated by removing all the surrounding tissues. The contact with the plant through the stem center sufficed for differentiation in response to localized auxin (A). No response was seen when auxin surrounded the tissue completely (B) and no flow could be expected. An outlet for auxin flow (C) resulted in pronounced differentiation. The diagrams show cleared material from experiments performed on bean and *Coleus* stems.

Feedback Accounts for Both Determined and Labile Polarity

These facts, and especially that shown by Fig. 4D, demonstrate that tissue polarity, as expressed by vessel differentiation, can be a stable or determined property. The details of the new vessels near wounds, a growing bud or a source of auxin (Fig. 3), however, show that differentiation at right angles to tissue polarity occurs readily, though at first the original polarity is evident in cell shape (Fig. 2C). Differentiation of vessels at all possible angles to polarity is most clearly demonstrated in plants with only narrow bridges connecting the shoot and the root (Fig. 6A, B). The reorientation of differentiation shown in Fig. 6A, has been shown recently to be expressed also in the expected new polarity of the transport of radioactive auxin (Gersani and Sachs, *in preparation*). These facts show that polarity can be either determined or labile (22), and raise the question of the conditions where the different expressions are found.

An experiment offering an answer is illustrated in Fig. 6C, D. When a polar path is present, vessels (and other vascular channels) form along it, even though it may lead to no functional contacts with the root. Under certain conditions this polar path causes the death of the shoot. When no polar alternative is available, however, new contacts with the root, at various angles with original tissue polarity, form readily and the shoot always survives the wound. This, and the other facts mentioned above, suggest not only that auxin flow and the resulting vessel differentiation depend on tissue polarity, but also that polarity itself can be induced by auxin flow. In relatively mature tissues, convenient for experimental work, polarity always exists, but it can be changed when cuts prevent normal transport and diffusion towards a connection with the roots starts a flow in a new direction (22). This suggestion, of a positive feedback relation between tissue polarity and auxin flow, accounts both for the determined nature of polarity (Figs. 4D and 6C) and for its lability (Fig. 6A, B).

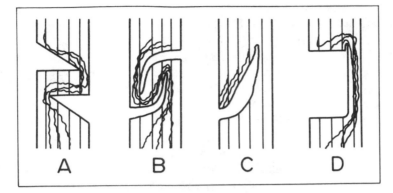

FIGURE 6. Vessel differentiation in response to wounds. New vessels (wavy lines) form readily at right angles to the original shoot-to-root polarity (A) and, for short distances, even opposite this polarity (B). When a polar alternative is available, however, it is followed by differentiation even when this does not lead to functional contacts (C). This is not due to any problem in the formation of contacts with the roots when the polar alternative is removed (C). There is diagrammatic representation of only a few of the vessels in cleared stem tissue of various plants. In bean seedlings, for example, the new vessels are seen clearly within three days.

Canalization of Induction to Cell Files

It is now possible to ask what controls the pattern of cells that compose a vessel (Fig. 2A–C). If vessels are induced by auxin flow and the oriented transport of this very same flow is a differentiated response of the tissue, then the vessels could represent a *canalization* of the response to defined strands of cells (21). Such canalization would be expected if the polarization response of the cells is a gradual process, for then the signals carried by a large cross-section of the tissue would be concentrated in a few cells belonging to strands that become most specialized for transport. It is these strands that would continue to be induced and would finally mature as vessels that consist of dead cells (which do not transport auxin). The differentiation of a cell file as a vessel would then require no predetermination, and indeed auxin application indicates that this does not exist. Once a cell has started to differentiate as an efficient, oriented transporter, however, further flow would occur preferentially through it, leading to continued differentiation. At the same time, it would also induce differentiation in cells along the same file, by efficient drainage from those above and localized supply of inductive signals to those below. The same response would inhibit similar differentiation in neighboring cell files by diverting away the inductive flow. This is, therefore, an economical control that accounts for cell determination, intercellular induction and intercellular inhibition on the basis of one system. This hypothesis has been analyzed mathematically (15) and is consistent with the parameters of transport, to the extent that they are known. The following are the main types of experimental evidence that support it:

1. The induction of vessel differentiation is a gradual process, requiring signal flow throughout most of the time differentiation takes place. The removal of an ex-

ternal source of auxin at various times has variable effects, presumably because of auxin accumulation in the tissue. It is possible, however, to change the configuration of auxin flow by adding sources at various times after the beginning of the experiment. A convenient system is illustrated in Fig. 7, where the additional source present only in B changes the effect of an initial one present in A (21, 22). this additional source can be added at various times, and it is found that the differentiation seen in A is labile to change for as long as two days after the initial auxin is added. Since the entire process of differentiation is complete, in the first new vessels, about two and a half days after auxin application, it could not be a process that is triggered by an early flow and then continues without further orienting signals.

2. There have been extensive studies of the transport of radioactive auxin (5). Though these have not been related to the canalization hypothesis of vessel formation, they do show that the capacity to transport varies with the tissue, and best transport occurs through the differentiating vascular system. Furthermore, it has been found that auxin treatments maintain the capacity of cut tissues to transport radioactive auxin (5). There is some more direct evidence (21 and *unpublished*) that auxin can also induce new channels for its own active transport at an angle to the original polarity of the treated tissue.

3. All stem tissues form in correlation with the presence of a new leaf primordium directly above them. After a short initial stage of development, however, the removal of leaf primordia influences only the development of the vascular system (21). These results are in accordance with the hypothesis that a primordium first influences the entire tissue and its effects become canalized to discrete strands of cells.

4. In cut plants, differentiation induced by auxin is oriented so that it connects to nearby vascular strands (Fig. 7A). This effect of existing vascular strands is

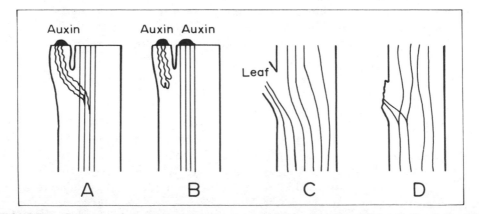

FIGURE 7. Controls of contacts between vascular strands. New differentiation induced by auxin (indicated by wavy lines) always connects to existing vascular tissues (A) unless they, too, are supplied by an auxin source (B). Strands formed in relation to the differentiation of leaves above do not connect directly to those of an older (and lower) leaf (C). Such connections are formed, however, if the leaf is removed at a primordial stage, after its connecting strands have been determined (D). Diagrams of experiments performed on pea seedlings; for references see (21).

generally prevented if they are supplied with a source of auxin (Fig. 7B). These results are in accordance with the hypothesis that vascular systems differentiate as the best channels for auxin so that they orient a new flow towards themselves unless they are already overloaded (21). This principle accounts, or at least suggests an explanation, for the location of many of the vascular contacts found in intact plants. It is further supported by the results of organ removal experiments (for example, Fig. 7C, D).

Axial Rather Than Polar Differentiation

The suggestion that vascular strands result from canalized inductive flow is thus supported by major experimental results. It does, however, make a prediction that is not in accordance with plant structure and this leads to an important modification of the hypothesis. Since flow must have a direction, it might be expected that all vascular strands should have a definable polarity, from the source of the inductive signals in the shoot to their sinks in the root tips. The general form of the vascular system of plants is, in fact, that of a drainage system, so that a polarity can be readily assigned to all major strands (Fig. 8A). The leaves of many ferns and most seed plants, however, are supplied by complex vascular networks that can not be understood as an expression of a polar flow (Fig. 8B). These networks include, for example, vascular triangles where any assignment of direction leads to opposite polarities occurring within one strand (Fig. 8C).

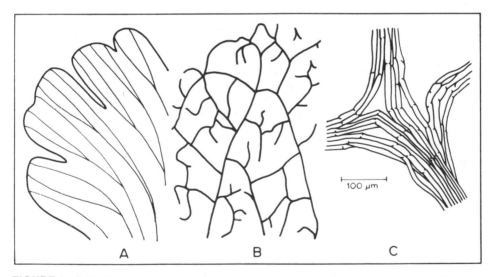

FIGURE 8. Polarity of leaf vessel systems. In many ferns, such as *Adiantum* (A), a clear shoot-to-root polarity can be assigned to all parts of the vascular system. Such polarity is not clear in the networks common to seed plants, such as pea (B). Large vessel connections at the base of pea and other leaf networks (C) show that vessels of opposite shoot-to-root polarity can differentiate next to one another, within the same strand. Thus no consistent polarity could account for vessel triangles (D). Projection microscope drawings of cleared material.

A solution to this problem is suggested by large transverse strands, to which no clear polarity can be assigned, found in various stems (21) and by similar structures at the base of many leaves (Fig. 8C). These strands include many vessels and the polarity of individual vessels can be readily defined by their connections with a leaf and the roots. The important fact is that vessels with opposite polarities are found next to one another, as is also shown by curved vessels (Fig. 9). This suggests that a strand that has no definable polarity can be formed by inductive signals moving in opposite directions. Observations on the development of such strands and experiments using application and removal of auxin sources (21) indicate that these opposite flows occur at different times, the direction changing more than once during differentiation.

This suggests that small apolar strands in leaves, often including only one vessel, are formed in response to inductive flows whose direction is reversed during the time differentiation takes place. Supporting evidence comes from the occurrence of such strands only in tissues where growth measurements and stomata maturation show that development is unsynchronized and could be the basis for changes in the direction of flow. The importance of this suggestion is that cells would specialize, at early stages of differentiation, to transport along a given axis while the direction may change according to source-sink relations of the signals. Since there is clear evidence for tissue-determined polarity of transport only for auxin, this modification of the canalization hypothesis explains how other signals could be involved in the induction of biological strands.

Two Open Problems

It is thus possible to suggest one control for the pattern of the cells that compose vessels and for the general pattern of the vessels in the plant. Two of the major problems that this hypothesis raises will be mentioned. The first is the cellular complexity of the vascular system, which can include not only the vessels considered above but also the sieve tubes, composed of files of living cells that lack nuclei, and at least three other distinct cell types, one of them distinguished by continued divisions. Auxin alone orients and induces the differentiation not only of vessels but of sieve tubes and entire vascular systems as well. How one signal, consisting of a simple molecule, specifies the organized differentiation of different cell types is not known (for possible hypotheses see 21).

A second general problem, about which a little more can be said, concerns the nature of the early cellular events induced by signal flow. There is evidence that the response of the cells themselves, and not only of the tissue as a whole, is to the flow of the signals. For example, a long-term response to intercellular gradients could not account for the differentiation of vessels in the form of sharp bends and closed rings (Fig. 9A, B) but they can be understood as a positive feedback response to canalized flow. This suggests that the individual cells respond as flow meters, a possibility not generally considered in developemental studies. A response to flow, however, is possible in various ways:

1. It could be mediated by intracellular gradients that may be both a necessary consequence and a cause of continued flow. Such internal gradients have been

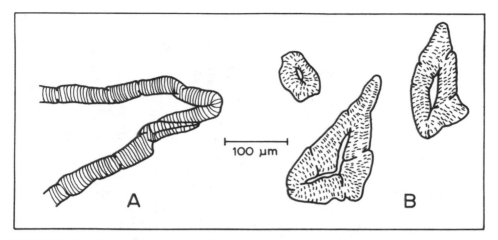

FIGURE 9. Vessel shapes that could be a response to a long-term inductive flow but not to an inductive gradient. Vessels with sharp bends (A) are commonly found in intact plants. Circular vessels (B) form where endogenus or exogenous auxin flows along tissue polarity and no outlet to the roots is available (as in Fig. 6C). Projection microscope drawings of cleared material from a bean node (A) and a cut piece of auxin-treated turnip (B).

calculated on the basis of available knowledge concerning the cellular mechanism of polar auxin transport (16). It is not known, of course, how internal gradients could control cellular differentiation, but this possibility at least transfers the problem to relatively familiar grounds.

2. Signal flow could directly or indirectly orient ultrastructural elements in cells. Electron microscopy supplies no evidence that the endoplasmic reticulum, in plant cells capable of further differentiation, is oriented in any consistent way. There is no doubt, however, that microtubuli are so oriented, and their effect is expressed in the orientation of cell divisions (17) and of cell wall thickenings, both with a definite relation to the axis of vascular differentiation. The wall structures, normally at right angles to the axis of vessels (Fig. 10A, B), are parallel to this axis in the first vessels formed above wounds or in other cases of tissue reorientation (Fig. 10C). Thus the reorientation of the elements that control the pattern of the wall structures need not occur during the differentiation of new vessels. This conclusion is also supported by the effects of colchicine treatments, which prevent cell division and completely disrupt the pattern of cell wall thickenings (Fig. 10D), but do not prevent new vessel differentiation (6).

3. The earliest changes induced by auxin flow could be in the localization of specific proteins in the cell membranes. Such localization may well be the structural basis for polar auxin transport (5). A flow of ions related by a positive feedback to protein localization in external membranes may be major control of oriented biological development (8). This directed movement of the proteins, due to self-electrophoresis, could not be operative in a response to auxin flow. It is possible, however, the auxin flow and protein localization are coupled in some indirect way, for example as a result of intracellular auxin gradients, or that auxin movement through the cells is associated with that of a small molecule left behind when

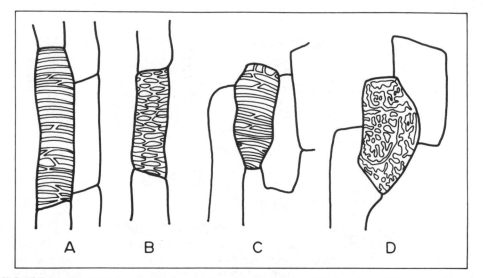

FIGURE 10. Wall thickenings of vessel members. Normal thickenings of vessels in growing tissues are in various forms, but at right angles to the vessel axis (A and B). In the first regenerative vessels near a wound the thickenings still conform to the original tissue polarity (C) even though the arrangement of the cells does not (Fig. 6A). Colchicine treatments disrupt wall thickenings completely (D) but do not prevent vessel differentiation. Projection microscope drawings of cleared sections of bean seedlings.

auxin crosses the cell membrane. Such molecules could activate specific proteins in the region of passage. All this is, of course, completely hypothetical.

The Spacing Pattern of Stomata

Measurement of Spacing

The distribution of stomata on the surface of most leaves gives the intuitive impression of relative uniformity with an absence of contiguous or almost contiguous stomata [Fig. 2D, E; (1)]. This is, therefore, an example of a "spacing pattern" (29) and the deviations from a random distribution must depend on a control of the location of differentiation. Stomata spacing patterns have the advantages for research that they can be readily observed, both in the mature state and during development, and their lack of precision indicates that they could be amenable to experimental manipulations. The intuitive impression of spacing, however, must be first checked by objective methods, especially since stomata in direct contact are not completely absent but only rare. Quantitative methods have been based largely on techniques of ecological work (2, 10, 18) and they compare predictions of random distributions with actual measurements. The following three examples (Fig. 11) illustrate parameters that have been used for this purpose:

1. The ratio of observed and random distances between closest neighbors (Fig. 11A). The use of this ratio, known as the parameter R, was developed for ecological work (2). A large random sample of distances between a stoma or any other

spaced structure and its closest neighbor can be readily measured and the expected value for a distribution governed by chance can be calculated from the density. A ratio that differs significantly from unity indicates order or pattern and the value for ideally spaced units is about 2.15. Common results for stomata distribution are, as expected, around 1.5. This measure of order has the advantage that it gives a clear quantitative result and requires only simple measurements. In its present form, however, it has two disadvantages. Distances in all different directions are pooled together, and as biological structures generally have a clear polarity, this pooling is, at best, a loss of information. The second disadvantage is that the calculation of the random value is based on the distribution of points with no area. Stomata, and other specialized cells, occupy a nontrivial part of the surface (Fig. 2D, E) and the impossibility of their overlapping one another should be taken into account.

2. Squares of a given size can be placed at random on the pattern (10, 18). The distribution of such squares with different stomata numbers can be readily counted (Fig. 11B) and compared with that expected for a random arrangement of stomata. This being the Poisson distribution

$$e^{-m}\left(1; m; \frac{m^2}{2!}; \frac{m^3}{3!} \cdots \right)$$

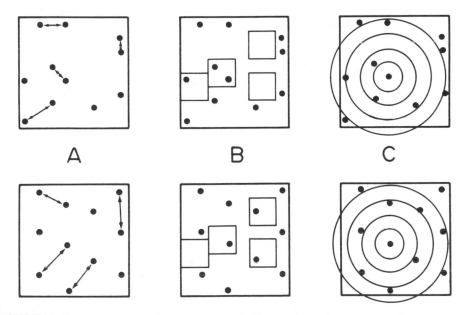

A　　　　　　　　B　　　　　　　　C

FIGURE 11. Three methods of measuring orderliness of spacing patterns. The top distributions could be random, while the bottom ones are spaced. (A:) The distances between closest neighbors are greater in the spaced than the random distribution. The ratio between averages of such distance, the parameter R, is a measure of order. (B) A larger number of randomly placed squares will have an average number of units in a spaced than in a random distribution. (C) The frequency of finding another unit in an area of given size will depend on the distance from randomly selected units in a spaced but not in a random distribution. Such frequencies can be plotted by super-imposing the distribution around many randomly chosen units (Fig. 12).

where m is the average number of stomata or other observed units in a square of the size chosen. This method has the advantage that pattern may occur on different scales, and these could be revealed by counts made with areas of various sizes. The disadvantages of the previous method apply here too, and the problem of the size and absence of overlap might cause actual mistakes. In addition, it is generally impossible to find the large uniform areas required for the measurements; disturbances caused by veins, for example, are harder to avoid than for the previous method.

3. Radial distribution functions (11, 20). These are plots of the frequency of finding all other stomata as a function of the distance from a central stoma (Fig.

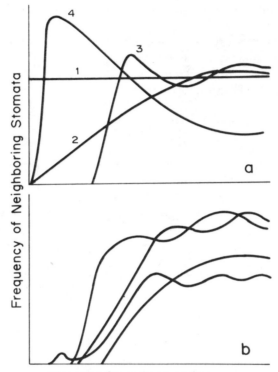

FIGURE 12. Stomata frequencies as a function of distance from randomly chosen stomata. (A) Possible hypothetical curves. Curve I shows no dependence of stomata frequency on distance. This would be found for a random distribution, and deviations from such a horizontal line imply an orderly component or pattern. Curve 2 would be found if each stoma had an inhibitory effect on the formation of additional stomata that declined gradually with distance. Curve 3 would imply that there is a minimal distance between close stomata and the exact location beyond this distance could depend on chance. Curve 4 shows stomata found near one another at higher than random frequencies. This means an arrangement in groups, as in Fig. 17A. (B) Examples of curves measured for actual stomata distributions. Though these and other, unpublished, curves show many unexplained variations, the general shape is that of curve 3 in (A), showing a minimal distance between neighboring stomata. The drawings are from published curves (14,20,23), and the distance to the first maximum varied between 40 to 160 μm, the maximal density being between one and 10 per 10^4 μm².

11C). For statistical results, measurements from a large number of central stomata, randomly selected, have to be superimposed. The deviations of the curves obtained from a horizontal straight line show a correlation between the presence of a stoma and differentiation events in its neighborhood (Fig. 12A). The method involves no assumptions about the size of a stoma or any other observed unit. Separate measurements for different directions can be readily made, relative to the axis of the stoma or any other reference. The main advantage, however, is that information is obtained not only on the degree of order but also on the distance over which any hypothetical control mechanism operates (Fig. 12).

Measurements of the distances between neighboring stomata proved, as expected, that stomata are not randomly spaced. Plotting the distribution functions for about a dozen plants confirmed this result and showed that the major, if not the only, correlation between the existence of a stoma and events in its neighborhood is the presence of a minimal distance between close stomata (Fig. 12B). The presence of stomata beyond this minimal distance appears to depend only on chance and not on any correlation with the stomata that are somewhat further away. This result was not expected and not the only one possible, for one could conceive of inhibitory effects of a stoma on its surroundings that decline gradually, up to considerable distances. The role of chance in determining whether a stoma will form in a "permissible" location is in accordance with the intuitive impression of stomata distribution (Figs. 1D and 2D, E) and the results of measurements showing that this distribution, though not random, is far from complete order.

Possible Controls of Minimal Distances

It follows that there must be controls preventing stomata from forming very close together, and the nature of these controls is the essential developmental question to arise from the observation of the spacing pattern. Before considering the evidence, it may be useful to define major possible groups of controls:

1. The pattern may depend on events that occur outside the epidermis. The cells may become unequal, and only some of them capable of forming stomata, as a result of inductive effects of another pattern found in the underlying tissues (1, 11, 20).

2. If controls are found in the epidermis, they could involve interactions between cells. These could be inhibitory effects of a cell on the formation of stomata in its immediate neighborhood (1, 12). They could also, however, have the complementary form that epidermal cells cause the formation of a stoma in their center. it is also possible that a stoma, once formed, induces the growth of other cells that separate it from its neighbors.

3. The interactions may be internal, and lead to a relatively orderly process of development of a stoma formed together with its neighboring cells, a "cell lineage" mechanism. If, for example, all mother cells divided twice in the same direction before differentiating as stomata, a perfect spacing pattern could be formed, with no stomata separated by fewer than two cells (1, 20).

The three suggested controls can be defined to cover all possibilities, for they are the answers to two dichotomous questions (are controls inside or outside the tissue and, if inside the tissue, do they involve intracellular or intercellular processes). These controls are not, however, mutually exclusive, and it is possible that more than one of them is effective. The essential question, therefore, is what is the relative contribution of the various controls.

Evidence for Cell Lineage Controlling Pattern

It is now necessary to consider testable predictions that could be based on these hypothetical controls. Induction by underlying tissues suggests a search for such an inductive pattern and for the time of its initial appearance. Stomata are, in fact, associated with the presence of air chambers and a causal relation between the two has been suggested. The late appearance of these chambers is not convincing evidence that they do not determine stomata location, for their early detection depends on the method of observation. A physiological presence, that could have inductive effects, might not be observable by methods now available. The correlation between stomata and the chambers is not, however, precise, either at early stages of development or in the mature state (11, 20). There is no other structure of the underlying tissues that is correlated with stomata distribution. All this is negative evidence, and can not be conclusive, for there could always be undetected structures. It does suggest, however, that the main controls of stomata distribution should be sought within the epidermis.

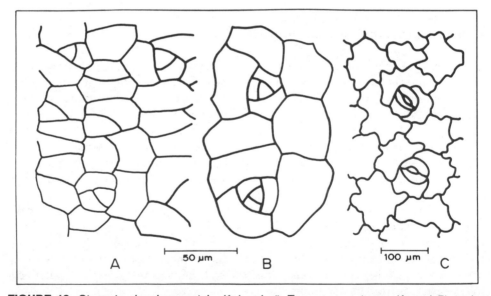

FIGURE 13. Stomata development in *Kalanchoë*. Two young stages (A and B) and a small mature region (C). Stoma development starts with an unequal division of relatively large cells (A). This is followed by a spiral of unequal divisions (A and B). As a result, two neighboring stomata (C), may develop from neighboring cells and yet be separated by a minimal distance. Measurements (Figs. 11C and 12B) show that this minimal distance is the major, if not the only, orderly aspect of stomata distribution. Projection microscope drawings of nail-polish copies of lower leaf surfaces.

Predictions concerning development differ for controls of pattern by interactions between stomata and neighboring cells and by cell lineage. Interactions would cause the earliest observable stages of stomata formation to be spaced; though the surface may continue to grow, the relative distances could be determined at the same time as the stomata. Cell lineage, on the other hand, would suggest that the earliest stages of development occur in neighboring cells and that the distance between the closest stomata forms from the very same cells as the stomata themselves. Observations of various stages of stoma development in some plants (Fig. 13) leave no doubt that cell lineage can be a major control of stomata spacing. The earliest observable stage of development, unequal cell division, occurs in neighboring cells, and this is a general rule in many plants (14, 20, 23). The stomata form together with neighboring cells which, in some cases, surround them completely (Fig. 13C). This produces a minimal distance between any two neighboring stomata; measurement of the distribution functions (Fig. 12B) show that this is the major or only orderly component of stomata distribution. This correspondence of predictions based on the developmental processes and measurements

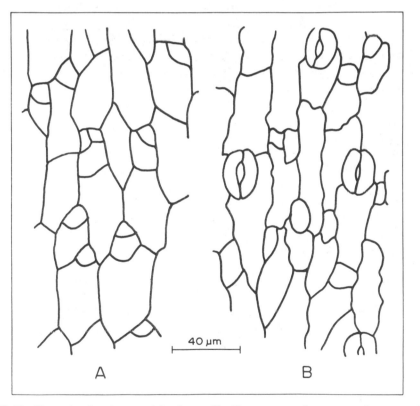

FIGURE 14. Stomata development in *Anagallis*. Two young stages are shown and the mature state appears in Fig. 2D. Stomata development is variable, but the same divisions also form cells toward the base of the leaf. The consistent polarity of the unequal divisions prevents stomata from being formed next to one another. Projection microscope drawings of peeled epidermis.

of the mature state proves that cell lineage can be the major factor in stomata spacing (20).

This conclusion leads to an examination of its validity for different cases of stomata spacing. In very many plants, the cells formed together with the stoma do not surround it from all sides (Fig. 14). Yet in such plants stomata formed from neighboring cells are only rarely in direct contact. A reason that this is not more common, however, is readily seen in the polarity of the developmental process (Fig. 14). Each stoma may, in fact, be "exposed" on one side, but since the initial unequal divisions and the developing complexes of stomata and neighboring cells all have the same orientation, stomata are in contact only with the epidermal cells formed together with the next stoma along the file of cells (14).

In very many plants, however, a stoma is not surrounded by the products of the same mother cell and neighboring complexes do not necessarily have the same polarity. A further complication is that in these plants, including most of the broad-leafed dicotyledons, arrangement of the cells both in the mature epidermis and in developmental stages of young leaves show that development can vary, even for adjoining stomata [Figs. 2D, E and (15)]. This variation is expressed in the locations, relative orientations, and number of cell divisions leading to a mature stoma. It follows that the process of development can not be considered to be determinate and it is possible for intercellular controls to modify cell lineage after the initial unequal cell divisions in neighboring cells. This does not contradict the conclusion reached above, about the importance of cell lineage but suggests that the controls may be complex and involve a number of different processes.

Where development is variable it can not be reconstructed from stages seen in different leaves or even in different parts of the same leaf. Evidence concerning modifications of development by interactions of neighboring stomata, therefore, should be sought in descriptions of the changes in specific developing stomata. In other words, development has to be studied by continuous, nondestructive methods. Since stomata are surface structures, this is possible using epi-illumination microscopy (20), though it is limited to convenient plants where the surface is not covered by hairs, where the borders between cells can be readily seen etc. The main problem, however, is that such studies are laborious, and have been carried out to date only on limited regions of leaves of *Pisum sativum* (20) and *Vinca major*. An example of the results (Fig. 15) shows that interactions between neighboring developing stomata may influence the orientation of divisions, and thus the precise location of maturation, rather than the occurrence of initiation. The available evidence, however, is not sufficient to supply anything more than hints or indications. Much more data will be required before a reliable estimation of events due to chance can be separated from expressions of orienting intercellular controls.

Problems Suggested by Comparative Studies

Stomata development and pattern vary remarkably in different plant groups and even a brief survey indicates that comparative studies, taking advantage of "experiments" carried out by evolution, could be rewarding in indications of possible controls of two-dimensional differentiation patterns. This is illustrated by the exam-

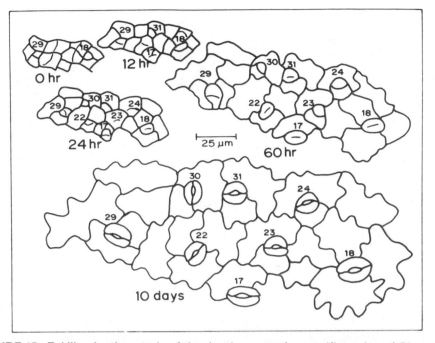

FIGURE 15. Epi-illumination study of the development of a specific region of *Pisum* epidermis. This is an unpublished part of the study reported in (20). The epidermis was drawn twice a day so local changes and cell divisions could be followed. Very recent divisions were recognizable, and they were marked by lines that do not quite join previous cell walls. The starting time (0 hr) and the numbers of the stomata are arbitrary. The last stage had been fully mature for some time before it was drawn. Note that stomata development varies greatly in number, rate, and relative orientation of divisions, even in neighboring cells. This indicates that there could be controls that modify stomata development during the entire time it takes place.

ples of Figs. 16 and 17. The formation of a stoma mother cell is followed, in many monocotyledons, by oriented divisions that form the adjoining cells (Fig. 16B). This suggests that a developing stoma can induce neighboring developmental events that form parts of the distance separating it from other stomata. The spacing pattern discussed above is not the only stomata distribution; in many plants stomata are in groups, and these can be quite distinct (Fig. 17A). One possibility is that these patterns are formed by a two-stage control: cells competent and incompetent to form stomata are first separated, followed by the formation of the spacing found within the groups. Finally, stomata are not the only structures on the epidermis, and the different patterns are certainly related [Fig. 17B (1)]. Observation of the formation of such complex patterns suggests that early unequal divisions may lead to the formation of one structure, such as hairs. Later divisions, when the state of the leaf has changed, form stomata by the same initial developmental processes (1). This possibility, however, is little more than an intuitive guess.

A last open question concerns the density of the stomata. A minimal distance between the closest stomata, the only measured aspect of their pattern (Fig. 12B), sets an upper limit to this density. Simple calculations, however, show that this

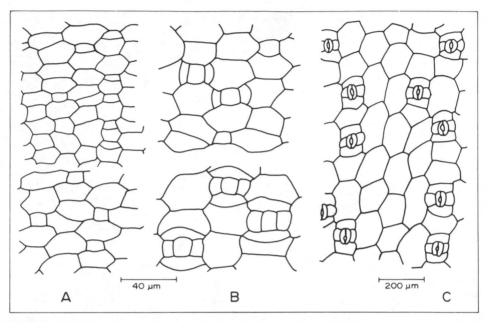

FIGURE 16. Stomata development in *Tradescantia*. Four developmental stages (A, B) and the mature state (C) show that stoma formation involves major changes in the relative shapes of cells and coordinated unequal division, presumably induced by the stoma mother cell which is the first to appear. Such developmental stages are the best available evidence for interactions between a developing stoma and neighboring cells. Projection microscope drawings of nail-polish copies of young leaf bases (A, B) and a mature leaf (C).

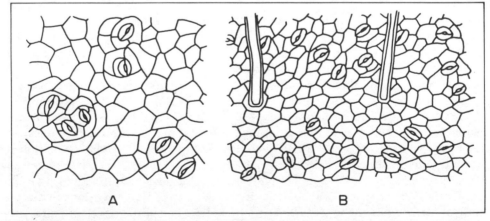

FIGURE 17. Examples of variations of stomata distribution. Groups of stomata in *Begonia* (A) and relations of hair and stomata distributions in *Pelargonium* (B) show the types of questions and possibilities raised by comparative surveys of epidermal structure. Projection microscope drawings of epidermal peels.

limit is rarely reached. Density could then depend on the random frequency of unequal divisions leading to stoma formation in the young epidermal cells competent to undergo these divisions. Differences in stomata density, such as those between the upper and lower epidermis of the same leaf, could then result not only from different minimal distances but also from frequencies of unequal divisions and the length of time they can occur (14).

The picture to emerge from this appears confusing and a summary hypothesis may be useful: An important factor in determining the exact location of stomata (and many other biological spaced structures) is chance. There is, however, a control preventing the regular occurrence of stomata closer than a minimal distance, typical of the species and environmental conditions. Though the inhibition of initiation by neighboring stomata could be a simple and efficient control (1, 12), it is cell lineage, or the development of a stoma from the same cell as the surrounding epidermis, that plays the largest role. This lineage, however, often requires a general polarizing influence to orient events in neighboring cells. Direct observations of development also show that lineage is not a rigid program but rather a set of rules capable of many changes, especially in the orientation of cell divisions (17). Stomata spacing, therefore, resembles vessel differentiation in depending on controls that operate during a relatively long period. Comparison of epidermal structure in the plant provides many variations of potential interest.

Discussion

It is now necessary to consider whether the subjects of this chapter suggest any ideas or principles applicable to other biological systems. The contents of this volume, including this discussion, show that a conceptual framework for working hypotheses about expressions of pattern formation is now available, and this is a major advance over the situation in the not too distant past (26, 28). The elaboration of a general, comparative theory need not involve a detailed physical and molecular knowledge of the controls involved, especially since they may well differ in major biological groups. What may be most needed at present are parameters of relative timing and distances that could serve as necessary constraints for hypotheses and quantitative models. Major questions concern the timing of pattern specification, possibly by a configuration of controlling morphogenes, relative to later stages of its expression by elaborate differentiation. In the processes of specification, it is possible to conceive of a separation (19) between interaction of the parts of the pattern (spatial controls) and the dependence of any part on its past development (temporal controls). Concerning spatial controls it is necessary to know, for example, the dependence of interactions on distance and the types of information that they convey. Temporal controls may maintain states that have been achieved, determine competence to respond to spatial signals and also specify elaborate programs of development. The following are major points, concerning these and similar problems, based on the study of vessels and stomata spacing summarized above:

1. The formation of pattern can be gradual and need not involve stepwise changes of the controls of differentiation. This is true for vessel formation, where it

is suggested by experimental evidence and by observations that the same hormonal signals control very different stages of differentiation. The evidence from epi-illumination studies of stomata formation (Fig. 15) also indicates a gradual process under constant control. This contrasts with the all-or-none decisions often assumed for these and for other patterns whose specification occurs at earlier stages, less amenable to observation and experimentation. Though the time scale may be different, however, all-or-none events, that do not allow for feedback relations and correction of mistakes, considered below, may be the exception rather than the rule.

2. It is often stated that the major question is how a uniform space becomes patterned (26). For plants, at least, this is not a common problem though it can be simulated in experimental conditions. Normally the formation of a space and its patterning occur together, as may be seen clearly for stomata development. Growth itself is coupled to other differentiation processes and is an expression of the patterning of the developing tissue.

3. The specification of pattern by morphogen distribution and its expression by differentiation are not necessarily separate processes. In vessel formation, the earliest stages of differentiation appear to be concerned with signal transport, and it is this differentiation that canalizes the signals and is a necessary element in their patterning. For stomata the situation is less clear, but the unequal divisions might compartmentalize essential morphogens that control the continuation of stomata formation. This means that the response to controlling signals is required for their continued influence, so pattern becomes dependent on feedback relations and not on unidirectional inductive processes. Even in the induction of vessels and other vascular tissues by leaves there is a strong element of feedback, for the leaves do not develop unless their signals are channeled away and "in return" they are supplied by the substances transported by vascular tissues. It follows from these feedback relations that morphogen patterning may depend on the differentiation it controls. This "differentiation dependent" pattern formation (19) means that differentiation processes of signal formation, transport, and change are major factors in patterning. Expressions of pattern after it is fully specified need not be excluded, of course, and they are likely even in vessel formation, since its last stages include cell death and the disappearance of directional transport. In animals where there is strong evidence for a "prepattern" (29), the controls may be different. This, however, could be a relative difference; the essential, unknown processes of prepattern formation would still be relatively gradual and would involve feedback relations between unknown cell differentiation and the distribution of signals.

4. In the cellular patterns considered here, the specification of differentiation need not be detailed and include precise differences between neighboring cells, as it does in hypotheses of positional information (29). These precise differences arise, instead, from the "differentiation dependent" localization of the signals. An important part of specification, however, includes the orientation of differentiation processes. This is true not only of vessels, where it is to be expected, but also of stomata spacing. Orientation is an elementary expression of pattern, and is likely to be a common early component of its formation. The specification of orientation requires not only a molecule that can be recognized by the cells, but also a gradient or a flow of such molecules. Oriented responses may be more specific, when differ-

ent signals are compared, than those of gene expression and may be useful for the study of the controls of pattern.

5. An important role has been found, as expected, for temporal controls, or the effect of the developmental history of the cells. In the cases considered here, however, they are less distinct from spatial controls than might have been expected and, furthermore, they may depend on positive feedback relations that are not directly related to controls of gene expression. Thus in vessel formation, the change in the transport capacity of the cells and its feedback relation to flow stabilize differentiation and promote its continuation by processes that could occur at the level of the cell membranes. In stomata formation, the situation is less clear, but unequal divisions may be both a result and a cause of the localization of morphogens. This localization would then be a central component of temporal controls. The common implied assumption that determinate aspects of differentiation are directly related to the controls of gene expression at the level of nucleic acids or proteins might not be always correct (8).

6. Positive feedback can create pattern, but it does not in itself determine exactly where specialized differentiation starts. For the patterns considered above, the answer appears to be that exact early locations depend on chance: precisely which cells transport a greater amount of signal may depend on fluctuations that are reinforced by the developmental controls, and the statistical measurements of somata distribution show that unequal divisions of a given cell are not correlated with the events in the rest of the tissue. The role of chance events is supported by the appearance of the mature patterns, which follow definite rules without being precise in any given detail. This dependence on chance events that are constrained by later feedback controls has the advantage that it is very economical in the specification of pattern. It is in accordance with the general appearance of biological patterns, not precise and yet remarkably free of mistakes of functional significance.

These generalizations point to major differences between the controls suggested here and most other hypotheses about controls of pattern formation. A similar suggestion, however, is that of self-electrophoresis determining polarity (8), where a gradual, positive feedback is the basis of oriented development. The obvious suggestion emerging from the work discussed above is that what applies to vessels and stomata could be useful, at least for working hypotheses, to patterns in other organisms and perhaps to development in general. Though in animals cell movement offers possibilities for additional controls, it might be specifically proposed that blood veins and nerves are oriented by the flow, rather than the gradient, of the appropriate growth factors, and that many spacing patterns are formed not by inhibitory interaction but rather by the development of complex structures that can be in direct contact with one another.

General References

BÜNNING, E.: Die Entstehung von Musten in der Entwicklung von Pflanzen. *Hand. Pflanzenphysiol.* **15** (1): 383–408 (1965).

SACHS, T.: The development of spacing patterns in the leaf epidermis. In: The clonal basis of *development*. (eds.), Subtelny, S., Sussex, I.M. 36 Symp. Soc. Devel. Biol., pp. 161–83. New York, Academic Press (1978).

SACHS, T.: The control of patterned differentiation of vascular tissues. *Adv. Bot. Res.* Vol. 9 (ed.), Woolhouse, H. W. pp. 151–262, London: Academic Press (1981).

SINNOTT, E. W.: *The problem of organic form*. New Haven: Yale University Press (1963).

References

1. Bünning, E.: Die Entstehung von Mustern in der Entwicklung von Pflanzen, *Hand. Pflanzenphysiol.* (ed.) Ruhland, W. Vol. XV (1) pp. 383–408, Berlin: Springer (1965).

2. Clark, P. J., Evans, F. C.: Distance to nearest neighbor as a measure of spatial relationships in population, *Ecology* 35: 445–53 (1954).

3. Esau, K.: *Plant anatomy*, 2nd ed., New York: Wiley (1965).

4. Fahn, A.: *Plant anatomy*, 3rd ed., Oxford: Pergamon (1982).

5. Goldsmith, M. H. M.: The polar transport of auxin, *Ann. Rev. Plant Physiol.* 28: 439–78 (1977).

6. Hammersley, D. R. H., McCully, M. E.: Differentiation of wound xylem in pea roots in the presence of colchicine, *Plant Sci. Lett.* 19: 151–6 (1980).

7. Jacobs, W. P.: The role of auxin in the differentiation of xylem around a wound, *Am. J. Bot.* 39: 301–9 (1952).

8. Jaffe, L. F.: The role of ionic currents in establishing developmental pattern, *Phil. Trans. R. Soc. Lond. B. Biol. Sci.* 295: 553–66 (1981).

9. Jost, L.: Über Beziehungen Zwischen der Blattentwicklung und der Gefässbildung in der Pflanze, *Bot. Zeitung (Berl.)* 51: 89–138 (1893).

10. Kershew, K. A.: *Quantitative and dynamic ecology*, London: Edward Arnold (1964).

11. Korn, R. W.: Arrangement of stomata on the leaves of *Pelargonium zonale* and *Sedum stahlii*. *Ann. Bot. (Lond.)* 36: 325–33 (1972).

12 Korn, R. W.: A neighboring-inhibition model stomata patterning, *Dev. Bio.* 88: 115–20 (1981).

13. Larson, P. R.: Development and organization of primary vascular system in *Populus deltoides* according to phylloxaxy, *Am. J. Bot.* 62: 1082–99 (1975).

14. Marx, A., Sachs, T.: The determination of stomata pattern and frequency in *Anagallis*, *Bot. Gaz.* 138: 385–92 (1977).

15. Mitchison, G. J.: A model for vein formation in higher plants, *Proc. R. Soc. Lond. B. Biol. Sci.* 207: 79–109 (1980).

16. Mitchison, G. J.: The dynamics of auxin transport, *Proc. R. Soc. Lond. B. Biol. Sci.* 209: 489–511 (1980).

17. Palevitz, B. A., Hepler, P. K.: The control of the place of division during stomata differentiation in *Allium*. I Spindle reorientation, *Chromosma (Berl.)* 46: 297–326 (1974).

18. Pielou, E. C.: *An introduction to mathematical ecology*, New York: Wiley-Interscience (1969).

19. Sachs, T.: Patterned differentiation in plants, *Differentiation* 11: 65–73 (1978).

20. Sachs, T.: The development of spacing patterns in the leaf epidermis. In: *The clonal basis of development*, 36th *Symp. Soc. Dev. Biol.* (eds.) Subtelny, S., Sussex, I. M., pp. 161–83. New York: Academic Press (1978).

21. Sachs, T.: The control of patterned differentiation of vascular tissues, *Adv. Bot. Res.* Vol. 9, (ed.) Woolhouse, H. W., pp. 151–262. New York: Academic Press, (1981).

22. Sachs, T.: Polarity changes and tissue organization in plants. In: *Cell Biology 1980/1981* (ed.) Schweiger, H. G., pp. 489–96, Berlin: Springer (1981).

23. Sachs, T., Benouaiche, P. A control of stomata maturation in *Aeonium*, *Isr. J. Bot.* **27**: 47–53 (1978).

24. Sinnott, E. W.: *Plant morphogenesis*. New York: McGraw Hill, (1960).

25. Thimann, K. W.: *Hormone action and the life of the whole plant*. Amherst: University of Massachusetts Press (1977).

26. Turing, A. M.: The chemical basis of morphogenesis, *Phil. Trans. R. Soc. Lond. B. Biol. Sci.* **237**: 37–72 (1952).

27. Vöchting, H.: *Über Organbildung in Pflanzenriech*. Bonn: Zweiter Teil, Emil Strass, (1892).

28. Waddington, C. H.: Ultrastructure aspects of cell differentiation, *Symp. Soc. Exp. Biol.* **17**, 85–97 (1963).

29. Wolpert, L.: Positional information and pattern formation, *Curr. Top. Dev. Biol.* **6**: 183–224 (1971).

Questions for Discussion with the Editors

1. *Cell and tissue response to auxin (and presumably other plant hormones) apparently depends on the development of "competence" to respond. Do you expect that pattern specification for competence arises completely independently of the regulation of auxin appearance?*

A generalization I tried to convey is that pattern formation can be profitably considered to result from gradual feedback relations rather than cause-and-effect controls of switch reactions. This would lead to the expectation that competence is elicited by the continued presence of signals (hormones) and might even be expressed by a relay production of the same signals. An example might be the suggestion explained above, that canalization of differentiation results from a feedback relation between auxin flow and the capacity (or competence) of the cells to transport auxin. For embryonic parts of shoot apices, where new tissues and organs are constantly formed, there is evidence that auxin synthesis and the capacity to respond to auxin by growth appear together.

2. *Do you believe that "reaction-diffusion" models of the type described in Meinhardt's chapter could be applied to plant systems?*

Models such as Meinhardt's are valuable because they demonstrate the patterning possibilities of positive feedback coupled with spatial correlations. The ideas are at least as valid for plants as for animals, and Meinhardt has discussed both leaf veins and stomata distribution. My feeling is, however, that present knowledge does not provide enough constraints for useful working hypotheses at a molecular level. The theory needed at present would define not models but questions with alternative answers that can all be considered. The most useful experiments would provide concrete answers to questions such as (a) to what distances are spatial signals effective, and for how long must they be present to elicit a differentiation response? (b) Do cells respond differently depending on precise concentrations and small gradients? (c) What are the earliest expressions of pattern? (d) Are patterns the result of rigid developmental programs or, as in plant veins, the stability of the final state that can be reached in various, even random, ways? Many other such basic questions can be readily suggested.

CHAPTER 16

Competence, Determination, and Induction in Plant Development

Carl N. McDaniel

THE DEVELOPMENT of a multicellular organism involves numerous processes that lead to a multiplicity of cell, tissue, and organ types. Specific tissues, organs, or structures result because the component cells become different in terms of their biochemistry, physiology, and morphology. The process by which cells become different is called cellular differentiation. The development of a population of cells into a particular pattern of functional organization is a result not only of the differentiation of the component cells but also of the spatial arrangement and the interactions among the component cells. An understanding of development will entail not only a comprehension of cellular differentiation but also an understanding of how cellular differentiation is integrated into high orders of cellular organization and interaction.

Developmental biologists have employed the operationally defined concepts of competence, determination, and induction to describe certain developmental processes and states relating to the development of tissues, organs, and structures. In general, induction occurs when a signal gives a unique developmental response from competent tissue. Competence is exhibited if a cell/tissue/organ is exposed to a signal and it responds in the expected manner. Determination is shown if a cell or group of cells exhibits the same developmental fate whether grown *in situ*, in isolation, or at a new place in or on the organism. In almost all situations, these developmental states have been examined in *populations* of cells and not in individual cells.

The concepts of competence, determination, and induction were initially used in animal embryology and subsequently have been extensively employed in describing various aspects of animal development. Although botantists have made limited use of these terms, they appear to be appropriate for describing some developmental phenomena in plants (10–12, 18, 20, 30). In this chapter, I will consider the general usage of determination, competence, and induction and the application of these terms to plant development.

To illustrate these three terms, I will briefly describe two classic animal systems, neural induction in amphibians (5) and imaginal disc development in *Drosophila* (23). During amphibian gastrulation, the invaginating mesoderm comes into contact with ectoderm. The mesoderm induces the overlying, competent ectoderm to develop into neural tissue. This induction process precedes the development of the ectoderm into neural tissue. Thus, once induced, the ectoderm will develop into neural tissue when it is removed and grown in isolation from other embryonic cells or when it is grafted to a new place on the embryo. After induction, the ectoderm is said to be determined for the development of neural tissue. This determination or commitment for the production of neural tissue is stable enough to withstand considerable environmental perturbations but should not be considered to be irreversible. The relationship among the three concepts can be clearly seen. Induction of competent ectoderm by some unknown signal produced by the mesoderm results in ectoderm which is determined for neural development. In amphibian neural induction, the expression of the determined state follows immediately after the induction event and does not appear to involve additional developmental signals.

In *Drosophila* imaginal disc development, the sequence of events appears to be different. It has been shown that about three hours after fertilization, groups of cells become determined as specific imaginal disc progenitors. Induction and competence are usually not considered or at least not discussed in the context of this determination process. The determination event leads to a stable state that is clonally propagated, i.e., all daughter cells inherit the state. The final expression of the determined state is revealed when ecdysone, the insect molting hormone, acts on competent imaginal discs to bring about the formation of an adult fly. It is of interest to note that imaginal discs become competent to respond to ecdysone sometime in the late second or early third instar larva, days after the initial determination event. In this example, the sequence of events appears to be determination, competence, induction, and expression of the determined state.

These two examples indicate that when different developmental systems are considered, the usage of developmental terms may not be synonymous. Determination is a case in point. In the imaginal disc example, determination refers to a commitment for a limited program of development but the expression of this commitment is latent depending upon additional developmental processes. In neural induction, determination refers to the expression of a limited program of development apparently without subsequent external inputs. It is not clear as to whether determination in neural induction merely elicits a specific developmental pattern like ecdysone elicits metamorphosis in previously determined imaginal discs or whether it involves programming for the developmental pattern as well as elicitation. Thus determination in imaginal discs appears to mean programming but not expression while in neural development it means expression and perhaps programming.

As one considers other developmental systems, it becomes apparent that it has not been possible to experimentally separate competence from determination in most instances. *Drosophila* imaginal disc development appears to be one instance where the two processes are cleanly separated. One might legitimately ask if the *Drosophila* example is an anomaly and if these two terms refer to the same developmental event. I do not consider the *Drosophila* example to be an exception and hypothesize that determination could be the programming of a cell or tissue for a specific program (i.e. pattern specification) of differentiation while competence is the linking of this program to the reception of a developmental signal. If this idea is correct, these events could be temporally separated and thus subject to experimental separation. The ease of experimental separation would depend upon how tightly the two processes are linked. Unfortunately we have very little knowledge of the various components that establish and permit the expression of specific developmental programs. Thus the separation of determination and competence is currently an impossible task in most animal systems. However, as discussed below, there have been some attempts to separate these two developmental events in plants.

Organ Determination

Leaves

Leaves are lateral appendages initiated for each species in a predictable pattern. They are formed as a bulge that is radially symmetrical and as development proceeds they become bilaterally symmetrical. Ferns have been ideal experimental systems because the development of a leaf covers a period of several years (27). Thus on a single apical meristem, there are many leaf primordia of various ages with each at a different developmental stage. *In vivo* surgical studies have shown that if the area which is to form the next leaf primodium is physically separated from the apical cell area, it will not form a leaf, but rather a shoot apical meristem. *In vitro* culture of young leaf primordia has produced similar results. However, as older and older primordia are cultured, the probability that they will develop as leaves goes from about 0.0 to 1.0.

Kuehnert (14) has studied this restricting of developmental potential using leaves from the fern *Osmunda cinnamonea*. His studies show that older leaf primordia act as inducers to bring about the determination of young primordia as leaves (Fig. 1). In fact, extracts from older leaves can substitute for the older leaves. By definition, these young primordia must be competent to respond to the inductive signal from older primordia but it is not clear as to when competence is acquired. The evidence also indicates that the developmental potential of a primordium becomes restricted in stages (9). It goes from being capable of producing an entire shoot to being able to produce a leaf and an axillary meristem, and finally it can only produce a leaf. As discussed earlier, the apical area also plays a role in leaf development and during *in vivo* leaf development inductive signals may come from the apex as well as from older leaf primordia.

Angiosperms initiate leaves more rapidly than ferns with the time period being measured usually in days (27). The earliest primordia that have been successfully cultured developed into leaves. Thus leaf determination has occurred very early in

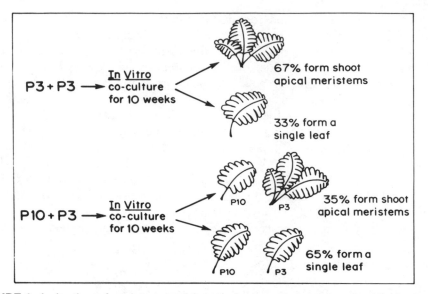

FIGURE 1. Induction of undetermined fern leaf primordia by determined leaf primordia. The fern apex contains many leaf primordia at various stages of development. These primordia are numbered by age with P1 being the youngest visible primordium. When cultured, the various primordia have different developmental fates. P3 primordia cultured alone develop as shoot meristems 75% of the time while P10 primordia always develop as leaves. When two P3 primordia are co-cultured, the primordia form shoot meristems 67% of the time and single leaves 33% of the time. When a P3 primordium is co-cultured with a P10 primordium, the incidence of leaf formation by the P3 primordium is increased to 65%. These results indicate that determined leaf primordia influence undetermined leaf primordia to develop as leaves and not as shoot meristems. Data from reference (14).

primordium development making it impossible to study the determination process in angiosperms with the techniques employed to study fern leaf development.

Apical Meristems

INITIATION OF APICAL MERISTEMS

The root and shoot apical meristems are organized during embryogeny. There is essentially no information on the developmental events that regulate or are responsible for the establishment of these structures. Studies on the production of apical meristems in culture have indicated that hormones may be involved in these processes. The classic work of Skoog's laboratory (26) on the *in vitro* initiation of tobacco apical meristems showed that different ratios of exogenous auxins and cytokinins lead to different types of meristem organization (Fig. 2). A high auxin to cytokinin ratio leads to root formation while a low auxin to cytokinin ratio leads to shoot formation. These external hormone concentrations are in some way interpreted by the cultured cells such that apical meristems are formed. Recently several reports have focused on the early developmental events associated with the or-

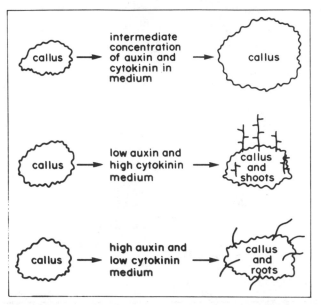

FIGURE 2. Hormonal control of apical meristem initiation in tobacco callus. The ratio of auxin to cytokinin establishes the type of response: roots develop in the presence of high auxin and low cytokinin, and shoots develop in the presence of low auxin and high cytokinin. Intermediate concentrations of auxin and cytokinin favor callus growth. Data from reference (26).

ganization of meristems in callus (a poorly organized mass of dividing cells) cultures of alfalfa, *Medicago Sativa* cv. "REGEN S," (29) and in leaf-callus cultures of *Convolvulus arvensis* (4).

Alfalfa callus requires a pretreatment with hormones for subsequent organogenesis on hormone free medium. A four-day pretreatment with high levels of kinetin and low levels of the synthetic auxin, 2,4-dichlorophenoxyacetic acid (2,4-D), leads to root formation when the callus is subsequently placed on hormone-free medium while a pretreatment of four days with low kinetin and high 2,4-D concentrations leads to shoot formation (Fig. 3). Apical meristem initiation will not occur on the hormone-containing media and small pieces of callus (less than 105 mm in diameter containing 1 to 12 cells) are not capable of responding to pretreatment by undergoing organogenesis (incompetent for shoot/root induction). Small, incompetent pieces of callus become competent for organogenesis if they are allowed to grow to a larger size.

Leaf explants of *Convolvulus* are capable of producing callus, roots, or shoots with a unique medium favoring each type of development. These media differ from the basal medium only in terms of the hormone composition. The process of shoot meristem production has been investigated in several clones or genotypes. The evidence presented indicates that the researchers may have temporally separated the acquisition of competence from the process of induction which occurs on shoot induction medium (SIM). Analogous to neural induction in amphibians, competent tissue become determined to form shoots by the induction process.

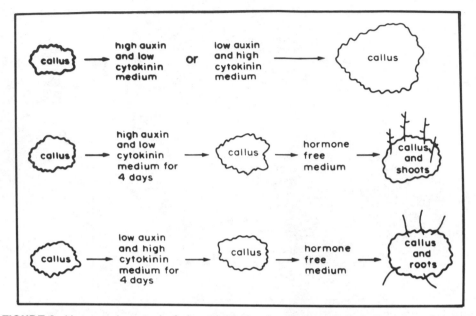

FIGURE 3. Hormonal control of shoot initiation in alfalfa callus. Callus placed on high-auxin and low-cytokinin medium for four days and then onto hormone-free medium will produce shoots. If the initial medium has high cytokinin and low auxin, then roots develop. Callus maintained on either hormone containing medium will not produce roots or shoots. Data from reference (29).

These observations can be illustrated by considering the response of one genotype (Fig. 4).

Genotype 30 must be maintained on SIM for 14 days before transfer to hormone-free basal medium and the subsequent production of shoots. A shorter amount of time of SIM leads to the production of few or no shoots. If cultured on callus-inducing medium for various periods of time and then transferred to SIM, the time required in SIM for induction to take place is shortened from 14 to about 11 days. These data can be interpreted to indicate that acquisition of competence requires about 3 days while induction requires about 11 days. Different genotypes have different competence acquisition and induction times; e.g., genotype 23 requires about 10 days for competence acquisition and 3 days for induction.

In *Convolvulus* the determination for shoot production appears to be quite stable and not readily redirected into root formation. That is, once the induction period is completed, shifting to root induction medium does not prevent shoot formation. However, roots will form from cells not determined for shoot meristem production. Unpublished histological data indicates that at the time of shoot meristem determination no meristems are present (Christianson, *personal communication*). Some meristemiods (areas of rapid cell division) and perhaps a tunica-corpus organization (cell organization found in shoot meristems) is observed in some cases but no leaf primordia can be detected. Thus determination appears to precede morphological expression.

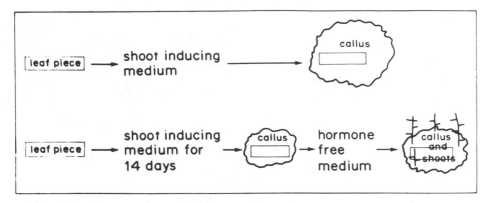

FIGURE 4. Hormonal control of shoot initiation in leaf explants of *Convolvulus arvensis*. Leaf pieces of genotype 30 are placed on shoot inducing medium for a period of 14 days and then placed on hormone-free medium or shoot-inducing medium. Those calli placed on hormone-free medium produce shoots within several weeks while those calli on shoot-inducing medium continue to produce callus. Data from reference (4).

Once organized, angiosperm apical meristems are very stable. There are no reports of a shoot apical meristem transforming itself into a root apical meristem or vice versa (10). On the other hand, it is common knowledge that the derivatives of each in many plants have the capacity to form the other type; i.e., shoot tissue can form roots and root tissue can form shoots. Apical meristems in lower plants may exhibit more plasticity than those of angiosperms as was shown by the work of Wochok and Sussex (31, 32) on *Selaginella willdenovii*, a member of the *Lycopsida* (Fig. 5). At the base of each leaf in this plant, there are two angle meristems which can develop either as a root or as a shoot meristem. Normally, the dorsal meristem forms a shoot and the ventral meristem forms a root. However, if the dorsal meristem is exposed to an appropriate concentration of auxin at an early point in its development, it will form a root. Failure to be exposed to auxin will lead to development as a shoot. It would appear that shoot/root apical meristem formation involves at least two distinct steps. First, organization as an apical meristem and then establishment of meristem type. However, the establishment of meristem type in *Selaginella* is not irreversible. If a root meristem is cultured in a medium lacking auxin, then it may reorganize as a shoot meristem. Thus auxin not only causes an angle meristem to be determined as a root meristem but its presence is required to stabilize and maintain root meristem expression.

FLOWER FORMATION

In many angiosperms, the shoot apical meristem has the capacity to express two patterns of organization. One, the production of leaves, can be considered to be a pattern of unlimited growth while the other, production of a flower, can be considered to be a pattern of limited growth because it eventually terminates the growth of the meristem. The transition from vegetative to reproductive growth is of tremendous practical importance and certain aspects have been extensively stud-

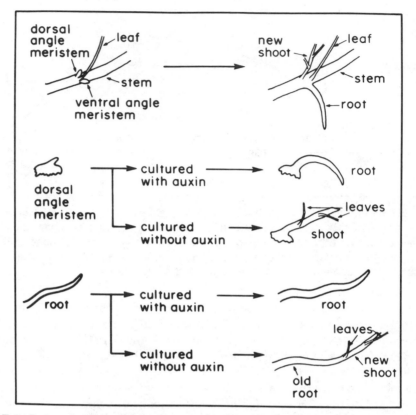

FIGURE 5. Determination of apical meristem type in *Selaginella willdenovii*. The dorsal angle meristem normally develops into a shoot. This results because it is not exposed to auxin in its early development, but if it develops in the presence of auxin, it will form a root apical meristem. Organization as a root apical meristem is not irreversible since a root apical meristem cultured in auxin-free medium can reorganize as a shoot apical meristem. Data from references (31, 32).

ied. Floral development has primarily been analyzed with photoperiodic plants using physiological and morphological approaches. Photoperiodic induction of flowering has been extensively studied and is mediated via the pigment phytochrome. An inductive photoperiod causes the leaves to transmit an unknown signal(s) (floral stimulus) to the shoot apical meristem(s) which responds by forming a flower. The floral stimulus may be hormonal in nature but whether it is a single substance, several substances, or some other arrangement is not known despite a tremendous research effort. There are a host of photoperiodic types; short-day plants, long-day plants, long-short-day plants, long-day plants which first require a chilling period, day-neutral plants, etc. The flowering literature has been reviewed (1, 2) and I have proposed a model for shoot development that considers the regulation of flowering in some detail (18). I will not consider the broader aspects of the flowering literature but rather will consider several investigations of floral development where competence, determination, and induction appear to be important.

If a plant is to flower, the floral stimulus must be produced and transported to the meristem, and the meristem must respond to the stimulus by changing the pattern and nature of the lateral structures produced (1, 2). In photoperiodic plants, this means that the leaves are competent to respond to the photoperiod and are thereby induced to change their state of development such that they now produce the floral stimulus. Following the logic presented in the introduction, competent leaves are induced by a light signal to produce the floral stimulus. This new state of differentation, production of the floral stimulus, can be quite stable in the absence of further photoperiodic induction and many persist until the leaf dies. In addition to the leaves being competent, the meristem must also be competent to respond to the floral stimulus. If a photoperiodic plant does not flower even though it is placed in an inductive photoperiod, incompetent leaves or incompetent meristems may be responsible.

Bryophyllum daigremontianum is an example of the first situation, incompetent leaves. This plant is a long-short-day plant that has a long juvenile phase. Zeevaart (33) grafted juvenile plants onto flowering plants and the juvenile plants flowered as rapidly as mature plants. Thus the juvenile apical meristems are capable of receiving and responding to the floral stimulus but the juvenile leaves appear to be unresponsive to the inductive photoperiod.

The second situation, incompetent meristem, is illustrated by some plants which require vernalization ["The acquisition or acceleration of the ability to flower by a chilling treatment." (3)]. It has been proposed that there are at least two classes of plants in terms of how chilling changes the developmental state of the meristem (Fig. 6, also see 18). In one class, vernalization changes the type of leaves produced by the meristem from photoperiodically insensitive to photoperiodically sensitive. In this situation, the meristem appears to be always competent to respond to the floral stimulus. In the second class, the meristem becomes competent to respond to the floral stimulus as a result of the cold treatment.

It is the shoot tip which is vernalized, and the vernalized state can be stable and clonally propagated. This condition appears to be similar to determination as it is used to describe Drosophila imaginal disc development. A meristem can remain vernalized over a long period of time and this developmental state is not revealed until the appropriate development signal (the photoperiod acting on the leaves in one case and the floral stimulus acting on the apical meristems in the other) is received.

The induction of a competent meristem leads to a meristem which is determined for flower formation. In some cases, this determination has been shown to be stable and the meristem can be cultured in vitro to produce a flower much the same as a determined leaf primordium can be cultured to produce a leaf (27). Thus, analogous to the amphibian example considered earlier, induction leads to determination and the expression of the determined state.

Studies of two day-neutral species, Nicotiana tabacum cv. Wisconsin 38 and Helianthus annuus, have provided data that casts some light on the developmental events that may occur prior to floral induction. In sunflower, Helianthus annuus, the terminal shoot meristem normally produces about 16 leaf pairs followed by a shift to the development of reproductive structures (8). Grafting of shoot tips from plants of various physiological ages (plants which had produced different numbers of leaf pairs) onto seedling stocks does not increase the total number of nodes produced by the grafted meristem. The total number of leaf pairs (produced before grafting + produced after grafting) is always about 16. However, grafting seedling terminal meristems onto the tops of stocks of increasing physiological age causes a reduction in the number of leaf pairs produced by the grafted shoot tip.

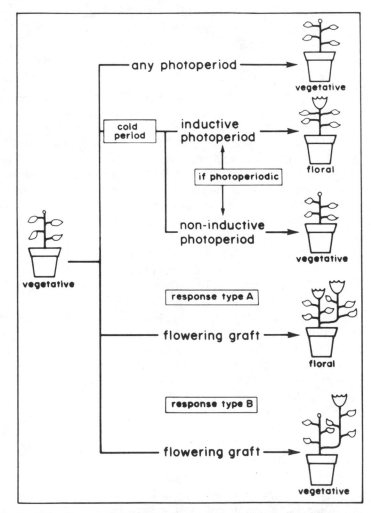

FIGURE 6. Vernalization and two hypothetical response types. Under normal circumstances, the plant requires a cold treatment before it will flower. After the cold treatment, it may also have a photoperiodic requirement. For response type A, the requirement for a cold treatment can be eliminated by grafting a flowering shoot onto the unvernalized plant. The "graft transmissible" characteristic indicates that the meristem can respond to the floral stimulus but the leaves produced by an unvernalized meristem are insensitive to the photoperiod. For response type E, vernalization is not "graft transmissible" indicating that the meristem cannot respond to the floral stimulus. Vernalization acts on the shoot tip or meristem in both instances: in type A vernalization enables the meristem to produce leaves which are subsequently competent to respond to the inductive photoperiod while in type E the meristem itself becomes competent to respond to the floral stimulus.

For example, a seedling grafted onto a stock which has produced 6 leaf pairs will produce about 11 leaf pairs while a seedling grafted onto a stock which has produced 9 leaf pairs will produce about 8 pairs. Rooting shoot tips of plants of various physiological ages does not increase or decrease the total number of leaf pairs produced by the meristem. The shoot tips of plants which had produced 3 leaf pairs at the time of rooting produced an additional 13 leaf pairs while ones that had produced 9 leaf pairs before rooting produced 7 after rooting. These data indicate that the amount of prereproductive growth can be reduced but not increased (Table 1). Thus the sunflower terminal meristem is programmed for a limited growth pattern in the young seedling; i.e., it is determined for floral development. The expression of this determined state is usually preceded by a period of growth, but by placing the meristem in a "physiologically older" environment, this growth period can be reduced and in the extreme case almost eliminated (8). It is possible that competence to respond to the floral stimulus is acquired by the meristem during this growth period or after being grafted to the top of another more mature plant. If this interpretation is correct, the the sequence of developmental events would be as follows: determination for reproductive development, competence for floral induction, floral induction, and expression of the floral pattern of development.

In day-neutral *Nicotiana tabacum*, the terminal shoot meristem develops into a single flower after giving rise to the cells which produce the rest of the shoot

TABLE 1. Floral determination in the terminal meristem of *Helianthus annuus*.[1]

TREATMENT	NUMBER OF LEAF PAIRS PRODUCED		
	Before Rooting	After Rooting	Total
Rooting of shoot tips	0	17	17
after production	3	13	16
of different numbers	9	7	16
of pairs.			
leaf	Before Grafting	After Grafting	Total
Grafting of shoot tips	0	16	16
from plants of various	4	13	17
ages onto seedling stocks.	8	9	17
	On Stock	Produced by Scion	On Stock & Scion
Grafting of seedling shoot	3	15	18
tips onto plants with	6	11	17
various numbers of leaf	9	8	17
pairs.			

[1]This sunflower species produces about 16 leaf pairs before forming an inflorescence. This total number of leaf pairs cannot be increased by rooting or grafting shoot tips from plants which have produced various numbers of leaf pairs. Thus the meristem is determined for floral development prior to or soon after germination. However, grafting a seedling shoot tip onto the top of an older stock can reduce the number of leaves produced by the seedling meristem. The fact that the amount of vegetative growth can be reduced indicates that the expression of the determined state can be accelerated by altering the environment of the shoot tip. Data from (8).

system. The conversion from vegetative to floral growth pattern is precisely regulated and endogenously controlled. This is evidenced by the fact that under a given set of environmental conditions, the number of nodes produced by the terminal meristem is uniform (19). Normally a terminal meristem will develop into a flower after producing about 35 nodes but by continually rooting the shoot, the meristem will remain vegetative indefinitely (17). Experiments with axillary buds (16) and with *de novo* bud formation from internodal tissues (13, 28) indicate that the cells of the meristem undergo stable developmental changes which lead to flower formation.

In *Nicotiana tabacum*, axillary buds below the inflorescence are arrested after the production of 7 to 9 leaf primordia and at the time of anthesis of the first flower, they appear to be morphologically vegetative as indicated in dissection and serial sections. Decapitation, rooting, and grafting experiments were used to measure a bud's developmental potential (number of nodes produced by a meristem before production of a flower). These experiments clearly demonstrate that there are two populations of axillary buds in terms of their developmental potential (16) (Fig. 6). Buds in population A develop in accordance with their new environment thereby showing no "memory" of their original position. For example, buds from the fourth node below the inflorescence produced about 21 nodes *in situ* but about 38 nodes when rooted and about 34 nodes when grafted to the base of the main axis. Buds in population B develop according to their original position. For example, buds from the second node below the inflorescence produced about 19 nodes *in situ*, about 18 nodes when rooted and about 22 nodes when grafted to the base of the main axis.

On a single plant, the populations are not intermixed along the main axis. Rather, B buds are always directly below the inflorescence and separated from A buds by a single internode. For example, the following numbers of nodes were produced by the five buds directly below the inflorescence from a single plant when their buds were rooted and grown to maturity; first bud—18 nodes, second bud—17 nodes, third bud—37 nodes, fourth bud—38 nodes, fifth bud—38 nodes. B buds exhibit the same growth potential when grown *in situ*, when grown independently, and when grafted to and grown at a new place on the plant. Thus they fit the classical definition of developmentally determined tissue. Since their growth is restricted to producing a limited number of leaves and then a flower, it is reasonable to say they have undergone an early determination event in floral development. That is, like the cells in the *Drosophila* embryo which have become determined several hours after fertilization, these determined meristem cells must proliferate and undergo additional changes (perhaps the acquisition of competence) before the determined state is expressed.

Studies of *de novo* meristem organization in *N. tabacum* demonstrate that determination for floral development occurs not only in an organized meristem but also at the tissue or cell level (13, 28) (Fig. 7). These studies and others have shown that cells from cultured internode segments or tissue explants of day-neutral *N. tabacum* will form *de novo* floral meristems if the explants are from or near the inflorescence area. Similarly treated explants taken from more basal areas of the plant form, in almost all cases, only vegetative buds. The data on axillary bud development and *de novo* meristem organization indicate that the shoot meristem is, at

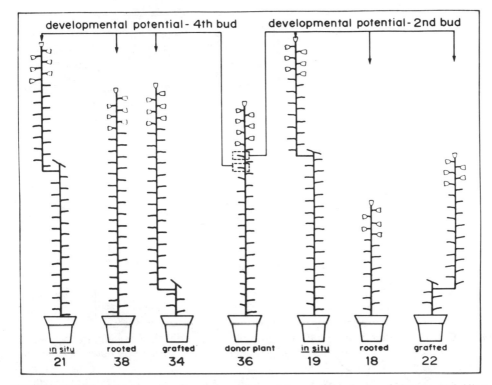

FIGURE 7. Determination for floral development in auxillary buds of day-neutral *Nicotiana tabacum* cv. Wisconsin 38. The developmental potential of an auxillary bud can be established by forcing it to develop and then counting the number of nodes produced below the terminal flower. The numbers underneath the pots indicate the number of nodes on the donor plant or on the stem produced by the auxillary bud after the indicated treatment. The developmental potential of the second vegetative bud below the inflorescence is the same *in situ*, when rooted, and when grafted to the base of the plant. The fourth vegetative bud exhibits a different growth pattern *in situ* and when rooted or grafted. It develops in response to its new environment when grafted or rooted. These results indicate that the second bud is determined for floral development while the fourth bud is not. A line indicates a leaf and a goblet indicates a terminal flower or a floral branch. Data from reference (16).

first, neither determined nor competent for floral development and that these developmental states are subsequently acquired.

These several examples indicate that plant tissues, like animal tissues, become competent for induction and a result of induction undergo a specific pattern of development. Fern leaf development and floral induction appear to be analogous to the example of neural induction in amphibians where induction of competent tissue leads to the expression of a specific development fate. Vernalization of shoot apical meristems appears to be similar to *Drosophila* imaginal disc development where determination results in stable programming for a future developmental fate but the expression of this state requires an additional developmental signal. Studies of floral development in day-neutral plants indicate that important devel-

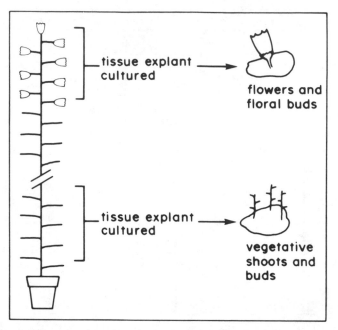

FIGURE 8. Floral determined and undetermined internodal tissues/cells in day-neutral *Nicotiana tabacum*. Explants from the inflorescence will produce *de novo* floral meristems while explants from the base of the plant will produce *de novo* vegetative meristems. Thus the cells from the two ends of the shoot system have distinctly different developmental potentials: those from the inflorescence being determined for floral development while those from the base are not. Data from references (13, 28).

opmental events may occur in the shoot meristem prior to floral induction and these events may be analogous to those required for vernalization.

Cell Determination

It seems logical that competence, determination, and induction be expressed not only at the tissue and organ level but also at the cell level. One approach is to study the development of cell types. Sachs (24) has studied both vascular strand and stomata development. His analysis of these two types of cell differentiation is that they appear to be autocatalytic. He calls their development "differentiation dependent" meaning that the "processes of the differentiation are responsible for the orderly distribution of the signals which in turn control continued differentiation" (24). Thus there may be a continuum of signals rather than a single induction event followed by determination and subsequent expression.

Does this indicate that the concept of determination and the events which precede determination are phenomena associated with populations of cells but not single cells? This is an interesting possibility and would indicate that plant cells may follow two sets of instructions, those related to tissue/organ organization and those related to cell differentiation. It is of interest to note that in alfalfa, competence for shoot meristem organization is size dependent as discussed earlier. The general problem of pattern formation and cell differentiation in plants is considered in depth in Chapter 15.

Control and Stabilization of Developmental States

Positive Feedback Loop

In order for one to study competence and determination, these states or processes must be stable enough to withstand some manipulation. Stability has certainly been a problem in the study of plant development. Some plant cells are totipotent and may express this potential upon manipulation. Many plant cells will divide to form callus when manipulated or the cells may fail to grow. This lack of stability may explain, in part, why investigators have had difficulty examining the early events of plant development in depth.

How might stability of a developmental state be achieved? A model system has been provided by the work of Meins and colleagues on cytokinin habituation in tobacco (20). The requirement of cultured tobacco cells for exogenous cytokinin in the culture medium can be lost and the cells are then said to be cytokinin habituated. The loss of dependency upon cytokinin can be controlled by manipulating the culture environment. Habituation is clonally propagated but reversible. Meins proposes that cytokinin habituation results from a positive feedback relationship that is maintained by cytokinins either inducing their own production or blocking their own degradation. It is possible that cells and tissues may employ similar positive feedback relationships to stabilize states of development like determination and competence.

The relevance of cytokinin habituation in the development of the tobacco plant is not known. However, different tissues exhibit tissue specific phenotypes in relation to habituation; e.g., pith tissues are not immediately habituated when put in culture while stem-cortex tissues are (21) and pith tissues from different parts of the plant exhibit different capacities for habituation (22). Small explants (less than 20–30 mg) do not habituate. This last observation is quite interesting since single habituated cells can be cloned and maintain their habituated state. Is it possible that habituation initially occurs in groups of cells but once initiated it can be stabily expressed by individual cells?

Role of Hormones

The work of Meins and colleagues has indicated one means of stabilizing a developmental state—a positive feedback loop. It is of interest to note that his model involved cytokinins, one class of plant hormones. Initiation of meristem organization, *in vitro* at least, certainly appears to involve the classic plant hormones, auxins and cytokinins. Auxins, cytokinins, ethylene and gibberellins have been reported to be involved in the flowering process, and each hormone has been shown in some species to stimulate and in other species to inhibit the flowering process (1, 2).

Sawhney and Greyson (25) in their studies of stamen development in a stamenless mutant of tomato have demonstrated that early in stamen development gibberellic acid (GA) treatment would permit normal stamen development. As the abnormal development proceeds further, GA treatment becomes less and less effective. They also reported that treatment with indoleacetic acid leads to carpel formation. Thus developmental expression is hormone-type dependent with the responses being developmental-stage specific. In another example, the development *in vitro* of spines or leaves by an axillary bud of the cactus, *Opuntia polyacantha* is dependent upon the hormonal composition of the medium (15). The cytokinin, benzylaminopurine, promotes axis elongation and leaf production while gibberellic acid leads to spine production without axis elongation.

The few examples and the studies mentioned earlier clearly show that the classical plant hormones are intimately involved in many aspects of plant development and are certainly involved in induction processes. Their role in such phenomena as competence and determination is not clear. A serious problem in understanding the involvement of plant hormones in a developmental process is that each hormone is apparently involved in numerous processes. Additionally the very limited knowledge on the mechanism of action of the various hormones makes this situation even more challenging.

What Is Being Controlled?

Control of the axis of cell elongation appears to be critical in the initiation of apical meristems or of leaf primordia as well as in the conversion from vegetation to floral growth. Thus, expression of one of these determined states involves a change in the orientation of cell elongation which in turn may be correlated with the plane of cell division. Green and co-workers have considered reorientation of cell elongation axes (polarity shifts) as they are related to the initiation of lateral organs (7). The general picture that emerges is that organ initiation involves an abrupt shift in polarity followed by a smoothing of this shift into a radially symmetrical organization and the concurrent growth along the new axis to form the bulge characteristic of organogenesis. It is clear that these sorts of investigations provide information relevant to the problem of organ determination in plants since the control of the timing and the patterns of these polarity shifts establishes the basis for morphogenic change. Thus, when a meristem shifts from producing leaves to producing sepals, petals, etc. the timing and pattern of primordia initiation has been altered. Obviously, it is at this level that control has been exerted.

In order to understand processes like floral determination one must know how polarity shifts are controlled. We also recognize that as development proceeds for a petal or a leaf, morphological differences appear as a result of unique patterns of cell division and elongation. In addition to morphogenic patterns within organs, the component cells exhibit different biochemical properties, e.g., pigments and unique proteins. Thus patterns of cell differentiation are integrated into patterns of morphogenesis.

A major challenge in developmental biology is to link morphogenic expression to genetic information. It is certainly possible that we could have a complete understanding of the organization of DNA and of the expression of structural and regulatory genes but fail in our ability to understand competence, induction, and determination and their relationship to gene expression and morphogenesis. Green (6) has considered this issue and has suggested that the recommended sequence of analysis is from the phenotypic end to the DNA. For example by starting with an analysis of cell elongation we should be able to walk back in developmental time through a sequence of events to regulation of the tubulin gene which is the starting point for the orientation of cellulose fibers in the cell wall. The challenge we face is to bridge this gap between the regulation of gene expression and the expression of morphogenic patterns.

Conclusion

Predictable patterns result because the development of organs, tissues, and cells in multicellular plants is position dependent. However, we do not know how positional information is communicated, recognized, or remembered. We certainly cannot describe with any degree of completeness the development of any cell, tissue, or organ nor do we understand the regulation of the developmental processes involved. The operationally defined terms competence, determination, and induction have been successfully employed to describe aspects of animal development which are not well understood. Botantists are beginning to use these terms more extensively. The material presented here indicate that these terms are certainly applicable and useful for describing plant development.

Acknowledgments

I thank Dr. M. Hanna, Dr. H. Roy, and Ms. S. Singer for their comments on the manuscript and NSF for their support (PCM 8204491).

General References

STEEVES, T.A., SUSSEX, I.M.: *Patterns in plant development.* Englewood Cliffs: Prentice Hall (1972).

WAREING, P.F.: Determination in plant development. *Bot. Mag. (Toyko) Special Issue No.* 1: 3–17 (1978).

McDANIEL, C.N.: Shoot meristem development. In: *Positional Controls in Plant Development* (eds.), Barlow, P., Carr, D.J. in press, Cambridge: Cambridge University Press (1983).

References

1. Bernier, G. Kinet, J., Sachs, R.M.: *The physiology of flowering: I. the initiation of flowers*. Boca Raton: CRC Press, (1981).

2. Bernier, G., Kinet, J., Sachs, R.M.: *The physiology of flowering: II. transition to reproductive growth*. Boca Raton: CRC Press (1981).

3. Chouard, P.: Vernalization and its relations to dormancy. In: *Annual review of plant physiology*, vol. 11 (eds.), Machlis, L., Briggs, W.R. pp. 191–238. Palo Alto: Annual Reviews Inc. (1960).

4. Christianson, M.L., Warnick, D.A.: Competence and determination in the process of *in vitro* shoot organogenesis. *Dev. Biol., in press* (1982).

5. Grant, P.: *Biology of developing systems*. New York: Holt, Rinehart and Winston, (1978).

6. Green, P.B.: Biophysics of axial growth. In: *Positional controls in plant development* (eds.), Barlow, P., Carr, D.J. in press, Cambridge: Cambridge University Press (1983).

7. Green, P.B., Poethig, R.S.: Biophysics of the extension and initiation of plant organs. In: *Developmental order: its origin and regulation* (eds.), Subtelny, S., Green, P.B. pp. 485–509, New York: Alan R. Liss (1982).

8. Habermann, H.M., Sekulow, D.B.: Development and aging in *Helianthus annuus* L. Effects of the biological *milieu* of the apical meristem on patterns of development. *Growth* 36: 339–349 (1972).

9. Haight, T.H., Kuehnert, C.C.: Developmental potentialities of leaf primordia of *Osmunda cinnamomea* VI. The expression of P1. *Can. J. Bot.* 49: 1941–1945 (1971).

10. Halperin, W.: Organogenesis at the shoot apex. In: *Annual review of plant physiology*, vol 29 (eds.), Briggs, W.R., Green, P.B., Jones, R.L. pp. 239–262, Palo Alto: Annual Reviews Inc. (1978).

11. Henshaw, G.G., O'Hara, J.F., Webb, K.J.: Morphogenetic studies in plant tissue cultures. In: *Differentiation in Vitro* (eds.), Yoeman, M.M., Truman, D.E.S. pp. 231–251, Cambridge: Cambridge University Press, (1982).

12. Hicks, G.S.: Patterns of organ development in plant tissue culture and the problem of organ determination. In: *The botanical review*, vol. 46 (ed.), Cronquist, A. pp. 1–23, Bronx: N.Y. Botanical Garden (1980).

13. Hillson, T.D., LaMotte, C.E.: *In vitro* formation and development of floral buds on tobacco stem explants. *Plants Physiol.* 60: 881–884 (1977).

14. Kuehnert, C.C.: Developmental potentials of leaf primordia of *Osmunda cinnamomea*. The influence of determined leaf primordia on undetermined leaf primordia. *Can. J. Bot.* 45: 2109–2113 (1967).

15. Mauseth, J.D.: Cytokinin- and gibbereleic acid-induced effects on the determination and morphogenesis of leaf primordia in *Opuntia polyacantha* (Cactaceae). *Am. J. Bot.* 64: 337–346 (1977).

16. McDaniel, C.N.: Determination for growth pattern in axillary buds of *Nicotiana tabacum* L. *Dev. Biol.* 66: 250–255 (1978).

17. McDaniel, C.N.: Influence of leaves and roots on meristem development in *Nicotiana tabacum* L. cv. Wisconsin 38. *Planta* 148: 462–467 (1980).

18. McDaniel, C.N.: Shoot meristem development. In: *Positional controls in plant development* (eds.), Barlow, P., Carr, D.J. in press, Cambridge: Cambridge University Press, (1983).

19. McDaniel, C.N., Hsu, F.C.: Position-dependent development of tobacco meristems *Nature (London)* **259**: 564–565 (1976).

20. Meins, Jr. F., Binns, A.N.: Cell determination in plant development. *Bio. Sci.* **29**: 221–225 (1979).

21. Meins, Jr. F., Lutz, J.: Tissue-specific variation in the cytokinin habituation of cultured tobacco cells. *Differentiation* **15**: 1–6 (1979).

22. Meins, Jr. F., Lutz, J., Binns, A.N.: Variation in the competence of tobacco pith cells for cytokinin-habituation in culture. *Differentiation* **16**: 71–75 (1980).

23. Nothiger, R.: The larval development of imaginal disks. In: *Results and problems in cell differentiation*, vol. 5 (eds.), Ursprung, H., Nothiger, R. pp. 1–34, Berlin: Springer-Verlag (1972).

24. Sachs, T.: Patterned differentiation in plants. *Differentiation* **11**: 65–73 (1978).

25. Sawhney, V.K., Greyson, R.I.: Interpretations of determination and canalization of stamen development in a tomato mutant. *Can. J. Bot.* **57**: 2471–2477 (1979).

26. Skoog, F., Miller, C.O.: Chemical regulation of growth and organ formation in plant tissue cultured *in vitro*. *Sym. Soc. Exp. Biol.* **11**: 118–131 (1957).

27. Steeves, T.A., Sussex, I.M.: *Patterns in plant development*. Englewood Cliffs: Prentice Hall (1972).

28. Tran Thanh Van, M.: Direct flower neoformation from superficial tissue of small explants of *Nicotiana tabacum* L. *Planta* **115**: 87–92 (1973).

29. Walker, K.A., Wendeln, M.L., Jaworski, E.G.: Organogenesis in callus tissue of *Medicago sativa*. The temporal separation of induction processes from differentiation processes. *Plant Sci. Lett.* **16**: 23–30 (1979).

30. Wareing, P.F.: Determination in plant development. *Bot. Mag. (Tokyo) Special Issue No.* **1**: 3–17 (1978).

31. Wochok, Z.S., Sussex, I.M.: Morphogenesis in *Selaginella*. III. Meristem determination and cell differentiation. *Dev. Biol.* **47**: 376–383 (1975).

32. Wochok, Z.S., Sussex, I.M.: Redetermination of cultured root tips to leafy shoots in *Selaginella willdenovii*. *Plant Sci. Lett.* **6**: 185–192 (1976).

33. Zeevaart, J.A.D.: The juvenile phase in *Bryophyllum daigremontianum*. *Planta* **58**: 543–548 (1962).

Questions for Discussion with the Editors

1. *Could you be more specific regarding where the concept of "positional information" fits into your scheme of competence, determination, and cell differentiation?*

The notion of "positional information" implies that different locations have unique characteristics which are "sensed" by the cells and this "sensing" somehow makes cells in one location different from cells in another location. These "differences" are eventually reflected in the patterns of development expressed by various cells or population of cells. Competence and determination are developmental states which make various cells or populations of cells "different." We do not know the mechanisms or processes involved in making cells competent and/or determined but we do know that "position" within an organism or relative to groups of cells is important. Thus the concepts of positional information, competence and determination have considerable common ground in that they relate to what cells will develop into and are intimately associated with relative positions. Perhaps determination and/or competence are the results of "sensing" and responding to a "positional" input.

2. *Please speculate on what "developmental competence" may mean at the molecular level. For example, what molecular changes might accompany this development of competence of* Convolvulus *leaf explants to respond to school induction stimuli?*

Developmental competence is the capacity to respond to a physical or chemical signal with a predictable pattern of development. In molecular terms this means that the cell or population of cells has receptor molecules which are "changed" by the signal. This "change" is in some way chemically or physically linked to the predicted phenotypic expression. For example, hormone H acts on competent cells causing them to synthesize protein P. In molecular terms hormone H binds to receptor molecule R producing an H-R complex which can now bind to a unique chromatin site S. Binding of H-R to S turns on gene G which codes for protein P. Our understanding of the regulation of a morphogenic process like the formation of a *Convoluvulus* shoot meristem from a callus is very primitive. However, one could speculate that competence either involves the production of receptors for the inducing hormones or the linking of the reception process to the process(es) which controls the phenotypic expression of shoot meristem organization.

CHAPTER 17

Patterns and Problems in Angiosperm Leaf Morphogenesis

R. Scott Poethig

LEAF MORPHOGENESIS IS an attractive system for an analysis of pattern formation in plants because leaves are morphologically simple, determinate organs, are readily accessible to observation and manipulation, and display tremendous variation in shape. Much of this variation in shape is due to environmental factors and a great deal of effort has been expended trying to indentify these factors and characterize their effects. Numerous mutations affecting leaf shape also exist, and have served as valuable tools for developmental analysis. This analysis is facilitated by the availability of genetic markers useful for cell lineage analysis and by the ease with which leaves can be manipulated and grown to maturity in vitro. For these and other reasons, leaf morphogenesis is one of the most intensively investigated phenomena in plant development.

But, in spite of its many advantages as an experimental system, the mechanism of leaf morphogenesis is still largely unknown. A complete explanation of the mechanism of a morphogenetic process must include a description of the cellular and subcellular processes that generate changes in shape and describe how these processes are regulated in space and time. Although classical dogma holds that the primary cellular determinants of leaf shape are spatial variation in the amount and orientation of cell division and cell expansion, the relative importance of these parameters has never been determined. In fact there is still some question as to whether cells can be considered the fundamental units of plant development. Some investigators believe that plant shape depends on processes that operate at the supracellular level, and that it is independent of the amount and orientation of cell division, cell size, and cell shape. How the spatial organization of a leaf arises is even less understood, in part because it has received so little attention.

There are a number of reasons for this state of affairs, not the least of which is technical. Until recently all studies of the cellular basis of leaf morphogenesis were conducted using sectioned specimens. This approach not only imposes restrictions on the amount of tissue that can be examined, but on the orientation in which it is viewed. As a result, it precludes a thorough quantitative analysis of the pattern of cell division within a leaf. In addition, the use of killed specimens makes it difficult to follow temporal changes in cell behavior and correlate these changes with the dynamics of leaf growth. To do this requires methods for following leaf growth and cell behavior simultaneously, methods which have only recently been developed. Spatial interactions in plant development have traditionally been defined by examining the effects of various types of surgical interventions. This technique has been useful for ascertaining the state of determination of a leaf primordium and has provided some clues about the mechanism of pattern formation. But it is of somewhat limited value because it provides negative rather than positive information, and because it is too crude to reveal the nature of local interactions.

In recent years, however, several new and more quantitative approaches to leaf morphogenesis have been taken. From these studies has emerged a picture of this process that differs radically from more classical interpretations. In this paper, I will summarize some new information about the cellular basis of leaf shape, and briefly consider earlier work on the phenomenology of leaf morphogenesis. Two issues will be dealt with in particular: the role of cellular determination and cellular differentiation in pattern formation, and the role of cell division in determining the pattern of growth. The deficiencies in our understanding of leaf morphogenesis will soon become obvious. Due to the lack of experimental studies dealing with the developmental organization of the leaf, it is impossible at present to find strong support for any particular model of pattern formation. Nevertheless, the picture that has emerged from recent studies of leaf morphogenesis provides a new framework for such models and suggests some reasonable working hypotheses about the cellular mechanism of this process.

The Developmental Morphology of the Leaf

The variation in leaf morphology between, as well as within, taxa makes it difficult to present general description of the developmental morphology of the leaf. The pattern described here is the one typical of the species that will be discussed later in this paper.

Leaves arise at regular intervals and in predictable positions around the flank of the shoot apical meristem (Fig. 1). A leaf primordium first appears at the surface of the shoot apex in the form of a lateral bulge called the leaf buttress. Soon thereafter, the primordium rises above the apex as a dorsiventrally or dorsally flattened cone generally referred to as the leaf axis. The boundary between the primordium and the shoot apex is established gradually. Almost as soon as the primordium emerges, a crease forms between it and the apex and it ceases to advance toward the apex. The lateral and abaxial margins of the primordium usually stop expanding somewhat later. In some cases, these boundaries are established simultaneously

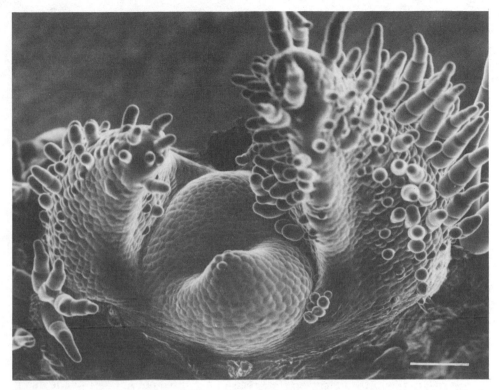

FIGURE 1. A scanning electron micrograph of a tobacco apex with four leaf primordia. In this species, leaves are initiated at about 1-day intervals. The youngest leaf primordium is at the buttress stage and is visible as a lateral bulge in the apex. the largest primordium has a well-defined lamina and midrib. The adaxial, or dorsal, side of the leaf is the side toward the apex, while the abaxial, or ventral, is the side away from the apex. Notice that in tobacco epidermal hairs are initiated very early in leaf development. Scale bar = 100 μm.

with the adaxial boundary; in others the edges of the primordium eventually encircle the apex.

The leaf blade, or lamina, arises from discrete regions along the lateral margins of the leaf axis, called marginal meristems. The outline of the lamina is its most variable feature and, from the standpoint of pattern formation, perhaps the most interesting. Spatial variation in the amount of lateral growth within the lamina occurs immediately after, if not during, its initiation. This variation may involve something as dramatic as the production of lobes and sinuses or simply involve a slightly higher rate of lateral expansion near the middle or base of the leaf. Although the growth pattern established at this stage often persists throughout development, significant changes in the pattern can also take place. In the nasturtium, for example, leaves begin as lobed primordia but eventually become circular in shape as the growth of the sinuses overtakes that of the lobes (23). A similar change in growth pattern occurs in tobacco leaves of the *broad* (*br/br*) genotype.

Here the base of the lamina initially grows at a slower relative rate than the middle of the lamina; but near the end of development the relative growth rate of the base increases dramatically so that the shape of the lamina changes from being elliptical to being ovate (14).

In addition to regional variation in rate and/or orientation of growth within the lamina, there is also variation in the duration of growth. Leaves mature basipetally. Cell division ceases at the tip of a leaf where the leaf is about $1/10$ its final size; by the time the leaf is $1/2$ its final size it is no longer meristematic. In the same way, expansion ceases first at the tip of the lamina and last at its base.

The Phenomenology of Pattern Formation

The Determination of the Leaf Primordium

The fate of a leaf primordium is determined gradually, but within a short time after its initiation. Several stages in this process have been defined on the basis of the effects of microsurgical manipulations. These stages involve the determination of (1) the site of the primordium, (2) the polarity and bilateral symmetry of the axis, and (3) the pattern of lateral growth.

The site of a leaf primordium becomes fixed about one plastochron before it appears at the surface of the shoot (a plastochron being the interval between the appearance of successive leaf primordia). Prior to this stage, the position of the primordium can be shifted by appropriate incisions in the shoot apex (16). The determination of the position of the primordium is closely associated with the first histological signs of leaf initiation—namely, a change in the orientation of cell division in the subepidermal layer of the meristem. Coinciding with, or soon after the determination of the site of leaf initiation, the primordium becomes destined for determinate growth. Although fern leaf primordia can be induced to form shoots if they are isolated from the rest of the shoot immediately after initiation, this transformation has not been accomplished in the case of Angiosperms. It seems reasonable to conclude, therefore, that in this latter group of plants determinate growth is the earliest feature adopted by a leaf primordium. The polarity and symmetry of the primordium become fixed after it appears at the surface of the shoot apex. At the buttress stage, a tangential incision between the apex and the primordium causes the primordium to form a short, cylindrical organ rather than a dorsally flattened axis (18). Operations performed on the primordium itself, however, have little effect on its morphology. In the case of pea, Sachs (15) found that removing up to $1/2$ of the leaf buttress did not prevent it from forming a normal leaf. Bisecting the primordium along its midline, on the other hand, often caused it to form two normal leaves. Thus both the polarity and the symmetry of the leaf are undetermined at the buttress stage. By the time the pea leaf is 30 μm long it has lost some of its regenerative ability. At this stage, most incisions produce unusual and unpredictable results. When the primordium is 50–70μm in length, it has become even more determined. Incisions made at this stage generally result in the loss of the part of the leaf operated on. Ablation of the apical part of the primordium, for example, results in a half leaf whose basal region is reasonably normal. Thus at an early stage, local regions of the leaf develop the capacity for autonomous growth. This important result demonstrates that leaf morphogenesis is controlled primarily by local conditions rather than by specialized organizing sites.

Although the basis for the determination of plant organs is still unknown, stable changes in cell state do not seem to be involved in this phenomenon. Most experiments used to demonstrate cellular determination in animals cannot be performed with plants, but the behavior of plant cells in culture, and the results of cell lineage analysis suggest that, in plants, determination is a result of the overall chemical and physical organization of an organ rather than a cell autonomous property. In many species, tissue from leaves at virtually any stage of development readily gives rise to undifferentiated callus, shoots, or roots under appropriate culture conditions. The conditions required to induce different types of morphogenesis vary with the age of the leaf, but this seems to be more a reflection of its meristematic state than its state of determination. This common observation has led to the general assumption that the fate of a plant cell is governed primarily by its environment, not its history. The histological fate of leaf cells in genetic mosaics supports this conclusion. In tobacco the epidermal layer of the shoot apex usually only contributes to the epidermis of the leaf. But occasionally epidermal cells contribute cells to the mesophyll, usually quite late in leaf development. Lineages resulting from these events can be observed in genetic mosaics in which the subepidermal layer of the leaf carries an albino mutation. Instead of having the morphology of epidermal cells, the cells in these lineages are indistinguishable from those in the surrounding mesophyll (17).

Green has often emphasized (see 6) that morphogenetic processes in plants are analogous to mathematical integration, where mathematical integration is broadly described as an operation that is defined in terms of immediate conditions and carries forward the product of earlier operations. In the same way, every step in the formation of a plant organ depends on the state of the organ at the point in time, and in some way changes this state. Because plant cells are surrounded by a wall of cellulose, what has already been produced cannot be undone; it can only be added to. When an organ is wounded before it has undergone cellular differentiation and acquired some degree of physical order, a new organ can be created out of the remaining undifferentiated tissue. (Because the location and symmetry of the new organ is often altered relative to the original structure, it is likely that this process involves a complete reconstruction of the original pattern, rather than a replacement of missing parts, as seems to be the case in some animals.) But as soon as the organ undergoes some form of cellular differentiation, then it only becomes increasingly more complex. It is unable to return to a less organized state unless it is removed from the plant and placed in culture. By this operational definition, the organ has become determined. Although its ultimate shape may still be subject to chemical or environmental modification, it is incapable of regenerating a part that was originally constructed under different conditions.

Wolpert's positional information model of morphogenesis (see chapter 1) postulates that pattern formation occurs in three steps: (1) the establishment of spatial variation in the distribution of a morphogen or morphogens, (2) the specification of stable cell states according to the distribution of cells within this field, and (3) the interpretation of this state by a cell according to its history and its interactions with other cells. This model cannot be applied to plant morphogenesis because the behavior of plant cells is constantly modulated by their environment, and in turn changes this environment. There is no evidence that plant cells receive positional cues at one time in development and act on these cues at some later time. The process of pattern formation in plants is inseparable from, and may in fact be dependent on, growth and differentiation. Pattern probably arises as certain types of cell behavior become reinforced by the products of this behavior (see Chapter 15). This type of autocatalytic interaction has been postulated to exist in many systems (see Chapter 3). The point made here is that in plants the expression of pattern at a cellular and morphological level probably coincides with, and is critical to the pattern generating mechanism operating at a subcellular level.

Factors that Affect the Shape of the Lamina

The shape of the lamina is influenced by the genotype, age and nutritional status of a plant, and by a variety of physical factors. The effect of these factors is illustrated by the phenomenon of heteroblasty, a term that refers to variation in leaf shape within a single plant. Every species exhibits heteroblasty to some extent. The first leaves a seedling produces often are small and relatively simple in shape. As a plant matures, leaf size increases and the shape of the lamina changes, often becoming more complex. Plants with prolonged vegetative phases eventually begin to produce a series of similar leaves. But in many determinate plants, such as corn and tobacco, leaf size and shape vary continuously from one node to the next. In addition to the variation in shape associated with the location of a leaf along the stem, dramatic changes in leaf shape may occur when a plant is induced to flower or in response to environmental factors such as the amount of illumination or, in the case of aquatic plants, whether or not a leaf is submerged.

The phenomenon of heteroblasty is at once interesting and problematic insofar as an analysis of pattern formation is concerned. Clearly it is important to determine how the various factors that affect leaf shape actually operate. But without an understanding of the basic mechanism of leaf morphogenesis this effort is essentially futile. In order to determine how this mechanism operates, it is necessary to distinguish genetically determined features of leaf shape from environmentally determined ones. In this regard, the work of Harte and her colleagues deserves special mention. These investigators have taken advantage of several narrow-leaf mutants of the snapdragon, *Antirrhinum majus*, in a detailed study of the effect of leaf position, genetic factors, and environmental conditions on leaf shape. Because *Antirrhinum* has a simple, entire leaf, the pattern of leaf expansion can be conveniently represented by the curve relating the logarithm of leaf length to the logarithm of leaf width, while the shape of a mature leaf is represented by the index leaf length/width (10, 12, 13).

In both normal and mutant plants, leaf shape varies according to the position of a leaf on the stem (10). Leaves become relatively narrower up to nodes 4–6 and then remain constant in shape until flowering. In addition, all genotypes respond similarly to an increase in photoperiod by producing relatively narrower leaves.

The response to temperature is more variable. The shape index of the normal variety and a few mutants show no correlation with a change in temperature. In contrast, leaves of the mutants *ang*, *ma*, and *sten* become narrower at high temperatures while *amb*, *arr*, and *imm* produce broader leaves under these conditions. The significance of this temperature effect is unclear, since it may reflect the nature of these mutations rather than the nature of the functions affected by these mutations. That fact that genotypic variation in leaf shape is independent of the effects of leaf position and photoperiod suggests that these latter factors act on general metabolic functions rather than on leaf specific ones.

Although narrow-leaf mutations, leaf position and photoperiod have similar effects on mature leaf shape, they influence the pattern of leaf expansion in different ways (11, 12). After a certain stage in its development, the relationship between the length and width of a leaf is expressed by the function $y = bx^k$, or $\log y = \log b + k \log x$, where y is leaf length, x is leaf width, k is the slope of the straight line relating $\log y$ to $\log x$, and b is the value of y when x equals 1. This equation, called the allometric function, provides two constants useful for characterizing leaf growth: k, which represents the relationship between the rate of growth in the length and width of the leaf, and b, which indirectly represents the initial shape of the leaf. If $k = 1$, then the shape of the leaf remains constant throughout its development. If $k < 1$, then the shape index decreases as the leaf elongates (meaning that the leaf becomes relatively broader), and if $k > 1$, then the shape index increases with an increase in leaf length (meaning that the leaf becomes relatively narrower). Thus, when $k \neq 1$, another important factor determining leaf shape is the size of the leaf at maturity.

The photoperiodic effect is due primarily to a change in k; long days generally produce an increase in k and thereby decrease the breadth of the leaf. Narrow leaf mutants also respond to photoperiod in this fashion, although they generally exhibit a much larger change in k than the normal variety. In contrast, the variation in shape associated with leaf position results from a change in the pattern of growth early in development. All the leaves on a plant have the same value of k, but their b values differ because the initial rate of lateral expansion decreases with an increase in leaf position. As a result, upper leaves are relatively more narrow than lower leaves when they enter the allometric growth phase. After this point, their pattern of expansion is essentially identical. The increase in leaf size associated with an increase in leaf position has no effect on leaf shape when environmental conditions dictate that $k = 1$. Under other conditions, however, leaf size plays an important role in determining leaf shape.

Genotypic effects on leaf shape can be attributed to variation in both k and b. Most mutant have a $k > 1$, which implies that late in development leaf width increases at a slower rate than leaf length. Mutants also differ from normal in that their initial rate of lateral expansion is slower than normal, as seen in their relatively high b values. These mutations therefore act during the entire course of leaf morphogenesis.

In summary, the shape of the lamina can be attributed to three parameters: (1) the shape of the primordium at the start of allometric growth, (2) the relationship between the growth rates in various dimensions of the leaf during this allometric phase, and (3) factors that determine the final size of the leaf. These parameters help define the stage at which a particular factor acts. The cellular parameters that mediate this activity will be discussed in the following section.

The Cellular Basis of Leaf Morphogenesis

The Intiation of the Leaf Axis

A problem of major interest in leaf morphogenesis is the number and fate of the cells that give rise to the leaf primordium. Early workers believed that a leaf is derived from a few stem cells (usually called *initial* cells) located in specific regions of the primordium. The axis was generally assumed to originate from initial cells located at its tip, while the lamina was thought to be derived from one or two rows of cells located along its margin. This model encouraged the belief that the apex and margin of the leaf control leaf morphology, despite the lack of experimental evidence for this conclusion.

The cell lineage of the leaf primordium has been re-examined recently with the help of genetic mosaics. Two types of mosaics have been used in these studies: chimeras and radiation-induced mosaics. As applied to plants, the term chimera refers specifically to mosaics in which mutant cell lineages originate in the shoot apical meristem. The most stable chimeras are those in which one or two of the three cell layers in the apex is completely mutant—periclinal chimeras. Periclinal chimeras have been particularly useful for determining the number of cell layers that contribute to a leaf, and the degree of flexibility in their pattern of cell division. However, they do not provide much information about the behavior of individual cells within a primordium. For this purpose, it is better to use mutations induced by ionizing radiation. In an appropriate genetic stock, such mutations can be induced at high frequency throughout the meristematic phase of development.

The fate of the cell layers that contribute to a leaf is most easily seen in periclinal chimeras in which the subepidermal layer of the shoot meristem carries an albino mutation. A schematic diagram of a leaf from such a plant is shown in Fig. 2. In this particular case, the epidermal layer of the shoot contributes exclusively to the epidermis of the leaf, the subepidermal layer of the shoot contributes to the subepidermal layer of the leaf and to a variable number of more internal layers, while the third cell layer of the shoot forms the core of the midrib and lamina. In other species, particularly monocots, the epidermal layer of the shoot usually makes a substantial contribution to the internal tissue of the lamina. The existence of these three topologically separate, genetically different lineages demonstrates that a leaf must be derived from at least three cell layers of the apical meristem. This is very likely an underestimate, however, because only layers that contribute to the lamina are readily visible in chimeras. A tobacco leaf, for example, is derived from at least four cell layers, but the fourth layer is restricted to the core of the midrib and is only visible under certain circumstances (14).

The size of the somatic sectors induced prior to the initiation of a leaf demonstrates that a leaf primordium encompasses several cells in each of the cell layers from which it is derived. In corn, where the leaf encircles the shoot, sectors in-

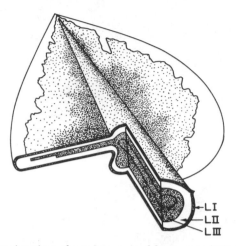

FIGURE 2. A schematic drawing of a tobacco leaf from a periclinal chimera carrying an albino mutation in the subepidermal layer (L–II) of the apical meristem. Although the epidermis (L–I) is genetically green, it does not contribute to the color of the leaf because chloroplasts are poorly developed in this layer. All of the mesophyll tissue at the margin of the lamina and a variable number of subepidermal layers elsewhere in the leaf are derived from the L–II. The intensity of color in the lamina varies in relation to the number of mesophyll layers derived from the green, L–III layer of shoot.

duced just before leaf initiation occupy about $1/45$ of the width of the leaf (Poethig, *unpublished data*), while in tobacco the smallest sectors induced at this stage are about $1/15$ of the width of the midrib (Fig. 3). These data imply that leaves arise from approximately 45 cells in the horizontal dimension of the corn apex and 15 cells in tobacco. The vertical dimension of a leaf primordium is more difficult to determine from clonal data, but it is clear that in both corn and tobacco it encompasses at least two, and probably three cells in this dimension. Thus, in both of these species leaf primordia arise from somewhere around 100–200 cells.

The pattern of cell division during leaf initiation has been extensively studied in the succulent *Graptopetalum paraguayense* by Green and his colleagues. Underlying this investigation is the assumption that the shape of the axis is due primarily to the transverse alignment of cellulose in its epidermis. Without this lateral hoop reinforcement, leaf cells would expand uniformly in all directions producing a spherical structure rather than one that is basically cylindrical. The essential phenomenon in generation of the leaf axis is therefore viewed as the establishment and stabilization of a circumferential pattern of cellulose within the epidermis of the primordium (4).

In *Graptopetalum*, leaves arise from a meristem in which both the cell pattern and the orientation of cellulose microfibrils in the epidermis is highly polarized (Fig. 4) (4, 5). Cells lie in parallel files with cellulose running perpendicular to these files. During leaf initiation, both features change: cellulose becomes circumferentially aligned around the primordium and cell files become radially oriented relative to the leaf apex. This pattern evolves gradually. Rather than have a few "stem" cells construct the pattern *de novo*, the plant reworks the existing pattern, saving some parts and modifying the rest. Only cells whose polarity is inappro-

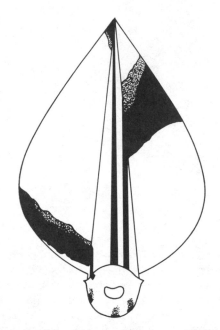

FIGURE 3. Clones in the subepidermal layer of a tobacco leaf irradiated 3 days before initiation. This particular stock was heterozygous for 2 cell-autonomous, semi-dominant chlorophyll mutations and was yellow-green in color. The sectors illustrated here were dark green and resulted from the loss of either one of the two mutant genes. Clones run in a continuous band from the dorsal subepidermal layer (black) to the ventral subepidermal layer (stippled). As originally reported by Dulieu (2), sectors near the lateral margin of the midrib contribute to the base of the lamina and those near the middle of the midrib contribute to the distal part of the lamina. This means that cells in the transverse dimension of the leaf buttress form longitudinal sections of the axis rather than radial sections, implying that the leaf is not derived from a small group of initial cells at its tip. The number of cells in the transverse dimension of the primordium—calculated by dividing the width of the petiole by the width of the smallest sectors in the petiole—is approximately 15.

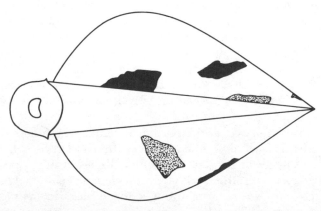

FIGURE 4. Clones in the subepidermal layer of a tobacco leaf irradiated prior to the initiation of the lamina (axis = 100 μm). At this stage, clones are usually confined either to the dorsal (black) or ventral (stippled) subepidermal layer, and none run from the margin to the midrib.

priate to the new organ undergo a change. In the regions with appropriate polarity, both the cell division pattern and the orientation of cellulose remain constant. Significantly, cell lineage does not seem to be important in this process. Within any given lineage, cell behavior varies considerably so that, although all cells eventually come to share a common transverse polarity, they achieve this polarity at different times and in different ways (6). Furthermore, cells that participate in the initiation of a growth center sometimes contribute both to the leaf and to the stem. The boundary between these organs becomes established after growth ensues.

Current thinking about the mechanism of these polarity shifts is still speculative. Nevertheless, it is worthwhile to mention some of the biophysical factors that seem to be involved. It is now generally accepted that the orientation of newly synthesized cellulose microfibrils depends on the orientation of microtubules in the underlying cortical cytoplasm. These two structures almost always share the same orientation during normal cell development, and when microtubules are disrupted by various agents, cellulose deposition occurs abnormally. Therefore, in the search for the mechanism of polarity shifts, attention has focused on factors that might influence microtubule behavior.

The orientation of microtubules is a cell-autonomous trait. Indeed their orientation in neighboring cells may differ by as much as 90° (9). This implies, of course, that cell polarity is directly dependent on intracellular factors rather than on some global agent. Such agents undoubtedly play a role in determining the overall polarity of the organ, but they apparently act through cellular intermediaries. The nature of these intermediaries is still unknown, but the phenomenology of polarity determination provides some interesting clues (6). Two factors have

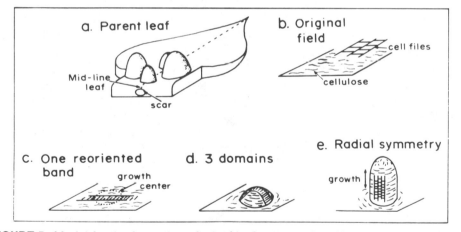

FIGURE 5. Model for the formation of a leaf in *Graptopetalum*. (A) A new shoot regenerates at the base of a detached leaf. The following figures show the changes that occur during the formation of the midline leaf. (B) The initial orientation of cell files and cellulose at the site of the midline leaf. (C) Prior to the emergence of the leaf, a transverse band of cells undergoes a 90° shift in cellulose orientation. (D) The new leaf encompasses 3 domains, one having a new polarity and two which retain their original polarity. (E) As the leaf grows, the orientation of cellulose in these domains is smoothed into a continuous radial pattern and cell division becomes transversely oriented relative to the new axis. Redrawn from Green and Lang (5).

been correlated with the orientation of microtubules in a cell after mitosis: the orientation of the new cell wall and cell shape. For the most part, microtubules parallel the new cell wall. In some instances, however, their orientation is most accurately predicted by the orientation of the long axis of the cell (5). The explanation for the correlation between microtubules and the new cell wall probably resides in the pre-prophase band. The pre-prophase band (PPB) is a thin ring of microtubules that appears briefly at the periphery of a cell prior to mitosis in a position that is later intersected by the new cell wall. Although the PPB disappears during mitosis, it seems reasonable to assume that microtubule organizing centers remain at that site. Microtubules synthesized after mitosis may take advantage of these pre-existing sites, and therby become aligned parallel to the new cell wall. Why microtubules sometimes prefer to become aligned along the longitudinal axis of a cell is more difficult to explain. It has been noted that this orientation minimizes their degree of curvature and hence may be energetically favorable (6). But this does not explain why microtubule organizing centers become localized along this axis.

Taking for granted that these two parameters are somehow involved in determining cell polarity one must then ask how the overall pattern of cellulose in the primordium is regulated. In particular, we need to know (1) the global signals responsible for pattern regulation, (2) how spatial variation in these signals arises, and (3) how a cell knows when its polarity is appropriate, i.e., the mechanism of pattern stabilization. Possible answers to these questions have been discussed recently and will not be elaborated here (6). Current hypotheses center around the idea that cell shape and the orientation of the pre-prophase band are governed to a large extent by the strain rate pattern. This pattern is described by the magnitude and orientation of the major and minor axes of strain at various points in a primordium. Factors that might influence this pattern include the boundaries of the primoridum, local variation in cellulose alignment and the distribution of "growth substances." Figure 5 illustrates how these parameters are envisioned to interact. This diagram makes one important point; namely, that the steps in morphogenesis are likely to be mutually dependent. Thus the orientation of cellulose affects the orientation of strain which ultimately feeds back on cellulose orientation via its influence on cell division and/or cell shape. The loop logic in this system serves both a stabilizing function and as an efficient mechanism for generating complex patterns. Given this type of system, it should be possible to encode the mechanism of leaf morphogenesis in a few rules of cell behavior coupled to a set of initial conditions.

The Initiation of the Lamina

The lamina arises along the lateral margins of the axis soon after leaf initiation. Responsibility for organizing the lamina is often attributed to a single row of subepidermal cells called submarginal initials. According to this view, submarginal initials establish and perpetuate the layered organization of the lamina by virtue of a highly ordered pattern of cell division, and are the ultimate source of all the internal tissue in the lamina. However, clonal patterns in genetic mosaics provide no support for this interpretation. As demonstrated by the structure of periclinal chimeras, the lamina is derived from at least three layers of cells in the leaf axis, and the internal tissue of the lamina originates from at least two of these layers (Fig. 2). The behavior of individual subepidermal cells was examined by irradiating tobacco leaves prior to the initiation of the lamina (Fig. 6). None of the resulting sectors spanned the entire width of the lamina, as would have been expected if a single row of cells was responsible for all the subepidermal tissue in the lamina. Instead, sectors were restricted to the upper or lower subepidermal layer, and were frequently found isolated from the midrib and leaf margin in an intercalary region of the lamina. The existence of such isolated sectors suggests that in any transverse section of the axis, the primordium of the lamina encompasses at least six subepidermal cells: three cells on the dorsal side of the axis and three cells on its ventral side. The most interesting sectors are those that abut the margin of the lamina. If submarginal cells served as the initial cells of the lamina, these sectors should have been oriented perpendicular to the margin. Instead they often ran parallel to it. Thus instead of contributing to the lateral expansion of the lamina, submarginal cells contribute primarily to the longitudinal extension of the leaf margin.

This brings up the question: Is there any evidence for the existence of a marginal meristem in tobacco? The term meristem is used here to refer to a specialized group of cells characterized by prolonged and/or rapid cell division. The answer to this question depends on the structure being considered. It is clear that the lamina

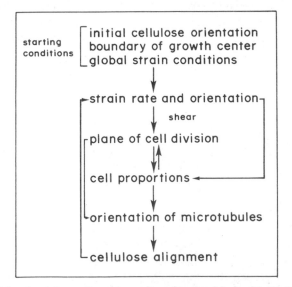

FIGURE 6. Model for the interacting parameters involved in leaf initiations. The starting conditions consist of the state of the cells at the time of initiation and any new information provided them by the factors that initiate leaf growth. These factors define the initial rate and orientation of strain in the primordium. Strain acts on the orientation of cell division directly via shear (new cell walls tend to be aligned in the direction of minimum shear) and indirectly via its effect on cell proportions. The orientation of the plane of cell division, in turn, determines the orientation of microtubules in two ways—directly via the location of the preprophase band, and indirectly via cell proportions. Cellulose is deposited parallel to cortical microtubules and feeds back to influence the strain pattern.

is derived from a marginal meristem of the leaf axis. All this means is that the lamina originates from the activity of a group of cells located along the margins of the axis rather than by the expansion of the entire axis. On the other hand, there is no reason to believe that cells at the margin of the lamina make a substantially greater contribution to its growth than cells in other regions. As indicated both by the frequency of clones in irradiated leaves (14) and by the frequency of mitotic figures in histological specimens (1), the rate of cell division is no higher at the leaf margin than elsewhere in the lamina. Although cell division persists longer at the margin than near the midrib, this difference is only a matter of a few hours, certainly less than one cell cycle. In short, the distribution and orientation of cell division in the lamina of a tobacco leaf does not support the hypothesis that the margin of the lamina plays a critical role in leaf morphogenesis. The basis for leaf shape must be sought in the overall pattern of cell division within the lamina.

The Expansion of the Lamina

Whereas the shape of the leaf axis results from a uniform, highly polarized pattern of expansion, the shape of the lamina (i.e., its outline) arises from spatial and temporal variation in the amount and orientation of growth. The role that these processes play in determining leaf shape may be illustrated by comparing the development of the broad leaf of a tobacco plant and the long, narrow leaf of a grass. Both leaf types exhibit a basipetal growth pattern; growth ceases first at the tip of the leaf and last at the leaf base. The basis for their difference in shape lies in the orientation of growth. Because growth is unpolarized (or only slightly so) throughout the tobacco leaf (Erickson, *personal communication*), variation in the rate and duration of growth along the length of the leaf produces corresponding variation in the width of the lamina. Hence there is a significant increase in the width of the lamina towards the middle and base of the leaf. On the other hand, because the growth of a grass leaf is highly polarized, the prolonged growth of the leaf base results in the elongation of the leaf rather than an increase in the width of this region. Therefore, at a gross level, leaf shape can be attributed to an interaction between the orientation of growth and the amount of growth in various regions of the lamina.

As pointed out earlier, the cellular basis of these growth parameters is usually considered to be spatial variation in the orientation and amount of cell division and cell enlargement. This interpretation assumes that leaves go through two independent stages of development—a meristematic stage, during which their expansion is under the control of cell division, followed by a maturation stage during which they expand in the absence of cell division. The pattern of expansion during this second stage is usually considered to be independent of the pattern of cell division in the preceding stage.

This interpretation can be criticized from two opposite standpoints. First it should be pointed out that even though cell expansion eventually occurs in the absence of cell division, the amount and orientation of this expansion may be determined by the polarity and the number of cells in a particular region of the leaf— factors that ultimately depend on the orientation and amount of cell division. In contrast to this proposal, Haber and Foard (7, 8) suggest that cell expansion is completely independent of cell division even during meristematic growth. According to this model the spatial pattern of expansion in a leaf is governed at a supracellular level with very little regard to cell number, cell shape or cell size.

The hypothesis that cell division plays a minor role in leaf morphogenesis is based on two types of observations. Haber's (7) analysis of leaf growth in γ-irradiated wheat seeds showed that even in the absence of cell division that polarity of growth of an irradiated leaf was identical to that of an unirradiated control. Although irradiated leaves never developed fully, the slope of the allometric curve relating leaf length to leaf width was normal. On the basis of this experiment Haber concluded that cell division was required to perpetuate growth, but that the orientation of cell division "plays no role in the manner in which leaves change shape" (7). In order to determine whether the amount of cell division is a crucial factor in leaf morphogenesis, Haber and Foard (8) compared the rate of increase in the cell number of a tobacco leaf to its polarity of growth. Noting that the polarity of growth remained constant during the period when the rate of increase in cell number was declining, they concluded that the extent of cell division, as well as its orientation, was of secondary importance in leaf morphogenesis.

Although these observations demonstrate that leaf shape is to some extent independent of cell division, they do not justify Haber and Foard's interpretation. In considering the significance of the developement of the first leaf in irradiated seedlings, it is important to realize that this leaf was several hundred microns long at the time of irradiation and that the polarity of growth and cell division prior to this stage is unknown. Clearly, if the cell polarity of the leaf had already been established then one cannot conclude that cell division is irrelevant to the pattern of expansion after irradiation. It is significant that normal leaves exhibit a constant growth polarity after this stage. That is, the rate of change in leaf length is related to the rate of change in leaf width in a linear fashion. Although one can conclude from this experiment that the maintenance of an established pattern of expansion does not require ongoing cell division, the possibility that cell division is involved in establishing this pattern remains to be tested. Furthermore, the fact that the polarity of leaf expansion remains unchanged in the face of a decline in the overall rate of cell division does not eliminate spatial variation in the amount of cell division as a determining factor in leaf morphogenesis. In order to study the effect that the rate and duration of cell division have on leaf shape it is necessary to examine the behavior of localized regions of the lamina. Haber and Foard used leaf width and leaf length—two local parameters—to represent the shape of the leaf, but measured the rate of cell division over the entire lamina. Because of the basipetal pattern of leaf development in tobacco, the overall rate of cell division may decline while the rate of cell division in a localized region of the lamina remains constant.

The shape of a mature leaf is more closely correlated with the number of cells in the lamina than with variation in either cell size or cell shape. Within any given layer of the lamina, cells are approximately the same size (22) and their shape bears no relationship to the shape of the lamina (19). The correlation between cell number and leaf shape may, of course, be strictly coincidental. It could be argued that cell division and cell expansion are completely independent but are controlled in parallel so that under normal conditions no variation in cell size or shape arises.

However, there is no conclusive evidence for this argument and it is clear that at some level growth must be linked to cell division. Cells do not become infinitely large, nor do they become infinitely small. Under a particular set of physiological conditions cells expand to a characteristic size. Consequently, if there are no physical restraints of growth, an increase in cell number leads to an increase in tissue volume. It seems reasonable to assume that this is the case in leaf development, and that as a consequence the orientation and amount of cell division are critical factors in determining leaf shape.

During the expansion of the lamina, changes in the pattern of cell division are closely correlated with changes in the pattern of growth. In *Tropaeolum peregrinum*, for example, during the initiation of lobules the mitotic index in the region of the lobule is significantly higher than in adjacent regions (3). Soon thereafter, the mitotic index becomes uniform throughout the lamina. This is consistent with the pattern of growth. During lobule initiation the length of the lobule increases dramatically in relation to the length of the adjacent sinus. However, following this period the ratio of these dimensions becomes relatively constant, signifying that their growth rates are equal. In *Tropaeolum majus*, lobes formed early in development later disappear due to the more rapid growth of the intervening sinus regions (23). Indirect evidence suggests that the rapid growth in these regions is also associated with a relatively high rate of cell division. A third example of the correlation between the rate of cell division and leaf shape is provided by leaf morphogenesis in tobacco of the *broad* genotype (14). In this particular genotype, the base of the lamina initially exhibits the lowest relative rate of expansion in the leaf. Similarly, the mitotic index of this region, and its sensitivity to irradiation (which is directly related to the rate of cell division) are also relatively low. Following the emergence of the leaf from the bud, the relative growth rate of the basal fifth of the leaf, its mitotic index and its sensitivity to irradiation all increase dramatically, both in absolute terms and in relation to the rest of the lamina. The relationship between the mitotic index and the relative rate of expansion after this stage is still unknown. However, the fact that sensitivity of the leaf base to irradiation fluctuates in concert with its relative rate of expansion throughout the rest of the meristematic phase of development suggests that the rate of cell division does so as well. Following the cessation of cell division in any given region of the leaf, the relative growth rate of that region drops rapidly. The correlation between these processes is again most obvious in the basal part of the lamina. Here the rate of cell division drops to zero within two days. One day after the cessation of cell division the relative rate of growth drops almost as dramatically, and then declines more slowly as the leaf matures.

The correlation between the rate and duration of cell division and the dynamics of leaf growth does not, of course, prove that these parameters determine leaf shape. In fact the exact temporal relationship between changes in the pattern of cell division and changes in the pattern of expansion is still unclear. Finding that every change in the rate or direction of expansion is preceded by an appropriate change in these cellular parameters would provide additional support for the hypothesis that cell division plays a primary role in leaf morphogenesis. In this respect, it is significant that, in tobacco, the relative rate of expansion drops immediately after the cessation of cell division. Further research is required in order to determine whether a change in the rate of cell division precedes an increase in the relative rate of expansion.

At present there is no evidence that distinguishes unequivocally between the hypothesis that cell division directly determines the orientation and amount of expansion within the lamina and the hypothesis that its acts solely to perpetuate a pattern of growth established at a supracellular level. To do so requires methods that allow one to specifically modify either cell division or cell expansion without totally preventing either of these processes. For this purpose, mutations can be quite useful. Although it may be difficult to determine the molecular basis of a mutant phenotype, the phenotype can at least be assumed to have a single cause. Moreover, because many mutations modify developmental patterns rather than disrupt them, they may provide clues to the regulatory circuitry of development.

Analysis of the cellular basis of the narrow leaf mutations in *Antirrhinum* mentioned earlier supports the hypothesis that cell division is the primary determinant of leaf shape (11). Although the mature cell size of some mutants differs significantly from normal, this difference is not large enough to account for their leaf shape. As in the case of normal leaves, the ratio of leaf length to leaf width is correlated with the number of cells in these dimensions rather than with cell size. The basis for this difference in cell number was studied by measuring the growth rate, cell size, and rate of cell division in isolated root tips grown in liquid media. This approach has the technical advantage that roots are morphologically simpler than leaves and therefore easier to analyze, and also made it possible to examine these parameters under uniform conditions and in the absence of the plant body. Most of the mutant roots grew more slowly than normal but did not exhibit any difference in mature cell size. Several differed from normal, however, in their rate of cell division and in the duration of division as a function of distance from the root tip. Using autoradiography, Viell-Maek (21) found that the *macer* mutation prolongs the cell cycle and delays cell differentation relative to wild type. *Loriformis* mutants, on the other hand, exhibit a more rapid cell cycle than normal, but undergo precocious cellular differentiation. These changes have the same net effect, namely, to reduce the number of cells produced per area per time. Thus, since the duration of leaf growth is apparently controlled independently of the amount of cell division, this reduction in the rate of cell production may explain why these mutants have narrow leaves. These observations indicate that leaf shape is determined in part by parameters shared by all the cells in a plant. The problem remains to determine which parameters give the leaf its species-specific shape. Much of the natural variation in leaf shape within and between species is due to only a few genes. Major variation in leaf shape in tobacco, for example, is controlled by only three loci (20), while in *Tropaeolum* two major genes determine the difference between lobed and circular leaves (23). Because these genes primarily affect the spatial pattern of growth rather than the extent of growth, they probably represent functions specific to leaf morphogenesis. A comparison of the pattern of cell division in these mutants should reveal whether a change in the pattern of cell division always accompanies a change in the pattern of growth. If so, this will provide additional evidence that leaf shape depends on regulated patterns of cell division.

Conclusions

Leaf morphogenesis is an ideal system to use for an analysis of pattern formation in plants, not only because leaves are relatively simple structures, but also because leaf morphogenesis is amenable to a variety of experimental approaches. Although the data are still too fragmentary to suggest a comprehensive model for pattern formation in leaves, some important features of this process are clear. The results of microsurgical studies demonstrate that early in its development a leaf primordium undergoes a progressive restriction in its developmental capacity. This phenomenon reflects the degree of cellular differentiation within the primordium, and is a property of the primordium as a whole, not of its constituent cells. While individual cells may acquire stable functions (such as the ability to synthesize hormones), they do not appear to become determined to form a specific part of the leaf. The shape of a leaf is a result of the amount and orientation of growth in local regions of the primordium. The mechanism regulating this spatial pattern remains a mystery. However, the identity of the cellular parameters involved in leaf morphogenesis is beginning to emerge. Recent studies suggest that the orientation, rate, and duration of cell division play a primary role in determining leaf shape, although the amount and orientation of expansion in the lamina is to some extent independent of these parameters.

In this paper I have tried to emphasize how little we actually know about leaf morphogenesis and the need for new research on this problem. A genetic approach is only one of many ways to attack this problem, although certainly one of the most powerful. As more investigators become aware of the deficiencies in our understanding of pattern formation in plants and the availability of techniques with which to study this problem, we can only hope to see new interest in this long neglected area of plant biology.

General References

DALE, J.E.: Cell division in leaves. In: Cell division in higher plants Yeoman, M.M., (ed.), pp. 315–345, New York: Academic Press (1976).

GREEN, P.B.: Organogenesis—a biophysical view. *Ann. Rev. Plant. Physio.* 31: 51–82 (1980).

SINNOTT, E.W.: *Plant morphogenesis.* New York: McGraw-Hill (1960).

References

1. Dubuc-Lebreux, M.A., Sattler, R.: Développement des organes foliacés chez *Nicotiana tabacum* et le problème des méristèmes marginaux. *Phytomorphology* 30: 17–32 (1980).

2. Dulieu, H.: Emploi des chimères chlorophylliennes pour l'étude de l' ontogénie foliare. *Bull. Sci. Bourgogne* 25: 13–72 (1968).

3. Fuchs, C.: Ontogenèse foliaire et acquisition de la fome chez le *Tropaeolum pere-grinum* L. I. Les premiers stades de l' ontogenèse du lobe median. *Ann. Sci. Nat. (Bot.)* **16**: 321–390 (1975).

4. Green, P.B., Brooks, K.E.: Stem formation from a succulent leaf: its bearing on theories of axiation. *Am. J. Bot.* **65**: 13–26 (1978).

5. Green, P.B., Lang, J.M.: Toward a biophysical theory of organogenesis: birefringence observations on regenerating leaves in the succulent *Graptopetalum paraguayense* E. Walther. *Planta* **151**: 413–426 (1981).

6. Green, P.B., Poethig, R.S.: Biophysics of the extension and initiation of plant organs. In: Developmental order: its origin and regulation (eds.) Subtelny, S., Green, P.B., pp. 485–509. New York: A.R. Liss (1982).

7. Haber, A.H.: Nonessentiality of concurrent cell divisions for degree of polarization of leaf growth. I. Studies with radiation-induced mitotic inhibition. *Am. J. Bot.* **49**: 583–589 (1962).

8. Haber, A.H., Foard, D.E.: Nonessentiality of concurrent cell divisions for degree of polarization of leaf growth. II. Evidence from untreated plants and from chemically induced changes of the degree of polarization. *Am. J. Bot.* **50**: 937–944 (1963).

9. Hardham, A.R., Green, P.B., Lang, J.M.: Reorganization of cortical microtubules and cellulose deposition during leaf formation in *Graptopetalum paraguayense*. *Planta* **149**: 181–195 (1980).

10. Harte, C: Phänogenetik der Blattform bei *Antirrhinum majus* L. I. Variabilität des Formindex in Abhängigkeit von Genotyp und Umwelt. *Biol Zbl.* **98**: 21–35 (1979).

11. Harte, C. Maek, A.: Genabhängigkeit des Wachstums pflanzlicher Meristeme untersucht an der Entwicklung isolierter wurzeln verschiedener Geotypen von *Antirrhinum majus* L. *Biol. Zbl.* **95**: 267–299 (1976).

12. Harte, C., Meinhard, T.: Phänogenetik der Blattform bei *Antirrhinum majus* L. II. Das allometrische Wachstum bei verschiedenen Umweltbedingungen. *Biol. Zbl.* **98**: 203–219 (1979a).

13. Harte, C., Meinhard, T.: Phänogenetik der Blattform bei *Antirrhinum majus* L. III. Das allometrische Wachsum bei Blattformmutanten. *Biol. Zbl.* **98**: 285–305 (1979b).

14. Poethig, R.S.: The cellular parameters of leaf development in *Nicotiana tabacum* L. Ph.D. thesis, Yale University (1981).

15. Sachs, T.: Regeneration experiments on the determination of the form of leaves. *Israel J. Bot.* **18**: 21–30 (1969).

16. Snow, M., Snow, R.: Experiments on phyllotaxis. II. The effect of displacing a primordium. *Phil. Trans. Roy. Soc. B.* **222**: 353–400 (1933).

17. Stewart, R.N., Burk, L.G.: Independence of tissues derived from apical layers in ontogeny of the tobacco leaf and ovary. *Am. J. Bot.* **57**: 1010–1016 (1970).

18. Sussex, I. M.: Morphogenesis in *Solanum tuberosum* L.: Experimental investigation of leaf dorsiventrality and orientation in the juvenile shoot. *Phytomorphology* **5**: 286–300 (1955).

19. Tenopyr, L.A.: On the constancy of cell shape in leaves of varying shape. *Bull. Torrey Bot. Club* **45**: 51–76 (1918).

20. Van der Veen, J.H.: Studies on the inheritance of leaf shape in *Nicotiana tabacum* L. Excelsior (Oranjeplein, The Hague) pp. 112, (1957).

21. Viell-Maek, A.: Genetische Einflüsse auf Wachstum und Differenzierung des Wurzelspitzenmeristems von *Antirrhinum majus* L. und einigen Mutanten. *Biol. Zbl.* **97**: 701–714 (1978).

22. von Papen, R.: Beitrage sur Kenntnis des Wachstums der Blattspriete. *Bot. Arch.* **37**: 159–206 (1935).

23. Whaley, W.G., Whaley, C.Y.: A developmental analysis of inherited leaf patterns in *Tropaeolum. Am. J. Bot.* **29**: 195–200 (1942).

Questions for Discussion with the Editors

1. *Do you believe that leaf cell walls are too rigid, and cell polarity not sufficiently "cell autonomous" to warrant the use of simpler model systems (e.g., cell cultures, Fucus eggs, etc.) for understanding the mechanism which lead to the specification of cell division planes?*

An analysis of the factors that determine the orientation of cell division would certainly benefit from the use of a simpler system than leaf development. However, this is not because of any inherent difference between the structure of leaf cells and other cell types (although such differences may exist), but because of the inherent complexity of the leaf as a whole. The fact that a higher plant leaf is a composite of several tissues and is embedded in the shoot apex probably has more significant implications for experimental analysis than do any hypothetical properties of their cell walls. But, it should be pointed out that while simpler systems offer the possibility of investigating the factors that determine the plane of cell division, they do not necessarily hold the answers to questions about the role that cell division play in morphogenesis. These are quite separate problems. In order to study the role of cell division in morphogenesis, what is needed are systems that undergo relatively complicated morphogenesis, but can still be observed and experimentally manipulated. Some promising systems are the moss *Funaria*, the water fern *Azolla* and the succulent *Graptopetalum*.

2. *Please comment on the possibility that the "positional information" model (e.g., J. Theor. Biol. 77: 195), although perhaps not applicable to leaf morphogenesis, might account for pattern specification in early plant embryogenesis, where environmental stimuli play a minimal role.*

The major argument against the application of the positional information model to plant development is not that plant morphogenesis is highly susceptible to environmental modification, but that the process of pattern specification coincides with the process of pattern interpretation; i.e., there is no evidence that patterns are first encoded in stable, invisible cell states. The fact that normal embryogenesis is more insulated from environmental stimuli than in many other organogenetic phenomena in plants does not make this a special case. In several species, pollen grains and somatic cells can be induced to regenerate plants via a pathway identical to normal embryogeny. Thus pattern formation during embryogenesis does not depend on factors unique to the zygote or embryo sac; nor are cell states irreversibly fixed during embryogenesis, otherwise somatic cells would be unable to regenerate a complete plant.

SECTION V

Vertebrates

CHAPTER **18**

Axis Specification in Amphibian Eggs

George M. Malacinski

DURING OOGENESIS the amphibian egg becomes polarized along its animal/vegetal (A/V) axis. That polarity can easily be seen in the superficial pigmentation pattern of virtually all species of amphibian eggs, which typically are darkly pigmented in the animal hemisphere and lightly pigmented in the vegetal hemisphere (Fig. 1). The pigment polarity is complimented by a polarization of internal components. For example, the smaller, less densely packed yolk platelets occupy the animal hemisphere while the larger, more densely packed platelets reside in the vegetal hemisphere (Fig. 2). Around the A/V axis the egg is radially symmetrical; that is, cross sections which pass through the A/V axis are identical at virtually all longitudes around the egg's equator.

After the egg is activated by sperm penetration, the A/V axis is preserved, but its radial symmetry is superceded by the emergence of bilateral symmetry. In some anuran eggs (e.g., *Xenopus laevis*), the sperm entrance site is visible as a dark spot and conveniently (for the experimenter) marks the future ventral region, while in several urodele (e.g., *Ambystoma mexicanum*) eggs a surface pigmentation alteration—the gray crescent—marks the future dorsal area (Fig. 3a, b). Having located either of those two surface markers, the experimenter can then deduce the remaining features of the spatial pattern (anterior and posterior; right and left) of the future embryo.

Historically the egg's animal/vegetal and dorsal/ventral (D/V) polarities have been studied as separate phenomena. Since the A/V polarity is established during oogenesis and the D/V polarity does not emerge until after fertilization, a temporal distinction between those polarities is certainly warranted. Alternatively, however, they may be considered more simply as different organizational states of the egg cytoplasm. Both polarities will be shown later in this review to be dramatically affected by novel gravity orientation. This suggests that they arise from a reorgani-

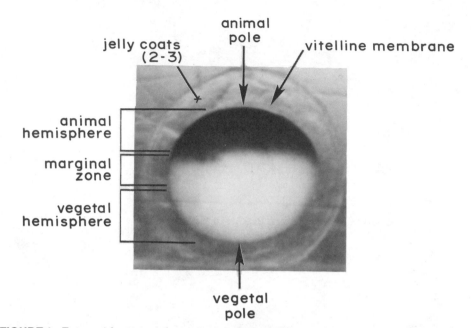

FIGURE 1. External features of a typical amphibian (*Xenopus*) egg (side view). The animal hemisphere is darkly pigmented. The vegetal hemisphere, which normally faces gravity, lacks appreciable surface pigmentation. Closely adhering to the egg surface is the vitelline membrane that is surrounded by 2–3 jelly coats.

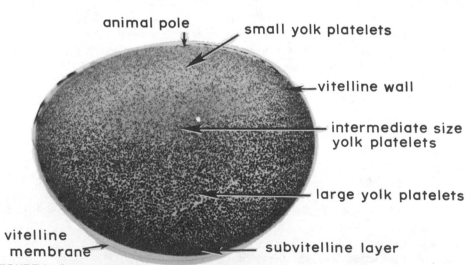

FIGURE 2. Cross section of a naturally rotated *Xenopus* egg reveals the major yolk components. Opposite the sperm entrance point, several minutes prior to the appearance of the first cleavage furrow, the vitelline wall can be observed. It consists of a wedge of large yolk platelets interspersed with smaller platelets along the inner surface of the cortex.

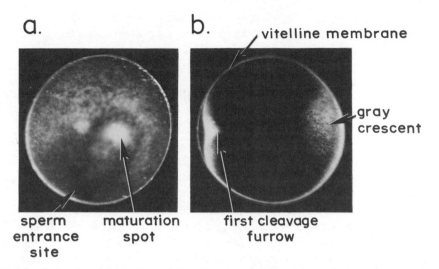

a.

b. vitelline membrane

gray crescent

sperm entrance site   maturation spot   first cleavage furrow

FIGURE 3. Following fertilization of the *Xenopus* egg a prominant pigment patch in the animal hemisphere marks the sperm penetration site. The maturation site that includes the extruded second polar body is also usually distinctly visible (a). Some amphibian eggs occasionally display a distinct gray crescent marking the future dorsal side. The axolotl egg shown in (b) has begun to form the first cleavage furrow.

zation of preexisting egg cytoplasmic components, rather than from the generation of distinctly different phenomena.

Nevertheless, those polarities will be considered separately in this review. A/V polarity will be considered first, followed by a brief discussion of the fertilization reaction. Then D/V polarity will be described, and finally in the Concluding Remarks an attempt will be made to interpret both polarities in terms of a single unified concept.

Animal/Vegetal Polarity

The major cytological feature of an amphibian egg's A/V polarity can be easily discerned, as mentioned previously, in histological cross sections. Consequently, substantial background information on the cytology of the typical amphibian egg is available (see reviews in 2, 22). Yolk platelets constitute the major cytoplasmic component. They make up approximately 50% or more of the egg's total protein and are distributed in a nonrandom pattern throughout the egg cytoplasm. In a typical egg (e.g., *Xenopus*, the animal hemisphere contains mostly smaller size (2–4 μ) platelets. In the entire vegetal hemisphere, the larger size (10–15 μ) platelets reside. An "intermediate" zone of platelets in the size range of 3–8 μ occupies a less well-defined "transitional zone" between the animal and vegetal regions (22). Lining the vegetal hemisphere cortex is a "subvitelline layer" that contains substantial numbers of both small and intermediate sized platelets (20, 21). Not only do the platelets vary in size, they display substantial density differences as well. Density gradient centrifugation separates the *Xenopus* platelets into three broad density categories (26). The most dense platelets are usually either of medium or large size, while the least dense platelets are generally small. Along the A/V axis the platelets are stratified according to density. The most dense platelets are predominantly localized in the vegetal hemisphere, whereas the least dense platelets are localized in the animal hemisphere. However, occasionally large (presumably low density) platelets are observed in the animal hemisphere, while small (low density?) platelets can be observed in the vegetal hemisphere (26). The stratification of yolk platelets is rigidly maintained in unfertilized eggs. Once fertilized, however, an inverted egg will display a substantial (but not complete) shift of its yolk platelets. The manner in which the A/V polarity of major cytoplasmic components (e.g., yolk) is rigidly maintained in unfertilized eggs but becomes subject to alteration in inverted fertile eggs is not at all understood. Perhaps a cytoskeleton maintains the A/V polarity in randomly oriented unfertilized eggs until the egg rotates in response to activation. Once rotated, under natural conditions, a mechanism which preserves the A/V stratification is no longer needed since the amphibian egg develops naturally with its vegetal hemisphere (containing large, more dense platelets) facing gravity.

In addition to yolk, a variety of cytoplasmic inclusions can be seen in histological cross sections of amphibian eggs. Glycogen granules, lipid droplets, and mitochondria are clearly visible. The extent to which those components are stratified according to buoyant density along the animal/vegetal axis is unknown. Both the animal and vegetal hemispheres, for example, contain glycogen granules.

Some components, such as the germplasm, appear to be located in discrete sites. The germplasm is localized in the vegetal hemisphere, close to the vegetal pole. Curiously, as will be discussed later, it does not redistribute to the hemisphere which faces gravity like much of the yolk does following inversion of the fertilized egg. In contrast to the germplasm, soluble proteins appear to be more or less uniformly distributed throughout the egg cytoplasm. Careful inspection of two-dimensional gel electrophoresis patterns of proteins extracted from animal versus vegetal hemispheres revealed that each of the approximately 300 soluble proteins detected on such gels is distributed almost evenly through the egg cytoplasm along the A/V axis (D. Meuler, *personal communication*).

It remains to be firmly established whether cytoplasmic mRNAs are uniformly distributed along the A/V axis. Preliminary results of Carpenter and Klein (3) and Capco and Jeffrey (4) strongly suggest that a stratification does indeed exist. Elucidation of the molecular mechanisms maintaining such a stratification should be interesting. Whether a passive situation such as differential buoyant densities of ribonucleoprotein particles, or more active phenomena such as localized attachment to specific regions of a cytoskeleton or endoplasmic reticulum are involved provide exciting opportunities for speculation.

The organization of those major cytoplasmic components along the animal/vegetal axis is illustrated in Fig. 2. The invariant and apparently rigid cytoplasmic organization which characterizes A/V polarity is striking. It is, therefore, not at all surprising that the maintenance of that cytoplasmic organization has historically and intuitively been considered to be a prerequisite for normal embryogenesis.

Only recently has a direct test been provided of the hypothesis that the cytoplasmic stratification generated during oogenesis contributes directly to pattern formation.

Figure 4 illustrates a relatively simple experimental protocol employed by Neff et al. (20) to analyze the relationship between A/V stratification and pattern formation. The principle upon which that protocol is based—exposure of fertile eggs to novel gravity orientation—is certainly not new. As long ago as the turn of the century, Born (1), Schultze (28), Penners and Schleip (24, 25), to mention a few, attempted to employ egg rotation and/or inversion as an experimental probe. Several aspects of the scheme described in Fig. 4 are, however, novel: (1) the effects of novel gravity orientation on eggs which were *never* permitted to rotate (Fig. 1a) following activation were examined; (2) large number of eggs from single spawnings were employed in each experiment; and (3) new data which contradict longstanding notions concerning the significance of the oogenetically derived A/V polarity were collected. The results obtained from the experimental protocol in Fig. 4 are worth reviewing in detail.

First, and perhaps most important of all, many of the eggs which were prevented from undergoing the classical "rotation response" to egg activation survived to the swimming tadpole stage and were, in many instances, indistinguishable (morphologically) from control embryos. The physical rotation of the egg is, therefore, not a prerequisite for normal embryogenesis as generally believed by the last two or three generations of embryologists! Clearly, the rearrangements in the egg cytoplasm that normally accompany rotation (discussed later) can either be dispensed with or compensated for in inverted prefertilization orientation (PreFO) eggs produced by the protocol shown in Fig. 4. Figure 5 reveals some major cytoplasmic rearrangements which can be detected in normal orientation, 90° off-axis, and inverted PreFO eggs. Although a major proportion of larger yolk platelets are displaced in 90° off-axis and inverted eggs, a complete rearrangement of platelets is never achieved. By the eight-cell stage, which includes the horizontal cleavage furrow that partitions the hemisphere that opposes gravity (OpG) from the G hemisphere (which faces gravity), substantial amounts of both large and small platelets

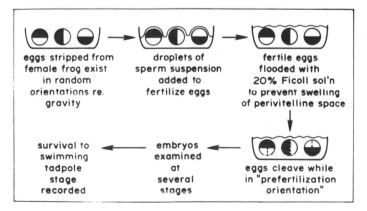

FIGURE 4. Protocol employed by Neff et al. (20) to prevent eggs from undergoing the natural rotation response to activation. Eggs remained in their "prefertilization orientation" (PreFO) throughout early development.

can be seen in unusual locations. The smaller OpG blastomeres of an inverted eight-cell embryo contain, for example, large numbers of medium and large yolk platelets. The reverse is also the case. Larger blastomeres often contain smaller sized platelets. The so-called "vitelline wall," (a wedge of large yolk platelets which lines the cortex on the future dorsal side) described by Pasteels (23) can usually be observed in control (natural orientation) eggs. It is assumed by recent authors (e.g., 10, 13) to be an integral component for establishing the D/V axis. However, it is most certainly disrupted in inverted PreFO eggs (Fig. 5c). Yet those eggs develop bilateral symmetry and proceed through axial structure morphogenesis in a seemingly normal fashion! The role yolk platelet shifts have been postulated to play in bilateral symmetrization should therefore be critically re-examined.

The behavior of other cytoplasmic components has also been monitored in inverted PreFO eggs. The germplasm does not shift from its original vegetal hemisphere location to the G hemisphere. An as yet unidentified yellow pigment found in the egg's internal cytoplasm does, however, shift completely from the original vegetal hemisphere to the G hemisphere of inverted eggs. It appears, therefore, that some cytoplasmic components which were originally stratified along the A/V axis (e.g., yellow pigment) shift dramatically in inverted eggs. Other components (e.g., yolk platelets) partially shift when an egg develops "upside down." Finally, some components (e.g., germplasm) apparently do not shift at all. Remarkably,

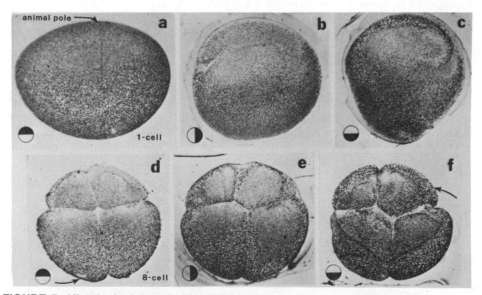

FIGURE 5. Histological examination of normal orientation (a), 90°-off axis (b), and inverted (c) *Xenopus* PreFO eggs reveals the distribution of major yolk components. Inverted eggs display a major, yet incomplete, shift of the large platelets from the original vegetal hemisphere to the G hemisphere. At the 8-cell stage (d–f), the incomplete yolk shift remains visible. A consequence of that incomplete shift is the appearance of large yolk platelets in the neural tube of tailbud stage inverted embryos. In the inverted egg (c) and 8-cell stage embryo (f), the germplasm (arrows) can be observed in its original location close to the original vegetal hemisphere cortex. Modified from (20).

however, the inverted amphibian egg is able to undergo normal development up to stage 47/48, and probably beyond (20). Those observations will certainly have to be taken into account when developing theoretical models to understand the role A/V polarity plays in pattern formation.

A detailed analysis of morphogenesis in 90° off-axis and inverted PreFO eggs provided novel observations regarding the programming of several of the early events of embryogenesis. The first cleavage furrow, for example, can bissect the egg in virtually any direction. Although inverted PreFO eggs usually form the first furrow in the original vegetal (OpG) hemisphere, occasionally it even forms around the egg's equator. The small/large blastomere pattern that characterizes the cleavage stages is reversed in inverted eggs (17). Smaller blastomeres develop in the lightly pigmented (OpG) hemisphere (Fig. 1). The site of involution, although usually located in the marginal zone, sometimes formed in the region initially identified as the egg's vegetal pole. Occasionally it even formed at the egg's original animal pole. As expected, those permutations in the normal pattern of cleavage generate embryos with dorsal sides which are frequently lightly pigmented while the ventral structures are often darkly pigmented. Nevertheless, except for pigmentation, pattern formation is remarkably normal. That is, normal swimming tadpoles can develop from PreFO inverted eggs. Table 1 summarizes several observations that emerged from the study of PreFO eggs.

If the typical amphibian egg is inverted after it has undergone its natural rotation response to activation, but before the first cleavage furrow has formed, a completely different result from that observed with PreFO eggs is obtained. The morphogenetic features listed in Table I can usually also be recognized in eggs which were inverted shortly after natural rotation. But when permitted to develop at normal laboratory temperature (18–22°C), such eggs invariably arrest during gastrulation. They never succeed in initiating neural morphogenesis. Analyzing primary embryonic induction in such embryos proved to be very instructive. Just prior to the mid-gastrula stage at which developmental arrest usually begins, grafts of prospective embryonic organizer and prospective neural ectoderm were made into normal orientation host embryos (7). The former graft (organizer) provided a test of inherent "inducing activity" of the dorsal lip cells while the latter graft (neural

TABLE 1. Characteristics of morphogenesis in inverted PreF0 eggs (summarized from Neff et al., [20]).

Egg pigmentation pattern	Surface pigment remains largely localized in original animal hemisphere
First cleavage furrow	Can bissect any plane, including equatorial
Small/large blastomere pattern	Usually reversed completely
Site of involution	Can form at any location on egg's original embryonic surface
Embryonic pigmentation pattern	Dorsal/ventral patterns usually reversed
Axial structure morphogenesis	Often normal
Primordial germ cell development	Frequently normal

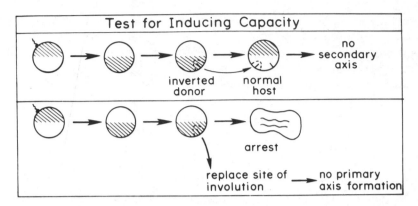

FIGURE 6. Grafting protocol employed to test the inducing capacity of the dorsal lip. The lip from an inverted embryo was either grafted onto the future ventral side of a normal host (top), or the host lip was discarded and replaced with a lip from a normal embryo (bottom).

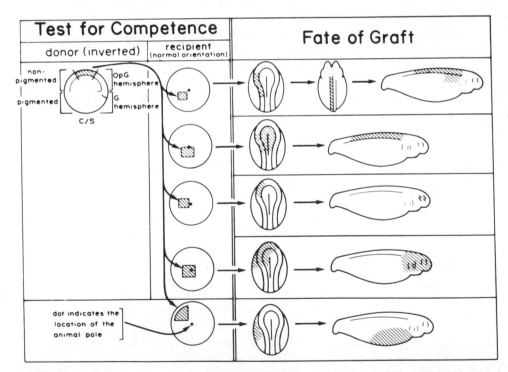

FIGURE 7. Grafting protocol designed to test the competence of the original vegetal hemisphere of inverted embryos to participate in neural development. The prospective neural ectoderm of an inverted embryo was grafted into various regions of the animal hemisphere of a normal orientation host embryo. At the tailbud stage, graft participation in normal ectodermal differentiation was observed.

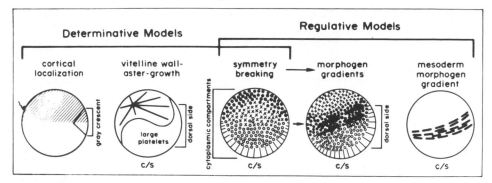

FIGURE 8. Types of models which have been proposed to account for dorsalization of the amphibian embryo.

ectoderm) tested "competence" of the prospective responding tissue. Figures 6 and 7 illustrate the experimental procedures. Although the inverted embryo's dorsal lip area lacked inducing activity, the prospective neural ectoderm was fully competent. Apparently a reversal of the pattern for neural competence took place. The program for induction, however, was not reversed since the dorsal lip consistently failed to promote neural induction in the test system (Fig. 6). Thus an uncoupling of "competence" from "organizer" activity occurs when eggs are inverted after they have undergone their natural rotation during activation. That observation should figure prominently in the development of models for the significance of the polar distribution of major cytoplasmic inclusions.

Models for A/V Polarity

How best to define A/V polarity? Perhaps in an elementary way such as "the stratification from the animal to vegetal pole of cytoplasmic components which is built up during oogenesis." What is the function of that polarity? For early pattern formation (e.g., cleavage and involution), perhaps simply to insure a high frequency of orderly development following egg activation. The significance of that polarity, especially with regard to the distribution of yolk platelets and other inclusions (e.g., germplasm) has, however, been overemphasized. Even after the cytoplasm organization is perturbed by egg inversion, substantial pattern regulation can ensue. For later pattern specification, A/V polarity is probably most significant for primary embryonic induction; that is, for insuring that cells in the animal hemisphere develop the capacity to respond to the action of the primary embryonic organizer. Likewise, A/V polarity generates primary organizer activity in marginal and vegetal zone cells.

For model building, the following two observations should be kept in mind. First, neither competence nor organizer activity track with other prominent cytoplasmic inclusions (e.g., yolk platelets or germplasm) during novel gravity orientation. This means that predicting the behavior of morphogens, especially those that might be expected to stratify according to buoyant density, is not straightforward. Second, the development of competence can be uncoupled from the development of primary organizer activity. This means that simple single gradient models (see Chapter 3 and 32) should explain how uncoupling of competence from organizer activity occurs.

In this author's opinion, the data just described suggest that A/V polarity is presently best understood in terms of a model in which the egg cytoplasm is viewed as being organized into a set of discrete physical "compartments" or "domains" (21, see also Fig. 9). Yolk platelets are organized into at least four compartments,

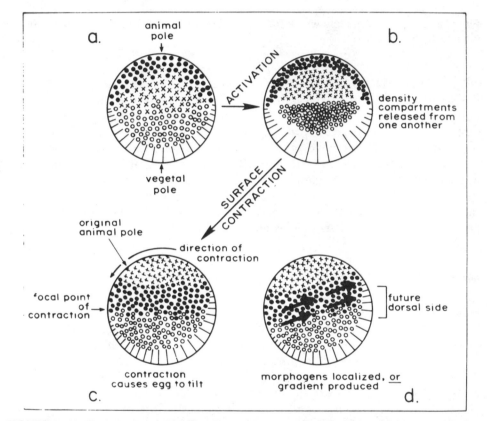

FIGURE 9. Unified model of density compartments to explain A/V and D/V axis specification. The cytoplasm of the unfertilized egg is viewed as containing several distinct "compartments" which differ from one another in intrinsic buoyant density and possibly also viscosity (a). Those compartments are, however, not stratified according to density from the animal to vegetal pole. Rather, in the above illustration the (•) compartment is more dense than the (x) compartment, and the (o) compartment is more dense than the (-) compartment. The (-) compartment, like the subvitelline layer (20, 21), is anchored to the egg cortex. Needless to say, the symbols in the above illustrations do not imply the content of individual compartments (such as yolk platelets, organelles, etc.). Prior to egg activation those compartments are stabilized, perhaps by a cytoskeleton. Once activated, the previously stable compartments become free to shift (b) and to assume natural buoyant density relationships (c). Accordingly, the various compartments relocate, and in the process

new compartment neighbor interactions emerge. The activation stimulus (e.g., sperm penetration, electrical activation, pricking, etc.) provides a direction to the contraction, and the egg tilts. The release of the compartments and the focal point of the cortical contraction serve to "break" the radial symmetry of the egg. The new compartment interactions are visualized as generating novel metabolic activities by the late blastula stage which could either (1) produce localized morphogenetic information, or (2) generate a morphogen gradient (as shown by the arrows in d). The diagrams represent an extreme case of compartment shifts. Similar end results (c) could be achieved if some of the compartments merely mixed, rather than completely shifted. Likewise, only 2 compartments rather than the 4 illustrated need to differ in density for the model to be formally correct. The compartment shifts (b and c) account for known cytoplasmic rearrangements (e.g., Fig. 5), while the morphogen gradient accommodates data on the development of primary embryonic organizer activity (19) and morpholaxis (see Chapter 20).

---

including the small, intermediate, large, and subcortical yolk masses. Numerous compartments which contain soluble components such as the yellow pigment referred to earlier also probably exist. Most of those compartments could easily be considered to contain developmental information. In some cases, that information may exist in the form of classical cytoplasmic localizations (e.g., germplasm compartment), while in other cases, a compartment may consist of a self-contained morphogen gradient system.

During egg inversion, those compartments are displaced to various extents. The yellow pigment compartment provides a paradigm for a compartment which shifts completely in PreFO eggs. The germplasm compartment apparently does not shift at all. During a relatively brief inversion, the compartment responsible for neural induction competence shifts from the original animal hemisphere to the OpG hemisphere. The compartment that specifies primary embryonic organizer activity, like the germplasm compartment, apparently does not shift. Only when eggs are inverted for relatively longer periods (e.g., PreFO eggs) does the embryonic organizer compartment shift to the G hemisphere.

Viewing A/V axis specification in terms of physical "compartments" or "domains" offers several advantages. First, recent observations concerning the reversal of developmental competence (7) and the displacement of cytoplasmic components (21) are easily accommodated. Second, subdivsion of the egg cytoplasm into a series of compartments permits several types of mechanisms, including determinative as well as regulative ones, to be active in the egg's vast expanse of cytoplasm at the same time. That is, A/V polarity need not be exclusively explained by only one type of mechanism, such as a reaction-diffusion system (see Chapter 3). Third, attempts at manipulations which serve to either rearrange one or another compartment or to mechanically isolate individual compartments are encouraged. Fourth, a reasonable mechanism—cytoskeleton—can be postulated to hold compartments together.

This compartment concept should not be confused with either of two other concepts which have been important in the field of experimental embryology. "Compartments" or "domains" are not necessarily directly equivalent to classical cytoplasmic localizations. Within a single compartment, one or more morphogenetic determinants may be contained. Perhaps some compartments contain complete reaction-diffusion systems. Likewise, compartments as described here should not be confused with the gene expression compartments which have been postulated to account for cell lineages in insect embryogenesis (see Chapter 13).

Sperm Penetration, Egg Activation, and the Establishment of D/V Polarity

Entrance of spermatozoa stimulates a series of alterations in the egg's structure, function, and developmental program. The earliest changes are those which are associated with the egg surface. Fusion of the sperm membranes with the egg plasma membrane triggers the breakdown of the cortical granules. Cortical granule exocytosis begins (in monospermic anurans) at the site of sperm entry and proceeds as a wave over the entire egg surface. As the contents of the granules are released to the egg surface, the vitelline membrane which previously adhered very tightly to the egg surface is elevated. The swelling of the mucopolysaccharides released through cortical granule breakdown creates the perivitelline space around the egg. That space is normally of major importance to early pattern formation, for within that space the egg is able, for the first time in its existence, to orient itself according to the natural force of gravity. Consequently, within a short time after the perivitelline space is generated, the egg rotates so that its animal hemisphere opposes gravity and its vegetal hemisphere faces gravity. That rotation is probably facilitated by contractions in the egg cortex, especially in the animal hemisphere region.

The scenario just described has been well documented for a variety of anuran eggs, including *Xenopus, Rana, Bombina*, etc. Similar details are not available for urodele eggs. Since urodele eggs are typically polyspermic, differences probably exist in the nature of the egg surface alterations which are stimulated by sperm entrance. Unresolved, for example, is the issue of whether urodele eggs even contain cortical granules (9). Sperm penetration in anuran eggs is normally confined to the darkly pigmented animal hemisphere. As mentioned previously, it is occasionally marked as a dark spot which represents the displacement of pigment granules around the entrance site (Fig. 3a). Urodele eggs often display multiple sperm penetrations over the entire surface of the egg, including the region surrounding the vegetal pole.

Unlike anuran eggs which display a classical electrical fast-block to polyspermy, urodele eggs do not exhibit a rapid membrane depolarization in response to sperm penetration. Urodele eggs, however, contain a very effective mechanism for inactivating all but one sperm pronucleus (31), so adverse consequences do not normally result from multiple sperm penetration. Conversely, anuran eggs apparantly lack the ability to inactivate supernumary sperm nuclei if the electrical block malfunctions (12).

Both anuran and urodele eggs display cortical contractions. Considering the differences in sperm penetration (monospermy—anurans; polyspermy—urodeles), it is not surprising that different types of surface contractions have been described for the two types of eggs. The activation contraction of anuran eggs appears to involve a contraction of the egg surface toward the sperm entrance site. That contraction may actually serve to move the sperm nucleus closer to the egg nucleus in preparation for syngamy. Urodele eggs apparently undergo a shift of the entire animal cap towards the side of the egg which will eventually develop a ventral orientation (reviewed in 16). Despite those potentially significant differences between anuran and urodele contractions, they may actually serve at least one common purpose: rearrangement of internal cytoplasmic components (e.g., yolk) in a manner which, under natural conditions, prepatterns the future dorsal/ventral axis of the embryo (see next two sections).

Nevertheless, the differences in mechanism of sperm penetration and subsequent cortical contractions between anurans and urodeles is worth emphasizing. In anuran eggs, there is solid evidence that sperm penetration plays a direct role in establishing the dorsal/ventral polarity of the egg (10). In urodele eggs the direction of the D/V axis is probably not under the direct control of the sperm penetration path.

Various other events follow sperm penetration, including the completion of meiosis and the emission of the second polar body, one or more surface contractions ["post-fertilization waves" (14)], sperm aster growth, etc. Several of those events, if not all of them, may participate in generating the bilateral symmetry which is established between sperm penetration and first cleavage.

Dorsal/Ventral Polarity

As mentioned previously, in some anuran eggs (e.g., *Xenopus laevis, Rana temporaria*, etc.), the sperm entrance site normally marks the future ventral side of the egg. In various urodele eggs, the gray crescent is likewise a reliable indicator of early bilateral symmetry since it appears on the future dorsal side of the egg. The relatively early appearance (before first cleavage) of those surface indicators reveals that the establishment of bilateral symmetry is actually initiated, at least in anuran eggs, by the penetrating sperm. As will be discussed in the next section, a "symmetry (i.e., radial symmetry) breaking process" may, in formal terms, account for the sperm's role in polarity establishment in monospermic anuran eggs.

Although the future dorsal side (invagination site) of the embryo usually develops on the side occupied earlier by the gray crescent, the dorsalization process can easily be uncoupled from the apparent controlling influence of either sperm penetration or the gray crescent. Briefly tilting the fertile egg 90° off its natural A/V axis overrides the normal influence of sperm penetration. The site of involution forms on the side of the egg which opposed gravity during the rotation regardless of the location of the gray crescent (27, 5). With that simple manipulation, it is even possible to locate the site of involution on the side of the egg where the sperm entered (which, under normal circumstances, would be the future ventral side) (27). That manipulation is effective during the developmental period from fertilization up to and including the 2-cell stage (10). At later stages the simple "tilting" manipulation is no longer effective in relocating the prospective D/V axis.

Those observations, as well as comparative studies on anuran and urodele eggs (6), lead to two important conclusions: (1) although the development of the prospective D/V axis probably begins with sperm penetration, it is labile to gravity manipulation (i.e., tilting) up to at least the 2-cell stage; and (2) despite the differences in fertilization described earlier, anuran and urodele eggs respond similarly to simple gravity perturbations. D/V polarity is, therefore, probably established in a progressive fashion, beginning with sperm penetration (at least in anurans) and being fixed certainly by the time the first horizontal cleavage furrow (8-cell stage) has formed.

.It it not clear whether bilateral symmetry is important to embryogenesis prior to gastrulation. In fact, the gray crescent pigment change is usually obscured during mid-cleavage. In most species of eggs it is very difficult, if not impossible, to distinguish dorsal from ventral side in the mid- and late blastula stages. Therein lies a major technical, if not also conceptual, problem with the study of egg dorsalization: there simply are no reliable indices of bilateral symmetrization other than the transient early surface markers (sperm penetration point—anurans; gray crescent—urodeles). Only when involution begins—at the early gastrula stage—can the experimeter routinely distinguish dorsal from ventral sides. During the interval from the 2-cell to the early gastrula stage, more than a dozen rounds of mitosis have taken place and 12 hr to 2 days (depending upon the species) have gone by. Yet the experimenter has no recourse but to wait until the dorsal lip makes its appearance before a reliable assignment of the egg's dorsal vs. ventral orientation can be undertaken.

The lack of any indices of egg D/V polarity during that long interval is considered by this author to be the single most important limitation of the amphibian egg for studies on symmetrization. To recall, animal/vegetal polarity reversal can be recognized immediately by the external features of the cleavage pattern. Internally, in inverted eggs, yolk platelet distributions can be conveniently monitored. No similar assessment of "what is D/V polarity in the cleaving embryo," however, can be made.

The changes characterizing the transition from a radially symmetrical egg to a bilaterally symmetrical embryo are no doubt very subtle. It is not clear exactly when the D/V polarity is finally fixed. Various types of perturbations, including simple egg rotation, the application of a mild heat gradient, centrifugation, distortion of shape, etc., have been reported over the years to be capable of altering the egg's D/V polarity. Such perturbations are summarized in Table 2. Those treatments are remarkably diverse and are, in some cases (e.g., egg rotation), effective only relatively early after sperm penetration. In other cases (e.g., shape distortion), the treatment is effective at all stages. No unifying underlying mechanism emerges—at least to this author—as a likely explanation to account for the effects of all of those perturbations. Rather, the observations included in Table 2 suggest that D/V polarity is probably established in the uncleaved egg rather passively by cytoplasmic rearrangements. Progressively, however, the D/V polarity becomes fixed by as yet unknown mechanisms. Only once gastrulation is underway is the D/V polarity finally fixed, once and for all.

The significance of D/V polarity is profound. Although the subtle mechanism(s) for specifying D/V polarity is unknown, the involution of a small group of cells in the marginal zone initiates a set of spectacular cell movements and cell interactions which result in the irreversible establishment of the embryo's body axis. They are so dramatic they can be seen with the naked eye. Cells around the involu-

TABLE 2. Techniques for perturbing d/v polarity.

TREATMENT	PROCEDURE	EFFECTS	REFERENCE
Heat gradient	Place egg in narrow heat gradient	Dorsal lip forms on warm side	(11)
Oxygen starvation	Restrict oxygen supply to one side of egg	Dorsal lip forms on high oxygen side	(15)
Ultraviolet irradiation	Irradiate uncleaved egg at vegetal hemisphere	Dorsal lip site rearranged	(18)
90° rotation	Tilt egg 90° off-axis	Dorsal lip forms on side which faced "up"	(27)
Centrifugation	Spin eggs prior to first cleavage	Relocation of Dorsal lip or formation of twin lips	(10)
Shape distortion	Remove vitelline membrane	High frequency of twinning	a

[a]Chung, unpublished observation.

tion site move inside the embryo. The mesodermal layer of cells which spreads inside the embryo then interacts with the overlying prospective neural ectoderm and differentiates into various tissues and organs (e.g., notochord, somites, proneophros). Various patterns become specified in each of three tissue layers (ectoderm, mesoderm, and endoderm). Those patterns themselves display various anterior/posterior and dorsal/ventral polarities, and as embryogenesis proceeds the egg's original D/V polarity becomes a mere footnote to the embryo's history.

Models for D/V Polarity

The search for the cytoplasmic determinants for amphibian egg polarity has been unrelentless. During the past century, various components of the egg cytoplasm have been assigned important roles in establishing D/V polarity. Some of the more recent proposals are included in Table 3. In this reviewer's opinion, each of those ideas has serious flaws and some are now known to play no role in dorsalization. The proposals represented in Table 3 could have failed for any of several reasons, including the following: First, D/V polarity may actually be established by a very subtle and/or complex mechanism which defies straightforward identification and characterization with present technology; second, three grenerations of embryologists may have erred by oversimplifying their conceptual approach to the problem.

This latter point concerning oversimplication warrants further discussion. A full understanding of the amphibian D/V polarity will be forthcoming only when the following three considerations are addressed in a satisfactory manner:

1. The molecular nature of the morphogenetic information which specifies D/V polarity. As two extremes, one could consider whether there exists in the egg a signal molecule for D/V polarity such as a diffusable morphogen, or whether a cytoskelton is erected in the egg cytoplasm for the purpose of pre-patterning a self-assembly process.

TABLE 3. Candidates for d/v polarity establishment.

Egg Component	Reference	Comment
Vitelline wall	(23)	Proposed in early days, probably because of dramatic yolk shift observed toward future dorsal side
Gray crescent cortex	(8)	Appealing idea, but never independently verified
Special proteins in dorsal cortex	(29)	Never rigorously established
Sperm aster	(10)	Aster growth probably, however, completed prior to establishment of D/V polarity
Dorsal yolk-free cytoplasm	(30)	Can be uncoupled from dorsalization by egg rotation

2. The manner in which that morphogenetic information becomes organized in the egg cytoplasm to polarize the embryo.
3. The mechanism of action of the morphogenetic information.

With regard to the first point, in a formal sense two opportunities for analysis exist. Bioassays could be developed for guiding purification and characterization schemes for morphogenetic determinants. Alternatively, mutant genes which generate defects in internal cytoplasmic organization could be collected and analyzed. The prospect for successfully employing either of those strategies with amphibian eggs in the near future is considered by this author to be, quite frankly, dim. Bioassays are difficult to interpret, especially if they generate negative results (i.e., "no effects"). Appropriate maternal effect mutant genes for egg polarity have simply never been recognized. Dealing with the third point—mechanism of action of the morphogenetic information—must naturally await progress on the first point.

From the above perspective, it is easy to understand why most attention has been directed toward the organization of the morphogenetic information which specifies D/V polarity. The wide variety of perturbation methods which alter D/V polarity (Table 2) has tempted embryologists, if for no other reason, because they can be so easily carried out. The information provided by those perturbation methods, however, is necessarily directed toward understanding how the morphogenetic information is organized rather than identifying the molecular nature of that information. Needless to say, it is expected that studies on the organization of the egg cytoplasm will lead into studies on the biochemistry of the morphogenetic information. Perhaps studies on cytoplasmic organization will generate suggestions, for example, that small molecular weight (diffusable) morphogens are the active components for dorsalization.

A compilation of the types of models which have been proposed—at various times during the long history of the D/V polarity problem—to account for bilateral symmetrization is included in Fig. 8. Each fails to be completely satisfactory for one or another reason. The cortical localization model proposed that morphogenetic information is localized in the gray crescent cortex. It was claimed that cortex grafts transferred the dorsalization information to the ventral side (8). Those claims have, however, been essentially disproved (19).

The vitelline wall model was originally proposed by Pasteels (23) and has recently been resurrected by Gerhart et al. (10). As was the case for the cortical localization model, its simplicity provides most of its appeal. Sperm entrance initiates the growth of asters away from the sperm entry site, causing a shift of some of the egg's large yolk platelets upwards on the future dorsal side to form the "vitelline wall." As the sperm asters grow, they fill the animal hemisphere and trigger the postferitilization contraction wave which amplifies the dimensions of the vitelline wall (Ubbels, personal communication). That vitelline wall is considered to represent localized morphogenetic information for specifying D/V polarity. Unfortunately, two recent observations are not consistent with that straightforward aster growth-vitelline wall model. First, inverted PreFO eggs display a remarkable redistribution of yolk platelets. The vitelline wall of an inverted egg is obliterated when the large yolk platelet compartment shifts downward in reponse to gravity. Yet pattern formation, including the specification of the D/V axis, occurs in a normal fashion (Table 1). Second, artificially activated eggs display the beginnings of vitelline wall formation despite the absence of sperm asters (Neff and Wakahara, unpublished observation).

It appears, therefore, that neither of the simple cytoplasmic localization (determinative) models stand up to rigorous testing. The regulative models offer features which circumvent some of the limitations of the strictly determinative models. A detailed mathematical treatment of one such model has been offered by Weyer et al. (32). It is depicted in Fig. 8 as the "mesoderm morphogen gradient model." In formal terms, that model is largely satisfactory. It does, however, contain some shortcomings. First, it lacks a precise definition of the manner in which the gradient sources are initially established. Second, like most morphogen gradient models it defies experimental verification with present technology. Third, it predates the recent information on egg rotation, inversion, etc. (Table 2), which should be addressed in up-to-date modeling efforts.

Considering the inadequacies of each of the models just discussed, it is appropriate to ask "Which features should a fully satisfactory and comprehensive model display?" Following is a list of some of the most important aspects of dorsalization which need to be accommodated:

1. The remarkable number of perturbations (Table 2) which can affect D/V polarity early (e.g., rotation) or late (e.g., collapsing of the egg suprastructure).
2. The apparent disparity between anuran egg dorsalization (in which the sperm entrance site normally plays a key role) and polyspermic urodele egg dorsalization (which does not require sperm penetration cues). Both types of eggs, as mentioned earlier, dorsalize in a similar fashion after egg rotation (6).
3. The specification of the mesoderm which arises from internal cells in anurans (e.g., *Xenopus*) and from superficial cells in urodeles (e.g., *Ambystoma*). The mesoderm presumably plays a key role in dorsal midline axis formation; its specification is probably the most important consequence of egg dorsalization.

In this author's opinion the most reasonable model for dealing with those three points is the "combined symmetry breaking and morphogen gradient" model depicted in Fig. 9. During oogenesis the locations of the various cytoplasmic compartments (discussed in Models of A/V Polarity) are held in place, probably by a cytoskeleton. The density differences of the various compartments make the egg cytoplasm potentially unstable. Thus in a mature egg, the compartments are viewed as being poised to shift but are actually restrained from doing so by a rigid cytoskeleton. Fertilization leads (by an unknown mechanism) to a reduction in the viscosity of the egg cytoplasm. As a result, the compartments shift. In anuran eggs, the direction (but not the force) for that shift may come from the growing sperm aster. In urodele eggs, the direction is random, but nevertheless unidirectional once initiated. The first step in dorsalization is therefore visualized as a breaking of the radial symmetry of the egg which results from the gravity driven shift of the various cytoplasmic compartments.

The radial symmetry of the unactivated egg can be "broken" by any number of stimuli. Under natural conditions sperm penetration and cortical contractions play such a role in anuran eggs. However, electrical activation, rotation, or probably various other treatments can substitute for sperm penetration (21). The nature of the stimulus that triggers the symmetry breaking process in urodele eggs is, however, unknown. Eventually, as the cleavage furrows partition the egg vertically (2- and 4-cell stages) and horizontally (8-cell stage), the bilateral symmetrization that emerged from the rearrangement of cytoplasmic components becomes relatively fixed. No longer is D/V polarity sensitive to either rotation or centrifugation. The result of the early events of symmetrization is the organization of the egg cytoplasm for the future activity of diffusable morphogen gradient sources and sinks which come into play during blastulation. Those morphogen gradient components in turn specify the polarity of the mesoderm which ultimately defines the location of the site of involution of gastrulation.

The strengths of such a unified model are readily apparent: anuran and urodele eggs are accomodated by a single model; the progressive manner in which mesoderm polarity is fixed during blastulation is understood (19), as are the relatively late (e.g., blastula) perturbations that alter the site of involution (Chung, *unpublished observation*). Finally, several features of the model are testable. For example, a cytoskeleton can be searched for, and its rigidity before and after fertilization examined. Also, the physical identification of various density compartments should be possible.

The weaknesses of such a model are also clear: it represents a compromise between earlier models and it offers few (if any) opportunities for direct analysis of the nature of the diffusable morphogen. Finally, in common with virtually all polarity models, it fails to address the issue of how the putative diffusable morphogen acts.

Concluding Remarks

Whether A/V and D/V polarity should be considered in terms of a single unified model warrants a brief discussion. Perturbations to A/V polarity generated by novel gravity orientation are most easily understood in terms of a compartment model. Likewise, if the egg cytoplasm is viewed as a diverse set of cytoplasmic compartments, the first steps in D/V polarity establishment can also be readily understood: sperm penetration and cortical contractions lead to a physical reorganization of several of the cytoplasmic compartments which were inherited from oogenesis. Once organized in the appropriate fashion, individual compartments, whether designed for A/V or D/V specification, somehow interact to program the cleavage pattern, the site of involution, etc.

Unfortunately, the "compartment" or "domain" concept fails to precisely explain in molecular terms the composition of the various A/V polarity and D/V polarity compartments. A priori there is no reason for believing that all compartments function in a similar fashion. Perhaps some compartments sequester particulate cytoplasmic localizations such as the germplasm, while other compartments function as source/sinks of morphogen gradients. Indeed, after A/V and D/V pattern specification is completed (probably by the 2- or 4-cell stage) their respective compartments may diverge substantially as they engage in the metabolic preparations for early embryonic cell differentiation.

Acknowledgments

The stimulating discussion, especially in regard to the compartment hypothesis, provided by several of the author's colleagues—A. Neff, M. Wakahara, G. Radice, and R. Cuny—is gratefully acknowledged. Much of the data included in the illustrations was collected with the support of NASA grant NAGW-60.

General References

GERHART, J.C.: Mechanisms regulating pattern formation in the amphibian egg and early embryo, pp. 133–316. In *Biological regulation and development*. Vol. II, (ed.) Goldberger, R. New York: Plenum (1981).

MALACINSKI, G.M.: Sperm penetration and the establishment of the dorsal/ventral polarity of the amphibian egg. In *Biology of fertilization*, (eds.) Metz, C.B., Monroy, A. New York: Academic Press, *in press* (1983).

References

1. Born, G.: Uber den Einfluss der Schwere auf das Froschei. *Arch. Mikrosk. Anat.* **24**: 475–545 (1885).

2. Brachet, J.: An old enigma: The grey crescent of amphibian eggs. *Curr. Top. Dev. Biol.* **11**: 133–186 (1977).

3. Carpenter, C.D., Klein, W.H.: A gradient of polyA$^+$ RNA sequences in *Xenopus laevis* eggs and embryos. *Dev. Biol.* **91**: 43–49 (1982).

4. Capco, D.G., Jeffrey, W.R.: Regional accumulation of vegetal pole poly(A)$^+$ RNA injected into fertilized *Xenopus* eggs. *Nature* **294**: 255–257 (1980).

5. Chung, H.-M., Malacinski, G.M.: Establishment of the dorsal/ventral polarity of the amphibian embryo: use of ultraviolet irradiation and egg rotation as probes. *Dev. Biol.* **80**: 120–133 (1980).

6. Chung, H.-M., Malacinski, G.M.: A comparative study of the effects of egg rotation (gravity orientation) and UV irradiation on anuran vs. urodele amphibian eggs. *Differen.* **18**: 185–189 (1981).

7. Chung, H.-M., Malacinski, G.M.: Reversal of developmental competence in inverted amphibian eggs. *J. Embryol. Exp. Morph. in press*, (1983).

8. Curtis, A.S.G.: Morphogenetic interactions before gastrulation in the amphibian *Xenopus laevis*—the cortical field. *J. Embryol. Exp. Morph.* **10**– 410–422 (1962).

9. Elinson, R.P.: The amphibian egg cortex in fertilization and early development. *Symp. Soc. Dev. Biol.* **38**: 217–234 (1980).

10. Gerhart, J., Ubbels, G., Black, S., Hara, K., Kirschner, M.: A reinvestigation of the role of the grey crescent in axis formation in *Xenopus laevis*. *Nature* **292**: 511–516 (1981).

11. Glade, R.W., Burrill, E.M., Falk, R.J.: The influence of a temperature gradient on bilateral symmetry in *Rana pipiens*. *Growth* **31**: 231–249 (1967).

12. Grey, R.D., Bastiani, M.J., Webb, D.J., Schertel, E.R.: An electrical block is required to prevent polyspermy in eggs fertilized by natural mating of *Xenopus laevis*. *Dev. Biol.* **89**: 475–484 (1982).

13. Grinfeld, S., Beetschen, J.-C.: Early grey crescent formation experimentally induced by cyclohexamide in the axolotl oocyte. *Wilhelm Roux' Arch.* **191**: 215–221 (1982).

14. Hara, K., Tydeman, P., Hengst, R.T.M.: Cinematographic observation of "post-fertilization waves" (PFW) on the zygote of *Xenopus laevis*. *Wilhelm Roux' Arch.* **181**: 189–192 (1977).

15. Lovtrup, S., Pigon, A.: Inversion of the dorsal-ventral axis by unilateral restriction of oxygen supply. *J. Embryol. Exp. Morph.* **6**: 486–493 (1958).

16. Malacinski, G.M.: Sperm penetration and the establishment of the dorsal/ventral polarity of the amphibian egg. In *Biology of fertilization*, (eds.) Metz, C.B., Monroy, A., New York: Academic Press, *in press* (1983).

17. Malacinski, G.M., Chung, H.-M.: Establishment of the site of involution at novel locations on the amphibian embryo. *J. Morphol.* **169**: 149–159 (1981).

18. Malacinski, G.M., Brothers, A.J., Chung, H.-M.: Destruction of components of the neural induction system of the amphibian egg with ultraviolet irradiation. *Dev. Biol.* **56**: 24–39 (1977).

19. Malacinski, G.M., Chung, H.-M., Asashima, M.: The association of primary embryonic organizer activity with the future dorsal side of amphibian eggs and early embryos. *Dev. Biol.* **77**: 449–462 (1980).

20. Neff, A.W., Malacinski, G.M., Wakahara, M., Jurand, A.: Pattern formation in amphibian embryos prevented from undergoing the classical "rotation response" to egg activation. *Dev. Biol.* **97**: 103–112 (1983).

21. Neff, A.W., Wakahara, M., Jurand, A., Malacinski, G.M.: Experimental analyses of cytoplasmic re-arrangements which follow fertilization and accompany symmetrization of the *Xenopus laevis* egg. *J. Embryol. Exp. Morph. in press* (1984).

22. Nieuwkoop, P.D. Origin and establishment of embryonic polar axes in amphibian development. *Curr. Top. Dev. Biol.* **11**: 115–132 (1977).

23. Pasteels, J.: The morphogenetic role of the cortex of the amphibian egg. *Adv. Morph.* **3**: 363–388 (1964).

24. Penners, A., Schleip, W.: Die Entwicklung des Schultzeschen Doppelbildungen aus dem Ei von *Rana Fusca*. Teil I-IV. *Zeitschr. Wiss. Zool.* **130**: 305–454 (1928).

25. Penners, A., Schleip, W.: Die Entwicklung des Schultzeschen Doppelbildungen aus dem Ei von *Rana fusca*. Teil V-VI. *Zeitschr. Wiss. Zool.* **131**: 1–156 (1982).

26. Radice, G.P., Neff, A.W., Malacinski, G.M.: The intracellular responses of frog eggs to novel orientations to gravity. *The Physiol.* **24**: (suppl.) 579–580 (1982).

27. Scharf, S., Gerhart, J.C.: Determination of the dorsal-ventral axis in eggs of *Xenopus laevis*: Complete rescue of UV-impaired eggs by oblique orientation before first cleavage. *Dev. Biol.* **79**: 181–198 (1980).

28. Schultze, O.: Die Kunstliche Erzeugung von Doppelbildungen bie Froschlarven mit Hilfe abnormer Gravitationswirkung. Archiv fur Entwicklungs-mechanik des organismen *Wilhelm Roux' Arch.* **1**: 269–305 (1894).

29. Tompkins, R., Rodman, W.P.: The cortex of *Xenopus laevis* embryos: Regional differences in composition and biological activity. *Proc. Nat. Acad. Sci.* (USA) **68**: 2921–2923 (1971).

30. Ubbels, G.A., Gerhart, J., Kirschner, M., Hara, K.: Determination of dorso-ventral polarity in the anuran eggs: Reversal experiments in *Xenopus laevis*. *Anat. Micr. Morph. Exp.* **68**: 211 (abstract) 1979.

31. Wakimoto, B.T.: DNA synthesis after polyspermic fertilization in the axolotl. *J. Embryol. Exp. Morphol.* **52**: 39–48 (1979).

32. Weyer, C.J., Nieuwkoop, P.D., Lindenmayer, A.: A diffusion model for mesoderm induction in amphibian embryos. *Acta Biotheoretica* **26**: 164–180 (1978).

Questions for Discussion with the Editors

1. *Is it reasonable to say that you regard embryonic dorsality as the product of a sequence of events, rather than the result of a single dramatic event that precedes first cleavage?*

Definitely. According to the views expressed in my chapter, a sequence of events occurs. Each event in turn sets up the next event. For example, during oogenesis the density compartments are built up. Next, activation releases those compartments to shift. Once shifted, novel metabolic activity follows. Finally, either a set of morphogenetic determinants is set in place, or (more likely) a morphogen gradient is established. The action of the morphogen gradient (or determinants) probably leads to the gene expression patterns which permanently establish dorsal-side activities such as involution. Each event leads—in a stepwise fashion—to the next. Viewing dorsalization in that fashion accounts for the wide variety of perturbations which can modify dorsal axis specification. Those perturbations are considered—in the above scenario—to alter one or more of the individual steps.

In summary, dorsalization is not considered to be a *single* well-orchestrated event. Instead, it is the product of a series of steps, none alone being the sole determinative event of dorsalization.

2. *Couldn't animal/vegetal and dorsal/ventral polarity have absolutely nothing in common? Perhaps you have unfairly stretched the A/V egg inversion data by attempting to unify them in the compartment density model?*

The first point is the A/V and D/V polarity both represent organizational states of the egg cytoplasm that are sensitive to gravity perturbation. One polarity is, however, perpendicular to the other. In order to account for the reorganization of the egg cytoplasm for D/V polarity establishment without disturbing A/V polarity, a model integrating the two is called for. A second point concerns the ultimate significance of D/V polarity—to program involution and mesoderm function at gastrulation. That D/V activity would be meaningless to the embryo unless competence to respond to the inductive signal arises in the animal hemisphere in concert with the emergence of the inductive signal. Since induction (D/V) and responsiveness (A/V) are linked, a unified model is warranted.

Chapter **19**

The Early Amphibian Embryo— A Hierarchy of Developmental Decisions

Jonathan M.W. Slack

AMPHIBIAN EMBRYOS have been favorite material for the experimentalist for over 100 years and accordingly we know more about them than about any other kind of embryo. Their technical advantages are threefold: they are laid in jelly coats and remain accessible to experimentation at all development stages, they are large enough to permit microsurgery, and it is possible to obtain advanced development of small explants cultured *in vitro* in simple, buffered salt solutions.

The establishment of the polar axes of the embryo is discussed in Chapter 18; this chapter is concerned with the process of regional specification from the fertilized egg to the end of neurulation, by which stage the general vertebrate body plan is complete.

Forms of Developmental Commitment

It is important to understand that different types of experiments give different types of information about the process of regional subdivision. The *prospective significance* or *fate* of a region in normal development can be established by marking the cells at an early stage and then finding where they go and what they become at a later stage. A *fate map* thus shows what will become of each region of the embryo in the course of normal development. It does not, however, give any information about the times at which cells become committed to their fates.

Commitment can be assessed by two kinds of experiments: isolation and transplantation. Isolation means cutting out a small explant and allowing it to develop *in vitro* in a buffered salt solution. If it develops into structure A, then we say that it was *specified* to form A at the time of explantation. Transplantation means grafting to other regions of the embryo at the same or a different stage. If a region forms a structure B, no matter where it is grafted, then we say that it is *determined* to form B. According to this definition, determination is irreversible and clonally heritable on cell division. It is important to note that although a map of determined regions must by definition correspond to the fate map for the stage this is not the case for a map of specified regions. For example, the animal-dorsal region of the early gastrula is fated to become neural plate in normal development. But if isolated, it will form epidermis, its state of specification thus being pre-epidermis. If grafted to other regions it can become any ecto or mesodermal structure, its state of determination thus being ecto-mesodermal. At a somewhat later stage when neural induction has taken place, this region has become determined to form neural plate and will accordingly do so not only in isolation but also after grafting to other regions.

As will become clear, the process of regional determination is hierarchical and each determinative decision involves a particular level of the hierarchy. For example, the neural plate is determined by the late gastrula but the eye territory within the neural plate is not yet determined. It is unfortunate that we do not at present possess direct means of observation of the various determined states and must infer their existence from terminally differentiated cell types that arise as the final result of the process.

Normal Development

It will be assumed that the reader is familiar with the visible events of cleavage, gastrulation, and neurulation in the amphibian embryo. If not, it will be worth consulting a textbook of embryology before reading on.

It should perhaps be emphasized that there is no net growth during early development. All the cell divisions are cleavage divisions in which the daughter cells are half the size of the mother cell and so any increase in size of the embryo is due simply to uptake of water. This is true of all holoblastic eggs which develop independently of the mother but is a fact often forgotten by mammalian and chick embryologists whose organisms undergo extensive growth simultaneously with regional specification.

At the beginning of the period we are considering, the embryo consists of at least four regions. An animal-vegetal polarity is established during oogenesis and a dorso-ventral polarity after fertilization. Both of these regional specifications are presumed to involve some sort of segregation of cytoplasmic components although nothing is known about their biochemical basis. In some species, the dorsal side of the fertilized egg is identifiable by the so-called "grey-crescent." A deterministic sequence of cleavage divisions inevitably means that particular regions of the zygote cytoplasm will be included in particular blastomeres, hence regional inhomogeneities may become a source of differential cellular commitment. In normal development, the first two cleavages are vertical and the third is equatorial. However if embryos are lightly compressed under a glass plate it is possible to cause the third cleavage also to occur in vertical orientation (41). Since such embryos develop normally once the plate is removed it seems likely that the precise position of cleavage planes relative to cytoplasmic regions is not critical for subsequent developmental decisions.

By the end of the period we are considering, the basic vertebrate body plan is visible in the form of a number of cellular condensations: neural tube, notochord, somites, lateral mesoderm, tailbud, etc. There is as yet no visible differentiation of the cells which are still densely packed with yolk granules. Later development involves the histological differentiation of the organs and a number of cellular migrations which take place on the scaffolding of the basic body plan, such as those of neural crest and haemopoietic cells.

Fate Maps

Amphibians possess a fate map which is continuous from the fertilized egg onwards showing that there is no stage of random cell mixing. Most of the experimental work has been carried out by the technique of vital staining. In this method, a small piece of agar impregnated with the stain is pressed against a particular region on the surface of the embryo until the cells have become deeply colored. The embryo is then allowed to develop to a later stage, is fixed and dissected and the location of the stained cells noted. Internal regions can be stained by using agar spikes which are inserted to a certain depth and which stain all the cells around them. This account of the gastrulation movements is based on the classic work of Vogt (68) as added to by Pasteels (54), Keller, (28, 29), and other authors.

Figure 1 shows the trajectories of a number of marks applied to the dorsal meridian. Those near the blastopore move in sequentially through the dorsal lip and eventually end up disposed in a continuous ring around the long axis of the archenteron. The marks near the animal pole never reach the blastopore and so end up inside the neural tube. In Fig. 2 a similar experiment is depicted in which equatorial marks are used. These become greatly elongated as the marginal zone enters through the ring-shaped blastopore but dorso-ventral continuity is maintained in the mesodermal mantle as shown in the transverse section of the tailbud stage (the ventral marks are present only posterior to the section shown). Figure 3 shows a fate map of a urodele gastrula compiled from many such marking experiments. All the prospective regions are shown on the surface but it is probable that the mesodermal regions extend to the interior. In the anura, the prospective lateral and ventral mesoderm is entirely internal at this stage. The boundaries between the prospective regions are shown as sharp lines but in reality they are probably somewhat fuzzy because of local cell mixing.

Vital staining is only applicable to clumps of cells and because the stain fades and becomes diluted it is not possible to say that every cell follows a strictly predetermined trajectory, simply that small regions do so. Recently the method of intracellular injection with horseradish peroxidase has been applied to the *Xenopus* embryo by Hirose and Jacobson (19) and Jacobson and Hirose (26). Amphibian embryos are well suited to a passive marking technique of this sort since the label is not diluted by growth. The enzyme is reasonably stable after injection and unlike the vital dyes is not passed to neighboring cells. It can be detected in single cells at a

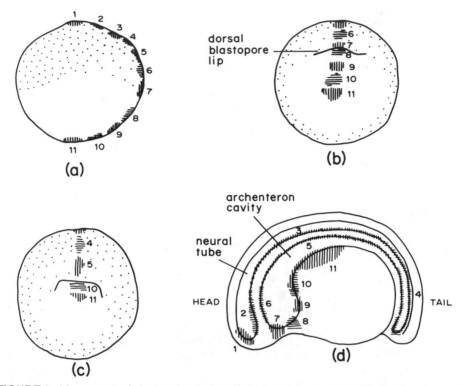

FIGURE 1. Movement of vital stain marks applied to the external surface of early urodele embryos. (a) shows the marks immediately after application to the blastula. (b) and (c) show the invagination of the marks through the dorsal lip (d) shows where they end up in the tailbud stage. After Vogt (68).

later stage by the very sensitive histochemical methods for peroxidase. This has yielded results for the central nervous system which are of higher resolution than the vital stain fate maps although not incompatible with them. The method has comfirmed suspicions that there is some cellular mixing at the edge of each clone, this being particularly noticeable in the ventral epidermis of the neurula in which there is extensive mixing of cells across the midline (author's unpublished results).

If it is shown that a clone remains coherent and ends up entirely within a particular structure at a later stage of development, this does not mean that the cell was committed to form that structure at the time it was labelled, since in embryos having a fate map, every structure must have some predictable prospective region at an earlier stage quite irrespective of any developmental decisions which may be made. However the converse is valid: if a clone crosses the boundary between two structures A and B, this means that the cell was not irreversibly committed to become either A or B at the time of labelling. So if there is a reasonable amount of cellular mixing locally some information about states of determination can be obtained by clonal analysis of cells which lie near prospective boundaries. At the time of writing, no such data has yet been published although we may expect it to appear soon since the problem is an important one and the techniques are now available.

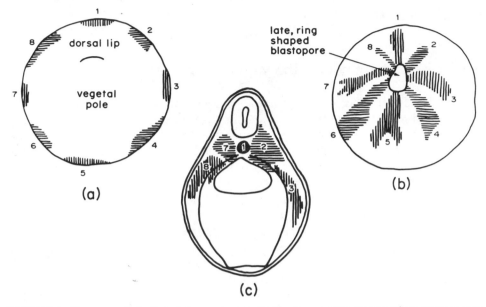

FIGURE 2. Movement of vital stain marks applied to the marginal zone of an anuran gastrula. The marks are originally equally spaced around the equator. They invaginate through the blastopore and end up as longitudinal stripes concentrated toward the dorsal side of the mesodermal mantle. Mark 1 is in notochord, 2 and 7 in somites, 3 and 8 in lateral plate. After Vogt (68).

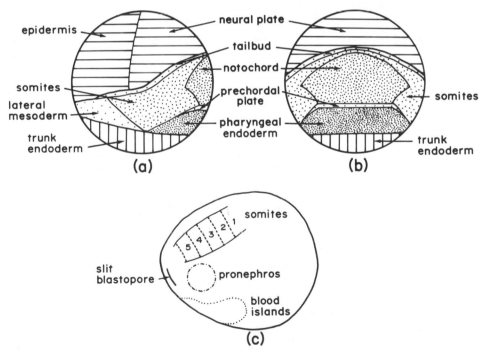

FIGURE 3. Fate maps of the early amphibian embryo compiled from vital staining data. (a) and (b) show the urodele blastula from the lateral and the dorsal aspects with the prospective regions in the surface layer indicated. After Pasteels (54). (c) shows prospective regions in the mesoderm of a urodele slit blastopore stage. After Yamada (70).

States of Commitment in the Morula and Blastula

The nuclei of early blastomeres do not appear to be restricted in potency since one nucleus can support the development of an entire embryo. This was first shown by Spemann in 1928 (62). A hair loop was used to constrict a fertilized egg of *Triturus* into a dumbell shape in which one half contained the zygote nucleus and the other half did not. Cleavage takes place only in the nucleated half but if the loop is not too tight a nucleus may pass through the constriction up to the 64-cell stage and initiate development in the previously enucleated half. These halves developed into complete embryos despite the fact that their nucleus had already undergone several rounds of DNA replication and mitosis. A similar result was obtained by Briggs and King (8) who injected blastula nuclei into enucleated host zygotes. This procedure has subsequently been used to "clone" amphibia since a number of embryos can be grown using nuclei from a single blastula (18, 33).

It is unfortunate, and to some extent surprising, that we have no information about the states of specification or determination of whole single cells in early amphibian development, but modern techniques of cell marking and micromanipulation should enable such data to be available soon.

There is however a great deal of evidence about the states of specification and determination of small regions of tissue. As explained above, specification is assayed by explanation of small regions followed by culture in buffered salt solutions, a method introduced by Holtfreter (22, 23). Amphibian embryos have the great advantage from the experimenter's point of view that every cell contains some yolk platelets and these can serve as its nutrient source until after the larva hatches. This means that various cell types will differentiate if explants are maintained in buffered salt solution for a sufficient length of time. The cells of the explant continue to cleave and so increase in number but since there is no external nutrient supply they do not increase in dry weight. Actually the advantages of this method are less than they appear at first sight, for three reasons.

First, explants usually give rise to more than one cell type. This is because histologically recognizable cell types are formed as a result of late local interactions rather than the interactions which partition the early embryo into regions. Fortunately in amphibian embryos, some cell types develop only from certain subdivisions of the body and so in the case of an explant it is often possible to assign it a state of specification on the basis of the cell types which are produced. Secondly, the assumption that the buffered salt solution is a "neutral" medium may perhaps be justified, but it is not so clear that it is a "simple" medium. Explants do not usually differentiate well unless they become surrounded by epidermis, either produced from the explant itself or supplied by the experimenter. So the microenvironment within the epidermal vesicle is an unknown quantity. Thirdly, single dissociated cells do not differentiate as well as clumps of tissue, even unwrapped with epidermis. Only at neurula and later stages will single cells differentiate reliably (16). So presumably something produced by the cells themselves is necessary for their proper development.

It is important to understand the drawbacks of this method not because it is without value, but because it has been of such decisive importance in understanding early amphibian development. It is largely because of this method that the dynamics of the early amphibian embryo are so much better understood than that of mammals or birds.

From morulae up to about the 64-cell stage explants from the animal hemisphere will develop into balls of ciliated epidermis while explants from the vegetal hemisphere fail to differentiate. From this stage onward, explants from the equatorial region (the marginal zone) will develop into mesodermal tissues indicating that a mesodermal region has been specified (42, 43). From the earliest stage at which this is possible, there is a difference between dorsal and ventral marginal zone. Explants from dorsal marginal zone form notochord and muscle, and those from ventral marginal zone tend to form mesenchyme and blood (Fig. 4). These are the tissues that arise from the dorsal and ventral mesoderm in the course of normal development.

Experiments carried out by P. D. Nieuwkoop and his coworkers (49) indicate that this mesodermal rudiment is brought into existence as a result of an interaction along the animal-vegetal axis (Fig. 5). Axolotl blastulae were divided into four zones along this axis with I and II drawn from the animal cap, III being the marginal zone, and IV the vegetal yolky region. When cultured alone, I and II tended to form epidermal bags and IV showed poor differentiation. But III developed into a deformed but not unreasonable complete embryo. When the regions were recombined and cultured it was found that the combination I + II + IV also yielded a reasonable version of a complete embryo (Fig. 5). When the orientation of IV was altered relative to I + II it could be shown that the dorsal structures developed on

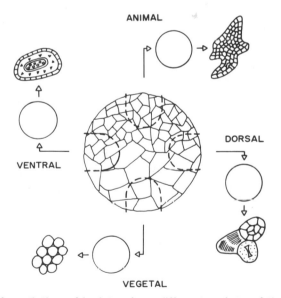

FIGURE 4. Self-differentiation of isolates from different regions of the amphibian blastula. Isolates from the animal cap produce only epidermis, isolates from the vegetal region fail to differentiate, isolates from the dorsal marginal zone produce notochord, muscle and neuroepithelium, while isolates from the ventral marginal zone produce epidermis, mesenchyme, and erythrocytes.

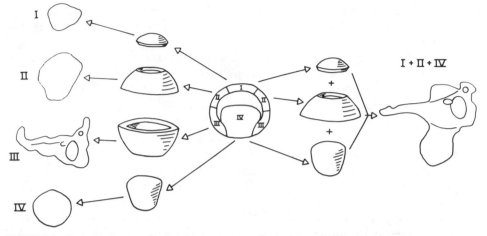

FIGURE 5. Development of annular fragments from the axolotl blastula. Only zone III can produce a reasonably complete embryo in isolation, but a similar embryo can be produced from zones I, II, and IV in combination.

the side which was dorsal in IV, irrespective of the orientation of I, II. This is despite the fact that the grey crescent itself is mainly confined to II and III. When the I and II/IV recombination was assembled from marked components it was established that the entire mesoderm and the pharyngeal endoderm were derived from the I and II section. So, it seems that even after the mesodermal rudiment has been specified, it can still be reconstituted from its surroundings, or if isolated, it can still reconstitute the remainder of the embryo.

The interaction is mimicked by two agents that are active on isolated blastula or gastrula ectoderm: lithium ion and the "mesoderm inducing factor." This effect of lithium is only one of several effects on early amphibian development (see below) and indeed on effects on early embryos in general. Masui (36) reported that isolates of *Triturus* gastrula ectoderm would develop into mesodermal tissues after treatment with lithium chloride. Nieuwkoop (48) is of the opinion that whole embryos or ectoderm-endoderm combinations can be vegetalized by lithium but isolated ectoderm cannot.

The existence of the "mesoderm inducing factor," here referred to as M-factor, was first inferred by Chuang (9) when he found that adult kidney contained a heat-labile mesoderm inducing activity as well as a heat-stable neural inducing activity (see below). Since then much effort has gone into the isolation of the active principle (66). The current M-factor is a purified protein of molecular weight 30,000 isolated from late (11–13 day) chick embryos. Blastula or early gastrula ectoderm will develop into mesodermal tissues after treatment and it is apparently also possible to determine which mesodermal tissue predominates by varying the time of treatment, or by combination with less pure preparations (1). Notochord and muscle are indicative of a dorsal mesoderm state of specification and mesenchyme and blood of a ventral mesoderm state.

So it seems that there are two agents which will achieve *in vitro* what the normal A-V axis interaction does *in vivo*: namely to transform ectoderm into mesoderm or even endoderm. It is thus somewhat disappointing that the biochemical investigations have not proceeded further than they have. The biochemical action of lithium is not known, although in adult brain tissue it is supposed to accelerate metabolism of noradrenaline and serotonin. Serotonin is present in the early amphibian embryo in small amounts (3), but shows no significant regional distribution and is unaffected by treatment with lithium (Forman and Slack, *unpublished results*).

The M factor is obtained from late chick embryos or other heterologous sources. It has been claimed that a similar activity can be extracted from early amphibian embryos but it has not been purified, and it must remain an open question whether it is really a similar protein or not. It seems unlikely that a 30,000 molecular weight protein would itself be a signal substance in an embryo as large as an amphibian where we might expect the time required for diffusion to be limiting. It is known that the amphibian blastula is coupled electrically, suggesting that gap junctions connect all cells together (59) so it is tempting to look for the signal substances among the small molecules which can pass through the junctions. It is possible that the M factor might affect the production or the removal of such substances.

In any case, it is clear that during its development from early cleavage to late blastula, the amphibian embryo acquires several distinctly specified regions. To begin with there are probably only four: the quadrants arising by passive partition of the zygote cytoplasm. By the late blastula, there is an ectoderm, mesoderm, and endoderm, and the mesoderm at least is divided into two or more zones in the dorso-ventral axis.

As emphasized above, states of specification may be reversible and in fact we know of various regulative properties possessed by the system of territories in the blastula. First we will consider the effect of changing the size of the embryo and then enquire which particular regions of the blastula are necessary for the reconstitution of the total pattern.

Pattern Regulation in the Morula and Blastula

The most straightforward way of changing the size of an embryo is by removing parts of the fertilized egg. Kobayakawa and Kubota (30) reported experiments in which fertilized eggs are slowly bisected by a glass rod into a nucleated and enucleated half. This procedure can be repeated to produce embryos having only a quarter of the normal volume. The enucleated fragments do not develop, but the nucleated ones do develop and produce normally proportioned embryos. Similar conclusions can be drawn from the classical constriction experiment involving separation of the first two blastomeres, at least one and usually both forming normally proportioned embryos. Cooke (12, 14, and Chapter 20) has made quantitative studies of somite number and of the number of cells in the subdivisions of the mesoderm and has shown that reduction of blastula volume by about 50% does not affect the proportions significantly.

It is also possible to create double-size embryos by the aggregation of two 2-cell stages (35). This procedure frequently yields embryos with multiple axes, but a single embryo can be formed when one embryo with medial cleavage is fused with one with frontal cleavage in such a way as to align the dorso-ventral axes. Such embryos are twice the normal size and are at least approximately normally proportioned.

Taking these experiments together, we see that it is possible to vary the volume of an amphibian embryo by a factor of 8, and hence vary each linear dimension by a factor of the cube root of 8, which is 2. Over this range, the size of each part of the body is scaled to the size of the whole and each part is composed of a different number of normal-sized cells from the usual. This conclusion is not affected by whether the volume change is made before the territories are established or while they are being established.

It is also possible to vary the cell size but keep the total embryo volume constant. Application of a pressure of about 6,000 psi around the time that the second polar body is expelled can cause its resorption and, often, its incorporation into the zygote nucleus along with male and female pronuclei. Such embryos then have three chromosome sets and are called triploid. Triploid cells are 50% larger than diploid ones but are otherwise normal and able to carry out mitosis. Pressure shocks, and also hot and cold shocks, can on occasion give rise to embryos of other ploidy ranging from haploid to heptaloid (17). These animals have been studied exhaustively by Fankhauser and Humphrey and it seems clear that although the embryos are of normal size, each structure within them is composed of a different number of cells from usual. In haploid embryos, there are twice as many cells as usual and in polyploids there are fewer than usual in inverse proportion to the increase in cell volume.

So the size of parts does not depend on the size of individual cells, nor does it depend on some absolute dimension specified in the developmental program; rather it depends on the size of the whole embryo. This suggests that the specification of territories depends on signals which extend across the whole extent of the embryo and can scale each territory to the amount of tissue available.

The second important aspect of regulation is the regional variation in regulative ability. In the early stage, it seems probable that it is necessary to include the dorsal region in order to obtain a complete embryo. This is suggested by the fact that separation of the first two blastomeres can produce two complete ½ size embryos if the first cleavage is medial, but often produces one complete ½ size embryo + one "belly piece" if the first cleavage is frontal (62).

In the blastula, the work of Nieuwkoop suggests that following division of the embryo along the egg axis only the prospective marginal zone (Zone III) can form a more or less complete embryo in isolation. However the combinations I–II–III and I–IV also form reasonably complete embryos, indicating that the zone III can be reformed from the extremes of the egg axis. These zones each comprise about ¼ the volume of the egg so presumably the size reduction is not in itself a sufficient reason for a failure of regulation, since we have previously seen that normally proportioned embryos can develop from ¼-sized fragments of fertilized eggs.

Several "two gradient" theories have been advanced to explain various aspects of early amphibian development (15, 57). The following description is an extension of these and an attempt to summarize the results in a form that suggests something about the underlying mechanism of the interactions.

1. The blastula is partitioned into territories arbitrarily labelled 1–4 in the egg axis and 1'–4' in the ventral-dorsal axis. Each territory is identified by two codings: e.g., (3,3') and left and right sides are mirror images with identical codings.

2. If the embryo is fragmented or rearranged the codings become re-established across the space available. Territories can be halved in size in each linear dimension but cannot be further reduced.

3. The extreme codings are the highest present in the fragment. High values can generate lower ones but not vice versa.

4. By commencement of gastrulation, the codings begin to control the behavior of the cells in each region.

This description is illustrated in Fig. 6. There is one region of the blastula

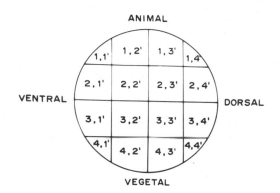

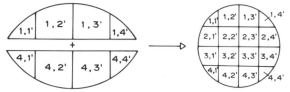

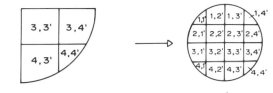

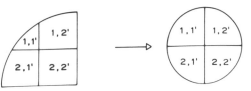

FIGURE 6. Formal description of the regulative properties of the amphibian blastula. The normal arrangement of territories is shown. In fragments, high codings can generate lower ones but not vice versa. A twofold reduction in linear dimensions is allowed for the size of each territory.

which has the highest codings in both axes and can therefore, in principle, generate all the other values in the embryo. This region is called the organizer.

The Organizer

No subject in embryology has been so misrepresented and misunderstood as the properties of the organizer. This is because there are several interactions involved in early amphibian development and a number of different assay procedures which can detect one or more of them at a time. In particular, the organizer should not be confused with "mesodermal induction" which is the regionalization along the egg axis discussed above, nor with "neural induction" which is discussed below.

The properties of the organizer must be understood in the context of grafting experiments on the early gastrula of which some of the most definitive are interspecies grafts carried out between *Triturus vulgaris* (= *taeniatus*) and *T. cristatus* [reviewed in (62)]. The former has more deeply pigmented eggs and so it is possible to identify at later stages which parts of the embryo are derived from the graft and which from the host. Three types of graft develop according to their new positions. These are dorsal ectoderm to ventral ectoderm, ventral ectoderm to dorsal ectoderm, and ectoderm to mesoderm. These results suggest that there is no irreversible determination of dorsal versus ventral ectoderm, and that, as in the blastula, it remains possible to vegetalize the ectoderm. They do not however show that these parts of the early gastrula are completely unspecified, or to use Spemann's term "indifferent," indeed we have seen above that the entire ectodermal region is specified to become epidermis at this stage.

The original organizer graft is a graft of dorsal mesoderm from the early blastopore into the ventral marginal zone (63). In favorable cases, this graft continues its invagination to form a second set of dorsal mesodermal structures, mainly notochord and prechordal plate, on the ventral side of the host. It also "captures" a certain region of host tissue and dorsalizes this so that it becomes the outer part of the secondary mesodermal axis (somites, kidney, lateral plate) (Fig. 7). The subsequent events of neural and endodermal regionalization follow from this, so that the end result can be the formation of a complete "secondary" embryo on the ventral side, whose parts are arranged in a relation of mirror symmetry to those of the original embryo. The two salient characteristics of the organizer are (1) that its own state of determination is not affected by its new position, and (2) that it can "dorsalize" the surrounding ventral mesoderm of the host to form a second embryonic axis.

An alternative method of performing the organizer graft was introduced by Mangold in which the graft is inserted into the blastocoel through a hole made in the ectoderm. This method is easier than implantation into the ventral marginal zone and much of the classical work on the organizer made use of it. Unfortunately it is difficult to control exactly where the graft ends up and so the result is a product of several processes which operate to different extents depending on the details of the experiment; namely vegetalization of ectoderm in contact with the graft, dorsalization of this induced mesoderm or of host mesoderm, self-differentiation of the graft, and, last but not least, neural inductions. More than any other single factor

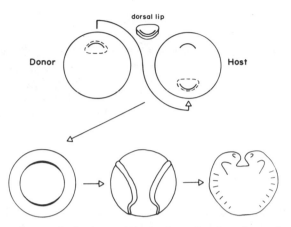

FIGURE 7. The organizer graft. A piece of tissue from the dorsal marginal zone is grafted into the ventral marginal zone and induces the surrounding tissue to participate in the formation of a double-dorsal embryo.

it is the complexity of this experiment which led to the idea that the organizer is an agent of neural induction rather than an agent of mesodermal dorsalization.

Recent studies of the organizer have been made by Cooke (10, 11, 13, 14) using *Xenopus laevis*. The capture by the graft of a certain invagination territory is rapid, within 1–2 hours. It is always smaller than that of the host organizer unless the host dorsal lip region is simultaneously removed. The number of cells in the mesoderm of double embryos is the same as that in control embryos which implies that the ventral mesoderm is dorsalized without any extra cell division taking place. Finally, the organizer not only captures the ventral part of the host but also compresses the pattern of the host mesodermal structures in the dorso-ventral axis, so the total number of distinctly determined zones in the double embryo is increased. Using explants cultured *in vitro*, Slack and Forman (60) showed directly that the ventral marginal zone could be dorsalized by co-culture with organizer tissue: explants that normally produced blood islands formed large masses of muscle after a certain period in contact with the dorsal lip region. It is thought that tissue from about 60 degrees on either side of the early blastopore has organizer activity (6, 37, 40). There is no space in this chapter for a discussion of models for pattern formation. However the properties mentioned suggest that the organizer is the source region for a dorso-ventral gradient responsible for the regional subdivision of the mesoderm. It is probably not a simple concentration gradient, the phenomena being better explained by one of the more complex gradient models (e.g., 38).

Regulation in the Gastrula

The regulative abilities of the early gastrula are less than those of the blastula, perhaps because there is now not much time left before large parts of the embryo become irreversibly determined. It is still possible to remove the ventral half of the early *Triturus* gastrula and obtain a reasonably well proportioned embryo from the dorsal half (55). However defects in the dorsal region often lead to defects in the later embryo unless they are very small.

The nuclei of gastrulae are not as effective at supporting development of enucleated eggs as those from blastulae, but some will still do so. The consensus among nuclear transplantationists is that the ability to support egg development declines gradually but that it is possible to obtain a few positive results even from nuclei of adult differentiated cells. The decline is thought to be due to factors other than an irreversible reduction in potency.

Formation of the Craniocaudal Pattern in the Course of Invagination

On the dorsal side of the embryo the region destined to invaginate may extend 30 to 90 degrees (depending on species) from the blastopore. The fates of regions of this dorsal marginal zone are, from the blastopore towards the animal pole, first pharyngeal endoderm, then prechordal plate, then notochord, and finally the dorsal part of the tailbud (53). The effects of removing the dorsal lip region at different stages of gastrulation are consistent with this fate map. Removal of the dorsal lip from early gastrulae leads to deficiencies in the head region, although such defects often regulate; removal of dorsal lips from mid-gastrulae leads to deficiencies in the trunk region, and removal of dorsal lips from late gastrulae leads to deficiencies in the tail (31, 58).

But the self-differentiation behavior of small explants from this region at early gastrula stages are rather different from the fate map. The region closest to the blastopore tends to form notochord and neural tissue and the region further away tends to form epidermis and pigment cells (24). The prospective trunk mesoderm only acquires a self-differentiation behavior of notochord/neural type by the mid-gastrula (27). Specification of the dorsal mesoderm does not, therefore, correspond to its fate until after invagination.

When grafted to the ventral marginal zone, an early dorsal lip can, as we have seen, form a complete secondary embryo with head, trunk and tail by a combination of self-differentiation and induction. But a dorsal lip from a mid-gastrula grafted to the ventral marginal zone of an early gastrula will form only trunk and tail, and a dorsal lip from a late gastrula only tail (61, 37). The dorsal lip itself contains both invaginated and uninvaginated tissue and it seems probable that the craniocaudal level of determination of each cohort of cells is achieved during invagination. This can be assayed by testing for neural induction activity by implantation into the blastocoel. In such experiments, the uninvaginated early gastrula dorsal marginal zone induces tail structures while invaginated early gastrula dorsal mesoderm induces head structures (64).

Gastrulae treated with lithium chloride tend to develop into embryos that have axial defects (32, 2). If early gastrulae are treated, the defects are concentrated in the head, while if later gastrulae are treated, the defects occur in the head and also more posteriorly. The primary effects are certainly on the mesoderm but gaps arise in the central nervous system as a consequence of the failure of neural induction. Hence cyclopia and anophthalmia are common head defects.

This evidence seems to suggest that the dorsal mesoderm acquires its craniocaudal state of determination on invagination[1]. Prior to invagination the self-differentiation behavior is still governed by the signals in the blastula (Fig. 6). After invagination, the archenteron roof (as it has now become) is determined to become head, trunk, or tail, and, as we shall see below, to induce corresponding neural structures from the overlying ectoderm. Future research on the mechanism of craniocaudal determination will probably focus on the changing sequence of microenvironments to which the invaginating cells are exposed.

Neural Induction

In the original organizer grant of Spemann and Mangold (63), one of the results, and the one on which most subsequent attention has concentrated, was that the graft induced a second nervous system from that part of the host ectoderm that came to lie above it at the end of gastrulation. In the absence of the graft, this tissue would have become ventral epidermis. Since it was also known that prospective epidermis and neural plate (dorsal and ventral ectoderm) could be exchanged in the gastrula and would each develop according to their new position, Spemann (62) argued that the nervous system was formed in *normal* development as a result of an inductive signal from the archenteron roof. This interpretation was strengthened by the discovery of Holtfreter (20) that the entire ectoderm could develop into epidermis if isolated from the mesoderm. This occurs in "exogastrulae" which arise if embryos are cultured in stronger salt solution than usual; the morphogenetic movements working to exclude the mesoderm and endoderm from the ectoderm so that the ectoderm becomes isolated as an empty sac.

In 1932 it was shown that killed dorsal lips would induce some neural tissue following implantation into the blastocoel (7). Because of the confusion between the organizer and the neural inductor properties of the dorsal lip this led to the famous "gold rush" for the biochemical basis of the organizer. It was found that many tissues from adult animals or indeed many purified chemical substances would induce neural tubes or patches of neuroepithelium from gastrula ectoderm. After a period of confusion, it become clear that the ectoderm was fairly delicately balanced between epidermal and neural pathways and that a variety of stimuli could tip the balance one way or the other.

Regional Specificity

One problem about much of the early work on neural induction was that the response consisted in the formation from gastrula ectoderm of nondescript patches of neuroepithelium sometimes rolled up into vesicles of neural tube. This made it unclear whether these reactions really resembled neural induction by the archenteron roof since the salient characteristic of the natural signal is regional specificity; different parts of the archenteron roof inducing different parts of the central nervous system. This specificity was shown by Mangold (34) who implanted explants from different craniocaudal regions of archenteron roof into the blastocoel of host *Triturus* embryos, and also by Holtfreter (21) who placed ectodermal explants on different regions of the surface of exogastrulae. Similar results were obtained by Horst (25) in combinations of these explants with gastrula ectoderm in isolation.

This implies that several distinct signals are involved in neural induction and this is usually recognized in diagrams which attempt to account for the course of development in terms of chains of induction (e.g., 44). However there is also evidence that there can be interactions within the neural plate itself. It has been known for a long time that explants of neural plate are themselves neural inductors. This is called homeogenetic induction (induction of like by like), and was

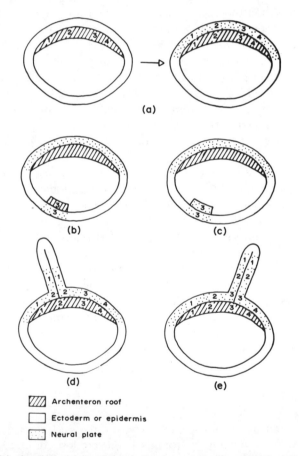

FIGURE 8. Neural induction. (a) In normal development, the neural plate is induced by the archenteron roof. (b) The archenteron can induce neural tissue from prospective epidermis. (c) Neural plate can induce further neural tissue of the same regional specificity from prospective epidermis (homeogenetic induction). (d, e) Serially ordered structures are induced in ectodermal folds which were implanted into the neural plate.

shown by Mangold (34) and Horst (25) to have some degree of regional specificity, in that prospective brain induced brain and prospective tail induced tail. Nieuwkoop (45-47) showed that if small folds of gastrula ectoderm are implanted into a neural plate, the basal parts of the folds develop into similar structures as are formed by their surroundings while the apical parts of the folds formed more anterior structures and that these tended to be arranged in the normal sequence. So anterior neural plate could induce only forebrain, but posterior neural plate could induce a complete sequence of structures.

These indications that interactions can occur within the neural plate, although perhaps only in abnormal situations, suggest that the signals from the archenteron roof may be representable as some sort of serial hierarchy as implied in Fig. 8 where they are shown as a series of numbers. This idea is implicit in the "gradient" models for neural induction put forward by Yamada (67) and Toivonen (70).

Biochemical Mechanism of Neural Induction

Biochemical studies on neural induction have been bedevilled by the complexity of the *in vivo* situation. As has been noted above, this makes the relevance of some of the assays which have been used for active substances rather questionable. It seems to be agreed that a "neuralizing factor" (N-factor) can be extracted from a number of adult tissues, such as liver, and that it is heat stable. On its own, it induces forebrain structures ("archencephalic induction"). The M-factor, which has already been discussed in connection with the regionalization of the blastula, will induce notochord, somite, and neural tube ("spinocaudal induction"). When N and M are applied together, structures appropriate to all craniocaudal levels may be induced (57, 66, 67).

Experiments have been carried out in which the inductor and the reacting tissue are separated by various types of filter. Saxen (56) showed that neuralization could take place across a Millipore filter. Tarin *et al.* (65) showed that it could take place across Nucleopore filters of small pore size (down to 0.1μ) but not across cellophane membranes. In these experiments, dorsal blastopore lip was used as the inducer so presumably the signal involved is the natural one. It is believed that 0.1μ Nucleopore membranes cannot be penetrated by cytoplasmic processes and so the implication is that the natural signal, as well as various unnatural ones, can be transmitted in the form of extracellular diffusible molecule(s).

Studies on the electrophysiology of the neural plate by Warner (69) suggest that neural differentiation in the axolotl is underway as early as the neural plate stage. The resting membrane potentials rise from about -15 mV characteristic of ectoderm to about -40 mV at this time. The region in the center of the "keyhole" of the neural plate develops the highest potential, and since the whole neural plate is electrically coupled, this implies that there is a current flow towards this region. Whether these early events are causes or consequences of the regionalization of the neural plate is not known, although there is evidence that a functional sodium pump is required for differention of neurons *in vitro* (39).

The Endoderm

The regionalization of the endoderm has not been the most popular of research topics but the results we have suggests that the territories become determined by the middle neurula under the influence of the adjacent mesoderm.

A fate map of the early neurula endoderm was produced by Balinsky (4) and used to examine the results of interchanging stomach and liver prospective regions (5). Normal embryos arose from these grafts indicating that determination had not yet occurred. The question was later examined by Okada using explants (50–52). He found that gastrula endoderm alone will not differentiate *in vitro*. Prospective pharyngeal endoderm would form pharynx in combination with archenteron roof tissue and intestine in combination with lateral plate mesoderm. Prospective gastric and intestinal endoderm would form branchial or oesophageal structures in the presence of head mesoderm and intestinal structures in the presence of lateral plate mesoderm.

Summary of Determinative Events in the Early Amphibian Embryo

The results which have been described in the present chapter may perhaps be brought together and summarized as follow (Fig. 9): Animal-vegetal polarity is established during oogenesis and dorso-ventral polarity is established after fertilization. Both of these apparently involve cytoplasmic localization, but we know nothing about the molecules which we presume to be segregated to the different regions.

The possibility of constructing fate maps from the stage of the fertilized egg onwards shows that there is no large scale mixing of cells. The gastrulation movements preserve the neighbor relations of cells in all positions except the ventrolateral blastopore lip where there is a rupture between mesoderm and endoderm.

The first inductive interaction occurs along the egg axis and results in the formation of the mesoderm. The process is mimicked *in vitro* by the action of the M factor on isolated ectoderm. As soon as the mesoderm is formed, the dorsal and ventral regions seem to be differently specified. The dorsal region is the organizer and emits an inductive signal which regionalizes the mesoderm in the dorso-ventral and perhaps also in the craniocaudal axes before and during gastrulation.

In the gastrula and neurula stages, the neural plate is induced by the action of the mesoderm on the overlying ectoderm. It is formed as an array of territories committed to become different parts of the nervous system. The endoderm is similarly regionalized under the influence of the mesoderm to form different parts of the visceral epithelium.

Up to the time at which the basic body plan has become determined, we therefore have at least two decisions involving cytoplasmic localization and at least five inductive signals, some of which have multiple outcomes and may be complex signals involving many biochemical substances.

Although the signals and responses involved in regional specification in early amphibian embryos are much better characterized than for other types of embryo,

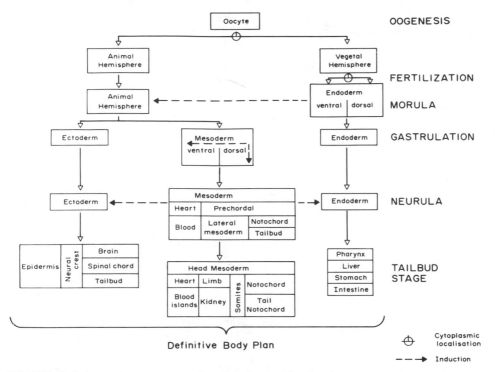

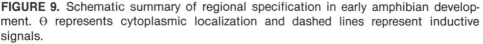

FIGURE 9. Schematic summary of regional specification in early amphibian development. θ represents cytoplasmic localization and dashed lines represent inductive signals.

we still have very little idea about the underlying mechanisms. We do not know either the identities of the chemical substances involved, nor the dynamic properties of the systems that must underlie such phenomena as size regulation or homeogenetic induction. The explanations of classical workers in this field must now be regarded as problems for solution.

Notes

[1] It should perhaps be mentioned that Spemann (1931) reported induction of head structures by late dorsal lips which is not compatible with this interpretation. However I have examined some of his slides held at the Hubrecht Laboratory and I am satisfied that this result can be explained by secondary interactions between the induced neural plate and that of the host.

Acknowledgments

This chapter is based on Chapter 3 of *From Egg to Embryo": Determinative Events in Early Development* by J. M. W. Slack which will be published by Cambridge University Press in 1983. I am grateful to the Press for permission to reproduce the figures.

General References

SPEMANN, H.: *Embryonic Development and induction.* New York: Hafner (1938, *reprinted* 1967).

NAKAMURA, O., TOIVONEN, S., EDS.: *Organizer—A milestone of a half century from Spemann.* Amsterdam: Elsevier North Holland (1978).

GERHART, J.C.: Mechanisms regulating pattern formation in the amphibian egg and early embryo. In *Biological regulation and development.* Vol. 2 pp 133–316 (ed.) R. Goldberger. New York: Plenum (1981).

References

1. Asahi, K., Born, J., Tiedemann, H., Tiedemann, H.: Formation of mesodermal pattern by secondary inducing interactions. *Wilhelm Roux' Arch. Dev. Biol.* **187**: 231–244 (1979).

2. Bäckström, S.: Morphogenetic effects of lithium on the embryonic development of *Xenopus. Ark. Zool.* **6**: 527–536 (1954).

3. Baker, P.C.: Changing serotonin levels in developing *Xenopus laevis. Act. Embryol. Morphol. Exp.* **8**: 197–204 (1965).

4. Balinsky, B.I.: Kinematik des endodermalen Materials bei der Gestaltung der Wichstigsten Teile des Darmkanals bei den Amphibien. *Wilhelm Roux' Arch. EntwMech. Orgs.*, **143**: 126–166 (1947).

5. Balinsky, B.I.: Korrelation in der Entwicklung der Mund und Kiemenregion und des Darmkanals bei Ambphibien. *Wilhelm Roux' Arch. EntwMech. Orgs.* **143**: 365–395 (1948).

6. Bautzmann, H.: Experimentelle Untersuchungen zur Abgrenzung des Organisationszentrums bei *Triton taeniatus. Wilhelm Roux' Arch. EntwMech. Orgs.* **108**: 283–321 (1926).

7. Bautzmann, H., Holtfreter, J., Spemann, H., Mangold, O.: Versuche zur Analyse der Induktionsmittel in der Embryonalentwicklung. *Naturwissenschaften* **20**: 971–974 (1932).

8. Briggs, R., King, T.J.: Transplantation of living nuclei from blastula cells into enucleated frogs' eggs. *Proc. Nat. Acad. Sci. USA* **38**: 455–463 (1952).

9. Chuang, H.H.: Induktionsleistungen von frischen und gekochten Organteilen (Niere, Leber) nach ihrer Verpflanzung in Explantate und verscheidene Wirtsregionen von Tritonkeimen. *Wilhelm Roux' Arch. EntwMech. Orgs.* **139**: 556–638 (1939).

10. Cooke, J.: Properties of the primary organization field in the embryo of *Xenopus laevis.* I. Autonomy of cell behaviour at the site of initial organizer formation. *J. Embryol. Exp. Morph.* **28**: 13–26 (1972).

11. Cooke, J.: Properties of the primary organization field in the embryo of *Xenopus laevis.* V. Regulation after removal of the head organizer in the normal early gastrulae and in those already possessing a second implanted organizer. *J. Embryol. Exp. Morph.* **30**: 283–300 (1973).

12. Cooke, J.: Control of somite number during morphogenesis of a vertebrate, *Xenopus laevis. Nature* **254**: 196–199 (1975).

13. Cooke, J.: Cell number in relation to primary pattern formation in the embryo of

Xenopus laevis. I. The cell cycle during new pattern formation in response to implanted organisers. *J. Embryol. Exp. Morph.* **51**: 165–182 (1979).

14. Cooke, J.: Scale of body pattern adjusts to available cell number in amphibian embryos. *Nature* **290**: 775–778 (1981).

15. Dalcq, A., Pasteels, J.: Une conception nouvelle des bases physiologiques de la morphogènese. *Arch. Biol.* **48**: 669–710 (1937).

16. Elsdale, T., Jones, J.: The independence and interdependence of cells in the amphibian embryo. *Symp. Soc. Exp. Biol.* **XVII**: 257–273 (1963).

17. Fankhauser, G.: The effects of changes in chromosome number on amphibian development. *Quart. Rev. Biol.* **20**: 20–78 (1945).

18. Gurdon, J.B.: *The control of gene expression in animal development.* Oxford: Oxford University Press (1974).

19. Hirose, G., Jacobson, M.: Clonal organization of the central nervous system of the frog. I. Clones stemming from individual blastomeres of the 16 cell and earlier stages. *Dev. Biol.* **71**: 191–202 (1979).

20. Holtfreter, J.: Organisierungsstufen nach regionaler Kombination von Entomesoderm mit Ektoderm. *Biol. Zentral.* **53**: 404–431 (1933).

21. Holtfreter, J.: Regionale Induktionen in xenoplastisch zusammengesetzten Explantaten. *Wilhelm Roux' Arch. EntwMech. Orgs.* **134**: 466–550 (1936).

22. Holtfreter, J.: Differenzierungspotenzen isolierter Teile der Urodelengastrula. *Wilhelm Roux' Arch. EntwMech. Orgs.* **138**: 522–656 (1938a).

23. Holtfreter, J.: Differenzierungspotenzen isolierter Teile der Anurengastrula. *Wilhelm Roux' Arch. EntwMech. Orgs.* **138**: 657–738 (1938b).

24. Holtfreter-Ban, H.: Differentiation capacities of Spemann's organizer investigated in explants of diminishing size. PhD thesis, University of Rochester, (1965).

25. Horst, J. Ter: Differenzierungs und Induktionsleistungen verschiedener Abschnitte der Medullarplatte und der Urdarmdaches von Triton im Kombinat. *Wilhelm Roux' Arch. EntwMech. Orgs.* **143**: 275–303 (1948).

26. Jacobson, M., Hirose, G.: Clonal organization of the central nervous system of the frog. *J. Neurosci.* **1**: 271–284 (1981).

27. Kaneda, T., Hama, T.: Studies on the formation and state of determination of the trunk organiser in the newt *C.pyrrhogaster. Wilhelm Roux's Arch. of Dev. Biol.* **187**: 25–34 (1979).

28. Keller, R.E.: Vital dye mapping of the gastrula and neurula of *Xenopus laevis* I. Prospective areas and morphogenetic movements of the superficial layer. *Dev. Biol.* **42**: 222–241 (1975).

29. Keller, R.E.: Vital dye mapping of the gastrula and neurula of *Xenopus laevis* II. Prospective areas and morphogenetic movements of the deep layer. *Dev. Biol.* **51**: 118–137 (1976).

30. Kobayakawa, Y., Kubota, H.Y.: Temporal pattern of cleavage and the onset of gastrulation in amphibian embryos developed from eggs with reduced cytoplasm. *J. Embryol. Exp. Morph.* **62**: 83–94 (1981).

31. Lehmann, F.E.: Entwicklungsstörungen in der Medullaranlage von Triton, erzeugt durch Unterlagerungsdefekte. *Wilhelm Roux' Arch. EntwMech. Orgs.* **108**: 243–282 (1926).

32. Lehmann, F.E.: Mesodermisierung des praesumptiven Chordamaterials durch Einwirkung von Lithiumchlorid auf die Gastrula von *Triturus alpestris. Wilhelm Roux' Arch. EntwMech. Orgs.* **136**: 112–146 (1937).

33. McKinnell, R.G.: Cloning. *Nuclear transplantation in amphibia*. Minneapolis: University of Minnesota Press (1978).

34. Mangold, O.: Über die Induktionsfähigkeit der verscheidener Bezirke der Neurula von Urodelen. *Naturwissenschaften* 21: 761–766 (1933).

35. Mangold, O., Seidel, F.: Homoplastiche und heteroplastische Verschmelzung ganzer Tritonkeime. *Wilhelm Roux' Arch. EntwMech. Orgs.* 111: 593–665 (1927).

36. Masui, Y.: Mesodermal and endodermal differentiation of the presumptive ectoderm of *Triturus* gastrulae through the influence of lithium ion. *Experientia*, 17: 458–459 (1961).

37. Mayer, B.: Über das Regulations—und Induktionsvermogen der halbseitigen oberen Urmundlippe von Triton. *Wilhelm Roux' Arch. EntwMech. Orgs.* 133: 518–581 (1935).

38. Meinhardt, H., Gierer, A.: Generation and regeneration of sequences of structures during morphogenesis. *J. Theor. Biol.* 85: 429–450 (1980).

39. Messenger, E.A., Warner, A.E.: The function of the sodium pump during differentiation of amphibian embryonic neurones. *J. Physiol.* 292: 85–105 (1979).

40. Minganti, A.: Transplantations d'un fragment du territoire somitique présomptif de la jeune gastrula chez l'Axolotl et chez le Triton. *Arch. Biol. Paris* 61: 251–355 (1949).

41. Morgan, T.H.: *Experimental embryology*. New York: Columbia University Press (1927).

42. Nakamura, O., Matsuzawa, T.: Differentiation capacity of the marginal zone in the morula and blastula of *Triturus pyrrhogaster*. *Embryologia* 9: 223–237 (1967).

43. Nakamura, O., Takasaki, H.: Further studies on the differentiation capacity of the dorsal marginal zone in the morula of *Triturus pyrrhogaster*. *Proc. Jap. Acad.* 46: 546–551 (1970).

44. Nakamura, O., Hayashi, Y., Asashima, M.: A half century from Spemann: Historical review of studies on the organizer. Chapter 1 in *Organizer: A milestone of a half century from Spemann*. Amsterdam: Elsevier/North Holland (1978).

45. Nieuwkoop, P.D.: Activation and organization of the central nervous system in amphibians. I. Induction and activation. *J. Exp. Zool.* 120: 1–31 (1952a).

46. Nieuwkoop, P.D.: Activation and organization of the central nervous system in amphibians. II. Differentiation and organization. *J. Exp. Zool.* 120: 33–81 (1952b).

47. Nieuwkoop, P.D.: Activation and organization of the central nervous system in amphibians. III. Synthesis of a new working hypothesis. *J. Exp. Zool.* 120: 83–108 (1952c).

48. Nieuwkoop, P.D. The formation of the mesoderm in urodelean amphibians. III. The vegetalizing action of the Li ion. *Wilhelm Roux' Arch. EntwMech. Orgs.* 166: 105–123 (1971).

49. Nieuwkoop, P.D.: The "organisation centre" of the amphibian embryo, its origin, spatial organization and morphogenetic action. *Adv. Morphogen.* 10: 1–39 (1973).

50. Okada, T.S.: Role of the mesoderm in the differentiation of endodermal organs. *Mem. Coll. Sci., Univ. Kyoto.* 20: 157–162 (1953).

51. Okada, T.S.: The pluripotency of the pharyngeal primordium in urodelan neurulae. *J. Embryol. Exp. Morph.* 5: 438–448 (1957).

52. Okada, T.S.: Epitheliomesenchymal relationships in the regional differentiation of the digestive tract in the amphibian embryo. *Wilhelm Roux's Arch. EntwMech. Orgs.* 152: 1–21 (1960).

53. Okada, Y.K., Hama, T.: Prospective fate and inductive capacity of the dorsal lip of the blastopore of the *Triturus* gastrula. *Proc. Imp. Acad. (Tokyo)*. **21**: 342–348 (1945).

54. Pasteels, J.: New observations concerning the maps of presumptive areas of the young amphibian gastrula (Ambystoma and Discoglossus). *J. Exp. Zool.* **89**: 255–281 (1942).

55. Ruud, G., Spemann, H.: Die Entwicklung isolierter dorsaler und lateraler. Gastrula-hälften von *Triton taeniatus* und alpestris, ihre Regulation und Postgeneration. *Wilhelm Roux' Arch. EntwMech. Orgs.* **52**: 95–165 (1922).

56. Saxen, L.: Transfilter neural induction of amphibian ectoderm. *Dev. Biol.* **3**: 140–152 (1961).

57. Saxen, L., Toivonen, S.: *Primary Embryonic Induction*. Logos Press, (1962).

58. Shen, G.: Experimente zür Analyse der Regulationsfähigkeit der fruhen gastrula von Triton, zugleich ein Betrag zum problem der Cyclopie. *Wilhelm Roux' Arch. Entw-Mech. Orgs.* **137**: 271–316 (1937).

59. Slack, C., Warner, A.E.: Intracellular and intercellular potentials in the early amphibian embryo. *J. Physiol.* **232**: 313–330 (1973).

60. Slack, J.M.W., Forman, D.: An interaction between dorsal and ventral regions of the marginal zone in early amphibian embryos. *J. Embryol. Exp. Morph.* **56**: 283–299 (1980).

61. Spemann, H.: Über den Anteil von Implantat und Wirskeim an der Orientierung und Beschaffenheit der induzierten Embryonalanlage. *Wilhelm Roux' Arch. EntwMech. Orgs.* **123**: 389–517 (1931).

62. Spemann, H.: *Embryonic development and induction*. New York: Hafner (1938, *reprinted* 1967).

63. Spemann, II., Mangold, H.: Uber Induktion von Embryonenanlagen durch Implantation artfremder Organisatoren. *Arch. Microscop. Anat. EntwMech.* **100**: 599–638 (1924).

64. Takaya, H.: Dynamics of the organizer. Chapter 2A in "*The organizer*" (eds.) Nakamura, O., Toivonen, S. Amsterdam: Elsevier, North Holland (1978).

65. Tarin, D., Toivonen, S., Saxen, L.: Studies on ectodermal-mesodermal relationship in neural induction. II. Intercellular contacts. *J. Anat.* **115**: 147–148 (1973).

66. Tiedemann, H.: Pattern formation in early developmental stages of amphibian embryos. *J. Embryol. Exp. Morph.* **35**: 437–444 (1976).

67. Toivonen, S.: Regionalisation of the embryo. Chapter 2 in "*The organizer*" (eds.) Nakamura, O., Toivonen, S. Amsterdam: Elsevier, North Holland (1978).

68. Vogt, W.: Gestaltungsanalyse am Amphibienkeim mit ortlicher Vitalfarbung. II Teil, Gastrulation und Mesodermbildung bei Urodelen und Anuren. *Wilhelm Roux' Arch. EntwMech. Orgs.* **120**: 384–706 (1929).

69. Warner, A.E. The electrical properties of the ectoderm in the amphibian embryo during induction and early development of the nervous system. *J. Physiol.* **235**: 267–286 (1973).

70. Yamada, T.: Der Determinationzustand des Rumpfmesoderms in Molchkeim nach der Gastrulation *Wilhelm Roux' Arch. EntwMech. Orgs.* **137**: 151–270 (1937).

Questions for Discussion with the Editors

1. *What do you perceive as the weaknesses of the gradient theories which have been proposed in the past to explain various aspects of early amphibian development?*

This question is not easy to answer since it is unlikely that many readers will be familiar with the theories in question. There are at least three and they are reviewed by Toivonen (1978). However in general, I would say that gradient models in this field have two principal shortcomings. First, it is clear that the body plan of the embryo is built up as the result of a sequence of interactions. A given model can only attempt to explain one process occurring at one stage and it is important to keep separate in one's mind, for example, the formation of the mesoderm and its dorso-ventral regionalization.

Secondly, as in other branches of embryology, the best evidence for gradients is the formation of mirror symmetrical duplications such as those arising from the organizer graft. But all that this implies is that there is some hierarchy of cell states and that the state of a cell can be altered by interactions with other parts of the embryo. It does not prove that there exists a concentration gradient of a single substance. In all probability, the signalling and responding systems will turn out to be rather complex and to involve several substances. Whether it will have been useful at the present stage in the history of science to classify such processes as "gradients" will only be known retrospectively. See (67).

2. *To what extent does early (post-fertilization) amphibian pattern formation emerge from pre-existing (maternal) vs. newly synthesized (zygotic) components?*

I would imagine that most or all of the substances involved in early regional specification would have been synthesized in the oocyte, but this does not necessarily mean that they have a significant regional location at this stage.

The animal-vegetal polarity becomes visible during oogenesis in terms of pigmentation, distribution of yolk platelets and location of the germinal vesicle. It cannot be reversed experimentally after fertilization and so it seems probable that the significant regionalization of components occurs during oogenesis.

By contrast, the dorso-ventral polarity is not associated with any visible feature prior to fertilization and it can for a while be reversed by a variety of stimuli. I am therefore inclined to think that in this case the significant regionalization of components occurs after fertilization.

CHAPTER **20**

Morphallaxis and Early Vertebrate Development

Jonathan Cooke

THE TERM MORPHALLAXIS was applied originally in the study of animal regeneration, early in the 1900s. It defined a process where structure appeared to be regenerated by remodeling over a wide area within remaining tissues, so that complete pattern was restored without any localized zone of special growth being responsible for the replaced parts. Such regeneration can be seen in adult planarians and urochordates, but the precise roles played in it by special reserve cell populations, cell migration, and the alteration of differentiation and growth remain imperfectly understood. Recently, the term has been extended to initial development to refer to those regulative systems, seen in early embryos of many (perhaps most) types, where cells' contributions to the final pattern of the body are set by their *relative positions* within a cell sheet or a mesenchyme during the period across which determination occurs. The proportions and completeness of the pattern are thus independent of growth schedules and of the size of individual examples. This contrasts with patterns of cell differentiation built up by directed lineage (24), or by the local control of growth and cell fate by interaction among neighbors i.e., epimorphic processes (16). Some far-reaching mechanism of intercommunication or signalling within the embryonic tissue is obviously implied by this positional phenomenon, but although the idea that there is essentially just one form of such mechanism is forceful and attractive (Positional Information; 38), we still lack evidence for this. Moreover there is, in evolutionary terms, less a priori reason to expect a universality of mechanism for control of multicellular patterns than for, say, the hereditary transmission of genetic information. In this sense, continued use of the term morphallaxis as if it referred to one as-yet-unelucidated mechanism may be disadvantageous.

I here propose to survey a developing system where relative position is determinative of cells' fates, but where the underlying principles may vary as between different dimensions of the pattern. I shall refer to quantitative observations and experiments, from which the relative plausibilities of current models for the regulative mechanisms can be assessed and lines of future work be defined that will further discriminate between them. The rationale for such work is that only from an advanced understanding at this level can a search for the molecular machinery involved be fruitfully undertaken.

The system is that which initially specifies the body plan in the early vertebrate (specifically, amphibian) embryo. Our own membership of the zoologically very homogeneous vertebrate group gives an incentive to attempt to understand the early development even when we currently lack some of the technical advantages evident for other groups, notably insects. Certain vertebrate systems have a classic status as experimental objects, probably because of ease of access for surgical manipulation. This has resulted in an extensive backlog of geometrical knowledge about the fate map of material and the cellular movements in undisturbed development, such knowledge being prerequisite for any approach to analytical rigor in experiments of the type to be described.

I first deal, in some detail, with work on the establishment and proportioning of the medial-to-lateral sequence of territories in the primary axial pattern of amphibian embryos. This occurs in the mesodermal cell sheet during gastrulation and neurulation (a 10-to 24-hour period), under the control of a mid-dorsal organizer region. I shall illustrate the positional information theory (38), a particular proposal for a universal principle of pattern control, by applying it to the normal and to experimentally observed, abnormal versions of this pattern. This will reveal quite profound problems for any of the model mechanisms whereby positional information might be realized. A different hypothesis is then proposed, that of "serial diversion of determination," which appears to accommodate the observed quantitative behavior and to predict the results of subsequent experiments which are described. I then briefly describe the morphallactic behavior of the other, antero-posterior dimension of pattern in the amphibian body plan, and draw attention to features that suggest that here quite detailed positional information exists, however this is controlled. I speculate, finally, that different principles may operate in pattern control along these two dimensions, and indeed in other early morphogenetic fields in vertebrate embryos. Some necessary, but as yet unperformed, experiments are outlined at various points in the discussion.

The Medio-Lateral Dimension

The Anatomy of the Developing Pattern

The amphibian mesoderm develops by immigration of cells from a surface or sub-surface equatorial zone, the marginal zone, of the blastula. The future dorsal mid-line cells lead the process but the zone from which tissue is recruited soon comes to be annular, so that a bilaterally sysmmetrical, cylindrical cell sheet develops with an extended dorsal side. Keller (20, 21) Nakatsuji and co-workers (31) have studied in fine detail the origin, fates, and behavior of the mesoderm cells during normal gastrulation and pattern formation. By the time the medio-lateral pattern of its future contributions to the body is being established, it is one or a few cells thick, and situated between the neurectodermal layer and the yolky endodermal mass. Cells are continually added to form the cross sections of more posterior levels of the cylinder (and thus body axis) from an annular zone around the closing blastopore.

The four territories comprising the archetypical vertebrate medio-lateral pattern are the mid-dorsal notochord (referred to as the medial extreme of pattern), the paired somite-forming columns, paired pronephric zones and the lateral plate mesoderm with blood-forming islands ventrally (called the lateral extreme of pattern). The tissue to be devoted to these structures is established as zones, of particular relative extents, within the cross-section of the mesodermal cylinder just described. The total field between the extremes is in the order of 100 cells and 1.5 mm within the cell sheet. Thus with four stripes and bilateral symmetry as trivial added features, this is an instance of the problem of the regulating "French flag" used by Wolpert (38) in his general treatment of the concept of pattern formation. Figure 1 illustrates the geometry of these events. The initial polarity for the pattern, i.e., the dorsal midline position and site of the first immigrating portion of the future mesodermal cylinder, is set from an "organizer" boundary that derives, in amphibians, from a material localization in the activated egg (17, 27). But with this exception, the evidence is that ooplasmic localizations, cell lineage, and the cell cycle, and even total cell number or tissue extent across a considerable range are of little consequence for the completeness and proportioning of the pattern attained. Position, by contrast, *is* determinative of the fate of material until quite late stages before the first visible expression of the determinations (see next section).

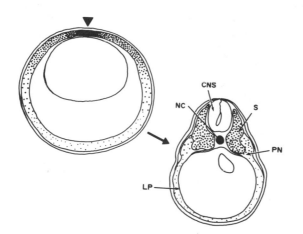

FIGURE 1. Schematic cross-sections of midgastrular and early larval stages of amphibian development. The middle, mesodermal cell layer and its four derivatives in the developed meiolateral pattern are stippled with decreasing density from middorsal (arrowhead) to ventral. Stippling density indicates differentiation tendencies at the early stage where the mesoderm is essentially a monolayered cylinder; see the later territories notochord (NC), somite (S), pronephros (PN), and lateral plate (LP), after the dorsal convergence movements and induction of the overlying nervous system (CNS). Except for dorsal midline positions, however, determination does not occur until relatively late stages (i.e., neurula and early tailbud).

Outline of Classical Findings and Concepts

CELL FATE DETERMINED BY POSITION

The term *determination* as used of embryonic tissue must always be defined operationally; that is, in terms of what has been done to the cells to test their state of commitment, or restriction in developmental potential. Unclearness about the implications of particular observations, in assessing the state of determination of parts of the embryo, has dogged vertebrate developmental biology ever since the period of its classic works. Thus one surprisingly often encounters views to the effect that the body plan of amphibian embryos is built up because the successive cleavage products, throughout the egg, inherit different subprograms in the form of "ooplasmic" precursor localizations, or because small groups of early cleavage products are set aside (e.g., by binary sequential decisions) as the only possible founders of parts of the pattern, as is seen in insect embryogenesis. It is appropriate here to summarize the evidence for three assertions, namely (a) that cell potential for most parts of the mesodermal body pattern is labile until late stages of gastrulation, (b) that the property of acting as a pattern-controlling boundary "the organizer" is restricted to a particular group of cells that normally develop the mid-dorsal territory, and (c) that cell-cycle related changes of differentiated state, and/or differential migration and adhesion among determined cells cannot make more than local and late-acting contributions to primary pattern establishment.

Despite Driesch's original characterization of the unknown mechanism of morphallactic development as a "harmonious equipotential system" (14), there are few embryo types which long remain in the condition where all large coherent parts, when isolated, will adjust the fates of their cells to form complete, normally proportioned editions of the body plan. Vertebrate embryos are indeed among the most highly regulative in this way, but from the onset of cleavage in amphibians, isolated pieces (e.g., halves) will only regulate and form substantially normal pattern if they include material from the restricted region, the organizer, whose fate is to contribute to dorsal mesoderm and head endoderm. As already mentioned, properties that are the precursors for this region are first localized within the egg near the onset of development. If an embryo fragment has the organizer region near one of the edges that must reheal to restore its geometric continuity (i.e., if a presumptive lateral, rather than a dorsal, half embryo is created), the achievement of normal pattern is unreliable from all stages and seldom found from operations as late as late blastula (5000 cell) stages. But this loss of regulative capacity does not mean that an advanced degree of determination has taken place among the individual cells of the embryo. Large mesodermal isolates cultured from much older (50,000 cell) gastrulae often develop according to presumptive fate only (i.e., no regulation), but even this does not imply determination among their cells at the time of explanation. Such observations tell us only that intercellular communication systems that normally regulate pattern (including any biomechanical as well as biochemical factors) have dynamics such that a spatially complete set of events can no longer be achieved from the starting conditions obtaining within the isolate.

No cellular "decisions" need have occurred, and such isolates, placed as part of another embryo that retains the appropriate dynamics (effectively, one that possesses the organizer), may deviate from their original fate in whatever direction is required for the production of normal pattern in the host.

Much about the communication system that regulates pattern might still be learned from a systematic exploration of the times at which each sector of the marginal zone other than the organizer loses the capacity to be assimilated into the normal pattern, when relocated to particular sites. Spemann and co-workers were responsible for a body of work which led to concepts, about the dynamics of such development, that have essentially been inherited by a majority of vertebrate embryology laboratories today (36, 30, 32). Their empirical discovery of the organizer and of the induction of the nervous system were undoubtedly achievements of crucial importance, but their theoretical legacy may have done much to retard the subsequent progress of understanding.

It was found that one specific cell group, when relocated to remote, presumptively ventral parts of a host embryo, was able to cause development of a second edition of the body pattern, centered on itself but sharing the host material with the pattern of original orientation. This group included the dorsal midline of the zone of mesoderm immigration (dorsal lip) and future head endoderm at early gastrula stages. The graft developed to make its normal, dorsal, and anterior presumptive contribution in the second pattern, while the extent of host participation was variable but could, under some circumstances, approach a 50% subdivision of the available material with the original pattern. To confine ourselves momentarily to the mesoderm, the real subject of this chapter, it can be deduced from Spemann's writings that he believed that the respecification of the fates of material, involved in this phenomenon as well as in that of regulation to wholeness by early dorsal or lateral half embryos (see above), was essentially on a geographical or relative positional basis. Any special stimulation of cell division and thus new tissue production played at most a minor role. Contemporary work has shown that these two major regulative adjustments of pattern, to the mirror-symmetric double dorsal and to the small and whole, indeed occur without detectable alteration in the schedule of cell multiplication anywhere in the participating material. Thus the final results are expressed in normal-sized (two patterns) and miniature (one whole pattern) mesodermal cell populations at early larval stages (7, 9). The schedule of cell division in gastrula and neurula mesoderm is very slow in relation to the progress of the cell sheet from the condition where material is pluripotent to that where the determined territories exist, since an average of some two cycles occur over this time. Spemann also knew from the very phenomenon of fate mapping i.e., the regular participation in particular later structures of particular coherent groups of cells marked at early stages, that *individual* cell migration and assortment was not instrumental in large-scale patterning. This has been confirmed (21) with more up-to-date and careful marking and following across the whole course of pattern determination, although a study at the level of individual hereditarily marked cells is still needed to assess the incidence of more localized cell mixing or possible sorting. The foregoing facts, and their logical exclusion of lineage and or ooplasmic localizations as underlying pattern formation, are summarized in Fig. 2.

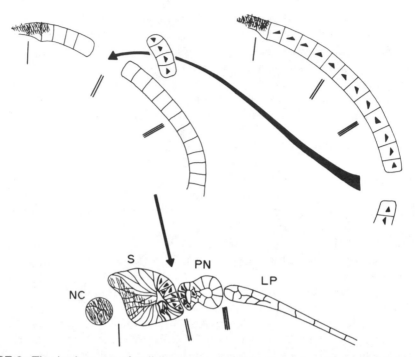

FIGURE 2. The irrelevance of cell lineage and cell sorting for pattern establishment. In the upper part of the diagram, a graft is made from a presumptively lateral position in a (cell-marked) donor mesodermal cell sheet, to a relatively dorsal site in a host. Shading indicates the restricted middorsal "organizer" regions, the only ones effectively determined at this (gastrular) stage. Some 24 hours and two cell cycles later, the formed pattern is observed in the host (lower part of diagram), showing descendants of the implanted cells to have remained as a largely coherent patch with few outliers, and to have been determined in harmony with their new positions in the cell sheet as a whole. Index lines mark the presumptive and actual frontiers between pattern territories. Without extensive cell sorting, the pattern could not be built up on a lineage basis within the number of cell generations available. Symbols as in Fig. 1.

The Inadequacy of the Induction Concept

Coupled with the organizer phenomenon was the discovery that the middle cell layer or mesoderm in the gastrula carried near its dorsal midline the influences responsible for positioning and regionalization of the neural plate in the overlying neurectoderm. The extra central nervous system accompanying the second body pattern after dorsal lip implantation was thus a natural result of incorporation of an organizer into the mesoderm beneath it, being a territory *induced* within the neurectoderm. As implied by Wolpert (38), it would have been well for developmental biology if the term, *induction*, had then been restricted to events like this evocation of the nervous system from neurectoderm by contact with mesoderm; that is, to the transfer of information determining a pathway of development between one cell layer and another. As it was, Spemann and his school expanded the concept to refer indiscriminately to all hypothetical signals, emanating from particular determined territories of patterns to evoke other determinations in neigh-boring material. The spatial organization of development came to be seen just as a sequential hierarchy of inductions, with failure to distinguish conceptually be-tween processes believed to occur along the plane of one cell layer or tissue (pattern formation, see below), and those occurring between two opposed layers or tissues and co-ordinating the patterns of differentiation between them.

This has remained the conceptual framework for most embryologists. Thus most textbook accounts of the grafted amphibian organizer phenomenon fail to dis-tinguish between the direct effect of the graft itself, which is to set a new boundary for the dynamic system organizing the pattern of mesodermal territories, and the accompanying induction of a second central nervous system with this new dorsal midline of pattern as a secondary expression.

We are here concerned only with events controlling primary pattern, which arise in the mesodermal layer. It can be seen intuitively that without some ancil-lary concept corresponding to negative feedback or self-limitation, models based on hierarchies of induction fail fundamentally to explain the salient features of morphallactic regulation, the size-independence of the completeness and propor-tions of pattern. Briefly, if cells on finding themselves to be developing as type A manufacture a specific substance which diffuses or is otherwise exported into the surroundings, and causes there the development of cell type B, and so on, there is no way that the system as a whole can take account of the overall space available in setting the spatial scale of the territories devoted to A, B etc. The kinetics of manu-facture, diffusion/export and breakdown of signals and of cells responses must be part of the species metabolic equipment, and not naturally susceptible to feedback modification against individual embryo size without a special mechanism. It is fur-thermore hard to see what such a feedback mechanism might be. Without it, the intrinsic property of sequential induction is to use up a particular amount of tissue or space for each particular territory induced, until either pattern is completed and the remaining space makes up the final territory, or the process runs out of tissue and pattern remains incomplete. Spemann, and most experimental embryologists writing since, while aware that vertebrate embryos avoid such incompleteness or variable proportioning over a wide, experimentally induced range of sizes at pat-tern formation, have been unable to confront this fact analytically.

The Testable Hypothesis that Positional Information Regulates the Pattern

Several biologists working with invertebrates, earlier this century, had tried to en-compass morphallactic regulation in development and regeneration within a clear theoretical framework. In recent years, Wolpert has proposed a clear and, in prin-ciple, testable concept in the hypothesis that *positional information* underlies the control of pattern (e.g., 38). He further proposed that this principal is universally used, and even that a universal mechanism underlies it in the development of regu-lative embryos across the animal kingdom. This vigorous and challenging notion has caused a wealth of new work, and undoubtedly stimulated the re-awakening of interest in the fundamental problem of pattern. It is currently in danger of degen-eration or "dilution" however in that wherever pattern develops, then cells are said to have interpreted "positional information" and the two phenomen are treated as almost synonymous. There is, accordingly, a certain impatience with the concept among experimental scientists. This is unfair, since although cast in abstract terms, the real scientific value of the hypothesis lies in its precise postulates about the form of the mechanisms, and thus its predictions which are potentially refutable in an appropriate test. It is not a *necessary* concept, since patterns could form and be regulated in other ways. I shall try to make clear the distinctive postulates and predictions of the positional information idea, before describing quantitative work on the regulative performance of the amphibian medio-lateral body pattern, which is aimed at testing it.

The abstract hypothesis is that normal pattern is achieved, despite any earlier disturbances which relocate cells within the field or create abnormally sized fields, because a communication system operates through the entire tissue. This system ensures that some cellular variable, whose local values set cells states to form the pattern territories, is distributed as a gradient with absolute boundary levels at the extremes of pattern and a particular, monotonic profile in the space between. The variable concerned might be a substance concentration or ratio of such concentrations or a range of cell surface states or configurations. Boundary states are set up by early localizations in the egg structure giving special properties to regions, or by symmetry breaking processes, and the consequent boundaries are the organizer regions seen in experimental embryology. Correct pattern depends on the appropriate interpretation of the gradient values by cells in terms of epigenetic action. There is then a family of more or less plausible hypothetical, but concrete physicochemical systems whereby such a positional gradient might be set up and regulated, bearing in mind that normal pattern depends upon normal boundary values for the signal state and a normal gradient profile between them (3, 18, 28, 37, 38).

Figure 3 shows the application of the formal idea to the pattern under discussion, and then the principal model gradient mechanisms that have been proposed. These are diffusion from a focal source with degradation elsewhere (Fig. 3c), and reaction/diffusion models. The latter propose that autocatalytic processes of short-range diffusion produce a particular state of cellular activation as a local peak near one boundary, but lead to a larger-range diffusion gradient of an inhibitor of their own activity. The profile of either activation or inhibitor level may be used as the positional gradient (Fig. 3d). Diffusional transfer of substances of appropriate molecular weight to carry certain biological specificity might be consistent with the formation of gradients on the space and time scales (10 hr—1mm) that are required. The possibility of a series of cell surface states modified by neighbor-interaction, as a positional signal, exists but has not been formally modeled due to our profound ignorance of the machinery involved. The amphibian mesoderm system offers no evidence from grafting experiments for more than one organizing boundary, the organizer of Spemann. Ventral tissue exerts no reciprocal effect to change the fate of its surroundings if grafted, but is assimilated into normal pattern if given time.

One area of test for the hypothesis concerns limits on the performance of these model physicochemical systems in producing complete and normal gradient profiles (thus complete patterns) under various abnormal conditions such as reduced size and bipolarity. Alteration of reaction kinetics and diffusibilities of the constitu-

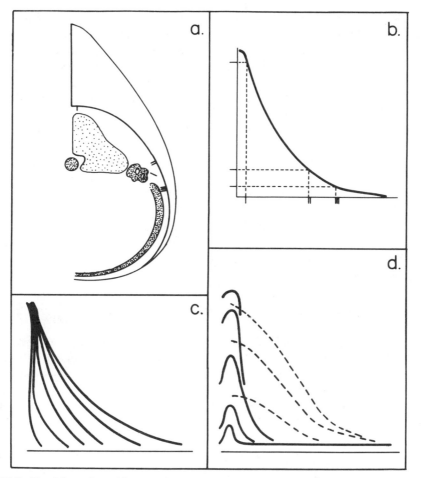

FIGURE 3. The idea of positional information in media-lateral pattern formation. (a) The formed mesodermal pattern *in situ* with a gradient in positional signal extending around the cell layer from dorsal to ventral. (b) The formal idea of the gradient extending through the earlier cell sheet, showing threshold levels that occur at positions that become the frontiers between territories, devoted to the pattern parts. The proportions of these territories, in terms of cell number, depend upon the completeness and shape of the gradient. (c) A source of fixed concentration of a morphogen at one boundary, and five timepoints on the way to equilibrium for a gradient maintained by diffusion/destruction elsewhere in tissue. Thus if the system is not growing around the time of pattern formation, such a gradient only exerts good control over proportions when utilized near its (late) equilibrium state. (d) A reaction/diffusion created positional gradient. The solid profiles represent five timepoints in the slow spread of an autocatalytic activated state of metabolism, first arising in tissue near one boundary. Between the second and third timepoints, the activation exceeds a threshold leading to the local production of a more diffusible morphogen which spreads into surrounding tissue to form a gradient (dashed profiles). This morphogen preserves its own gradient distribution by acting as an inhibitor of initiation or autocatalysis of the activated state in cells which have not yet passed an activation threshold, and thus limiting the spread of the activated region which alone produces it. At equilibrium, either the activation or the inhibitor profile could be used as positional information.

ent molecules in any particular such system, in relation to individual embryo size, is impermissible because implausible; regulating behavior must come, if at all, from the system dynamics. Does the performance of embryos under various circumstances match what plausible models can be shown to produce? We shall see (below) that the information from such experiments is very suggestive. But exploration of possible performances by model gradients from computer simulation or analysis is hardly exhaustive enough for quantitative observations about behavior of real patterns to be critical tests of the validity of positional information. A much stronger test concerns the postulate, distinctive to this theory, that there are *no* specific interactions between emerging pattern territories, in the control of their relative proportions. Such control is stated to occur automatically from the independent positioning of the divides between the territories according to positional signal levels in the gradient (Fig. 3a, b). This suggests quite critical experiments (see later section) in that specific, determined territories that do not include the organizer can be grafted into the cell sheet in which pattern is to be determined in somewhat younger embryos, and the effect upon sizes of the territories subsequently formed in those embryos can be measured.

The Pattern in Miniature Embryos and in Those with Two Organizers

Figure 4a shows the region of the long axis, in tailbud stage larvae, where the proportions and overall scale of the pattern can be assayed as soon as histodifferentiation has revealed the character of the component cells. Overall proportions of the total cells assigned to each of the four territories are estimated by counting the nuclei seen in each structure in a regular subset of the series of precisely transverse histological sections cut at standard thickness (9, 11). Meaningful comparative results are obtained without correcting to absolute cell numbers present in each tissue (1), because pilot investigations reveal that nuclear size and internuclear distance in the various tissues, in the principal planes of section, do not differ as between normal and experimentally small embryos from within an egg-batch (i.e., sibling embryos). That is, their tissues differ in overall scale (cell number), but not significantly in texture. The rate of cell cycle between the period of determination of pattern and that of pattern assay is either zero or slow in the various territories (8) and again there is no evidence for differences as between normal and experimental embryos during this interval, which lasts some 24 hr or 72 hr from the middle of the period of actual determination, in the fast-developing *Xenopus* and the slow-developing *Ambystoma*, respectively. The sampling involves several thousand cells per embryo, and repeat sampling of individual embryos gives very comparable results in terms of percentage of the sample population found in each structure. The control data have revealed that as expected, normal embryos, especially sibling sets, achieve rather constant proportions for the pattern. More remarkably, proportions are almost indentical in the two species used in the study, which represent the two major amphibian subclasses, and which differ about two-fold in the mean numbers of cells between medial and lateral extremes of pattern in cross-section (*Xenopus laevis*, the clawed frog, c. 70 cells, *Ambystoma mexicanum*, the axolotl c. 140 cells). The assessment of performance of the regulative system in this work is based on comparisons between synchronous, sibling sham-operated and experimental embryos in matched sets.

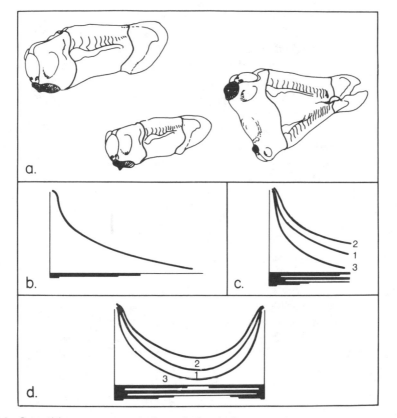

FIGURE 4. Quantitive assessment of regulative performance by real embryos. (a) Normal, experimentally small and double dorsal organizer grafted embryos seen in fronto-lateral view. Heavy lines mark cross-sectional positions between which the standard estimates of pattern proportions by cell counting are made. Somite boundaries and pronephros can be seen in outline beneath the skin of the assayed area. The head contains a small proportion of the mesodermal cells in a complex set of structures, while the tailbud is the site of delayed differentiation and the first true growth. (b) An arbitrary morphogen gradient profile, taken to specify the normal pattern in a normal-sized mesoderm (the four thicknesses of the baseline), by a particular physico-chemical system. Performance of regulation in small (c) and double-dorsal (d) embryos. Profile and baseline versions one and two in these diagrams show the performance to be expected of almost all models for diffusionally maintained morphogen gradients, on the positional information hypothesis of strict gradient interpretation. Profile and baselines versions three on the other hand, depict the regulative performance actually observed, and the profile of the gradient that would be implied, on a positional information mechanism for pattern specification. The essential discrepancy is that, empirically, there is little or no loss from territories far from the "gradient source," even though there is no evidence for control of the pattern from more than this one organizing boundary.

One completed study has assessed the response of the pattern specification system to the challenge created by two different abnormal circumstances. In the first test, presumptively dorsal part-embryos were made by removing ventral blastomeres (early cleavage products) from young blastulae and causing them to round up and heal. Morphologically small gastrulae and neurulae are produced which develop to tailbud larvae in which the mesoderm contains as little as 25% of the normal cell number, thus some 40% of normal cell numbers along the medio-lateral dimension. In the second test, late blastula stage embryos were provided with a second organizer (see later section), midventrally, by grafting a dorsal blastoporal lip from an early gastrula into a site made by removing a few cells from the ventral marginal zone (5). The new organizer diverts the development of some third or more of the total mesodermal cylinder into a new axial pattern having full cellular continuity with the "host" pattern at their adjoined lateral margins. Thus a double dorsal, mirror-image duplicated pattern is produced, involving a normal-sized cell population in total cross-section (7). These operations and their results are described elsewhere in full (9, 11). Here, I wish simply to indicate how the data, while not ruling out the possibility that a positional information mechanism controls the pattern, are very challenging to this assumption and prompt us to ask what other classes of mechanism could be at work.

In single patterns of reduced dimension down to 40% of normal cell number, regulation achieves proportions for the four territories that are statistically indistinguishable from normal. The normal precursor material for lateral territories (even including pronephros) has probably been entirely removed in the more extreme examples. It is therefore striking that in the very smallest such patterns I have examined, the percentage devoted to notochord and somite is smaller than normal rather than larger. "Over-regulation" appears to have occurred. In patterns of the second experimental type, the tissue lies between two organizing boundaries where notochords have differentiated, and normal cell numbers are involved. Regulation produces mesoderms where the total percentages devoted to the four territories, across the two opposed patterns, lie much closer to control values than might be expected. Since the presumptively lateral third of the cell sheet has been reorganized in relation to a new dorsal midline, loss of lateral representations and the joining of two imcomplete patterns might be expected. This is largely avoided, although lateral plate is statistically under-represented. Most strikingly and unexpectedly, the two patterns *both* contribute to this avoidance of lateral loss in the territory between them, since both show smaller cell populations than normal devoted to pronephros, somite, and notochord. The new organizer, mid-ventral in the original blastula/gastrula, has caused a scaling down of the positions of boundaries between territories in the original, host-controlled pattern. The phenomenon occurs most with respect to pronephros and somite territories, but to a significant extent even in "host" notochords, whose territory we believe to become fixed soonest after the onset of gastrulation.

These relationships and their implications for the positional information theory are represented in Fig. 4b–d. The curve of Fig. 4b represents the profile of a signal gradient whose levels specify normal pattern proportions at normal tissue size (shown along the baseline as cell numbers) on the positional information hypothesis for this system. A profile from a concave family is selected simply because plausible model gradient mechanisms characteristically produce such curves, but any monotonic form would do as well. The proportions for the pattern are known at some 4%, 40%, 10%, and 46% for notochord, somite, pronephros, and lateral plate/blood island respectively, with 10–20% coefficient of *relative* variation in each. We can thus mark, on the ordinate, threshold signal levels which would establish boundaries between territories so as to divide up the tissue space or cell number (baseline) in the appropriate ratios. The response empirically seen in small and in doubly organized mesoderms is shown as version 3 in terms of cell number along the baseline of Fig. 4c and d. If true regulation of an interpreted gradient were to underlie this response, therefore, the gradient profiles achieved would bear the relationships to the normal one that are shown in curves 3 of Fig. 4c and d, because of the new positions of boundaries. No gradient models having one organizer, as a signal source, with diffusion and degradation of signal elsewhere, can replicate such changes of profile in response to simulated equivalents of the biological experiments described. Few plausible models based on the more dynamic reaction/diffusion mechanisms (see later section) achieve the required behavior either. Profiles are produced which more resemble numbers 1 and 2 of the figure, which when interpreted would produce baseline pattern versions 1 and 2, corresponding to loss of more lateral regions of the pattern, and even an associated expansion of medical territories.

A particularly challenging feature of the present biological pattern is that a restricted territory, the pronephros, far from either of its borders, is both rescaled and appropriately positioned in small versions. A steeply nonlinear curve corresponding effectively to a peak of signal level near one border, the most usual reaction diffusion component that actually responds to change of field size by repositioning the peak shoulder (3, 18), would not be adequate here. The need to produce the implied regulation with respect to a profile that would show slope, thus carry positional information, across most of the field, imposes even greater stability and plausibility problems on such models. In fact, only very narrow ranges of parameters for the simulated systems result in anything like effective behavior (29). Certain "solutions" to the problem of steepening gradients by feedback from reduced overall scale of the system have been suggested. Thus it is proposed that by various ancillary signals which are functions of tissue size, the diffusion terms or those describing the kinetics between autocatalytic and inhibitory components of the morphogen system are appropriately adjusted to restore the necessary gradient at equilibrium. All such "solutions" could be designed to work, by a human engineer. For the biologist, however, the intellectual choice would seem to lie between continuing to use such models, meanwhile wondering about their plausibility for quite basic and primitive instances of biological regulation, and on the other hand, turning to other classes of model which suggest new experiments. A more detailed discussion of the implications of the results just described, for various positional information models, appears elsewhere (11).

The Hypothesis that Serial Diversion of a Wavefront of Determination Regulates the Pattern

Biological patterns like the one we are considering have two main features, and an appropriate model must explain both of them. The territories that are determined are correctly ordered in space, and follow the "rule of continuity" when one edition of the pattern joins another in tissue, in that territory B is seldom omitted to leave A and C adjoining. Pronephros *is* occasionally absent in small amphibian patterns, but this may be because its visible expression is subject to a threshold number of determined cells being created. In addition, the relative numbers of cells assigned to the territories are quite highly regulated. There follows the outline of a model which departs from the positional information idea but which would explain the data thus far. It is related to previous ideas of Rose (33, 34) and of Meinhardt (Chapter 3) and also marks a reversion towards the original conception by Child (4) of physiological gradients and dominance hierarchies controlling development.

The model involves a wavefront of cell determination passing across the available tissue, organized by a primitive gradient or by propagated cellular "activation" from the organizer, and a set of specific signals which divert determination from one course towards another according to a logical pattern. The organizer is assumed to control only an overall polarity within the embryo's tissue, this polarity being expressed as graded rates of physiological progress towards determination by cells, smoothly ranked from medial (dorsal) to presumptive lateral (ventral) extremes. Such control could be via a diffusion gradient of a signal state, declining smoothly with distance from source and setting local developmental rates, or else via early propagation of an activated cellular state, in the form of a wave that starts or accelerates progress towards determination, spreading from its origin at the autonomous organizer region. This spatial communication system need have no further, accurate informational function such as would be required of regulated positional information. It could therefore be a primitive and plausible diffusion-controlled or propagated signal, lacking regulative capacity against scale variation and maintaining only the direction and continuity of the wavelike sequence of development irrespective of precise position of material in relation to the boundaries of the pattern to be formed. If of gradient form, it could assume profiles 1 and 2 under the conditions corresponding to Fig. 4c and d, while if of propagated wavelike form, its transit time would be proportional to tissue distance traversed rather than of particular absolute duration in both normal and small embryos (see later section).

Normal ordering and relative extents of the determined territories are assumed to follow because of the logic governing access to the various determinations by the cells. An early biosynthetic product of cells which have advanced into each determined state is assumed to diffuse rather rapidly into the rest of the tissue and to prevent, at a threshold concentration, development towards that same state by "younger" i.e., slower-developing or as-yet-unactivated cells that receive it. Thus in normal circumstances (though see later section.) all cells that can develop fast enough move initially towards notochord determination. But a diffusable signal, produced specifically by cells that have become committed to the notochord state, builds up in the system as a whole until it attains a level that diverts less advanced cells towards the next state, somite.

Such a process might be termed the *serial diversion* of a wavefront of *determination*, to produce pattern. Each specific signal is assumed to be so diffusible or transportable within the cell sheet as a whole that its level or concentration is a negative function of the size of the mesoderm as well as a positive function of the numbers of cells producing it. It thus acts as a sensor of the proportion of the total tissue that consists of cells which have been switched to produce it. The logic of proportion control is somewhat similar to that of the regulation of a single frontier position by a reaction-diffusion system (3, 18, 28), and the whole mechanism is like a linked sequence of such events that avoids the regulatory problems associated with positioning all the frontiers by one informational gradient. Due to more rapid buildup of each specific signal to threshold, the wavefront of determination will be diverted in character more frequently, marking off smaller territories between frontiers, when the cell sheet is smaller than normal or is being simultaneously invaded by another wavefront originating elsewhere as in organizer grafting. If the organizer is in fact the source for a primitive gradient, rather than the origin for propagated activation then the gradient is assumed to produce a wavefront of development simply by having smoothly graded, at some prior time, the rates of maturation in cells across the embryo. The conceptual separateness of the wavefront of determination (spatially organized), and then the succession of territory-specific signals that monitor the proportions of each pattern part (rapidly spreading throughout the system with a size-related time-course), is crucial to the understanding of this hypothesis.

The positional information hypothesis is distinguished from the foregoing one by its strong proposition that there are *no* specific interactions between territories during pattern control. We have seen that in double dorsal patterns, even the two notochords and pairs of somite columns, far apart in cellular terms, can affect one anothers' sizes. If specific feedback signals that exert effects on determination are to account for this, they must be sufficiently mobile so that significant spread throughout the mesoderm occurs within fractions of the time-course of pattern determination. This has consequences that suggest experimental studies.

Firstly, a quantitative study of the time course of dissemination through cell sheets of labelled molecules of various characters and sizes is long overdue. Such a study could be carried out by following dissemination with time from small grafts of tissue heavily labelled with marked (preferably nonbiodegradable) molecules, that had been injected into a blastomere of the original donor embryo. We have currently no idea how this time course would compare with that for free diffusion in aqueous space of the same dimensions, because the effects of restricted communication between cells via gap junctions, and possible effects of active transport or mixing in intracellular space, could modify effective "diffusion" coefficients in opposite ways. Figures derived from such studies could offer plausibility estimates for a variety of models, but no more than this. The biosynthetic arrangements underlying actual signals used in pattern regulation could give their progress through tissue a character intermediate between that of diffusional spread (proportional to distance squared) and that of a propagated state or wave (proportional to distance). More detailed considerations of such possibilities are beyond the scope of this chapter.

Specific Suppression of Territories by Heterochronic Grafts; a Test of the Serial Diversion Hypothesis

Another study suggested by the serial diversion model involves transplantation of large pieces of already-determined territories, from older embryos, to ectopic locations in gastrulae whose own patterns have yet to be determined. With certain do-nor-host age relationships, such grafts should exert specific suppressant effects upon the percentage of the host mesoderm finally devoted to the homologous territory. Any significant observation of such a phenomenon would constitute strong evidence against positional information as the effective control mechanism for the pattern. Experiments have been designed to search for such effects and the results of a study in progress are outlined in this section.

Figure 5 shows a type of operation, the heterochronic graft, where a dorso-lateral stripe of neurula mesodermal mantle together with overlying neuroectoderm and posterior zone of recruitment at its blasopore margin, is grafted to a narrow mid-ventral slot in a younger gastrula stage sibling. Zones of mesodermal recruitment and neurectodermal layer are matched up in graft and host and heal into continuity. The developing mesodermal cylinder now has, in addition to an approximately normal-sized host-derived cell population, a mid-ventral strip of older tissue which proves to have maintained its advanced schedule of development and to be in fact determined, under these conditions, to give rise to somite alone or to somite flanked on one side by pro-nephric differentiations. The notochord territory has been excluded from grafts and none of this structure in fact develops, but a small neural plate or rodlike neural formation usually overlies the ectopic somite, since neural fold was part of the graft.

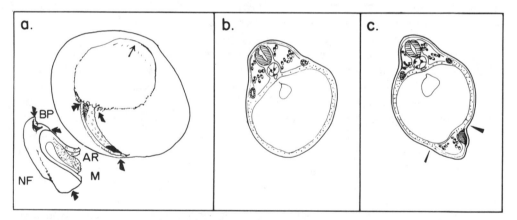

FIGURE 5. The heterochronic grafting operation. (a) The operation whereby a dorso-lateral sector of neurula, including neural fold (NF), presumptive somite mesoderm (M), and the archenteron roof (AR), joined at the blastoporal zone of mesodermal recruitment (BP), is grafted to a ventral site made by cutting a slot in the ectoderm back to the marginal zone in a mid/late gastrula host. Light arrow marks dorsal midline of the host (which is seen from the yolk-plug aspect), while heavy arrows show how cell layers are matched up; (b) control sham-operated, and (c) heterochronically grafted tailbud stage embryos in schematic transverse section, to show how pattern parts are histologically recognizable. The experimental embryo has in addition to the normal cross-sectional complement of cells, extending to the arrowheads, an ectopic midventral strip of somite in continuity with its ventral mesoderm, as well as a neural formation, extending along much of the body. Proportion of host mesoderm devoted to somite is specifically reduced in these regions. For key to mesodermal pattern and neural tissue in sections, see Fig. 1.

Figure 6 summarizes in histogram form the results of quantitative pattern analysis (see earlier section.) in experimental *Xenopus* embryos of this type, in relation to their sham-operated control siblings. Total cross-sectional cell number in *host* mesoderm is not significantly different from normal, but the proportion devoted to somite is quite significantly reduced, the effect being very variable but present in most embryos. This somite deficit is largely concentrated in those parts of the axis lying opposite the relatively posterior position of the ventral ectopic somite in the mesoderm. Pronephros proportions are much more variable among experimentals than in controls and, on average, somwhat lower. But analysis of individuals reveals that, in those experimental embryos where pronephric proportions are diminished right below the normal range of variation in controls, a pronounced pro-nephric differentiation has occurred in graft-derived structures ventrally. Axolotl embryos behave similarly in response to these heterochronic grafts, though there the specific response of somite-proportion is distributed more widely along the host's axis.

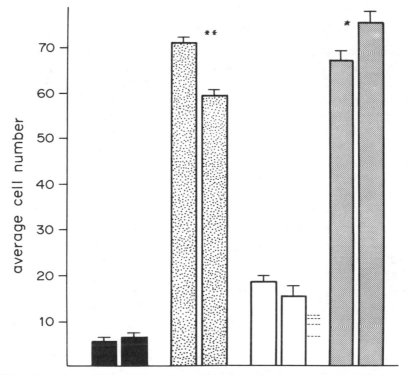

FIGURE 6. The effect of heterochronic grafting. Data from a population of heterochronically grafted *Xenopus* embryos with ectopic somite, and their matched sibling controls. Means of absolute cell numbers per pattern territory per section, within embryos, and their standard errors for the population, are shown. Control columns on left, experimental on right. Notochord, solid black; Somite, stippled; Pronephros, clear; Pronephros in four individuals showing ectopic graft-derived pronephric differentiations, dashed index lines: Lateral plate, hatched. Key: ** = difference significant at $p0.002$; * = difference significant at $p0.05$.

These results make it quite improbable, in my view, that a positional information principle underlies control of the medio-lateral dimension of the body plan. The alternative theory proposed here would, however, accommodate and even predict them. They would represent feedback effects exerted by specific signals of the sort that normally regulate the pattern, but in this case emanating from advanced representatives of the territories concerned, that have been joined to the cell sheet at one extreme of the pattern. Buildup of signal specific to a particular territory would be accelerated, if older cells of that territory had already primed the system with an appreciable level at the site in the host where the control of proportions (i.e., the positioning of a frontier) is taking place. The result would be abnormally early diversion of the character of the wavefront of determination, and thus an abnormally small proportion of the tissue space devoted to the territory in question.

Graded Differentiation Tendencies in Embryonic Mesoderm, and Other Difficulties for the Serial Diversion Idea

The Need for Experiments on Timing

There is one well-documented aspect of organization in blastulae and gastrulae that the model just discussed does not automatically account for. The differentiation tendencies of pieces of the uncommitted mesoderm or its precursors from early cleavage stages, isolated *in vitro* and allowed to continue development, are graded accordingly to their presumptive fates in the normal development of the medio-lateral pattern. Most of these explanation studies (19, 30) report the gradation to be such that material from presumptive medial territories adjusts to produce more complete pattern than is its normal fate (i.e., gives lateral territory representatives as well as medial ones), whereas lateral material produces lateral parts only. This last is not predicted by the serial diversion model in its simple form, which says that the formation of the normally more lateral differentiations is consequent on the prior development by other cells in the system towards earlier, normally more medial ones. The expectation is thus that all isolates in which differentiations are achieved will preferentially form representatives of medial territories, and that more lateral ones will only be achieved insofar as each explant maintains or reachieves a spatial gradation in intrinsic rates of maturation within its material, such as is normally set up by communication with the organizer. But differentiation patterns in explants can be produced without notochord. Furthermore, under particular conditions (2, 39) in whole embryos, notochordless body patterns can be found that are otherwise normally organized in the medio-lateral dimension. There is no reason, in such cases, to postulate the death of cells that would otherwise have become notochord; such cells have presumably contributed to the somite that now marks the medial extreme of pattern.

From limited published work and from informal exchange of information, it appears that the capacity for differentiation as isolated explants develops earlier in the mesodermal precursors of embryos of the frog type than in those of salamander type, and that its dorso-ventrally graded properties when explanted are somewhat less pronounced in the former type. This is reminiscent of another quantitative difference in the progress of pattern control as between the two types. If a salamander-type embryo is divided into a smaller dorsal and a larger ventral fragment when gastrulation activity has already started in the dorsal midline (the organizer), the dorsal fragment can continue development and regulate its fate to form pattern to a considerable degree. The ventral fragment however, although surviving for a long time, usually differentiates little except epidermis, yolky endoderm, and, after prolonged delay, a primitive lateral platelike mesenchyme with blood islands. In frog embryos subjected to the same procedure, the ventral fragment has already acquired the capacity to gastrulate and to develop mesodermal pattern to a large degree, though the medial extreme of pattern is missing.

The general trend is for all these partial patterns deficient in medial territories, however produced, to result from abnormally slow schedules of development and (presumably) determination in the cells. These are associated either with the early treatment of the whole embryo, or with the ambient conditions following grafting operations or the culture of isolates. Proper quantitative study of the time schedules associated with the development of all examples of abnormal patterns will be of prime importance in assimilating all the data into a coherent theory for pattern control in this system. However, the following outline might accommodate both the concept of serial diversion arrived at in a previous section and the phenomenon of patterns without medial parts mentioned above. The effect of the organizer is assumed to be one of "spreading activation," propagating into surrounding tissues at some time before its first expression as the spreading of gastrulation activity to each location. Such activation of progress towards gastrulation and the determined stage may start sooner and progress relatively more rapidly in frog-type than in salamander-type embryos, hence the systematically less graded character of development in frog isolates, and the less delayed and incomplete morphogenesis of the ventral fragments in these embryos. The degree of activation achieved locally may steeply affect the rate of cellular development, ensuring the gradient rate needed to effect a wavefront of determination susceptible to serial diversion. But such an organization would also mean that in cases of subnormal activation or attenuation of the apical (organizer) end of the activation gradient due to specific early treatments (e.g., 2, 17, 26), or perhaps loss of activation through early isolation in culture or tissue disruption, development is slowed to varying extents. The more "medial" determined states may become sequentially unavailable with prolonged passage of physiological time in cells, even in the absence of diverting signals from "older" cells elsewhere. Total isolation from activation may lead always to blood island/lateral plate development after a prolonged period. In short, the normal pattern developed *in vivo* may be the combined outcome of a (nonregulatable) gradient system in degree of activation, itself biasing the "first choice" state of determination at each location, together with the series of specific diverting signals which ensure proportionality and which prohibit the formation of *laterally* incomplete patterns in small and in organizer-grafted embryos. The specific signals would also underlie the results of heterochronic grafting.

The gastrulation movements themselves occur according to a pronounced graded time schedule that appears quite well controlled as between individuals. Presumptive blood island cells begin their immigration a set time after the presumptive notochord (organizer) region. The precise relation between the endogenous controls of rate in these early activities and those controlling onset of determination for pattern is unknown, but it will be of great relevance for the theories so far discussed to time the visible dorso-ventral gastrulation sequence in small and in organizer grafted double dorsal embryos in relation to their normal siblings. Embryos of both types are able to reduce the sizes of territories so as to complete pattern in the spaces available. Regulation of a true positional signal gradient, if successful, might be expressed in the gastrula stage as normal schedules of progression in absolute time, spanning the reduced dorso-ventral tissue spaces available. That is, little gastrulae and the two half fields of double gastrulae should each take a *normal* time for immigration activity to proceed across them from its onset mid-dorsally. By contrast, the primitive sort of activation system, necessary for pattern coherence on the serial diversion model, might be expected to behave like the profiles number 1 or 2 in Fig. 4c and d. If timing reflected the gradient profile, this would lead to a contrasting phenomenon at gastrulation; the completion of the blastoporal ring by rapid spread of the immigration activity in less than normal time. These simple, relevant but exacting experimental observations remain to be made.

The units of pattern I have called "territories" in this chapter do not correspond simply to determinations for development into cell types as recognized by the cytologist or the cell physiologist. They are, rather, particular geographical units where cells' possible development becomes restricted to formation of particular components of the final body, each component then contributing a multiplicity of cell types and structures. Somite contributes skeletal and muscle tissue, while lateral plate and blood-forming islands contribute an even greater variety of histological compartments in the finally differentiated frog. There are thus limits to an analogy between the concept of pattern formation by serial diversion of determination due to specific signals, and that of control over the functional sizes of different tissue systems in the later body by systemic "chalones" (e.g., erythropoietin, liver-chalone). At the time of pattern control, the embryo has no vascular system, but the cells of the entire mesoderm are probably in communication *via* gap functions allowing passage of molecules up to around 1,000 molecular weight, and *via* the restricted extracellular diffusion space between the sealed faces of ectoderm and endoderm. The postulated signals are thus more allied to traditional morphogens than to hormones or chalones, and could conceivably be their common evolutionary precursors.

The Antero-Posterior Dimension

The Regulation of Longitudinal Pattern

Pattern in the antero-posterior or cephalo-caudal dimension of the body plan is also subject to a morphallactic regulatory process at early stages, at least as regards anterior regions down to the level of the hind limb rudiments (see 10). Experimental embryos of the type discussed in an early section with medio-lateral pattern down to below half-normal scale, also show normal numbers of abnormally short somites (mesodermal segments) between unique markers such as ear vesicle and root of the tailbud. There is reduced cell number in the original long dimension of each somite at its formation, so that a normal complement of these structures can be made with the tissue available (6). These embryos also show reduced cell number in the long dimensions of the pronephros, the heart and the brain parts, the last presumably a reflection of shorter zones of inductive activity in the mesoderm.

In order to normalize longitudinal pattern to small size in this way, embryos must heal after size reduction by middle blastula stages. Transection of later gastrulae and neurulae in the transverse plane, or transposition of somitogenic tissue along the axis at these times (13), results in highly mosaic behavior. That is, pieces of embryos isolated or transposed at such stages differentiate, with respect to numbers of somites developed in each piece and the timing of their segmentation, like pieces of a photographic film exposed to a complex scene, then cut up, then developed. It therefore appears that although the visible processes demarcating successive segments from the somitongenic tissue occur in mesoderm which has been in its definitive position for some time, the original factors which scale the processes and which set their local timing to give a "wavefront" progressing down the body plan, must be subject to a regulatory process acting before or at gastrulation. The result of this early process must be some graded cell state along the presumptive axis, which sets the smoothly graded time-sequence for local events as well as the different determinations attained, and which transits the same absolute range of values from anterior to posterior pattern limits in normal and in small embryos. A system which can do this is indeed reminiscent of true positional information, if in fact the cells that are to make the pattern are laid out in coherent sequence in a sheet at the time their "values" are given them. But our understanding of the fate map for this pattern dimension, whereby tissue is recruited into the mesoderm in its antero-posterior sequence during gastrulation, is not complete. Extensive cell mixing could conceivably occur with respect to position along this axis as between blastula and gastrula stages, though this seems unlikely (20, 21).

Positional Signals, Nonequivalence, and Timing

Most embryologists feel that we are presently much further from understanding the mechanism of antero-posterior pattern regulation than that of medio-lateral pattern regulation in the vertebrate embryo. The positional information concept still seems helpful for the longitudinal axis, especially if it is broadened to include the idea of positional value being given by a mechanism that measures time spent by cells in the marginal zone before their recruitment into the mesoderm, rather than their physical positions within a spatial gradient of a signal (10). It is along the antero-posterior axis that the phenomenon referred to by Lewis and Wolpert as nonequivalence (25) is most evident in primary pattern formation. Thus the medio-lateral dimension of pattern described earlier, with its zones of unique histological character, appears similar at many antero-posterior levels. Within each of its longitudinally arranged zones, however, tissue undergoing similar initial behavior and differentiations, e.g., as somite or lateral plate, can be shown to be encoded with unique properties at each level. The most striking work demonstrating this has been carried out on bird, rather than on amphibian embryos. The tissues' early nonequivalence is revealed by the autonomous nature of the overall developmental sequences carried out by regions when transplanted elsewhere along the axis, but also, most obviously, in the specific physiological "age" in development at which the morphogenesis takes place in tissue from each level. Thus pre-segmental material from the neck somite region, transplanted to the thoracic or lumbar region, will ultimately produce there vertebrae of "neck" character (23). Similarly, lateral plate mesenchyme of fore- and hind-limb regions acquires very early its specific character, determinative of the type of limb produced when ectodermal jackets and mesoderms are cross combined.

Time, or rate as a continuous variable, seems to be fundamental to the expression and perhaps the genesis of pattern in the longitudinal axis. If experimentally small amphibian embryos having undergone morphallactic regulation are synchronized precisely with normal siblings at the early neurula stage (i.e., at the anterior onset of somite segmentation), their subsequent progress is parallel in terms of *numbers* of somites segmented and thus proportion of the axial pattern developed, rather than in terms of absolute cell numbers incorporated into somites and other axial structures. The latter arrangement would complete the development, by using up tissue, earlier in the smaller embryos, whereas in reality the same absolute time span is linked to the development of the complete pattern in both cases. Taken overall, the evidence is for a great multiplicity of states coded into tissue to give rise to this dimension of pattern. Such a multiplicity could not readily be regulated by a series of unique size-assessment/switching events such as were postulated for the relatively simple medio-lateral dimension of pattern. Instead, some property occupying a graded range between absolute boundary values is implied, as in positional information theory, although there are problems in considering how such information might be used in controlling numbers of similar structures (10).

A Note on Universality

There is no a priori reason why control of the body pattern in the mesodermal cell layer should not be achieved by different mechanisms in each of its two dimensions. Frankel (15) and chapter 7 has suggested that the regulation of cortical geometry in ciliated protozoans has such a dual character, as well as drawing out the parallel phenomena of organization as between this and pattern regulation in the early development of vertebrate embryos. A similar dualism may apply to pattern establishment in the insect blastoderm (35). In another early episode of vertebrate development, namely the establishment of antero-posterior limb pattern, the data are currently consistent with the idea of a true, interpreted positional gradient underlying the character of the skeletal elements (37). There is even evidence in the chick limb-bud to suggest that by contrast with the primary body pattern discussed in this chapter, growth control is correlated with pattern control so that the gradient mechanism for the latter may be simple and primitive, and yet avoid the difficulty of regulation for size because size is itself controlled accurately at an early stage (12). Should this prove to be the case, the way is open to reconsider yet other concepts for control of the relatively complex pattern that characterizes, say, the limb skeleton. These concepts come under the general heading of "pre-patterns."

Unlike positional information, or the serial diversion model outlined here, pre-patterns involve the concept of a spatial distribution in the morphogen levels that is isomorphic with the future pattern, i.e., a set of peaks and troughs of concentration, rather than being a simple gradient that requires a complex genetic response from the cells (38).

A problem with the physico-chemical model systems whereby such pre-patterns might emerge, has always been that the pattern produced is crucially dependent upon the scale of the system, for reasons similar to those whereby reaction-diffusion-controlled gradient profiles are similarly dependent. If limb rudiments in fact contrive to keep spatial scale constant during the initial specification of pattern within them, then some (though not all) aspects of these somewhat complex patterns may depend upon pre-patterning processes (see 11).

The possibility is emerging that, rather than a single universal form of process, there may be a selection from a small repertoire of available processes according to functional considerations connected with particular dimensions of pattern in particular embryos at particular times. The various classes of process touched on in this chapter are all, nevertheless, to be thought of as morphallactic if indeed the term is to continue in modern use. In none of them is the development of a particular spatial array of cells with specific potentialities intimately tied, *via* neighbor interactions, to the process of tissue production by cell division itself.

General References

HOLTFRETER, J., HAMBURGER, V.: Embryogenesis and progressive differentiation in amphibians. In (eds) Willer, E., Weiss, P., Hamburger, V. pp 230–297. Philadelphia: Saunders *Analysis of Development*, (1955).

WOLPERT, L.: Positional information and pattern formation. *Cur. Top. Dev. Biol.* 6: 183–224 (1971).

References

1. Abercrombie, M.: Estimation of nuclear population from microtome sections. *Anat. Res.* 94: 239–247 (1946).

2. Backstrom, S.: Morphogenetic effects of lithium on embryonic development of *Xenopus. Arck. Zool.* 6.2: 527–536 (1953).

3. Bode, P., Bode, H.R.: Formation of pattern in regenerating tissue pieces of hydra. I. Head—body proportion regulation. *Dev. Biol.* 78: 484–496 (1980).

4. Child, C.M.: *Patterns and problems of development.* Chicago: University of Chicago Press (1941).

5. Cooke, J.: Properties of the primary organization field in the embryo of *Xenopus laevis.* I. Autonomy of cell behaviour at the site of initial organizer formation. *J. Embryol. Exp. Morph.* 28: 13–26 (1972).

6. Cooke, J.: Control of somite number during development of a vertebrate, *Xenopus laevis. Nature, Lond.* 254: 196–199 (1975).

7. Cooke, J.: Cell number in relation to primary pattern formation in the embryo of

Xenopus laevis. I. The cell cycle during new pattern formation in response to implanted organizers. *J. Embryol. Exp. Morph.* **51**: 165–182 (1979).

8. Cooke, J.: Cell number in relation to primary pattern formation in the embryo of *Xenopus laevis.* II. Sequential cell recruitment, and control of the cell cycle during mesoderm formation. *J. Embryol. Exp. Morph.* **53**: 269–289 (1979).

9. Cooke, J.: Scale of body pattern adjusts to available cell number in amphibian embryos. *Nature, Lond.* **210**: 775–778 (1981).

10. Cooke, J.: The problem of periodic patterns in embryos. *Phil. Trans. Roy. Soc. B.* **295**: 509–524 (1981).

11. Cooke, J.: The relation between scale and the completeness of pattern in vertebrate embryogenesis; models and experiments. *Am. Zool.* **22**: 91–104 (1982).

12. Cooke, J., Summerbell, D.: Cell cycle and experimental pattern duplication in the chick wing during embryonic development. *Nature, Lond.* **287**: 697–701 (1980).

13. Deucher, E.M., Burgess, A.M.C.: Somite segmentation in amphibian embryos; is there a transmitted control mechanism? *J. Embryol. Exp. Morph.* **17**: 349–359 (1967).

14. Driesch, H.: *The science and philosophy of the organism.* 2nd ed. London: Black (1929).

15. Frankel, J.: Global patterning in single cells. *J. Theoret. Biol.* (1982, *in press*).

16. French, V., Bryant, P., Bryant, S.V.: Pattern regulation in epimorphic fields. *Science,* **193**: 969–981 (1976).

17. Gerhart, J.C.: Mechanisms regulating pattern formation in the amphibian egg and early embryo, in *Biological regulation and development,* Vol. 2. *Molecular organization and cell function,* (ed) Goldberger, Robert F. N.Y.: Plenum pp. 133–316 (1981).

18. Gierer, A., Meinhardt, H.: A theory of biological pattern formation. *Kybernetik,* **12**: 30–39 (1972).

19. Holtfreter, J., Hamburger, V.: Embryogenesis and progressive differentiation in amphibians. In *Analysis of development,* (eds) Willer, E., Weiss, P., Hamburger, V. pp. 230–297. Philadelphia: Saunders, (1955).

20. Keller, R.E.: Dye mapping of the gastrula and neurula of *Xenopus laevis.* I. Prospective areas and morphogenetic movements of the superficial layer. *Dev. Biol.* **42**: 222–241 (1975).

21. Keller, R.E.: Dye mapping of the gastrula and neurula of *Xenopus laevis.* II. Prospective areas and morphogenetic movements of the deep layer. *Dev. Biol.* **51**: 118–137 (1976).

22. Keller, R.E., Schoenwolf, G.C.: An S.E.M. study of cellular morphology, contact and arrangement, as related to gastrulation in *Xenopus laevis. Wilhelm Roux' Arch.* **182**: 165–186 (1977).

23. Kieny, M., Mauger, A., Sengel, P.: Early regionalization of the somite mesoderm as studied by the development of the axial skeleton of the chick embryo. *Dev. Biol.* **28**: 142–161 (1972).

24. Laufer, J., Bazzicalupo, P., Wood, W.B.: Segregation of developmental potential in early embryos of *Caenorhabditis elegans. Cell* **19**: 569–577 (1980).

25. Lewis, J., Wolpert, L.: The principle of non-equivalence in development. *J. Theoret. Biol.* **62**: 478–490 (1976).

26. Malacinski, G.M.: Identification of a presumptive morphogenetic determinant from the amphibian oocyte germinal vesicle nucleus. *Cell Diff.* **1**: 253–264 (1972).

27. Malacinski, G.M.: Chapter 18 of this volume.

28. Meinhardt, H., Gierer, A.: Generation and regeneration of sequences of structures during morphogenesis. *J. Theoret. Biol.* **85**: 429–450 (1980).

29. Murray, J.D.: Parameter space for turing instability in reaction diffusion mechanisms; a comparison of models. *J. Theoret. Biol.* (1982, *in press*).

30. Nakamura, O., Toivonen, S. (eds.): *Organizer; milestone of a half-century since Spemann.* Amsterdam Elsevier, North Holland (1978).

31. Nakatsuki, N., Gould, A.C., Johnson, K.E.: Movement and guidance of migrating mesoderm cells in *Ambystoma maculatum* gastrulae. *J. Cell Sci.* (*in press*).

32. Nieuwkoop, P.D.: The "organization center" of the amphibian embryo, its origin, spatial organization, and morphogenetic action. In: *Advances in morphogenesis.* New York: Academic Press **10**: 1–39 (1973).

33. Rose, S.M.: Polarized inhibitory control of regional differentiation during regeneration in *Tubularia. Growth*, **31**: 149–164 (1967).

34. Rose, S.M.: Differentiation during regeneration caused by migration of repressors in bioelectric fields. *Am. Zool.* **10**: 91–99 (1970).

35. Sander, K.: Specification of the basic body pattern in insect embryogenesis. *Adv. Insect Physiol.* **12**: 125–238 (1976).

36. Spemann, H.: *Embryonic development and induction.* Yale University Press (1938). Reprinted 1967, New York: Hafner.

37. Tickle, C., Summerbell, D., Wolpert, L.: Positional signalling and specification of digits in chick limb morphogenesis. *Nature, Lond* **254**: 199–202 (1975).

38. Wolpert, L.: Positional information and pattern formation. *Curr. Top. Dev. Biol.* **6**: 183–224 (1971).

39. Youn, B.W., Malacinski, G.M. Axial structure development in UV-irradiated (notochord defective) amphibian embryos. *Dev. Biol.* **83**: 339–352 (1981).

Questions for Discussion with the Editors

1. *Given your assumption that systems which display "morphalaxis" are not necessarily homogeneous in terms of underlying mechanism, is your proposed serial diversion mechanism for patterning of the medio-lateral dimension of the mesoderm unique to this system? Or, are there other systems to which it could apply?*

My hypothesis is that this principle is utilized for control of the primary body pattern, medio-lateral dimension, in the embryos of vertebrates and possibly other regulative types of embryo, but that other classes of mechanism (such as true positional information) are utilized at other stages of development which show morphallactic behavior. I propose that the diversion mechanism is associated with pattern control in systems where (a) there is only one organizing or "boundary" region from which a physiological gradient might be set up, rather than two complementary-such regions, and (b) size is not controlled in the individual while pattern establishment is occurring, so that pattern proportions achieved must be size-insensitive. The primary body patterns of embryos have this property, and I would expect a homology of mechanism, underlying what are perceptibly homologous episodes of development, on grounds of evolutionary conservatism.

2. *Would you expand on your comments about the possible evolutionary relationship between traditional morphogens on the one hand and hormones and chalones on the other?*

I am proposing that there exists a class of signal substances or cell states, involved in pattern formation, which function more by their changing levels or intensity with *time*, in the whole system, than by their local spatial variations as would gradient or pattern morphogens. Pattern is set up more by cells being informed of what has happened elsewhere in each system as it develops, than by local interpretation of a spatial variable. This idea leads one to the "hormone/chalone" principle whereby the functional sizes of mature tissues and organs are continuously controlled during life by feedback from the systemic levels of certain products. It could be that at a primitive level there was homology of the molecular machinery, as well as analogy of principle. A limit to the degree of homology concerns the absence of a circulatory system at stages of pattern formation: the morphogens must be "circulated" by diffusion.

CHAPTER 21

Mechanisms of Polarization and Pattern Formation in Urodele Limb Ontogeny: A Polarizing Zone Model

David L. Stocum and John F. Fallon

AMONG THE MOST FASCINATING MYSTERIES of ontogeny is how axial polarities are established during the course of embryogenesis. The unfertilized eggs of most radially and bilaterally symmetrical organisms exhibit a visible polarity in density of yolk and/or pigment along their future anterior-posterior axis, and in addition, the eggs of bilaterally symmetrical organisms acquire a visible dorsal-ventral polarity after fertilization. During organogenesis, each organ develops with one or more axes of asymmetry, which define its morphology. Asymmetry is the rule in animals; the only exceptions seem to be the radiolarian protozoans, which are spherically symmetrical.

One of the first and most productive systems used to analyze the development of axial polarity was the urodele prospective forelimb region. On the basis of heterotopic transplantation and axial reversal experiments, Harrison and colleagues (10, 27) concluded that the limb region is determined as limb and also that the anterior-posterior (AP) axis of this region is determined shortly after gastrulation begins. However, its dorsal-ventral (DV) and proximal-distal (PD) axes are not determined until much later, at tail bud stages. Harrison was interested in the molecular basis of axial polarity, and unsuccessfully sought x-ray diffraction evidence for the hypothesis that the axial polarity of the limb was the result of the tandem alignment of asymmetric protein molecules within the limb cells (9). To this day, the mechanism of polarization and pattern formation within the limb region have remained unknown.

In this essay, we wish to do two things: first, propose a mechanism for axial polarization and pattern development during urodele limb ontogeny which can be integrated with limb regeneration, and second, challenge the classical view on the sequence of limb axial determination.

Development and Regulation of the Forelimb

Figure 1 is a map of the Harrison stage-29 forelimb area of the *Ambystoma maculatum (punctatum)* embryo, identifying the prospective free limb and shoulder girdle tissues and also the prospective peribrachial flank tissues which are capable of regulating to form free limb and girdle after removal of these prospective tissues. The map makes a significant point: the region of tissue, which has been called the limb disc and which has been transplanted in experiments investigating axial polarity, is the 3.5 somite wide inner large circle of tissue. Hence, all grafts of axially reversed limb discs also contain peribrachial flank tissue next to the posterior border of the actual limb and girdle tissues. This fact will be important to the formulation of our polarization model.

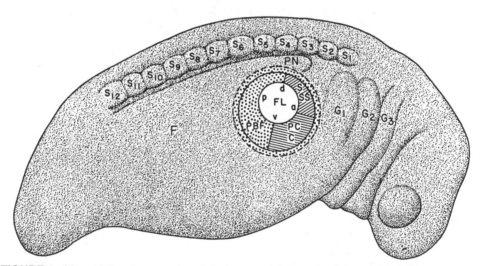

FIGURE 1. Map of the presumptive right forelimb and surrounding areas of the Harrison stage 29 *Ambystoma maculatum* embryo. The shoulder girdle develops from the region marked by hatching, while the free limb (FL) is derived from a 2–2.5 somite-wide area (clear circle). The AP and DV axes of the free limb are rotated 30–45° (clockwise for the right limb, counterclockwise for the left) from the corresponding body axes. The stippled area is peribrachial flank tissue (PBF) that does not normally take part in forelimb development. The inner large circle (3.5 somite diameter) encompasses the classical limb disc that has been used in transplantation experiments, even though much of its posterior tissue is not prospective limb. The outer large circle (5 somite diameter) represents the maximum size of the region that can regulate to form limb after actual presumptive limb tissue is removed. Outside this region, flank tissue (F) will form limb only when induced by other tissues such as ear or nasal placodes. S, SS = scapula and suprascapula of the shoulder girdle; C, PC = coracoid and procoracoid of the girdle; a, p, d, v = anterior, posterior, dorsal and ventral poles of the principal transverse axes of the limb; PN = pronephros; S_{1-12} = somites; G_{1-3} = gills. Originally published in (22).

The free limb area begins to thicken at stage 29, and becomes visible as a growing bud by stage 35–36. The limb bud is conical at first and points in a posterior-dorsal direction. At stage 40 the distal half of the bud flattens and undergoes a torsion that orients the radial-ulnar plane approximately 30° to the vertical. As the elbow joint and digits appear, the lower arm twists further to a vertical alignment (8).

It is commonly held that the limb rudiment is derived solely from the somatopleure of the lateral plate mesoderm in amphibians. In birds, however, wing myogenic cells are derived from the limb level somites (3, 4) and somite cells appear to be required for the development of day-11 mouse limb buds in vitro (1). Furthermore, it has been shown that limb buds develop from masses of neurula stage *A. maculatum* somite plates wrapped in ventral ectoderm and cultured in vitro (13). In view of these facts, it would be profitable to reinvestigate the question of a somitic contribution to the limb regions of urodeles, using marked cells.

The stage-29 limb disc exhibits considerable regulative ability. Any of its four cardinal halves can regulate to form a whole limb, although regulation is observed more consistently with posterior and dorsal halves (8). The prospective free limb and girdle tissues can be removed from the limb disc and the remaining peribrachial flank tissue can regulate to form a whole limb. The limb-forming capacity of flank tissue has been mapped by Harrison (8), who found the limit of this capacity to be located in a one-somite-wide ring of peribrachial flank tissue surrounding the limb disc (see Fig. 1). Flank tissue beyond this ring cannot regulate to form a limb after tissue extirpation, but it does have latent limb properties, because it can be induced to form limbs if stimulated by transplants of various foreign tissues such as ear and nasal placodes (e.g., 2).

Time of Determination of Axial Polarity

The classical test to determine the time at which the axes of the urodele forelimb become determined is to reverse one of the three cardinal axes of the limb disc at different stages of development while grafting the disc in either the orthotopic or heterotopic (flank) position. If the limb develops with asymmetry opposite to that of its side of origin, the polarity of the axis in question is not determined at the time of surgery. Using this test, Harrison and colleagues concluded that the axes of the limb disc are determined in sequence (27). That the limbs developing at the graft sites are actually derived from the donor discs was shown by heteroplastic exchange between *A. maculatum* and *A. tigrinum* (16). By this test, the AP axis is already determined in *A. maculatum* and *T. pyrrohgaster* by the late gastrula or medullary fold stages when the limb mesoderm becomes determined as limb (6, 30). The DV axis becomes determined between stages 33 and 35 in *A. maculatum* and becomes determined between stages 36 and 38 in *A. tigrinum* and *T. pyrrohgaster* (11, 27, 30). The PD axis has been investigated only in *A. maculatum* and becomes determined between stages 35 and 36 about the time the free limb region begins outgrowth (27).

Axially reversed limb discs interact with their surrounding flank tissue, and the nature of this interaction is different prior and subsequent to determination of their transverse axes. Axial determination is associated with duplication in axially reversed limb discs (27). Orthotopic reversal of the AP axis at any stage results in a primary limb which retains the handedness of origin, and the frequent production of mirror-image supernumerary limbs on the anterior and posterior sides of the primary. Similarly, orthotopic reversal of the DV axis after its determination at stage 35 results in a primary limb with the handedness of origin, and the frequent production of mirror-image supernumerary limbs on the dorsal and ventral sides of the primary. Identical results are obtained after the reversal of the AP or DV axes of larval and adult limb regeneration blastemas (31). When the DV axis of the limb disc is orthotopically reversed *prior* to its determination, however, no duplications arise, and a single limb forms with the handedness of the side of transplantation. Removal and replacement of limb discs in normal orientation always gives normal single limbs.

The same axial reversals in the heterotopic position give similar results with regard to the handedness of the primary limbs derived from the grafts. However, the results are different with regard to duplications. Reversal of the AP axis alone now gives only a very low frequency of duplication, but a high frequency of duplication when not reversed. Reversal of the DV axis alone prior to its determination results in a high frequency of duplication, the duplications arising on the anterior side of the primary. Reversal of the DV axis after its determination also results in a high frequency of duplication but the duplications now arise dorsally or ventrally, or both (32).

These results suggest that prior to its determination, the DV axis of a grafted limb disc can be repolarized by adjacent flank tissues so that it corresponds again to the DV axis of the body, but that during determination, the limb disc cells are imprinted with a positional memory which now leads to duplication when limb tissue interacts with flank tissue after axial reversal. The AP axis never shows repolarization after reversal because it is presumably determined by the late gastrula to neurula stage (6). These duplications will be discussed in the next section on axial polarization.

There is evidence which leads us to question whether the AP axis is actually determined prior to tail bud stages. First, Stultz (23) has shown that the AP axial polarity of the hindlimb region of tailbud stage *A. maculatum* is reversible. He grafted AP-reversed stage 30–35 (no later stages were tested) *A. maculatum* hindlimb areas homoplastically or heteroplastically (to *A. tigrinum*) to the flank, or in the orthotopic position. In the heterotopic position, the grafts developed with reversed AP polarity, but in the orthotopic position, they developed with normal AP polarity. The heteroplastic grafts developed with the species specific characteristics of the donor, indicating that this result was not due to graft resorption, and limb development from host tissues. Second, we return to the point made earlier, that grafts of what has been called the forelimb disc have contained a thick crescent of peribrachial flank tissue on their posterior side, but only a small amount on their dorsal side (Fig. 1). This fact suggests that the presence or absence of peribrachial tissue is correlated with whether or not an axis behaves as determined (AP axis) or undetermined (DV axis) prior to stage 35.

The importance of peribrachial flank tissue to forelimb development in fact has been demonstrated in several studies. Stage 29 limb tissue transplanted to the dorsal or ventral midline, (14, 30, 7) to the side of the head between eye and gill

(18) or cultured in vitro (34) fails to develop unless peribrachial flank tissue is included. That this developmental failure is not due to nonspecific effects related to the suitability of the transplantation site is shown by the fact that limb tissue grafted first to the head, then regrafted to the flank with a ring of head tissue around it also fails to develop (29).

These facts argue for a role of peribrachial tissue in the normal development and axial polarization of the urodele forelimb.

Mechanisms of Axial Polarization: The Polarizing Model

We propose a polarizing model that can account for both the genesis (or reinforcement) of axial polarity during limb ontogeny and the known facts about limb duplication after orthotopic or heterotopic transplantation of limb discs. In this model, posterior and dorsal peribrachial flank mesoderm represent high points of morphogenetic activity which specify AP and DV polarities in the adjacent free limb and girdle mesoderm by means of signals which spread across this mesoderm (Fig. 2).

There are several possibilities in regard to the nature of the polarizing signal (22). The polarizing zones could be sources of diffusable morphogens which form a concentration gradient across the limb mesoderm, each concentration specifying a positional value to be interpreted by the cell and transduced into the proper pattern of gene activity. The morphogens could diffuse through the extracellular spaces or from cell to cell via gap junctions. Alternatively, the polarizing signal could be in the form of local cell interactions, mediated through the organization

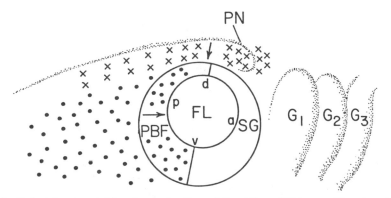

FIGURE 2. Polarizing model for determination of the AP and DV axes. The classical limb disc, containing limb tissue (FL), shoulder girdle (SG), and posterior peribrachial flank tissue (PBF) is outlined by the large circle. The Xs indicate tissue inferred to have a DV polarizing effect, the dots indicate tissue inferred to have an AP polarizing effect. The arrows show the direction of spread of the polarizing signals across the limb mesoderm in the normal embryo. The signals can spread in any direction from the polarizing tissue as shown by experiments in which dorsal polarizing tissue is rotated with respect to the limb tissue (28). PN = pronephros; a, p, d, v = poles of the AP and DV axes; G_{1-3} = gills. Originally published in (22).

of cell surface constituents. The polarizing zones, whose cell surfaces already are specified as posterior and dorsal, would interact with the cells adjacent to them to make surface changes specifying the normally adjacent positional value, and so on in a cascade of interaction across the limb tissue. The same kind of molecules and mechanisms may be involved in establishing both AP and DV polarities, or the molecules and/or mechanisms may be different for each polarity.

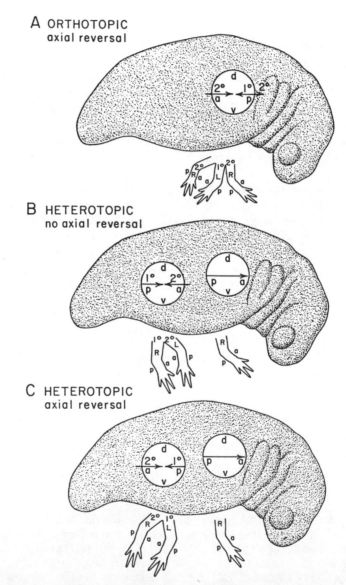

FIGURE 3. Diagrams indicating how the polarizing model accounts for the location and handedness of duplications after orthotopic and heterotopic transplantation of limb discs at stage 29. The upper diagrams of each set depict the grafting operation and the tissues from which the primary (1°) and secondary (2°) limbs are predicted to arise. The arrows show the origin, direction, and extent of penetration of the anterior-posterior polar-

izing effects of the flank tissue on the posterior side of the disc and the remaining flank tissue surrounding the disc. The dorsal-ventral axis is repolarized to normal at stage 29 whenever it is reversed by the operation and is pictured this way in the diagrams. The lower diagrams of each set show the symmetry relations of the fully developed, primary and secondary limbs. a, p, d, v = poles of the anterior-posterior and dorsal-ventral axes; L, R = left and right. (A) Orthotopic transplant of either a right limb disc to the right side, reversing the AP and DV axes, or a left limb disc to the right side, reversing only the AP axis. (B) Heterotopic transplant of a right limb disc to the right side, no axial reversal, or a left limb disc to the right side, reversing only the DV axis. (C) Heterotopic transplant of either a limb disc to the right side, with AP and DV axial reversal, or a left limb disc to the right side, reversing only the AP axis. The frequency of duplication after the latter transplantations is much lower than after any other, because the anterior edge of the grafted disc is in a flank region having little or no polarizing activity. Originally published in (22).

The frequency, location and handedness of duplicated limbs arising after orthotopic and heterotopic transplantations of urodele limb discs can be explained by assuming that a wide region of flank tissue posterior to the limb anlage possesses polarizing activity which falls abruptly at the posterior edge of this region (Fig. 3). The high frequency of duplication on both the anterior and posterior sides of primary limb discs grafted in the AP-reversed orthotopic position would be due to the fact that polarizing tissue is put in contact at both these loci with anterior tissue having limb-forming capability. The high frequency of duplication on the anterior side of limb discs grafted to the flank without AP reversal would be the result of placing the anterior edge of the limb disc within a flank region of high polarizing activity. No duplications would arise on the posterior side of these grafts because the polarizing tissue of their posterior edge is coincident with similar polarizing tissue. Duplications would be expected to arise at the observed low frequency on the anterior side of AP-reversed limb discs grafted to the flank because the anterior edge of the disc would sometimes lie within the region of lowest polarizing activity, but would lie completely outside the polarizing region in the vast majority of grafts.

The polarizing model predicts that the secondary limbs produced after grafting limb discs to the flank should all arise from the graft. This prediction has been borne out in experiments by Swett (26) in which limb discs were heteroplastically exchanged between A. *maculatum* and A. *tigrinum*, although Scharrer (15), on the basis of similar heterografting experiments using the same species, has claimed that some of these limbs can arise partly from host flank tissue. The model also predicts that secondary limbs produced on the posterior side of limb discs grafted in the AP-reversed orthotopic position should arise from respondent anterior peribrachial tissue, while those produced on the anterior side should arise from respondent anterior limb disc tissue. Scharrer (15) showed by heteroplastic grafting that secondary limbs produced in the orthotopic position can arise entirely from host or graft tissue, which is in general accord with the prediction. However, he did not make any specific correlation between the tissue origin of the secondary limbs and their anterior or posterior location, and this type of experiment should be repeated with such a correlation in mind. Nevertheless, Slack (17) has shown that secondary limbs formed after hetergrafting flank tissue adjacent to the anterior edge of the limb disc in an orthotopic position are always derived from the limb disc tissue, as predicted by the polarizing model.

Is there any direct evidence that dorsal and posterior flank tissues have special properties which specify axes in a dorsal to ventral and posterior to anterior direction? Yes, Swett (28) found that if dorsal peribrachial flank tissue alone was included with a stage 29 *A. maculatum* limb disc grafted to the flank with DV axial reversal, repolarization of this axis to harmonize with the DV axis of the body did not occur. When the same experiment was made by including ventral instead of dorsal peribrachial tissue in the graft, the DV axis was repolarized. However, dorsal peribrachial tissue transplanted ventral to a flank-grafted limb disc having normal DV orientation was unable to reverse the DV polarity of the disc. These results led Swett (28) to view the dorsal peribrachial tissue in these experiments as a physical barrier to other DV polarizing factors in the flank. From the discussion in Swett's paper, one would conclude that these factors are ventrally located. The interpretation of results is complicated, however, by the fact that the transplantations were made to the flank, the dorsal portion of which he found to have effects similar to that of dorsal peribrachial tissue, thus failing to isolate the dorsal peribrachial tissue as the sole variable in the experiments. Takaya (30) transplanted limb discs with attached dorsal or ventral peribrachial tissue to the ventral midline (where limb tissue of the same stage minus peribrachial tissue fails to develop), and found that the discs developed only when the dorsal tissue was included. This result suggests that the dorsal peribrachial tissue is indeed acting as a polarizer, specifying the DV axis, but these ventral midline experiments need to be repeated placing dorsal peribrachial tissue on the ventral side of the disc to see if this operation will reverse the DV polarity of the limb.

Evidence for a polarizing effect of a wide region of posterior flank tissue comes from experiments in which the spatial relationships of limb and flank tissue are altered. First, Swett (24, 25) showed that when the limb disc of a stage 27–34 *A. maculatum* embryo was separated into anterior and posterior halves by a thin strip of flank tissue, two primary limbs with the same asymmetry were formed. However, a mirror-image, secondary limb usually arose on the anterior side of the more posterior primary member, suggesting a selective repolarization of a portion of this member by the grafted flank strip. Second, the posterior edge of the limb disc (the portion which does not participate in the formation of the limb), as well as flank tissues from the 5th to the 8th somites, can stimulate the production of a mirror-image, secondary limb on the anterior side of a limb disc when in contact with that side (5, 17). The mesoderm, not the ectoderm, is the effective component. Taken as a whole, the data indicate that the ability of flank tissues to effect duplication is high from somites 5 (nonlimb tissue that is part of the classical limb disc) to 8, exhibiting an increase between these two points, and declining sharply under somites 4 and 9 (Fig. 4).

Simultaneous Determination of AP And DV Axes

The polarizing model suggests that the conclusion that the AP axis is determined as early as gastrula stages (6) is a misinterpretation created by the fact that in the

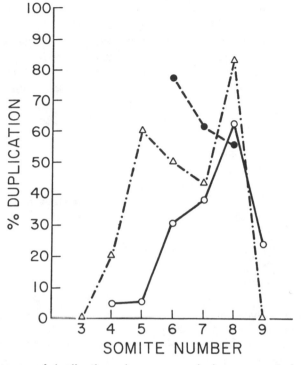

FIGURE 4. Frequency of duplication when progressively more posterior flank tissue is placed contiguous to anterior limb rudiment tissue, by (a) autoplastic grafting of the limb disc to progressively more posterior locations on the flank (o—o—o; *A. maculatum*), (5); (b) homoplastic grafting of the limb disc in the same fashion (•---•---•; *A. maculatum* (5); or (c) heterografting progressively more posterior strips of flank tissue (one somite wide) adjacent to the anterior border of the limb rudiment (Δ-•Δ-•Δ-•; *A. mexicanum*) (17). In the *A. maculatum* experiments, the limb rudiment was positioned so that its anterior border was at the flank levels indicated by the somite numbers. Note that the duplication frequency is highest from the fifth to the eighth somites, decreasing sharply under the fourth and nineth somites. Originally published in (22).

flank grafting experiments used to test this determination, the grafted tissue contains AP polarizing mesoderm and/or is embedded in it. Thus a limb disc with an undetermined AP axis would behave in reversal experiments as if this axis were determined, due to the action of its polarizing tissue, just as the DV axis behaves as if determined when reversed with attached dorsal peribrachial tissue (18, 28). There is evidence supporting this suggestion. We have already mentioned that the AP axis of the hindlimb region is reversible even at stage 35 (23). This may seem paradoxical since grafts of the hindlimb regions might also be expected to contain AP polarizing zone and thus behave as determined, but it should be pointed out that the exact location of limb and flank tissues comparable to those of the forelimb have not yet been mapped for the hindlimb.

Further evidence for this idea involves a relationship between limb and gill tissues. Wilde (35) carried out a series of ingenious experiments on stage 28-38 *A. maculatum* embryos, in which the positions of gills 1–2 and the limb disc were reversed by 180° rotation of the limb-gill complex (leaving gill 3 in place), thereby placing gill tissue anterior and posterior to the limb disc. The operations were done in such a way that the AP and DV axes of both limb and gill were either reversed or remained harmonic with respect to the body. From stages 28–33, when the AP axis is supposedly determined and the DV axis is not, limb development was inhibited. Normal limbs developed from transplants done at stages 35–38. Wilde (36) also conducted in vitro studies which demonstrated a similar time course for the acquisition of self-organizational capacity by the limb disc or bud in the presence of gill tissue. The suppressive effect on self-organization was proportional to the amount of gill tissue in the explants, and the limbs exhibited *radial* symmetry of what little differentiation there was, instead of the expected asymmetry.

Wilde (35) noted that the time at which the limb rudiment acquired the ability to develop in the presence of gill tissue coincided with determination of the DV axis. This fact suggests that outgrowth and morphogenesis of the limb cannot begin until the DV axis has become polarized. We might then conclude that the failure of pre-stage 35 limb discs to develop in these experiments is due to inhibition of DV axial determination of gill tissue, even though the AP axis is already determined. However, Wilde's diagrams show that a substantial amount of dorsal peribrachial (polarizing) tissue was included in his grafts, and that this tissue was not in contact with gill tissue. Therefore, the DV polarizing region of prestage 35 limb discs should be present in these experiments. We are then led to the conclusion that development of these discs fails *because their AP polarity has not yet been established* and is prevented by adjacent gill tissue from becoming established. The effect of the gill tissue would be to specifically inhibit the AP polarizing activity of the flank mesoderm in the posterior part of the limb disc. Specific inhibition of posterior disc tissue is indicated by the fact that transplantation of gill tissue to the *anterior* edge of the hindlimb rudiment does not inhibit hindlimb development (35).

The foregoing discussion not only leads to the conclusion that neither the AP nor the DV axes of the prospective limb tissues are determined prior to stage 35, but that both axes must be determined before limb outgrowth and morphogenesis can begin. This conclusion is also supported by the fact that stage 29 limb discs (which contain AP polarizing tissue) grafted to the ventral midline with attached *dorsal* peribrachial tissue alone will produce limbs. However, if such limb discs are grafted with attached *ventral* peribrachial tissue alone, they fail to develop (30). Recently, Slack (19) has made the very important finding that half or double half limb rudiments will not develop when grafted to the head unless *both* posterior and dorsal peribrachial flank tissue are present in the graft. In view of these facts, we propose the following two concepts as central to the understanding of limb ontogeny: first, the AP and DV axes of the prospective vertebrate limb mesoderm are determined together, and second, determination of both these axes is a prerequisite for the outgrowth and polarization of the PD axis.

We suggest that the test for AP axial lability which should be done is to axially reverse the prospective *free limb region* of the fore- or hindlimb disc before stage 35. According to our model, the expectation would be that this region should develop a limb harmonic to the side of its location. Such experiments, of course, should be conducted using grafts of marked cells.

Proximal-Distal Outgrowth

The parts of an amphibian limb bud or regenerating limb are laid down in a proximo-distal sequence by the mesoderm under the outgrowth promoting influence of the apical epithelium (21, 33). Hence, the dividing cells of the limb mesoderm must change positional value in a distal direction as the disc grows out into a bud. The limb is a three-dimensional structure with internal pattern; therefore the organization of the limb bud must be specified in three axes, and any model must account for all three axes. We propose that the transverse axes are first specified by the action of the dorsal and posterior polarizing regions and the limb mesoderm cells are imprinted with a memory of the positional values thus specified. At this point, the limb tissue becomes a self-organizing system able to grow out and lay down pattern in a PD sequence. How can the PD sequence of positional values be realized? We propose the following scheme: The combination of AP and DV positional values establishes the proximal boundary of the forming limb bud. We believe that the ectoderm has the intrinsic property of being the distal boundary of the developing limb. However, it is not precluded that the mesodermal cells acquire the ability to induce the distal boundary value in the ectoderm once their transverse positional values are set. In either case, we propose that the mesodermal cells do cause the overlying ectoderm (apical epithelium) to produce a mitogenic factor, thus accounting for the outgrowth-promoting properties of the apical ectoderm. The PD discontinuity between the proximal (mesodermal) and distal (ectodermal) boundary positional values is subsequently eliminated by intercalation of the mesodermal positional values which constitute the PD axis (12).

In establishing the PD limb positional value map by intercalation, the mesoderm and ectoderm cells interact in an integrated network of strictly local communications. At the same time, the mitogenic-promoting activity of the apical ectoderm may be the result of a long-range, diffusable signal.

Relationship of Limb Ontogeny to Regeneration

The limb regeneration blastema is formed by the dedifferentiation of stump tissues, and it redifferentiates a replica of the parts lost by amputation. The blastema is a polarized, self-organizing structure that undergoes autonomous redifferentiation and morphogenesis when isolated from its stump (20, 21). This self-organizing capacity is explainable in terms of the positional memory imposed on limb cells by the polarizing zones and by PD intercalation during limb bud outgrowth as discussed above. The intercalation process generates a sequence of positional values specifying pattern along this axis. The proximal boundary of the regenerate is then determined by the positional value of the cells at the level of amputation, and the distal boundary is imposed by the apical wound epidermis, which retains the distal boundary memory of the limb ectoderm (12). A question which may arise is whether polarizing regions necessary for specifying transverse axial asymmetry are present in the blastema. Our model predicts that such regions are not present or necessary, since positional memory of transverse polarity is built into prospective free limb cells during ontogeny prior to outgrowth of the limb bud.

Acknowledgments

We are grateful to the following colleagues for helpful discussion and criticism of the ideas that led to this manuscript: Eugenie Boutin, Susan Bryant, Jill Carrington, Bruce Carlson, Charles Dinsmore, Richard Goss, Malcolm Maden, Paul Maderson, James Nardi, Donene Rowe, B. Kay Simandl, David Slautterback, Patrick Tank, Stephen Thoms, William Todt, and Charles Wilde. We thank Sue Leonard for typing the manuscript. The drawings were done by Alice Prickett. This work was supported by NIH grant HD-12659 to D.L.S. and NSF grant PCM8205368 to J.F.F.

General References

HARRISON, R.G.: 1969. Relations of symmetry in the developing embryo. In, *Organization and development of the embryo.* (ed. Wilens, S.), New Haven: Yale University Press, pp. 166–214 (1969).

TAKAYA, H.: Experimental study on limb asymmetry. *Annot. Zool. Japan* **20**: (Suppl.) 181–279 (1941).

SWETT, F.H.: Determination of limb axes. *Quart. Rev. Biol.* **12**: 322–339 (1937).

References

1. Agnish, N.D., Kochar, D.M.: The role of somites in the growth and early development of mouse limb buds. *Dev. Biol.* **56**: 174–183 (1977).

2. Balinsky, B.I.: Weiters zur Frages der experimenteller Induktion einer Extremitätenanlage. *Wilhelm Roux' Arch. EntwMech. Org.* **107**: 679–683 (1926).

3. Chevallier, A., Kieny, M., Mauger, A.: Limb-somite relationship: origin of the limb musculature. *J. Embryol. Exp. Morph.* **41**: 245–258 (1977).

4. Christ, B., Jacob, H.J., Jacob, M.: Experimental analysis of the origin of wing musculature in avian embryos. *Anat. Embryol.* **150**: 171–186 (1977).

5. Detwiler, S.R.: Experiments on the transplantation of limbs in *Amblystoma. J. Exp. Zool.* **31**: 117–169 (1920).

6. Detwiler, S.R.: On the time of determination of the anteroposterior axis of the forelimb of *Amblystoma. J. Exp. Zool.* **64**: 405–414 (1933).

7. Finnegan, C.V.: Observations of dependent histogenesis in salamander limb development. *J. Embryol. Exp. Morph.* **11**: 325–338 (1960).

8. Harrison, R.G.: Experiments on the development of the forelimb of *Amblystoma,* a self-differentiating equipotential system. *J. Exp. Zool.* **25**: 413–459 (1918).

9. Harrison, R.G., Astbury, J.T., Rudall, K.M. An attempt at an x-ray analysis of embryonic processes. *J. Exp. Zool.* **85**: 339–363 (1940).

10. Harrison, R.G.: Organization and development of the embryo. (ed. Wilens, S.), New Haven: Yale University Press (1969).

11. Hollinshead, W.H.: Determination of the dorsoventral axis of the forelimb in *Amblystoma tigrinum. J. Exp. Zool.* **73**: 183–194 (1936).

12. Maden, M.: The regeneration of positional information in the amphibian limb. *J. Theoret. Biol.* **69**: 735–753 (1977).

13. Muchmore, W.B. Differentiation of the trunk mesoderm in *Amblystoma maculatum. J. Exp. Zool.* **134**: 293–313 (1957).

14. Nicholas, J.S.: Ventral and dorsal implantations of the limb bud in *Amblystoma punctatum. J. Exp. Zool.* **39**: 27–42 (1924).

15. Scharrer, E.: Uber den Ursprung spiegelbildlicher Verdoppelungen von Amphibienextremitaten. *Gesellschaft fur Morph. und Physiol.* Munchen, Sitzungsberichte. **40** Jahrgang 1930/31. pp. 66–68 (1931).

16. Schwind, J.L.: Heteroplastic experiments on the limb and shoulder girdle of *Amblystoma. J. Exp. Zool.* **59**: 265–295 (1931).

17. Slack, J.M.W.: Determination of polarity in the amphibian limb. *Nature* **261**: 44–46 (1976).

18. Slack, J.M.W.: Determination of anteroposterior polarity in the axolotl forelimb by an interaction between limb and flank rudiments. *J. Embryol. Exp. Morph.* **39**: 151–168 (1977).

19. Slack, J.M.W.: Regulation and potency in the forelimb rudiment of the axolotl embryo. *J. Embryol. Exp. Morph.* **57**: 203–217 (1980).

20. Stocum, D.L. Organization of the morphogenetic field in regenerating amphibian limbs. *Am. Zool.* **18**: 883–896 (1978).

21. Stocum, D.L., Dearlove, G.E.: Epidermal-mesodermal interaction during morphogenesis of the limb regeneration blastema in larval salamanders. *J. Exp. Zool.* **181**: 49–62 (1972).

22. Stocum, D.L., Fallon, J.F.: Control of pattern formation in urodele limb ontogeny: a review and a hypothesis. *J. Embryol. Exp. Morph.* **69**: 7–36 (1982).

23. Stultz, W.A.: Relations of symmetry in the hindlimb of *Amblystoma punctatum. J. Exp. Zool.* **72**: 317–367 (1936).

24. Swett, F.H.: On the production of double limbs in amphibians. *J. Exp. Zool.* **44**: 419–473 (1926).

25. Swett, F.H.: Transplantation of divided limb rudiments in *Amblystoma punctatum* (Linn.). *J. Exp. Zool.* **52**: 127–158 (1928).

26. Swett, F.H.: Reduplications in heteroplastic limb grafts. *J. Exp. Zool.* **61**: 129–148 (1932).

27. Swett, F.H.: Determination of limb axes. *Quart. Rev. Biol.* **12**: 322–339 (1937).

28. Swett, F.H.: Experiments upon the relationship of surrounding areas to polarization of the dorsoventral limb-axis in *Amblystoma punctatum* (Linn.). *J. Exp. Zool.* **78**: 81–100 (1938).

29. Swett, F.H.: The role of the peribrachial area in the control of reduplication in Amblystoma. *J. Exp. Zool.* **100**: 67–77 (1945).

30. Takaya, H.: Experimental study on limb asymmetry. *Annot. Zool. Japan*, **20** (Suppl.) 181–279 (1941).

31. Tank, P.W., Holder, N.: Pattern regulation in the regenerating limbs of urodele amphibians. *Quart. Rev. Biol.* **56**: 113–142 (1981).

32. Thoms, S.D. and Fallon, J.F.: Pattern regulation and the development of extra parts

following axial misalignments in the urodele limb bud. *J. Embryol. Exp. Morph.* **60:** 33–55 (1980).

33. Tschumi, P.A.: The growth of hindlimb bud of *Xenopus laevis* and its dependence upon the epidermis. *J. Anat., London,* **91:** 149–173 (1957).

34. Wilde, C.E., Jr.: Studies on the organogenesis *in vitro* of the urodele limb bud. *J. Morph.* **86:** 73–114 (1950).

35. Wilde, C.E., Jr.: Studies on the organogenesis of the urodele limb bud. *J. Exp. Zool.* **119:** 65–92 (1952).

36. Wilde, C.E., Jr.: Studies on the organogenesis *in vitro* of the urodele limb bud. II. The effect of size of the explant and the interrelation between gill and limb. *J. Morph.* **90:** 119–148 (1952).

Questions for Discussion with the Editors

1. *Are there only similarities between your polarizing model for urodele limb ontogeny and the "zone of polarizing activity" model which others have proposed to explain avian limb development?*

Our model is similar to the view put forth for the chick by Fallon and colleagues. These investigators have proposed the action of the polarizing zone is confined to the period of development before limb bud outgrowth. Others who investigate axial polarity in the chick limb, principally Wolpert and associates, postulate that the polarizing zone is required for normal antero-posterior limb polarization during the outgrowth of the limb bud. A third group, primarily Saunders and Iten and their colleagues, question whether such a zone has any role at all in limb development. Finally, we postulate both antero-posterior and dorsal-ventral polarizing zones; while chick workers have considered only the antero-posterior polarizer.

2. *Your challenge to the classical view on the sequence of limb axis determination appears to be well-founded. Yet "determination" has proved to be one of the most refractory concepts in embryology. What are the prospects for using cell cultures from, for example, tissue which flanks the limb discs, as a model system for studying the cell interactions which might be at the basis of axis determination?*

The prospects for using culture systems to test the polarizing hypothesis should be very good. For example, one could put prospective limb tissue and postulated polarizing tissue together in various orientations either in in vitro hanging drop cultures or in in vivo culture chambers hollowed out of larval dorsal fin connective tissue. Likewise, one could condition medium with polarizing tissue and test the effects of the medium when presented either directionally or randomly to prospective limb tissue.

CHAPTER **22**

Regeneration of the Axolotl Limb: Patterns and Polar Coordinates

Nigel Holder

AXOLOTLS ARE neotenic salamanders that spend all of their normal life as aquatic larvae. As is the case with all larval urodele amphibians, axolotls are capable of regenerating their appendages remarkably well. In recent years, they have become increasingly important as laboratory animals because they are easy to maintain and breed readily over the winter months. The limbs of the larvae are particularly well suited to studies of pattern regulation because normal limbs faithfully replace the limb parts distal to the level of amputation in an ordered histological sequence, and the skeleton, muscles, and skin provide clear anatomical markers for the results of tissue level manipulations (Fig. 1).

In this discussion, a theoretical framework for the control of pattern regulation will be analyzed in terms of a series of recent experiments. The intention of the discussion is twofold; the experiments to be outlined will reveal some of what we know of the cellular interactions occurring during distal outgrowth and will indicate how a theoretical framework for these interactions molds the design of experiments and evolves as new data concerning specific tenets of the model become available. Some of the basic features of the polar coordinate model have remained unscathed during the last six years, whereas others have been discarded or have been refocused. Prior to describing the series of experiments that will form the bulk of the discussion, the model itself will be outlined.

The Polar Coordinate Model

The polar coordinate model (6) is a formal representation of a two-dimensional coordinate array of positional values with a circumferential and a radial component (Fig. 2). The model was devised in an attempt to explain how pattern regulation

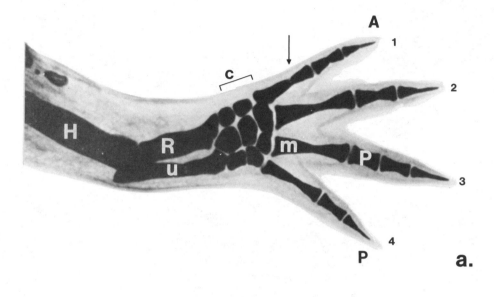

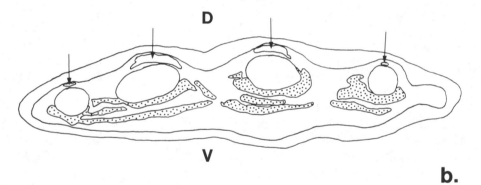

FIGURE 1. The axolotl forelimb bears numerous clear structural markers that can be used to analyse the results of tissue level grafting experiments. (a) The normal skeleton (magnification × 15) of the forelimb demonstrated by Victoria blue wholemount staining. Several skeletal markers are evident. For example, the third digit has three phalanges, digits three and four articulate with individual carpals and the ulna overlaps the humerus at the elbow joint. (b) A camera lucida drawing of a 10-micron wax section cut at the proximal metacarpal level (arrow on a) revealing the dorsal to ventral differences in the muscle pattern (magnification × 30). The ventral muscles (dotted regions) are a complex arrangement of four muscles which are virtually continuous across the anterior to posterior extent of the limb. In contrast, the dorsal muscles (arrows) are discrete semicircular shaped masses covering the dorsal side of the metacarpals (see 14–17 for a more detailed description of the muscles). H, humerus; R, radius; U, ulna; c, carpals; m, metacarpals; p, phalanges. The digits are numbers 1 to 4 from anterior (A) to posterior (P). On the section, D is dorsal, and V is ventral.

was achieved in structures which regenerated by the addition of new cells, as opposed to such animals as *Hydra*, which can regenerate in the absence of cell division by reorganizing the tissue which remains. Regeneration by the addition of new cells occurs in such animals as the cockroach and the fruit fly when tissue damage is incurred. This type of pattern regulation is termed *epimorphosis*. In its initial form, two basic rules were suggested to explain the epimorphic regulation of pattern when normally nonadjacent cells were confronted following surgical intervention. The first of these rules stated that a complete circumference of positional values needed to be present before outgrowth could commence. This complete circle rule was initially put forward in order to explain the appearance of supernumerary limbs that appear after axial misalignment of blastemas and stumps (see, for example, 5). In a sense, the rule was in conflict with the remainder of the model because it relied on a global realization within the limb that a complete asymmetrical circle was present before outgrowth could begin. The remainder of the model relies solely on local cell-cell contacts and the intercalation of new cells by cell division that restore circumferential or radial continuity after the confrontation of normally nonadjacent cells. The polarity of restoration of continuity by intercalation is controlled by the second rule which states that this always occurs by the shortest possible route (Fig. 3) (6, 3). In this way, the actual positional values intercalated

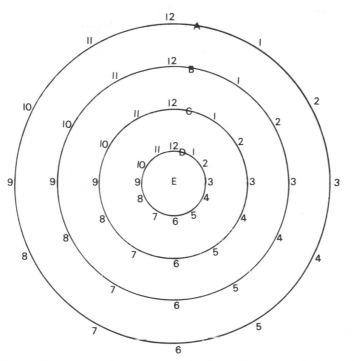

FIGURE 2. The polar coordinate model can be formally represented as a two-dimensional array of positional values. The circular sequence is represented by the numbers 1 through 12 although no discontinuity exists in the sequence. The radial sequence is represented by the letters A through E, with A being proximal and E distal.

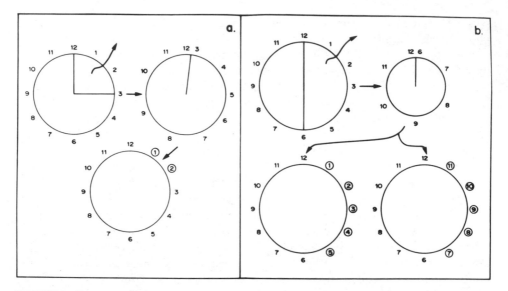

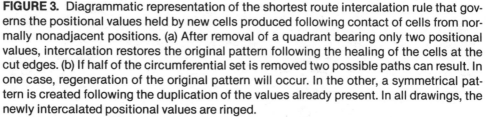

FIGURE 3. Diagrammatic representation of the shortest route intercalation rule that governs the positional values held by new cells produced following contact of cells from normally nonadjacent positions. (a) After removal of a quadrant bearing only two positional values, intercalation restores the original pattern following the healing of the cells at the cut edges. (b) If half of the circumferential set is removed two possible paths can result. In one case, regeneration of the original pattern will occur. In the other, a symmetrical pattern is created following the duplication of the values already present. In all drawings, the newly intercalated positional values are ringed.

should always be predicted except when maximum points of discontinuity are confronted when a choice of two paths of intercalation are evident (Fig. 3).

Soon after the model first appeared, numerous experimental results demonstrated that the complete circle rule was not required to explain pattern regulation. It is now clear that tissues bearing incomplete asymmetrical circles or symmetrical circles can commence outgrowth. This is evident from experiments concerned with both the outgrowth of stumps bearing symmetrical tissues (18, 11, 4) and from the structure of supernumerary limbs formed following ipsilateral blastemal rotations (see, for example, 15, 16, 17, 22).

In the first section of experimental results presented, the likely controls of normal outgrowth will be discussed. The possible interactions leading to the formation of supernumerary limbs will be outlined later on.

The Control of Distal Outgrowth

Following amputation of the limb, differentiated cells in the stump close to the amputation plane begin to divide and undergo a morphological transformation resulting in the accumulation of many cells that appear to be undifferentiated. This process occurs beneath a thin wound epidermis which covers the forming blastema throughout the process of pattern regulation. The initial accumulation of blastemal cells forms a mound known as the stage of medium bud (23). During the formation of the medium bud blastema, it is thought that cells from locally disparate circumferential positions contact and intercalate cells with intervening positional values (see Fig. 5a; 1, 3). These local interactions occur at points all around the circumference and the new cells produced take on the next most distal radial value to ensure that new pattern elements are produced only in a distal direction. The continuous growth of the blastema coupled with local contacts between cells in this way at the tip of the blastema replaces the lost pattern in a proximal to distal sequence. This simplistic way of thinking about how normal distal outgrowth may occur gains support from experiments in which symmetrical limb stumps are surgically created and amputated. From the results of these experiments, it is now clear that the amount of distal outgrowth involving the sequential production of new radial levels is intimately linked to both the nature of the circumferential sequence at the amputation plane, and the mode of healing that controls which circumferential cells actually make contact during outgrowth.

Symmetrical limb stumps have been created surgically in both the upper and lower limb regions (2, 11, 12, 21, 24, 27). The basic experiment involves cutting one half of the limb out from upper or lower limb positions and exchanging tissues from contralateral limbs (Fig. 4). Double anterior and double posterior constructions have been made in most cases, although double dorsal and double ventral constructions have also been examined. The amputation of such limbs produces a variation of structures ranging from extensive symmetrical regenerates to nothing at all.

The very fact that any regeneration is achieved is evidence that the complete circle rule is not upheld. Several other directives emerge from an analysis of the data from these experiments. It is clear from operations performed on the upper forelimb that double posterior constructions consistently regenerate more structure than double anterior constructions (11, 24). Furthermore, the degree of distal out-

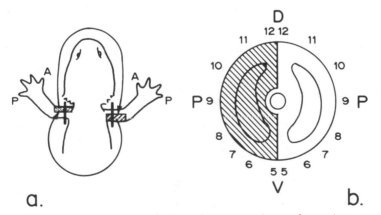

FIGURE 4. (a) Diagrammatic representation of the operation performed to create double anterior and double posterior symmetrical stumps. (b) An end-on view of the amputation plane of a double posterior stump. The graft is hatched and the theoretical distribution of circumferential positional values is biased so that the posterior half of the limb bears greater than half.

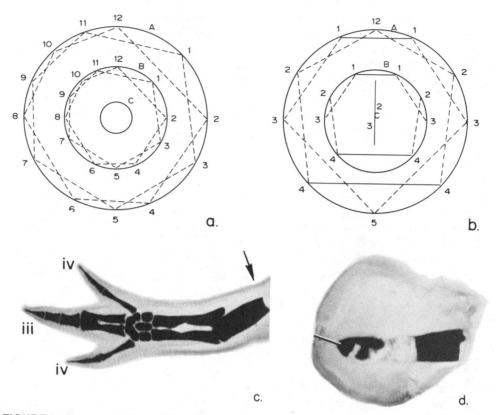

FIGURE 5. Representation of the theory of short arc intercalation. (a) Diagrammatic view of normal distal outgrowth following amputation at radial level A. Cells are thought to contact each other over short circumferential distances during initial blastema accumulation (dotted lines). Contacts between normally locally nonadjacent cells leads to intercalation of intervening cells whose new positional values are determined by the shortest route intercalation rule. (b) Short arc intercalation in a symmetrical, double-anterior upper arm stump in which the graft has healed for a month before amputation. The symmetrical arrangement of circumferential values and the allowance of nonproductive cell contacts leads to rapid loss of positional values and the cessation of regeneration. (c) Victoria blue stained whole mount of a double anterior limb amputated in the upper arm after one month of healing. This limb failed to regenerate any clear structure (magnification × 16). The arrow points to the level of amputation. (d) A double posterior, symmetrical limb regenerated following amputation of a double posterior upper arm in which the graft had healed for only five days (magnification × 18). In contrast to long healing time experiments, regeneration continues to a distal limb region and digits are formed. In such cases, the wound between the graft and host stump tissues is thought to prevent nonproductive cell contacts. The leader points to the regenerated cartilage.

growth is dependent upon the length of time that the initial graft is allowed to heal prior to amputation. The longer the graft healing time, the less the regeneration that is seen following amputation. This effect has also been detected in similar experiments in the thigh region of the axolotl leg (19). Two further conclusions have been drawn from these results. The reason for posterior tissues being more productive than anterior tissues in symmetrical combinations is assumed to be due to there being greater than half the number of circumferential positional values on the posterior side of the upper limb (Figs. 4 and 5). The role of tissue healing is thought to involve the structural constraint of contacts between cells on either side of the wound created during surgery (11).

This last point can be most clearly thought of in terms of the simple model discussed above for normal limb regrowth. It is thought that after long periods of healing, cells on the two sides of the wound can interact freely. In this case, cells will interact locally as in normal outgrowth and, because of the symmetry in the experimental cases, this will result in a number of cell contacts occurring between cells with the same or normally adjacent circumferential value (Fig. 5b). This type of contact will not stimulate intercalation. In addition, no local contact around the circle will result in the intercalation of the mid-line circumferential values at the next most distal level to that of amputation. The outcome of these two properties of symmetrical circles will be a tapered symmetrical outgrowth that forms from a blastema containing few cells (Fig. 5c). In contrast to this effect, which occurs after long periods of healing, short healing grafts regenerate far more structure. This is assumed to be due to the prevention of contacts between cells on either side of the central wound, and consequently the prevention of nonproductive cell interactions. The result of this will be more cells available for intercalation at the next most distal level. The result of amputating such short healing time limbs should be a distally tapering structure that regenerates to a more distal pattern level than the longer healing type of graft (Fig. 5d). In many cases, regenerates of this class distally transform to the digits.

Thus far we have examined only symmetrical outgrowths from upper arm stumps. Two further experiments have been performed on axolotls which examine the regenerative ability of symmetrical double posterior and double anterior lower arms. The first of these experiments (12) clearly suggested that lower arm symmetrical constructs behave differently from comparable upper arm constructions, in two ways. First, no clear healing time effect is seen in the axolotl lower arm. Amputations of symmetrical stumps immediately after surgery produces similar degrees of outgrowth to those amputated after a month of healing. Secondly, the degree of regeneration of double posteriors and double anteriors is comparable. In the second experiment, double posterior limbs that had regenerated from initial short healing time upper arm amputations of symmetrical double posterior constructions produce expanded patterns when amputated in the lower arm (11). These types of results strongly suggested that the lower arm was behaving differently from the upper arm in some fundamental way. Not only is the distribution of values around the circumference thought to be more even, but the constraints on contacts between the cells around the circumference seem to be different.

This point can be clearly demonstrated by examining how distal expansion of a double posterior limb amputated in the lower arm can occur. In the majority of cases, double posterior upper arms amputated immediately make symmetrical limbs bearing between three and six digits (Fig. 6a). When reamputated in the lower arm, limbs bearing up to seven digits regenerate with a normally anterior digit appearing in the middle of the pattern (Fig. 6b). In terms of shortest route intercalation, this can only occur if cells with the dorsal and ventral extremes of the circumferential set of values interact (Fig. 6c). The question at hand is why should this occur in the lower arm but not in the upper arm? The simplest conclusion, if local cell-cell contacts are required, is that the dorsal and ventral sides of the blastema are closer together in the lower arm than in the upper arm; that is, the blastemas formed in these two different limb regions are of different shape. The same conclusion can be drawn from the observation that symmetrical double posterior and double anterior lower arm stumps do not exhibit a healing time effect. This will occur if cells preferentially contact from dorsal to ventral sides of the blastema and not from anterior to posterior sides (Fig. 6c).

In conclusion, several points can be drawn from the results of amputating double anterior and double posterior limb stumps.

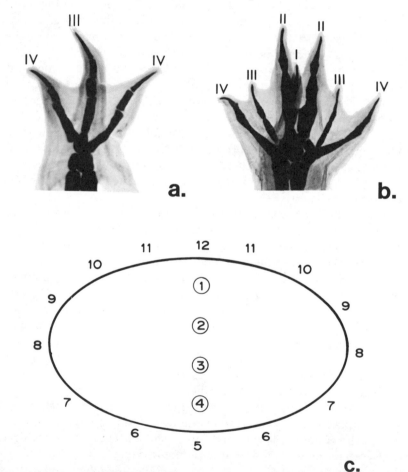

FIGURE 6. (a) A three-digit, double posterior limb formed following immediate amputation of an upper arm double posterior construction. (Magnification × 9). (b) A seven-digit, double posterior limb that regenerated following amputation of the limb shown in (a) at the forearm level. Four extra digits, including a digit 1, have been added in the midline of the pattern (magnification × 9). (c) Shortest route intercalation predicts such an event only if local contact between cells on the dorsal and ventral sides of the blastema can occur. Intercalated positional values are ringed.

1. It appears that the spacing of positional values around the circumference is not even in the upper arm of both newts and axolotls.
2. Interactions between cells can be modulated by physical constraints, such as lines of healing or by the shape of blastemal populations (this point will be demonstrated further below).
3. The amount of distal outgrowth (radial value production) is directly related to the number of circumferential positional values present at the amputation plane. In short, the amount of growth that is achieved during regeneration is intimately linked to the pattern formation process and the constraints on cell contacts.

Blastemal Shape: Its Control and Significance

Blastemal shape has recently been quantitatively described for different morphological stages of blastemas derived from different limb levels of both axolotl hindlimb and forelimb. It has clearly been demonstrated that upper arm and leg blastemas are round in cross-section whereas lower arm and leg blastemas are elliptical (8, 9). This observation is entirely consistent with the predictions made from the results of the experiments just discussed.

This analysis of blastemal shape was based upon a simple morphometric analysis. Limbs were amputated at the required level and allowed to regenerate to the stage of medium bud, medium to late bud, and palette (23). At the required stage, limbs were fixed, wax embedded, and serially sectioned. The distance of the main anterior to posterior and dorsal to ventral axes were measured at 100-μ levels from the distal tip and a simple ratio of these values calculated. This ratio, r, was then plotted for each blastemal stage at each level of amputation. These data for upper and lower leg (M/LB) and upper and lower arm (MB) are shown in Fig. 7 along with some representative sections.

It is clear from these graphs that the lower limb blastemas are significantly more elliptical than the round upper limb blastemas. It should be emphasized that this shape analysis was carried out subsequent to the surgical symmetrical limb stump experiments and indicate therefore that structural constraints can be predicted from such an approach. At the same time, however, blastemal shape may be used to examine the problem of pattern formation. If the control of shape can be established, then it should be possible to alter blastemal shape and the critical relationship between shape of tissue fields and pattern formation can be further examined. For this reason, a further set of experiments were performed in an attempt to elucidate the tissue controls of blastemal shape.

At the outset of these experiments, the simple questions to be examined involved the role of the stump in the control of blastemal shape. Clearly the stump is of maximum importance during the early phases of regeneration because all blastemal cells derive from it. However, it is already known that by the stage of MB the pattern regulation mechanisms of the regenerate are autonomously controlled by the blastema and are not influenced by the stump from which it derives (10, 20). We wished to know if the same was true of the control of blastemal shape. An indication that this was the case came from observing a change in shape from round to elliptical during outgrowth of a palette staged upper arm blastema (Fig. 7c). In this case, the flattening was occurring spontaneously within the blastema because it was forming on an upper arm stump. Further evidence for an autonomous blastemal control came from surgical experiments where specific tissues of the stump were removed prior to amputation to see if an abnormal stump would produce misshapen blastemas.

In these experiments, individual fore and hindlimbs had either the cartilage, muscle, or dermis removed. Removal of one bone, either the radius or ulna, from the lower forelimb and removal of stump musculature from the thigh region of the leg had little effect upon blastemal shape as compared with sham operated controls (9). These results suggested that the role of the stump in governing blastemal shape involved cryptic, level specific information rather than a purely mechanical role. This conclusion was supported by the removal of the dermis from the stump which caused a flattening of the blastemas derived from both the upper and lower limb levels.

The alterations in shape of blastemas from round to elliptical is consistent with the mode of healing and interaction that is predicted from the analysis of regeneration of symmetrical limb stumps. With regard to this particular point, an analysis of the number of cells separating the dorsal-ventral and anterior-posterior poles of blastemas and measurement of the dimensions of blastemas with reference to these axes demonstrates that these facets of blastema outgrowth are entirely consistent with the process of short arc intercalation and the proposed constraints on healing following amputation at different limb levels. Geraudie and Singer (7) have recently demonstrated, using the scanning electron microscope, that newt limb blastema cells have cellular processes that may be up to 50μ long. In the distal region of a palette staged blastema formed after upper arm amputation, the dimensions of and cell counts along the cardinal axes are consistent with the possibility of cells

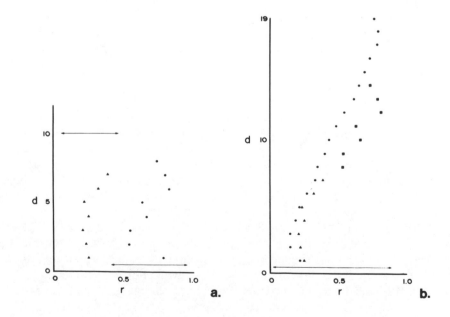

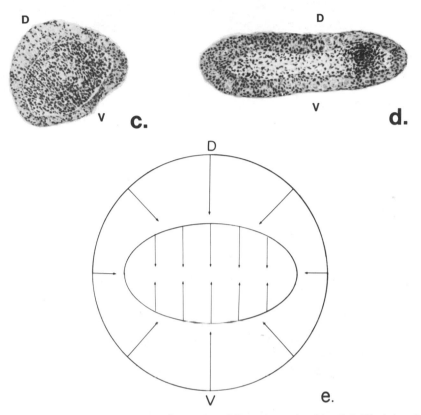

FIGURE 7. The shape of blastemas formed at different proximal to distal limb levels and at different stages. r = shape ratio; d = distance from the blastemal tip in 100 micron steps. D = dorsal; V = ventral. The arrows on the graphs represent 95% confidence limits for the range of r values for each blastema population. Each point represents the mean of between 3 and 5 separate blastemas at corresponding levels from the distal tip. (a) Upper arm (•) and lower arm (▲) medium bud shape ratios. The lower arm MB is markedly more elliptical than the upper arm MB blastema. (b) Upper arm palette staged blastema shape ratios (•). The blastema flattens distally after being initially rounded. This point is emphasised by placing the measurement of upper (■) and lower (▲) arm MB blastemas shown in (a) on the same graph. (c) A typical transverse section of a rounded upper arm MB blastema (magnification × 150). (d) A typical transverse section of an elliptical lower arm MB blastema (magnification × 150). (e) A hypothetical diagram of the healing modes in blastemas at upper and lower limb regions. In the upper arm (outer circle), the healing mode is radial. In the lower arm, the healing mode is preferentially dorsal to ventral.

from dorsal and ventral sides of the blastema contacting directly. For example, in the distal 700μ of such blastemas, which is the region forming the limb region distal to the elbow, the distance from dorsal to ventral extremes varies from 20 to 50μ (Fig. 8a), and are separated by between means of 7 and 26 cells (Fig. 8b). In contrast, the anterior and posterior poles are between 84 and 151μ apart (Fig. 8a) and are separated by between means of 25 and 67 cells (Fig. 8b). Thus the likelihood of cells contacting preferentially from dorsal to ventral sides of the blastema is higher

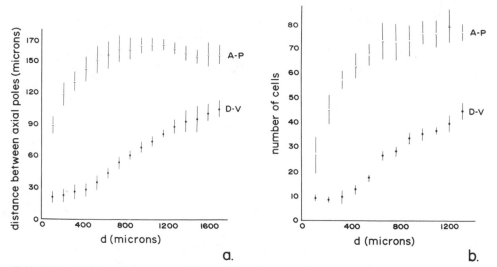

FIGURE 8. (a) Graphic representation of the distance (in microns) separating the poles of the cardinal axes in a palette staged blastema formed after amputation in the upper arm. The anterior and posterior (A-P) poles are clearly wider apart than the dorsal and ventral poles (D-V). d is distance from the distal blastema tip in microns. (b) The number of cells separating the cardinal axes in a palette staged blastema formed after amputation in the upper arm. In both graphs the points represent the means of 5 measurements and the bars represent standard errors.

than the likelihood of anterior and posterior cells contacting and the dimensions are consistent with a local cell contact model. Furthermore, dimensions and cell counts analyzed in upper arm MB blastemas are consistent with a radial healing mode.

Conclusions: The Importance of Moderators of Cell-Cell Contact

The data presented in this discussion are consistent with a short range cell-cell contact model controlling pattern regulation in the amphibian limb. It is clear, however, that trying to explain the results solely in terms of the formal representation of the polar coordinate model is inadequate and that secondary considerations, such as healing constraints and blastemal shape are needed to make the explanations more complete. This point is further emphasized by recent experimental findings where regenerating limbs bearing clear structural discontinuities have been demonstrated. These experiments will be briefly outlined as the conclusion to this discussion to demonstrate the problems that face attempts to elucidate the basis of pattern regulation in the urodeles.

The first demonstrations that limbs could form bearing structural discontinuities were made by Maden (14, 15) and Maden and Mustafa (16) who analyzed the muscle patterns in supernumerary limbs produced following ipsilateral blastema rotations in the axolotl. It is now clear that these extra limbs bear a variety of dorsal-ventral pattern characteristics despite the fact that the anterior-posterior pattern appears normally asymmetric. Four basic limb anatomies have been found. These include double dorsal and double ventral limbs, limbs with part of their pattern symmetrical (double dorsal and double ventral) and part asymmetrical, limbs with mixed handed patterns where half of the pattern is dorso-ventrally reversed compared with the other half, and normal asymmetrical limbs (16, 17, 22).

The supernumerary limbs comprising part symmetrical and part asymmetrical patterns or mixed-handed patterns have clear structural discontinuities within them. The formal polar coordinate model and any other model in which clear discontinuities should be smoothed out (see 25, and this volume for reviews of these models) fail to explain how such limb patterns can occur. The same is true of the recent descriptions of coordinate-free models based simply on rules of pattern continuity that do not discuss likely cellular mechanisms (13, 26). An additional problem is that these supernumerary limbs are derived from interactions between positionally mismatched blastemas and stumps and no way is currently available for elucidating the details of these interactions. For this reason, we have recently surgically created forearms that are anatomically equivalent to the mixed-handed supernumerary limbs (Holder, *unpublished results*). Such forearms were made by splitting the axolotl limb between digits 2 and 3 and separating one complete half of the forearm to the elbow between the radius and ulna. Anterior or posterior halves were then exchanged between left and right limbs and sutured into the vacant lower arm position such that the dorsal and ventral axes of graft and host limb tissues were opposed but the anterior to posterior axes were normal (Fig. 9a). Such constructions were allowed to heal for one month before amputation in the mid forearm position or amputated immediately. In both instances, such mixed-handed limbs regenerated essentially the same mixed-handed pattern that was removed. This fact was established by the analysis of muscle patterns in the regenerates (Fig. 9b). Thus in these surgically created cases, as in the supernumerary limbs, clear structural discontinuities in the limb are maintained during outgrowth and apparently no attempt is made to smooth out the discontinuity. No extra pattern elements are found in these regenerates. The faithful regeneration of mixed-handed limbs also occurs following amputation of such limbs formed following ipsilateral blastema rotation (16). Maden has recently suggested that the complete range of anatomies in supernumerary limbs initially formed following this operation can be explained by the fusion of initially discrete blastema populations around the graft-host junction (16). This idea may well explain the first formation of these limbs but no such fusion can occur following their amputation. The regeneration of mixed-handed and partly symmetrical, partly asymmetrical limbs has as yet only been observed following amputation of these structures in the lower arm. One clue as to the reason why they are able to regenerate faithful copies may be the elliptical shape of the lower arm blastema that was illustrated earlier in this discussion. It is possible that the preferential dorsal to ventral healing constraint within the forearm blastema prevents, or at least restricts, contact between cells across the anterior-posterior axis and thus allows the outgrowth of limbs bearing discontinuities. This is essentially the same explanation as that put forward to explain the lack of a healing time effect following amputation of double anterior and double posterior axolotl forearms (12). It is important, therefore, to examine the regenerative potential of mixed-handed tissues constructed in the upper arm where the radial

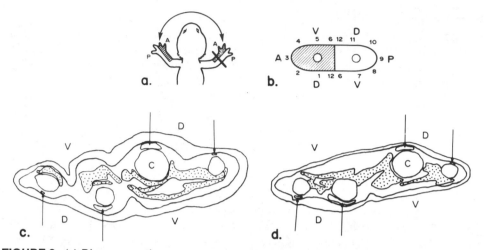

FIGURE 9. (a) Diagrammatic representation of the operation performed to create mixed handed lower arms. The line across the limb represents the amputation plane. (b) Representation of the amputation plane of such a limb. Hatched area is the graft. (c) Camera lucida drawing of a histological section cut through the proximal metacarpal level of a surgically constructed mixed handed limb (magnification × 30). (d) Camera lucida drawing of the regenerate formed following amputation of the limb shown in (c) after one month of healing. The level of section is comparable and the mixed handed pattern in the regenerate is closely similar to that initially formed following surgery. No extra structures were evident and a clear structural discontinuity still exists in the regenerated limb (magnification × 30). The ventral muscles in (c) and (d) are marked by dots and the dorsal muscles are arrowed. C, cartilage; D, dorsal; V, ventral.

healing mode should ensure that the discontinuities in the pattern are smoothed out by intercalation. In these experimental limbs, extra structures would be predicted.

This particular category of results reveals that we are as yet far from an adequate explanation for the cellular basis of pattern regulation in urodele limbs. It is clear, however, that the use of formal models that are clearly predictive remain essential as tools with which to further our knowledge (25). We must proceed with the hope that in the near future enough data will be available to allow the move to the analysis of pattern regulation at the cellular level. A clear understanding of this complex problem will only emerge when information from these different levels of organization can be held together. The urodele limb may be one of the systems in which this daunting task will be achieved.

Note Added in Proof

This chapter represents our understanding of pattern regulation in the axolotl limb as it was when the article was written in the summer of 1982. The results of subsequent research have shed doubt on some of the conclusions that are drawn in the chapter. The author therefore suggests that the reader consult two recent papers that discuss these results:

HOLDER, N., REYNOLDS, S.: Morphogenesis of the amphibian limb blastema: The relationship between pattern and form. *J. Embryol. Exp. Morph.* (1984, *in press*).

HOLDER, N., WEEKES, C.: Regeneration of surgically created mixed-handed axolotl forelimbs: pattern formation in the dorsal-ventral axis. *J. Embryol. Exp. Morph.* (1984, *in press*).

Acknowledgments

It is a pleasure to thank Susan Reynolds, Rosie Burton, Malcolm Maden, Ken Muneoka, David Wise, Nancy O'Rourke, Christine Dinsmore, Susan Bryant, Patrick Tank, Spyros Papageorgiou, Charleston Weekes, and Warren Fox for their advice, practical help, and collaboration during various stages of the work discussed in this chapter. I also thank Phil Batten for help with the figures, and the SERC for financial support for all aspects of the work performed in my laboratory.

General References

TANK, P.W., HOLDER, N.: Pattern regulation in the regenerating limbs of urodele amphibians. *Q. Rev. Biol.* **56**: 113–142 (1981).

BRYANT, P.J., BRYANT, S.V., FRENCH, V.: Biological regeneration and pattern formation. *Sci. Am.* **237**: 66–81 (1977).

References

1. Bryant, S.V.: Pattern regulation and cell commitment in amphibian limbs. In: *The clonal basis of development.* (eds.) Subtelny, S., Sussex, I. 36th Symposium of the Society for Developmental Biology, pp. 63–82. New York: Academic Press (1978).

2. Bryant, S.V., Baca, B.: Regenerative ability of double-half and half upper arms in the newt *Notophthalmus viridescens. J. Exp. Zool.* **204**: 307–324 (1978).

3. Bryant, S.V., French, V., Bryant, P.J.: Distal regeneration and symmetry. *Science* **212**: 993–1002 (1981).

4. Bryant, S.V., Holder, N., Tank, P.W.: Cell-cell interactions and distal outgrowth in amphibian limbs. *Am. Zool.* **22**: 143–151 (1982).

5. Bryant, S.V., Iten, L.E.: Supernumerary limbs in amphibian experimental production in *Notophthalmus viridescens* and a new interpretation of their formation. *Dev. Biol.* **50**: 212–234 (1976).

6. French, V., Bryant, P.J., Bryant, S.V.: Pattern regulation in epimorphic fields. *Science* **193**: 969–981 (1976).

7. Geraudie, J., Singer, M.: Scanning electron microscopy of the normal and denervated limb regenerate in the newt, *Notophthalmus viridescens. Am. J. Anat.* **162**: 73–87 (1981).

8. Holder, N.: Pattern formation and growth in the regenerating limbs of urodelean amphibians. *J. Embryol. Exp. Morph.* **65**: (suppl.), 19–36 (1981).

9. Holder, N., Reynolds, S.: Morphogenesis of the regenerating limb blastema of the axolotl. Shape, autonomy and pattern. In: *Limb development and regeneration, 3rd International Symposium.* (eds.) Caplan, A.I., Fallon, J., Goetinck, P., Kelley, R., MacCabe, J. New York: A. Liss, (1982, *in press*).

10. Holder, N., Tank, P.W.: Morphogenetic interactions occurring between blastemas and stumps after exchanging blastemas between normal and double-half forelimbs in the axolotl *Ambystoma mexicanum. Dev. Biol.* **68**: 271–279 (1979).

11. Holder, N., Tank, P.W., Bryant, S.V.: Regeneration of symmetrical forelimbs in the axolotl *Ambystoma mexicanum. Dev. Biol.*, **74**: 302–314 (1980).

12. Krasner, G.N., Bryant, S.V.: Distal transformation from double-half forelimbs in the axolotl *Ambystoma mexicanum. Dev. Biol.* **74**: 315–325 (1980).

13. Lewis, J.H.: Simpler rules for epimorphic regeneration. The polar coordinate model without polar coordinates. *J. Theoret. Biol.* **88**: 371–392 (1981).

14. Maden, M.: Structure of supernumerary limbs. *Nature* **286**: 803–805 (1980).

15. Maden, M.: Supernumerary limbs in amphibians. *Am. Zool.* **22**: 131–142 (1982).

16. Maden, M., Mustafa, K.: The structure of 180 degree supernumerary limbs and a hypothesis of their formation. *Dev. Biol.* **93**: 257–265 (1982).

17. Papageorgiou, S., Holder, N.: The structure of supernumerary limbs formed after 180 degree blastemal rotation in the newt *Triturus cristatus. J. Embryol. Exp. Morph.* (1982).

18. Slack, J.M.W., Savage, S.: Regeneration of mirror symmetrical limbs in the axolotl. *Cell* **14**: 1–8 (1978).

19. Stocum, D.L.: Regeneration from symmetrical hindlimbs in larval salamanders. *Science* **200**: 790–793 (1978).

20. Stocum, D.L.: Autonomous development of reciprocally exchanged regeneration blastemas of normal forelimbs and symmetrical hindlimbs. *J. Exp. Zool.* **212**: 361–371 (1980).

21. Tank, P.W.: The failure of double half forelimbs to undergo distal transformation following amputation in the axolotl *Ambystoma mexicanum. J. Exp. Zool.* **204**: 325–336 (1978).

22. Tank, P.W.: Pattern formation following 180 degree rotation of regeneration blastemas in the axolotl, *Ambystoma mexicanum. J. Exp. Zool.* **217**: 377–387 (1981).

23. Tank, P.W., Carlson, B.M., Connolly, T.G.: A staging system for forelimb regeneration in the axolotl *Ambystoma mexicanum. J. Morphol.* **150**: 117–128 (1976).

24. Tank, P.W., Holder, N.: The effect of healing time on the proximodistal organisation of double-half forelimb regenerates in the axolotl, *Ambystoma mexicanum. Dev. Biol.* **66**: 72–85 (1978).

25. Tank, P.W., Holder, N.: Pattern regulation in the regenerating limbs of urodele amphibians. *Q. Rev. Biol.* **56**: 113–142 (1981).

26. Winfree, A.T.: *The geometry of biological time.* New York: Springer (1980).

Questions for Discussion with the Editors

1. *You frequently mention the role cell-cell contacts might play in establishing, for example, the shortest route of intercalation in the regenerating blastema. Can you propose any direct tests which might be carried out to support the notion that cell-cell contacts provide the information employed to specify pattern?*

This is a difficult question to get at. Four approaches spring to mind. (a) To establish morphologically whether cells do contact over specific distances within the blastema, e.g., from dorsal to ventral blastema extremes in the lower arm blastema. This type of quantitation has now been performed and all predictions of specific healing modes are consistent with the data. (b) What are the structural features of cell contacts within the blastema? Do junctions exist between blastema cells? (c) Cell culture. Can different cell division kinetics of blastema cells be characterized in cell culture? Can such kinetics be altered in precise ways by confronting cells from different blastema positions in culture? If so, do cells need to contact for such alterations to occur? The problem here is the infancy of amphibian cell culture and the problems of regulation of cell behavior in culture. (d) Characterization of the blastema cell surface. If information lies on the cell surface can it be detected biochemically by, for example, antibody specificity or lectin binding.

2. *Where do retinoids and electric currents fit into your view of the limb regeneration problem?*

(a) Electric currents do not fit into my view of limb regeneration. The evidence for any role of currents in regeneration is tenuous at best and I can see no evidence for their role in patterning in multicellular fields in vertebrates. (b) Retinoids are useful probes that disturb patterns in vertebrate limbs in specific ways. As such, they may provide a valuable tool for studying pattern regulation. No evidence exists which suggests that vitamin A is any kind of normal patterning molecule.

CHAPTER **23**

Retinoids as Probes for Investigating the Molecular Basis of Pattern Formation

Malcolm Maden

As OTHER CHAPTERS in this book reveal, our current understanding of the molecular mechanisms of pattern formation is far from complete, to say the least. Research in pattern formation still involves the classical embryological methodology of tissue removal and tissue transplantation. Any subsequent disruption of pattern is observed, and from this, the rules of cellular behavior that the particular tissue under study follows during normal and abnormal pattern formation are deduced. Much data has been collected and many models have been proposed to explain these "rules," some involving detailed computer simulations of the behavior of hypothetical morphogens (6, 10).

However, with the possible exception of *Hydra* (1, 7) the molecules responsible for the way in which cells determine their position, establish polarity and form characteristic patterns have not been identified. This is partly because of the sheer complexity of pattern formation in higher organisms and partly because we have had no clues as to where in the cell to look for such molecules. Knowledge of a cell's position could, for example, be imparted by an extracellular concentration gradient of a morphogen, an intracellular concentration gradient of a morphogen, chemical differences between cell surfaces, differences in the extracellular matrix, etc. To begin biochemical investigations into any one of these areas without good reason would surely be a waste of time. One way to decide which is the most likely candidate is to alter the positional information (15) of cells in a reliable and controlled way. In the case of the limb, the system described here, we would like to make distal cells of the hand, for example, become proximal cells of the upper arm or vice versa. Fortunately, recent experiments with Vitamin A and its derivatives, the retinoids, have revealed that this is precisely what seems to be happening. The positional information of limb cells is being changed as the experiments described below reveal. Hopefully we now have a valuable probe with which to investigate positional information in the limb.

Vitamin A and Limb Regeneration in Urodeles

The limbs of axolotls are, upon amputation, capable of regenerating exactly those elements removed, whatever the level. If the hand is amputated only the carpals and digits are regenerated. If the lower arm is amputated, the radius, ulna, and hand are regenerated (Fig. 1a) and so on. However, when young axolotl larvae are

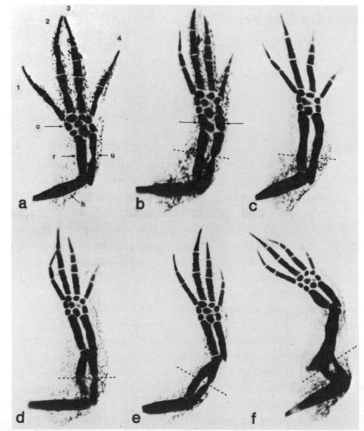

FIGURE 1. Regenerated axolotl limbs after amputation through the radius and ulna, stained with Victoria blue and cleared. Broken lines mark the amputation plane. (a) Control; h = humerus, r = radius, u = ulna, c = carpals, 1 2 3 4 = digits. (b–e) Limbs treated with 60 mg/l retinol palmitate for 12 days. (b) Three extra cartilages (arrows) have been intercalated between the carpals and radius and ulna. (c), An abnormally long radius and ulna have been regenerated. (d) A complete, extra radius and ulna has been produced in tandem. (e) The distal part of the humerus is now appearing too. (f) After 15 days in 300 mg/l retinol palmitate a complete limb has regenerated from the amputation plane.

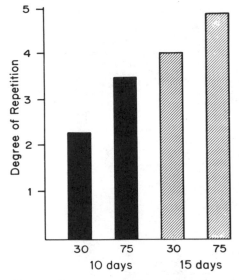

FIGURE 2. The effect of increasing concentration at two times of treatment on the degree of serial repetition. This scale is scored according to the six classes of limb in Fig. 1 with normal scoring 0, extra carpals (Fig. 1b) 1, extra long radius and ulna (Fig. 1c) 2, and so on up to a complete limb (Fig. 1f) scoring 5. 30 = 30 IU/ml retinol palmitate (120 mg/l), 70 = 70 IU/ml (300 mg/l) each for 10 or 15 days.

placed in a solution of retinol palmitate after amputation through the radius and ulna, instead of regenerating just those elements removed, a graded sequence of extra elements appears in the proximodistal axis. These range from extra carpal-like structures (Fig. 1b), an extra length of radius and ulna (Fig. 1c), an extra complete radius and ulna (Fig. 1d) to an extra distal humerus and lower arm (Fig. 1e). If the length of time that the animals are in retinol palmitate or if the concentration is increased, then the series continues (Fig. 1f) until a complete new limb regenerates from the cut ends of the radius and ulna (3).

A clear concentration effect is apparent (Fig. 2). At higher concentrations, the average degree of reduplication is higher, i.e., regeneration commences from a more proximal level. The time effect can also be demonstrated by placing animals in a solution of fixed concentration for an increasing interval after amputation. Here too, the level at which regeneration commences becomes gradually more proximal up to a maximum at 12 days (Fig. 3a). The complementary experiment to this—leaving increasing periods of time for the limbs to regenerate *before* placing them in Vitamin A reveals that the effectiveness decreases after the first week of regeneration (Fig. 3b). Thus, it is the early events of regeneration on which Vitamin A is acting—dedifferentiation and establishment of the blastema—precisely the time when we would expect pattern formation mechanisms to be operating (4).

It is important to emphasize that, as can be seen from the Figs., the anteroposterior axis (digit and carpal sequence) of the regenerates is perfectly normal. The dorsoventral axis, examined by studying muscle patterns in serial sections of such limbs, is normal too. Therefore Vitamin A seems to affect uniquely the proximodistal axis of regenerating axolotl limbs.

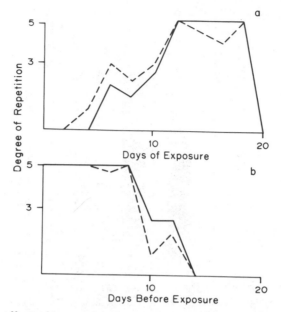

FIGURE 3. (a) The effect of increasing the time of exposure to 300 mg/l retinol palmitate on the degree of serial repetition (see Fig. 2 for explanation of this scale). The maximum effect is obtained with a time of 12–18 days. (b) The effect of delaying exposure to 300 mg/l retinol palmitate for increasing time after amputation. For the maximum effect to be obtained the limbs must be exposed to Vitamin A during the first 8 days after amputation. Broken line = forelimbs, solid line = hindlimbs.

The Effects at Other Amputation Levels

Similar concentration and time effects can be produced after amputation through the humerus or through the hand. In both cases, a complete limb can be regenerated from the cut (Fig. 4a, b). Interestingly, the concentration of retinol palmitate needed to produce this effect is lower from the humerus level than from the carpal level. That is, the amount of Vitamin A needed to induce the regeneration of a complete limb decreases as the amputation plane becomes more proximal. This gives us a valuable clue as to how Vitamin A might be acting, as will be discussed later.

The same phenomena can be observed in hindlimb regeneration from each level. At the appropriate concentration and time, a complete limb can be regenerated from amputation through the tibia and fibula (Fig. 4c) or tarsals (Fig. 4d). Note that in both these cases, a piece of girdle has also been produced. Therefore the effects of Vitamin A are not just confined to the limb itself, but extend into the girdle, a result noted later.

The Cellular Effect of Vitamin A

It can be seen by simple external observations of axolotl limbs that in high concentrations of Vitamin A no regeneration takes place. Paradoxically, only when the

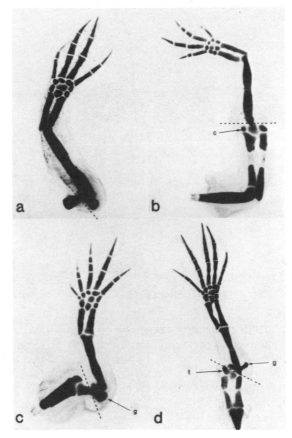

FIGURE 4. Cleared whole mounts of limbs treated with 300 mg/l (a–c) or 600 mg/l (d) retinol palmitate for 15 days. Broken lines mark the amputation planes. (a) Forelimb—amputation through the humerus. Proximal to the cut a small potion of humerus remains and distally a complete limb has regenerated. (b) Forelimb amputation through the carpals. Three carpals remain in the stump (c) and distally a complete limb has regenerated. (c) Hindlimb amputation through the tibia and fibula. Distally a complete limb plus part of the pelvic girdle (g) has regenerated. (d) Hindlimb amputation through the tarsals (t). Again a complete limb plus part of the girdle has regenerated.

animals are removed into fresh water does regeneration commence. This is shown in Fig. 5 with camera lucida drawings of control and experimental limbs. The effect is elicited via the suppression of DNA synthesis (Fig. 6a) and mitosis (Fig. 6b). However, once released from this inhibition, extra pattern is then regenerated.

Histologically, Vitamin A-treated limbs (Fig. 7a) show only one peculiarity compared to control limbs (Fig. 7b). Dedifferentiation progresses, an apial cap forms as usual, but in the absence of cell division the blastemal cells clump together in a tight, often eccentrically positioned, ball (Fig. 7a). Whether this clustering is significant for the subsequent abnormal pattern formation remains to be seen, but what can be demonstrated is that the inhibition of cell division *per se* is not responsible for Vitamin A effects in the following manner.

When limbs are denervated by cutting the nerves at the brachial plexus, re-

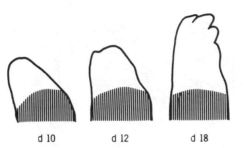

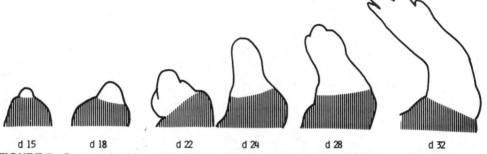

FIGURE 5. Camera lucida drawings of control (upper row) and Vitamin A treated (lower row) limbs after amputation through the radius and ulna. Hatching marks the stump, clear area the regenerate. Control—by day 10 post amputation a cone shaped blastema is present and by day 18 regeneration is virtually complete with the outline of the four digits present. Vitamin A treatment with 300 mg/l retinol palmitate for 15 days results in no regeneration during that period. Only when the limbs are removed from Vitamin A does regeneration commence.

generation is prevented. Here too cell division is inhibited (8). When the denervated limb becomes reinervated, regeneration commences after a delay period that can be manipulated at will by one or more subsequent redenervations. But when the limb finally does regenerate, the pattern is perfectly normal—only those elements removed are replaced. Thus the inhibition of cell division is not responsible for these Vitamin A effects.

The Effect of Different Retinoids

Some clues as to the mode of action of Vitamin A can be obtained by investigating the effects of its various analogues. Retinol acetate, retinol palmitate, retinol and retinoic acid have been compared. At the same concentration (0.12 mM) the effects in terms of the degree of repetition of elements vary across a 20-fold range (Fig. 8). Retinoic acid is the most active, followed by retinol, retinol palmitate, and last retinol acetate. It has been pointed out that this sequence of efficacy is the same as the sequence of net charge of the molecules. This suggests that their mode of action is, at least initially, at the surface of the blastemal cell since a highly charged molecule

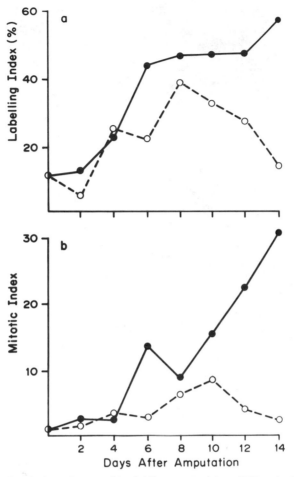

FIGURE 6. (a) Labeling index of control (solid line, crosses) and Vitamin A treated (broken line, circles) sampled every other day after amputation. Experimental limbs were treated with 300 mg/l retinol palmitate for the duration. The initial stimulation of DNA synthesis after amputation occurs in both control and experimental limbs, but after day 8 an inhibition by Vitamin A begins to take effect. (b) Mitotic index (per thousand cells) in control (solid line, crosses) and Vitamin A treated (broken line, circles) limbs. Mitosis is severely inhibited when control limbs start to rapidly divide after day 8.

would be more active at the cell surface and more capable of being inserted into the cell membrane than a less charged molecule.

Vitamin A and Limb Development

The experiments described above were performed on regenerating limbs and it is natural to ask whether the same effects can be demonstrated on developing limbs.

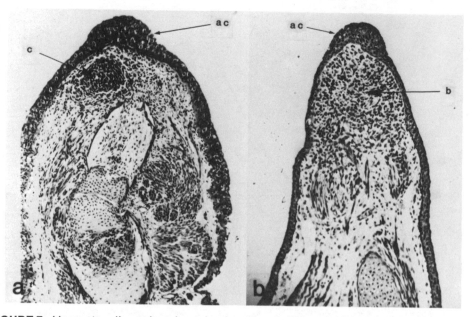

FIGURE 7. Haematoxylin and eosin stained sections of Vitamin A treated (a) and control (b) limbs. An apical cap (ac) is present in both, but instead of a rapidly dividing population of blastemal cells (b) present in control limbs, Vitamin A causes dedifferentiated cells to clump together in a tight ball (c).

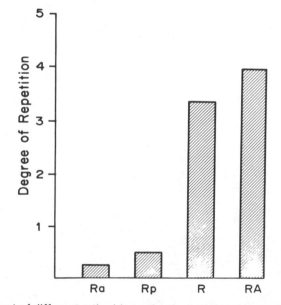

FIGURE 8. The effect of different retinoids on the degree of serial repetition (see Fig. 2 for details). Each retinoid was administered for 15 days at a concentration of 0.12 mM to fore-limbs amputated through the radius and ulna. Ra = retinol acetate, Rp = retinol palmitate, R = retinol, RA = retinoic acid. Clearly retinoic acid and retinol are the most effective.

The developing limb buds of chick embryos have been used for such studies (11, 14). Various differences in procedure have necessarily been introduced because simply adding Vitamin A to embryos causes dose-dependent lethality and those that do survive have normal limbs. Local application is therefore imperative and is effected by absorbing retinoic acid on to filter paper, newsprint, agar or silastin and implanting this into the limb bud.

In contrast to the results on the regenerating limb, in the chick no proximodistal effects can be detected. Instead when the implants are grafted to the anterior side of the limb bud, mirror-image duplications in the anteroposterior axis are produced (Fig. 9b). If the implants are placed at the posterior edge, either the limbs are normal or limb reductions result, but not duplications. Despite affecting a different axis, there are interesting similarities between the regenerating and developing limbs. Firstly, low doses of retinoic acid gives less of an effect which is manifest as reduced duplications. Instead of a 6-digit, 4 3 2 2 3 4 duplication (Fig. 9b) lower doses give only digits 2 2 3 4 or 3 2 2 3 4 (Fig. 9c). This is directly analogous to the dose dependent degree of serial repetition in the regenerating limb. Secondly, at too high doses, limb development is virtually abolished and severe reductions result (Fig. 9d). Again, this occurs in regenerating limbs.

Thus the cellular effects on the two systems seem to be the same, but the mor-

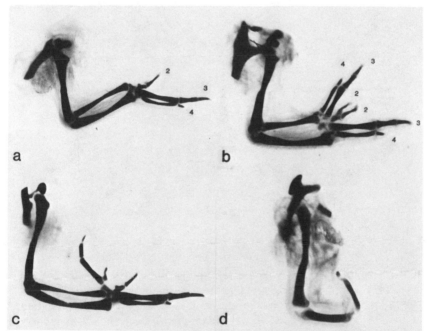

FIGURE 9. Cleared whole mounts of chick limbs stained with Alcian green. (a) Control limb with digits 2 3 4 marked. b–d limbs treated with a piece of newsprint soaked in retinoic acid and grafted to the anterior side of the limb. (b) A full anteroposterior duplication has been produced with a digital sequence of 4 3 2 2 3 4. (c) At lower concentrations of retinoic acid less complete duplications appear, this has digits 3 2 2 3 4 (from left to right). (d) At too high concentrations, severe reductions of the limb appear. This limb has only one digit.

phological results are different. In the chick limb, Vitamin A acts exactly like a graft of the posterior limb organizer (ZPA) which gives identical anteroposterior duplications (13). Their mode of action is similar too. Both a ZPA graft and Vitamin A need a minimum time of exposure of 12–24 hr with increasing time giving increased duplications (11). The dose effects of Vitamin A are very similar to the attenuation of ZPA grafts by γ-irradiation (9). Furthermore, when tissue adjacent to the locally applied Vitamin A is removed from the limb bud and grafted to the anterior edge of another limb, a mirror-image duplication results (11). Thus it is possible that Vitamin A is acting by inducing a posterior polarizing in the tissue next to the implant and the limb then behaves as if it had received a ZPA graft.

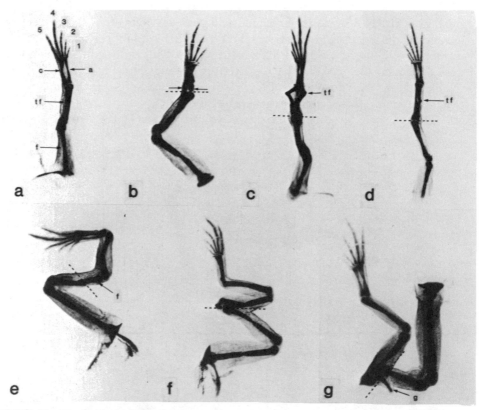

FIGURE 10. Hindlimbs of *Rana temporaria* stained with Victoria blue and cleared. (a) Control limb f = femur, tf = tibia and fibula, c = calcaneum, a = astragalus 1 2 3 4 5 = digits. (b–g) limbs treated with increasing concentrations of retinol palmitate (30–120 mg/l) for increasing time periods (2–9 days after amputation). Broken lines mark the amputation planes. (b) Two extra carpal-like elements (arrows) have regenerated in addition to the calcaneum, astragalus and 5 digits. (c) An abnormal segment of tibia and fibula (tf) has been intercalated. (d) A complete extra tibia and fibula (tf) is present. (e) The distal part of the femur (f) has now regenerated. (f) A complete limb has regenerated from the foot level. (g) A complete limb plus a pelvic girdle (g) has been regenerated from the distal amputation plane.

Vitamin A and Limb Regeneration in Anurans

One further system in which the effect of Vitamin A has been studied is the regenerating hindlimbs of tadpoles of *Rana temporaria* (5). *Rana* hindlimbs can be amputated at various stages throughout development and for a short period after development, and regeneration ensues (Fig. 10a) (2). For the effects of Vitamin A to be most pronounced, limbs are amputated through the shank just as the first indentations of digits appear and the tadpoles placed in various concentrations of retinol palmitate for varying lengths of time.

Remarkably, frog tadpoles seem to combine the effects of the previous two systems described in that *both* the proximodistal and anteroposterior axes are affected. Control limbs regenerated precisely those elements removed by amputation, that is, the calcaneum and astragalus and 5 digits (Fig. 10a). At low doses and short times of exposure, abnormalities in the proximodistal axis appear (Fig. 10b and c) and as the concentration is increased, the level at which regeneration commences becomes more proximal (Fig. 10d–g) in exactly the same way as the regenerating axolotl limb. The time and concentration effects of Vitamin A are also the same as in axolotls (Fig. 11), that is increasing the time of administration at any one concentration increases the degree of serial reduplication and conversely, increasing the concentration at any one time of administration increases the degree of serial reduplication.

At the higher concentrations and exposure times (9 days at 30 mg/1 or 6 days at 60 or 120 mg/1), effects previously described for the chick limb appear. That is, mirror-image duplications in the anteroposterior axis *in addition to* the serial du-

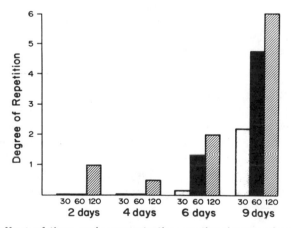

FIGURE 11. The effect of time and concentration on the degree of serial repetition of *Rana temporaria* limbs amputated through the foot. same scoring as Fig. 2 except that in addition an extra complete limb plus girdle (Fig. 10g) scores 6. Clearly, increasing time and/or concentration causes extra segments to be regenerated as in the case of axolotl limbs.

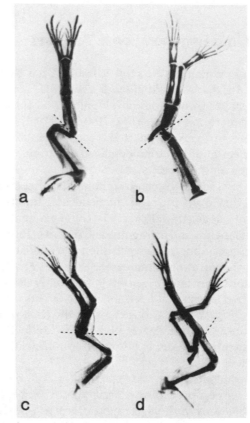

FIGURE 12. Anteroposterior and proximodistal reduplicated limbs of *Rana temporaria*. Broken lines mark the amputation planes. Limbs treated with 60 or 120 mg/l retinol palmitate for 6 or 9 days. Each example has regenerated a complete limb from the distal amputation level but also has mirror imaged in the anteroposterior axis. Note that in every case the anterior edges of the pair of limbs (digit 1) face each other and the posterior edges (digit 5) are outermost. In d, the limbs are crossed over.

plications (Fig. 12). What is also remarkable is that these anteroposterior duplications take exactly the same form as the chick limb bud duplicates despite the great differences in method of administration, developmental stage etc. In both cases, the duplications appear at the *anterior* edge, producing limbs with their posterior digits outermost and their anterior digits innermost. In the chick, the digital sequence is thus 4 3 2 2 3 4 (Fig. 9b) and in *Rana* 5 4 3 2 1 1 2 3 4 5 (Fig. 12).

Thus *Rana* limbs seem to combine the effects of both axolotl and chick limbs.

Hypothesis of Vitamin A Effects

From the results described above, it does seem that Vitamin A is changing the positional information of cells in both developing and regenerating limbs. More tissue is being produced than would be expected and in recognizable patterns. But to speculate on the precise biochemical effects of Vitamin A and thus on the form in which positional information is encoded in the limb would be premature.

Nevertheless, some progress can be made by drawing together the results and searching for a consistent scheme couched in only the most general terms. Considering first the effects on the proximodistal axis observed in axolotl and *Rana* limbs, they can be described as follows. Suppose there is a gradient of a substance, say P, along the proximodistal axis of the limb and the amount of P at any point along the axis determines what that cell will become (Fig. 13a, b). That is, a typical system of positional information as described by Wolpert (15). Now, if Vitamin A switches on the synthetic pathway of substance P in the blastemal cells at the level of the amputation, it will begin to accumulate. Since the amount of P determines the positional information of the cell then as P accumulates, blastemal cells will become more proximal. The longer Vitamin A is present, the more P will accumulate and the more proximal the cells will become. Similarly, the more Vitamin A that is present at any one time, the more proximal the cells will become (Fig. 13c). Clearly, to attain the maximum amount of P (shoulder level of the limb) will take longer or will need a higher concentration of Vitamin A from distal levels (lower starting amount of P) than from proximal levels (higher starting amount of P) and this is precisely what is found. Of course, other schemes are equally plausible such as the gradient being of Vitamin A itself, although that seems unlikely, or a gradient of opposite polarity with Vitamin A destroying the substance responsible for positional information.

Results on the developing chick limb have been explained by assuming that the localized Vitamin A implant induces adjacent cells to become posterior polarizing cells. The limb then behaves exactly as if it had received a ZPA graft. Since the characteristics of Vitamin A effects and ZPA grafting are very similar, as described above, this seems a highly plausible scheme. The mechanics of the hypothesis, involving a source-sink diffusion gradient across the anteroposterior axis, have already been proposed (13) and extensively tested (12) and we can simply adopt this explanation for use here (Fig. 14). However, this scheme depends upon the fact that Vitamin A is administered locally by an implant into the anterior part of the limb bud. The hypothesis would not predict a mirror-image duplicate to arise from the anterior side when Vitamin A is administered systemically—why should *only* those cells at the anterior edge be stimulated to become a new source of morphogen? Yet this is precisely what occurs in *Rana* limbs and so this cannot be adopted as a universal explanation.

An alternative hypothesis which can accommodate the *Rana* results is to consider the anteroposterior axis in terms of a Gierer-Meinhardt inhibitor-activator system with the peaks of activator and inhibitor at the posterior edge of the limb (6; and Chapter 3) (Fig. 15a). This establishes a positional information system similar in concept to the ZPA source-sink diffusion hypothesis, but different in mechanics. Now if we assume that Vitamin A destroys the inhibitor throughout the limb, then the first location at which the level of inhibitor falls below that of the activator will be at the anterior edge since here the inhibitor is at its lowest concentration (Fig. 15b). Such a reduction in the inhibitor will be sufficient to induce a new center of activation (Fig. 15c). Thus the precise location of the extra limb can be explained after a systematic administration of Vitamin A.

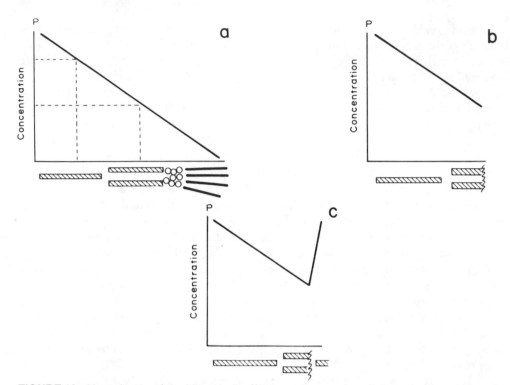

FIGURE 13. Hypothesis of the Vitamin A effects on the proximodistal axis of axolotl and *Rana* limbs. (a) Normal limb with the proximodistal level specified by the concentration of P at that level, that is, a typical positional information field. (b) After amputation through the radius and ulna the cells at the level of the cut know which position they are at by sensing the concentration of P. Thus they know from which level to commence regeneration. (c) As in b, but here Vitamin A has been applied and the concentration of P has risen to a maximum. Thus the cells at the amputation plane think they are at the proximal end on the humerus level and will mistakenly commence regeneration from that level.

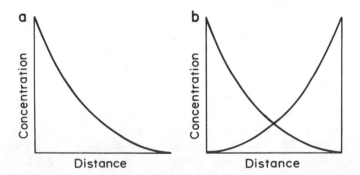

FIGURE 14. Hypothesis of the Vitamin A effects in the chick limb. (a) The normal anteroposterior organization is controlled by the concentration of a morphogen released by the zone of polarizing activity (ZPA) at the posterior edge and which diffuses across the width of the limb. (b) After implantation of a retinoic acid soaked graft to the anterior edge the cells in the vicinity are stimulated to become ZPA cells. They consequently start synthesizing morphogen and a mirror imaged gradient is set up which results in an anteroposterior reduplicated limb. However, this hypothesis cannot explain the *Rana* results.

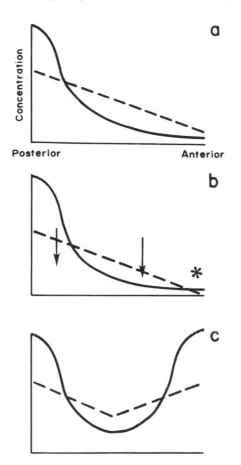

FIGURE 15. Hypothesis of the Vitamin A effects on the anteroposterior axis of *Rana* limbs, which can also explain the chick limb results. (a) The normal anteroposterior organization is by a Gierer-Meinhardt activator—inhibitor system. Solid line indicates rapidly diffusing activator; a broken line, slowly diffusing inhibitor. See (6) for details. The peaks of activator and inhibitor are at the posterior edge of the limb. (b) Vitamin A destroys the inhibitor which decreases in concentration throughout the limb (arrows). At the first point where the level of inhibitor falls below that of the activator (*), the anterior pole, a new source of activation is initiated. (c) Mirror imaged gradient with the new peak of activator established at the anterior pole. This system also explains the chick results (Fig. 14).

Conclusion

There is clearly great scope for hypothesizing on the results described here. Whether any or none are subsequently validated is largely immaterial. What is important is that Vitamin A is affecting the pattern formation of specific limb axes in a reliable and biochemically interesting fashion. It seems to be changing the positional information of cells in both developing and regenerating limbs. We can now begin biochemical studies to answer such questions as: where in the cell are the Vitamin A-induced changes taking place, what biochemical pathways are affected and ultimately what molecules are affected? Hopefully this will lead to the elucidation of the molecular basis of pattern formation, at least in the limb.

General References

MADEN, M.: Vitamin A and pattern formation in the regenerating limb. *Nature (Lond.)* **295**: 672 (1982).

SUMMERBELL, D., HARVEY, F.: Vitamin A and the control of pattern in developing limbs. *Proc. 3rd Int. Conf on Limb Dev. and Regen.*, (1983, *in press*).

TICKLE, C., ALBERTS, B., WOLPERT, L., LEE, J.: Local application of retinoic acid to the limb bond mimics the action of the polarizing region. *Nature (Lond.)* **296**: 564 (1982).

References

1. Berking, S.: Analysis of head and foot formation in *Hydra* by means of an endogenous inhibitor. *Wilmhelm Roux' Arch. Dev. Biol.* **186**: 189 (1979).

2. Maden, M.: Experiments on Anuran limb buds and their significance for principles of vertebrate limb development. *J. Embryol. Exp. Morph.* **63**: 243 (1981).

3. Maden, M.: Vitamin A and pattern formation in the regenerating limb. *Nature (Lond.)* **295**: 672 (1982).

4. Maden, M.: Vitamin A and the control of pattern in regenerating limbs. *Proc. 3rd Int. Conf. on Limb Dev. and Regen.* (1982a, *in press*).

5. Maden, M.: *The effect of Vitamin A on limb regeneration in Rana temporaria in preparation* (1983).

6. Meinhardt, H.: Models for the ontogenetic development of higher organisms. *Rev. Physiol. Biochem. Pharmacol.* **80**: 47 (1978).

7. Schaller, H.C.: Isolation and characterization of a low-molecular weight substance activating head and bud formation in hydra. *J. Embryol. Exp. Morph.* **29**: 27 (1973).

8. Singer, M.: On the nature of the neurotrophic phenomenon in Urodele limb regeneration. *Am. Zool.* **18**: 829 (1978).

9. Smith, J.C., Tickle, C., Wolpert, L.: Alteration of positional signalling in the chick limb by high doses of γ-irradiation. *Nature (Lond.)* **272**: 612 (1978).

10. Summerbell, D.: The zone of polarizing activity: evidence for a role in normal chick limb morphogenesis. *J. Embryol. Exp. Morph.* **50**: 217 (1979).

11. Summerbell, D., Harvey, F. Vitamin A and the control of pattern in developing limbs. *Proc. 3rd Int. Conf. on Limb Dev. and Regen.*, (1983).

12. Summerbell, D., Honig, L.S.: The control of pattern across the antero-posterior axis of the chick limb bud by a unique signalling region. *Am. Zool.* **22**: 105 (1982).

13. Tickle, C., Summerbell, D., Wolpert, L.: Positional signalling and specification of digits in chick limb morphogenesis. *Nature (Lond.)* **254**: 199 (1975).

14. Tickle, C., Alberts, B., Wolpert, L., Lee, J.: Local application of retinoic acid to the limb bud mimics the action of the polarizing region. *Nature (Lond.)*, **296**: 564 (1982).

15. Wolpert, L.: Positional information and the spatial pattern of cellular differentiation. *J. Theor. Biol.* **25**: 1 (1969).

Questions for Discussion with the Editors

1. *Do your observations on the retinoid stimulated reduplication of proximal structures modify—to any great extent—the polar coordinate model for limb regeneration?*

The observations of proximal respecification in regenerating axolotl limbs do not call for a modification of the polar coordinate model (PCM) because it is not particularly concerned with the proximodistal axis. The model simply says that limbs will distally transform if a complete circle is present without proposing any mechanisms for the process. The model is more concerned with the organization of the transverse axes, being a rival to Cartesian models. The generation of anteroposterior reduplications in *Rana*, therefore, has more relevance to the PCM. It can provide no rational explanation for such reduplications apart from implying the spontaneous generation of a new number 3 in the 9 position. This is a very unsatisfactory explanation, a better one is discussed in the text based on a Cartesian gradient system.

2. *Have rigorous searches for endogenous levels of retinoids or retinoid-like molecules in regenerating limbs been carried out?*

Searches for retinoids have not been performed in regenerating limbs, neither have they even been carried out in amphibians. Most retinoid research is, not surprisingly, carried out on mammalian tissues because of the therapeutic uses of retinoids in cancer treatment and dermatology. One presumes retinoids exist in amphibians since they are derivatives of Vitamin A and therefore one presumes that retinol binding proteins exist in the cells of amphibian tissues. But no one has demonstrated the fact.

CHAPTER 24

Pattern Specification in the Developing Chick Limb

Lorette C. Javois

ONE OF THE MOST INTERESTING problems in development concerns how the cells of a developing organism know their physical location and how this positional information is interpreted. How do the patterns of cellular differentiation we recognize as the adult organism arise? We know that the molecular changes which occur as cells differentiate into muscle, cartilage, and a variety of other cell types are not responsible for, but rather, are the result of pattern specification. For example, both an arm and a leg are composed of the same differentiated cell types, yet the overall pattern of each is distinctly different. It is believed that pattern specification relies on the acquisition of a "positional address" according to a cell's physical location in the embryo. Interpretation of this "positional value" leads to the appropriate cyto-differentiation. This concept of positional information was first suggested by Driesch (7) and more recently described in Wolpert's positional information theory (39 and Chapter 1).

The avian limb provides a model system in which to study this process of pattern specification during the development of a secondary embryonic field. The entire embryonic period of chick limb development is amenable to analysis and experimental intervention; precise patterns of structural elements are formed, and a cell marker is available. The ease of manipulation of limb bud tissue during development allows us to examine what factors are important in establishing the final pattern of structures. The similarities between the developing avian limb, the developing and regenerating amphibian limb, and the developing mammalian limb allow us to further investigate the possibility that common control mechanisms underlie limb patterning events in vertebrates.

Pattern specification in the developing chick limb will be considered here in two parts. In the first section, the epigenetic mode of patterning will be reviewed both in terms of the classical work which has provided the foundation of our knowledge, and the more recent experiments which suggest further insights. In the second section, models describing pattern specification will be considered. While an exhaustive review of all the models proposed to account for pattern specification in the developing chick limb is beyond the scope of this chapter, two current models are discussed. In particular, the applicability of the polar coordinate model in describing limb development is considered.

The Epigenetic Mode of Pattern Specification

A General Description of Limb Bud Outgrowth and the Pattern of Structures Formed

Limb buds arise as paired condensations of mesoderm covered by ectodermal "jackets" which protrude from the lateral body walls of the chick embryo at stage 16 for the wings and stage 17 for the legs (9). The appearance of the mesodermal bulge is due to decreased proliferation in the flank region possibly accompanied by decreased expansion of the limb region (30). The limb-forming mesoderm is of dual origin: somatic mesoderm of the body wall and somitic mesoderm which migrates into the buds from the ventro-lateral ends of somites 15–20 for the wings and somites 26–32 for the legs. Because quail cells can be distinguished histologically from chick cells on the basis of nuclear morphology, the contribution of the somatic and somitic mesoderm to the wing can be determined. By replacing chick somitic tissue at the level of the wing with quail somitic tissue, it has been demonstrated that the somatopleure (chick) gives rise to the skeleton, connective tissue of the muscles, tendons, vasculature, and integument while the somitic mesoderm (quail) gives rise to the intrinsic musculature of the wing. Similar experiments also indicate that the spatial deposition of somitic cells to individual muscles is dependent on positional information contained in the somitic mesoderm of the limb rather than in the somitic cells which migrate into the bud (3–5, 20).

In response to a mesodermal inductive signal, the ectoderm rimming the apex of the bud thickens to form the apical ectodermal ridge (AER). The AER in turn reciprocally induces mesodermal outgrowth and is necessary for continued axial elongation. By stage 21 (3.5 days), the dorsal side of the bud is rounded, the ventral surface flattened, and the posterior half is growing more than the anterior half. Cytodifferentiation begins in the proximal regions of the paddle-shaped bud by late stage 22 (4 days) and progresses distally. Myogenic cells come to occupy the dorsal and ventral regions of the bud while the prospective cartilage cells occupy the central portion. Differential growth and characteristic patterns of cellular necrosis give rise to the contours and interdigital spaces of the limb. By stages 36–38 (10 to 12 days), the skeletal elements, muscles, tendons, connective tissue, vasculature, and feather germs and/or scales characteristic of the adult wing or leg are present (Figs. 1, 2, 3; Table 1). Growth continues until hatching at 21 days. The final pattern of structures arises epigenetically through interaction of the various limb tissues (for review see 25).

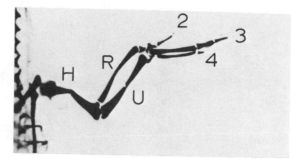

FIGURE 1. Normal chick wing skeletal pattern. Dorsal view of a normal right chick wing skeleton at 10 days. The normal wing skeleton consists of the humerus (H) that articulates proximally with the shoulder girdle and distally with the forearm. The forearm is composed of an anterior radius (R) and posterior ulna (U) that articulate proximally with the humerus and distally with the anterior radius and posterior ulna of the wrist. The hand is comprised of three digits, anterior digit 2, middle digit 3, and posterior digit 4.

The Ability of the Limb Bud to Self-differentiate

The limb is present as a bud protruding from the body wall at stage 17. Well before this, tissue comprising the limb field is capable of morphological and axial self-differentiation. As early as stage 11, mesoderm and ectoderm (or mesoderm alone) from the wing field grafted to the flank of an embryo will develop according to its original axial orientation. Before early stage 10, prospective wing bud mesoderm

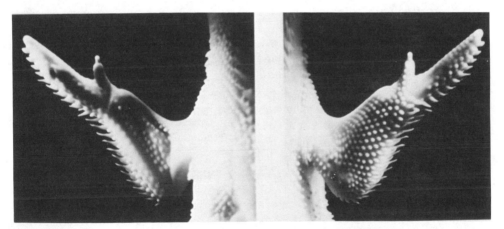

FIGURE 2. Normal chick wing integument. On the left is a ventral view of a normal right chick wing at 10 days and on the right is the dorsal view of the same wing. Note that there are more feather germs on the dorsal as opposed to ventral wing surface. There is also an antero-posterior asymmetry of feather germs across the dorsal surface of the wing. At the posterior edge of the wing, two rows of long secondary coverts begin just proximal to the elbow and extend to the wrist; primary coverts extend from the wrist to the distal tip of digit 3. Anteriorly there is a prepatagium (web) extending from the shoulder to the wrist.

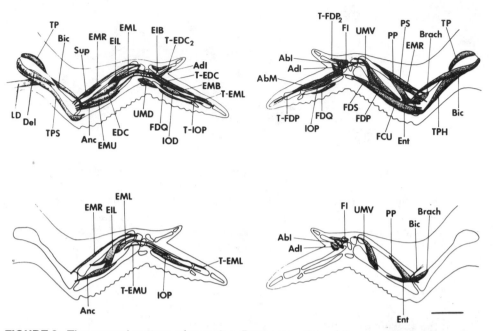

FIGURE 3. The normal pattern of muscles. Reconstruction of the normal muscle pattern of a right wing of a 10–12 day embryo. The left drawings show the dorsal muscles, and the right drawings show the ventral muscles. The bottom two drawings have had some of the overlying muscles removed. The muscles are stippled and the tendons are blackened. Abbreviations for muscles as in Table 1. Scale unit represents 1 mm (from 17).

must be grafted with adjacent segmental plate in order for development to occur. The segmental plate apparently influences the prospective limb tissue, imparting the capacity for self-differentiation. Prior to bud formation, studies have revealed no morphological or biochemical polarity in the prospective limb mesoderm, and even during early stages of outgrowth the bud is comprised of undifferentiated mesenchyme. Nevertheless, some inhomogeneity which imparts axial order must be present as indicated by the morphogenetic performance of the tissue under the experimental conditions described above (19).

The Role of the AER and the Generation of "Specification" Maps

Much of the classic work on the role of the AER during limb outgrowth was done by Saunders and co-workers (24). Saunders was the first to demonstrate that following removal of the AER, terminal wing parts failed to differentiate and that the level at which limb development was truncated depended on the stage of the bud at the time the AER was removed. These experiments illustrate that the presumptive wing parts acquire the ability to self-differentiate in a proximo-distal sequence, and the continued presence of an AER is necessary for complete limb outgrowth. That the AER provides the conditions necessary for axial elongation can be illustrated by grafting an extra AER to the dorsal or ventral surface of a wing

TABLE 1

NORMAL WING MUSCULATURE

		Origin	Insertion			Origin	Insertion
Dorsal stylopodium							
TP	Tensor propatagii	Medial surface dorsal clavicle	Tendon not formed	PP	Pronator profundus	Distal ventral humerus	Proximal midventral radius (no tendon)
Del	Deltoid	Coracoid	Tendon not formed				
TPS	Triceps pars scapularis	Scapula posterior to glenoid cavity	Proximal ulna (no tendon)	Ent	Entepicondylo-ulnaris	Common tendon with PP	Proximal midventral ulna (no tendon)
LD	Latissimus dorsi	Neural spines from some vertebrae	Humerus between TPS and TPH (no tendon)	FDS	Flexor digitorum superficialis	Distal ventral humerus	Common tendon with FDP, anterior ventral metacarpal digit 3
TPH	Triceps pars humeralis	Proximal posterior humerus (no tendon)	Proximal ulna (no tendon)	FCU	Flexor carpi ulnaris	Distal ventral humerus	Ulnari
				FDP	Flexor digitorum profundis	Proximal anterior ventral ulna	Common tendon with FDS, anterior ventral metacarpal, digit 3; Proximal branch to first phalanx digit 2 (T-FDP$_2$)
Ventral stylopodium							
Bic	Biceps	Coracoid and bicipital crest of humerus via tendinous sheet	Tendons to radius and ulna				
				UMV	Ulnimetacarpalis ventralis	Distal anterior ventral ulna	Radiale
Dorsal zeugopodium				**Dorsal autopodium**			
EMR	Extensor metacarpi radialis	Distal dorsal humerus	Common tendon with EIL, to radiale	EIB	Extensor indicis brevis	Anterior metacarpal digit 2	Dorsal first phalanx metacarpal digit 2
Sup	Supinator	Distal dorsal humerus	Proximal midanterior radius (no tendon); proximal to insertion of PS	EMB	Extensor medius brevis	Dorsal metacarpal digit 3 (no tendon)	Muscle applied to tendon of EML
EDC	Extensor digitorum communia	Distal dorsal humerus	Tendon branches to dorsal metacarpals of digits 2 and 3 (T-EDC$_2$) and 3	IOD	Interosseus dorsalis	Proximal metacarpals digits 3 and 4	First phalanx digit 3 common tendon with EDC
EMU	Extensor metacarpi ulnaris	Distal dorsal humerus	Tendon to dorsal posterior metacarpal digit 2	UMD	Ulnimetacarpalis dorsalis	Distal dorsal posterior ulna	Proximal posterior metacarpal digit 4
Anc	Anconeus	Distal dorsal humerus	Anterior middistal ulna (no tendon)				
EIL	Extensor indicis longus	Two areas: midradius; proximal ulna (no tendon)	Common tendon with EMR, to radiale	**Ventral Autopodium**			
				AbI	Abductor indicis	Anterior ventral proximal metacarpal digit 3	Anterior ventral distal metacarpal digit 2
EML	Extensor medius longus	Proximal radius	Tendon to phalanx digit 3	FI	Flexor indicis	Ventral metacarpal digit 2	Midventral first phalanx digit 2
				AdI	Adductor indicis	Anterior proximal metacarpal digit 3	Posterior first phalanx digit 2
Ventral zeugopodium				AbM	Abductor medius	Ventral proximal metacarpal digit 3	Ventral proximal first phalanx digit 3
Brach	Brachialis	Anterior ventral humerus	Brach (R): ventral proximal radius (no tendon) Brach (U): ventral proximal ulna (no tendon)	IOP	Interosseus palmaris	Ventral proximal metacarpals digits 3 and 4	Tendon to dorsal posterior first phalanx digit 3
PS	Pronator superficialis	Distal ventral humerus	Proximal midventral radius (no tendon)	FDQ	Flexor digiti quarti	Ventral posterior metacarpal digit 4 (just distal to insertion of UMD)	Phalanx digit 4

bud. This manipulation results in the induction and outgrowth of a second limb tip.

Several experiments indicate the action of the AER is not involved in the patterning process. For example, interchanging the AERs between a wing and leg bud does not affect the development and differentiation of the underlying mesoderm. Likewise, reversing the antero-posterior polarity of the AER on the bud does not affect pattern specification. The AER does not specify the proximo-distal level of differentiation achieved by the underlying mesoderm. This was demonstrated by capping the mesodermal core of a young wing bud with the apical ectoderm of an old wing bud, and *vice versa*. These operations resulted in the formation of normal wings. The architecture of the ridge is observed to conform to the age of the underlying mesoderm. For a more detailed discussion of the properties of the AER see the review by Saunders (25).

Recent work by Summerbell (32) and Iten (10) has expanded the earlier findings of Saunders. By examining the effect of removing AERs from a wider range of limb buds, what might be referred to as a "specification" map is obtained. Removing the AER is interpreted as stopping the specification of more distal positional values, suggesting that during the earlier stages of development only cells with proximal positional values are specified. As development proceeds cells with more distal positional values are generated. Hence, the AER removal experiments allow us to map the proximo-distal specification of cells during axial elongation. Iten not only confirmed the progressive proximo-distal specification of limb structures but also reported a high frequency of limbs truncated at the level of the forearm with only one forearm skeletal element, the ulna (10). When the AER was removed from stage 19 through 21 wing buds, wings with a partial or complete humerus resulted. AER removals from stage 22 and 23 wing buds resulted in wings with a complete humerus and partial or complete ulna but no radius. It was concluded that there is also a posterior to anterior specification of forearm structures with cells specified to form the ulna before the more anterior radius.

The Regulative Ability of Limb Bud Tissue

Different portions of the chick limb bud can be added or removed such that a limb with the normal complement of differentiated structures forms. These experiments demonstrate that at certain stages of development the presumptive fate of the cells is not fixed and regulation can occur (Fig. 4) (1). Dissociating limb bud mesenchyme, repacking it into a limb bud ectodermal jacket, and grafting it onto the somites of a host embryo can also result in the formation of a relatively normal limb (23, 40). While the limb bud does exhibit an extensive capacity to regulate following surgical intervention at early stages, there is a progressive loss of this ability in a proximal to distal direction at specific stages of development (28).

Just as the chick limb bud can regulate for surgical defects to produce a normal wing, it can also regulate to form supernumerary or extra limb structures following the apposition of normally nonadjacent limb bud tissue. One of the earliest examples of this form of regulation is illustrated by an experiment of Saunders and co-workers (29). They showed that when the distal third of a left limb bud was grafted to the contralateral limb stump opposing anterior and posterior tissue, supernumerary structures developed. Likewise, 180° rotation of the distal third of a limb bud on its base also resulted in the formation of extra limb structures (Fig. 5).

These experiments and others can be interpreted to indicate limb bud cells have information about their physical location within the developing bud and can respond to changes in their environment by regulating to form extra limb structures. While this regulative ability is extensive, with increasing age it diminishes. Iten and co-workers have recently asked whether limb bud cells lose the ability to recognize positional disparity or lose the ability to respond to positional disparity (14). To answer this question, they performed 180° rotations of limb bud tips between stage 21 and stage 24 wing buds. While rotation of stage 21 wing bud tips on their stumps resulted in wings with extra limb structures, rotation of stage 24 tips

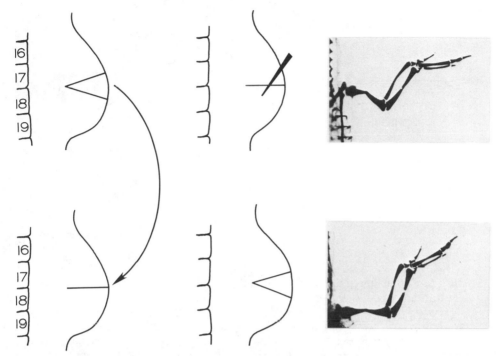

FIGURE 4. The ability of the chick wing bud to regulate for deletion or addition of tissue. Dorsal view diagrams of right wing buds and adjacent somites (16–19) illustrating the deletion (upper) and addition of extra (lower) wing bud tissue. At the right are the skeletons of typical wings resulting from these manipulations. These wings are normal (see Fig. 1).

on their stumps rarely resulted in supernumerary structures. The supernumerary limb structures which resulted from the rotation of stage 21 tips on stage 24 stumps, and *vice versa*, appeared to arise mostly from stage 21 wing bud tissue. These results suggest that stage 21 wing bud cells have the ability to recognize and respond to positional disparities while stage 24 cells can be recognized, but can not respond to positional disparities. The authors suggest that during development cells do not lose their ability to recognize or be recognized as normally nonadjacent neighbors; rather, as development proceeds they lose the ability to respond to the positional disparity created by the grafting operation.

Because of its extensive regulative powers, the developing chick limb can be systematically manipulated to determine the key factors involved in the process of pattern specification. The next section considers two models of pattern specification which have arisen as the result of a great body of experimental manipulations investigating the regulative capabilities of the developing chick limb bud.

Models of Pattern Specification in the Developing Chick Limb

Models are often formulated to describe experimental results. By using a model as a working hypothesis, one can make specific predictions about how a developing system may respond to an experimental manipulation. Several models have been pro-

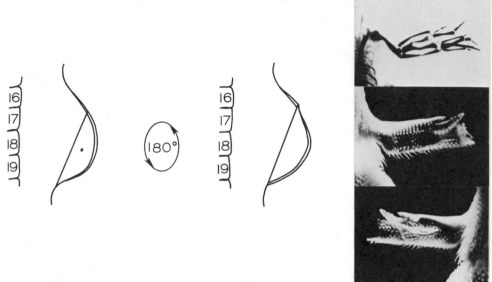

FIGURE 5. The ability of the chick wing bud to regulate and form extra limb structures. The left dorsal view illustrates a right wing bud and adjacent somites (16–19) and 180° rotation of the wing bud tip on its base. This manipulation opposes antero-posterior and dorso-ventral wing bud tissue. (Note the dorsal carbon mark is no longer visible after rotation of the tip.) Typical resulting limbs have supernumerary muscles, skeletal elements, and feather germs. The top view of the resulting wing illustrated to the right shows a dorsal view of the skeleton. Note the supernumerary skeletal elements in the forearm and hand. Based on a histological reconstruction of the muscles and tendons of this limb, the forearm consists of an anterior radius, posterior ulna, and two unidentifiable extra elements in between. The digits comprise three "hands": an anterior supernumerary right hand, a middle inverted right hand, and a posterior inverted supernumerary left hand. Beneath the view of the skeleton is a dorsal and ventral view of the integument of this wing. The integument can also be used to confirm the handedness of the digits [For further discussion see (15).]

posed to account for pattern specification in the developing chick limb. Two models in particular are considered here, the zone of polarizing activity/progress zone model and the polar coordinate model. Both are based on the concept of positional information. This concept assumes that cells are first assigned positional values according to their physical location in a developing system like the embryonic chick limb; they subsequently interpret this positional information and undergo the appropriate cytodifferentiation. Over the years models of pattern specification have suggested experimental approaches which have helped elucidate our present understanding of limb development. As additional knowledge has accumulated the models have been further refined, and certainly this evolution will continue as more details of pattern specification are worked out.

The ZPA/Progress Zone Model

From a historical perspective, the idea that a special region of the developing limb bud played a role in pattern specification during limb outgrowth was suggested by the work of Saunders and Gasseling (26). They observed that posterior limb bud tissue was capable of causing a mirror image duplication of posterior limb structures if it was grafted to a more anterior site in a host limb bud. Saunders and co-workers coined the term "zone of polarizing activity" (ZPA) for this special posterior tissue and mapped the location of the ZPA during limb outgrowth (21). The idea that a gradient of a diffusible morphogen could be used to tell a cell its positional address was first suggested by Wolpert (39). This concept was extended by Tickle and co-workers who proposed that a gradient of a diffusible morphogenetic substance with its source in the ZPA might be responsible for telling cells their antero-posterior position across the entire width of the developing limb (Fig. 6a) (36). About this same time, Summerbell and co-workers suggested that the distal 200–400 μm of mesenchyme subjacent to the AER comprised another special region of the growing limb bud, the "progress zone" (35). They hypothesized that the sequential proximal to distal pattern of limb structures arose during normal development as the result of an autonomous mechanism whereby cells measure the length of time that they spend in the progress zone, perhaps by counting the number of times they divide while in this zone (Fig. 6b) (34). Cells which leave the progress zone early in limb outgrowth would give rise to proximal limb structures while cells which undergo further cell divisions and leave the zone later would differentiate into more distal limb structures (Fig. 6c). In a further refinement, the roles of the ZPA and progress zone during limb outgrowth were linked by Summerbell's proposal that cells could only be told their antero-posterior position by the morphogenetic substance emanating from the ZPA while they were in the progress zone (33). Once they left the progress zone their positional values were fixed.

The biochemical nature of the polarizing activity has remained elusive. No subcellular preparation from the ZPA exhibits *in vivo* polarizing activity, and *in vitro* assays using ZPA tissue or extracts have not yet demonstrated the presence of a diffusible polarizing morphogen (22).

Since its inception the ZPA/progress zone model has undergone considerable refinement, and most recently it has been proposed that the ZPA is the source of two distinct signals, one which stimulates cell growth and the other which tells cells their positional address (6). It is no longer assumed the ZPA influences cells across the entire width of the limb bud; rather, it is suggested the ZPA specifies structures over a 200–300 μm range (31).

Some disenchantment with the concept of a ZPA in the developing limb has been expressed by those who originally defined it. Saunders and co-workers have found that the morphogenetic "activity" previously regarded as a unique property of the ZPA is also present in a number of non-limb tissues (25, 27). These data, as well as other recent findings, have prompted Saunders to suggest, as have others (see the discussion that follows), that interactions between individual cells may play a more active role in the process of pattern specification.

The Polar Coordinate Model

Just after the ZPA/progress zone model was proposed to describe chick limb development, another model, also based on the concept of positional information, was

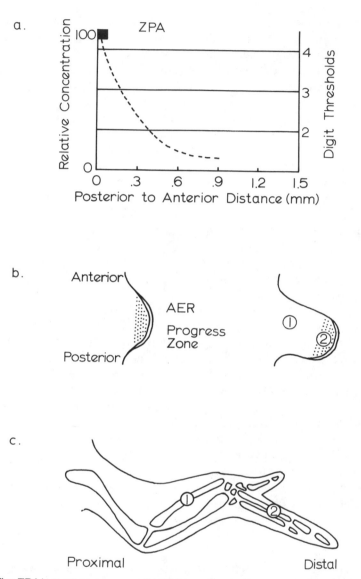

FIGURE 6. The ZPA/progress zone model. (a) Hypothesized concentration profile of a polarizing morphogen in a normal wing bud. The zone of polarizing activity (ZPA) located at the posterior edge of the wing bud is assumed to be the source of a morphogen that diffuses into the limb bud where it is broken down. The gradient that is established (---) specifies the antero-posterior pattern of structures. Posterior tissue exposed to a high concentration of morphogen gives rise to digit 4; whereas, more anterior tissue, exposed to a lower concentration, gives rise to digit 2. (b) Diagrammatic representation of wing bud outgrowth according to the progress zone model. Positional values along the proximo-distal axis are postulated to be specified by the amount of time a cell spends in the progress zone subjacent to the apical ectodermal ridge. Cells in region 1 have had their proximo-distal positional values specified; the positional values of cells in region 2 are not yet specified. (c) Following cytodifferentiation, the two regions ultimately give rise to different proximo-distal limb structures.

suggested to describe pattern specification. This model, the polar coordinate model, arose as a synthesis of experimental findings from studies carried out on regenerating amphibian and cockroach limbs and regenerating *Drosophila* imaginal discs (8). The polar coordinate model, a formal rather than mechanistic model, took the view that cells play a more active role during pattern specification. It suggested cells may be able to recognize their local environment and respond to changes in it, perhaps at the level of local cell-cell interactions. While this model was initially proposed to describe the process of regeneration, it was suggested it might also be applicable to limb development. It was this suggestion which led Iten and co-workers to examine chick limb development from the vantage provided by the polar coordinate model. As a working hypothesis, the model has allowed us to make specific and testable predictions about how the developing chick limb would respond to various experimental manipulations. Because the polar coordinate model describes pattern specification in terms of local cell-cell interactions and cell proliferation, it does not propose the existence of any special regions controlling development within the limb bud. Rather, individual cells have positional addresses and are able to recognize and respond to changes in their local environment. Iten's initial work examined the ability of various limb bud tissues to cause the formation of extra structures when grafted into host limb buds. It was concluded from this work that there did not appear to be anything special about the posterior edge of the limb bud (ZPA); anterior (non-ZPA) limb bud tissue was also capable of causing the formation of extra limb structures when grafted to more posterior host sites (11).

Bryant and co-workers have refined the polar coordinate model based on further experiments with regenerating limbs (2), and Iten and co-workers have continued to test the applicability of this model to the developing chick limb. In the following discussion normal chick limb outgrowth and the pattern regulation which occurs following experimental manipulations of limb bud tissue are reviewed in terms of the formal rules and predictions of the polar coordinate model. The same rules for regeneration can also be used to describe normal limb development and the formation of supernumerary structures following experimental manipulations (15).

Rules of the Polar Coordinate Model in Chick Limb Outgrowth

While the developing chick limb is a three-dimensional solid, the polar coordinate model proposes a two-dimensional positional information system. It does not directly address the specification of positional information in the peripheral as opposed to the internal dimension. However, evidence suggests peripheral limb mesenchyme contains positional information while the internal mesenchyme and overlying ectoderm do not (for further discussion see 15).

The polar coordinate model describes the positional address of a cell in terms of a two-dimensional grid (Fig. 7a). For the late embryonic or adult chick wing this grid can be depicted as a series of skewed concentric ellipses representing the wing viewed end-on. A cell's positional address is described by two values. One value represents the location of the cell on the circumference of an ellipse, this

a.

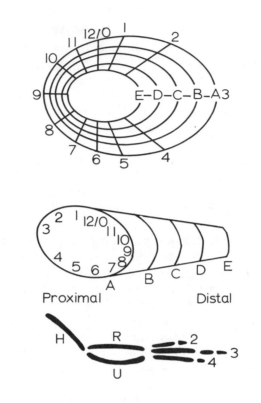

b.

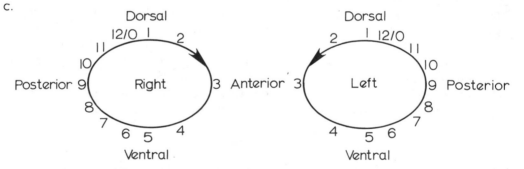

c.

FIGURE 7. A diagrammatic description of the polar coordinate model. (a) Grid representing the distribution of positional values in the late embryonic or adult limb proposed by the polar coordinate model. This grid represents an end-on view of the wing. The cells have their positional addresses specified with respect to their location on a circumference represented by the continuum 1 through 12/0 (there are no special boundry conditions; at 12/0 the sequence simply becomes continuous), and a radius represented by the values A through E. It is proposed that circumferential and radial positional values are unequally spaced, perhaps reflecting the position of "scorable" structures in the final wing outgrowth. (b) When this grid is expanded into a three-dimensional hollow ellipsoid, radial level A can be seen to correspond to the proximal shoulder region of the diagrammatic wing beneath, while radial level E corresponds to the distal digits 2, 3, and 4; (H) humerus; (R) radius; (U) ulna. (c) In an end-on view, the direction of increasing circumferential positional values specifies the "handedness" of the wing. A clockwise sequence specifies a right wing while a counterclockwise sequence specifies the mirror-image left wing.

value corresponding to the cell's antero-posterior and dorso-ventral position around the wing circumference. The other value represents the distance of the cell from the center of the grid, corresponding to the cell's proximo-distal location in the wing. In the three-dimensional limb, proximal structures of the wing (e.g., the shoulder) correspond to radial level A while distal structures (e.g., the digits) correspond to radial level E. Therefore, the grid expanded into a hollow ellipsoid represents the three-dimensional wing (Fig. 7b). Note also, the handedness of the wing is described by the direction of ascending circumferential positional values (Fig. 7c).

The polar coordinate model proposes that cellular behavior is governed by local cell-cell interactions and the general property of intercalation. According to the model, cells know their own positional addresses and can recognize their neighbors's positional addresses. If during development two cells with normally nonadjacent positional addresses confront each other, the cells recognize this disparity, proliferate, and eliminate the disparity via the shorter circumferential route. For example, if cells with positional values 8 and 10 interact, intercalation will occur producing a cell with positional value 9 (rather than cells with positional values 11, 12/0, 1, . . ., 7) (Fig. 8).

During early wing bud outgrowth, we propose wing bud cells have circumferential and *proximal* radial positional values. Nonadjacent cells around the circumference in the distal regions of the bud would interact, perhaps via cell processes. The positional disparity recognized through these interactions would result in intercalation. The model also proposes that progeny cells which arise during intercalation will adopt the next most *distal* radial positional values if their circumferential positional values are identical to those of pre-existing neighboring cells. This results in cells with the next most distal level of circumferential positional values (Fig. 8). At a predetermined proximo-distal level interactions between particular circumferential values lead to digit formation and distalization completes limb outgrowth.

A Polar Coordinate Model Description of Pattern Regulation

The polar coordinate model makes specific predictions about what will happen when limb bud tissue is grafted into a host wing bud confronting normally nonadjacent cells. The model predicts that the cells will detect a local positional disparity, and the donor and host cells will intercalate eliminating the disparity. If the intercalation gives rise to cells with positional values specifying "scorable" limb structures, extra limb structures will be observed in the resulting wing. Based on the model, we can predict whether supernumerary limb structures will arise and what these structures will be, as well as their asymmetry or handedness.

Many of the grafting experiments we have done to study pattern regulation during chick limb development support these predictions. (11–13, 16–18). Our work has led us to conclude that some of the key factors determining whether and what extra structures form following a grafting operation are the position of origin of the donor tissue, the position it is transplanted to in a host wing bud, and the orientation of the graft in the host. Other factors, such as the presence of an AER at the site of the confrontation, probably determine the degree to which extra limb

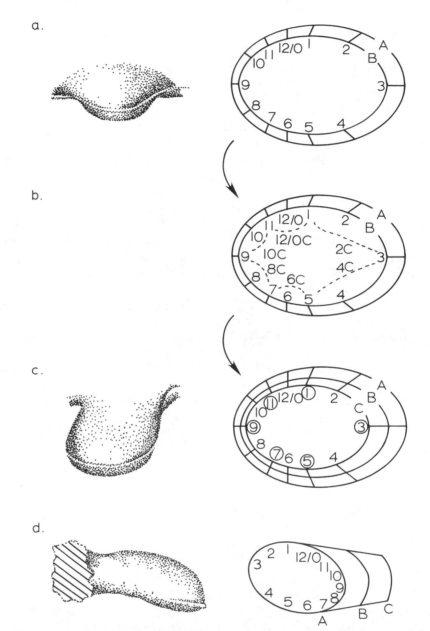

FIGURE 8. The process of distalization during normal wing bud outgrowth. (a) A distal dorsal view depicting an early stage wing bud (left) and a diagrammatic polar coordinate description of this wing bud (right). At this stage of development wing bud cells have circumferential and proximal radial positional values. (b) Diagrammatic polar coordinate description of the interactions which give rise to some of the cells with circumferential positional values of radial level C. Cells with normally nonadjacent positional values interact via short arcs (dashed lines) around the circumference. Intercalation produces progeny cells which eliminate the circumferential positional disparity recognized during these interactions. Because the intercalated cells have circumferential positional values identical to preexisting neighboring cells, they adopt the next most distal radial positional value,

radial value C. (c) A distal dorsal view depicting a later stage wing bud which has under-gone further distal outgrowth (left) and a diagrammatic polar coordinate description of this wing bud (right). Further intercalation (circled values) between the cells generated in (b) has completed radial level C circumferential positional values. This process of interca-lation and distalization, with some provision for stopping at the distal tip, continues pro-ducing a complete limb outgrowth. (d) A proximal side view of the wing bud depicted in (c) (left) and an expanded three-dimensional polar coordinate diagram depicting this wing bud (right). When expanded into a hollow ellipsoid, radial level A corresponds to the proxi-mal base of the wing bud while successive radial values represent more distal levels. The entire proximo-distal sequence is not represented, as the process of distalization is not complete.

structures develop into separate outgrowths. As cited, we have published detailed descriptions of the supernumerary limb structures resulting from the opposition of anterior/posterior, dorsal/ventral, and proximal/distal limb bud cells. A diagram-matic polar coordinate description of all these grafting operations is beyond the scope of this chapter. However, for the purposes of illustrating how key factors in-fluence the final pattern of a limb outgrowth, one experimental manipulation and its corresponding control has been selected to discuss in detail. Additional examples have been discussed by Javois (15).

Supernumerary Limb Structures Following Dorso-Ventral Opposition of Limb Bud Tissue

Previously, we described a detailed study demonstrating that dorso-ventral opposi-tion of wing bud tissue without associated antero-posterior positional disparity leads to the formation of extra muscles; these extra muscles arise from both the do-nor and host tissue and are the result of opposing dorsal and ventral mesenchyme rather than re-orienting the overly ectoderm (17). One of this series of grafting op-erations, and its control operation, is illustrated in Fig. 9. Polar coordinate dia-grammatic descriptions of these operations are presented in Figs. 10 and 11.

A comparison of Figs. 10 and 11 illustrates the dramatic differences in tissue interactions which the polar coordinate model predicts will occur following the control and experimental manipulations. In the control grafting situation (Fig. 10), a slight serial repetition of mid-wing bud circumferential positional values results from the addition of a donor wedge of mid-wing bud tissue. The slightly re-dundant circumferential sequence is restored to normal during the process of dista-lization, since only cells with normally nonadjacent positional addresses produce progeny with more distal positional values. Indeed, wings resulting from this con-trol grafting operation have normal patterns of limb structures. Thus, adding a do-nor wedge of tissue to a host wing bud at the same antero-posterior level from which it was removed does not result in autonomous development of the graft and the formation of extra limb structures.

When the same type of grafting operation is performed but dorsal and ventral mid-wing bud cells are opposed, the polar coordinate model again predicts that in-

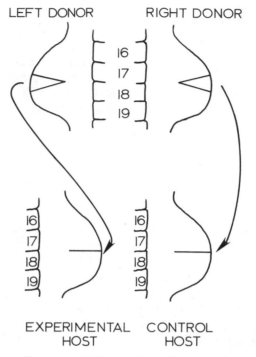

LEFT DONOR   RIGHT DONOR

16
17
18
19

16   16
17   17
18   18
19   19

EXPERIMENTAL   CONTROL
HOST   HOST

FIGURE 9. Dorsal view diagrams of left and right donor wing buds and right host wing buds with adjacent somites (16–19) illustrating the grafting of wedges of mid-donor wing bud tissue to slits at the same anterio-posterior level in the host wing buds. In the control grafting situation, no positional disparity is created by the operation, and resulting limbs have normal patterns of wing structures. In the experimental graft, however, donor and host dorsal and ventral tissue are opposed by inserting the wedge upside-down. Resulting wings have normal skeletal patterns, dorso-ventrally reversed integuments in the mid-wing region, and extra mid-wing muscles and tendons in the forearm and hand.

tercalation will eliminate the positional disparity as illustrated in Fig. 11. During the process of distalization, some of the intercalated cells will not produce progeny with more distal positional value because they recognize no positional disparity with their neighbors. This occurs in circumferential regions where intercalation has created areas of symmetry. Hence a "convergence" or loss of cells with certain circumferential positional addresses occurs as distal outgrowth proceeds. However, because of the extent of the original disparity created by the manipulation, the polar coordinate model predicts that the final limb outgrowth will still have extra cells with extra circumferential positional values localized in the mid-wing bud region. We propose that these cells give rise to the extra muscles observed in wings resulting from this experimental manipulation. Histological analyses of these resulting limbs reveal that the supernumerary muscles have not arisen randomly but are localized to the middle of the wing, reflecting the antero-posterior region of the wing bud in which the positional disparity was created. Also, the disruption of the muscle pattern is more extensive proximally than distally, perhaps reflecting the "convergence" toward a more normal circumferential sequence of positional values which the model predicts occurs during distalization.

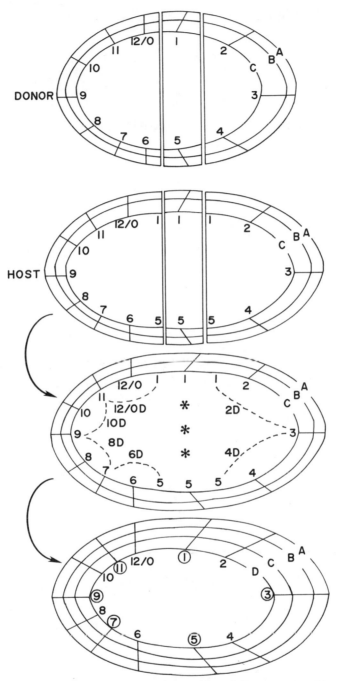

FIGURE 10. Diagrammatic polar coordinate description of the control grafting operation illustrated in Fig. 9. Diagrams illustrate the donor and host wing buds in end-on views and show the slight redundancy of positional values created by grafting tissue from the middle of the donor wing bud to the middle of the host wing bud. At the time of the grafting operation, only proximal radial levels are present. Because cells with identical positional values interact about the line of symmetry indicated by the asterisks but produce no progeny during distalization, interactions at radial level C (dashed lines) as described in Fig. 8, lead to the intercalation of a normal sequence of circumferential positional values at radial level D. A normal wing results.

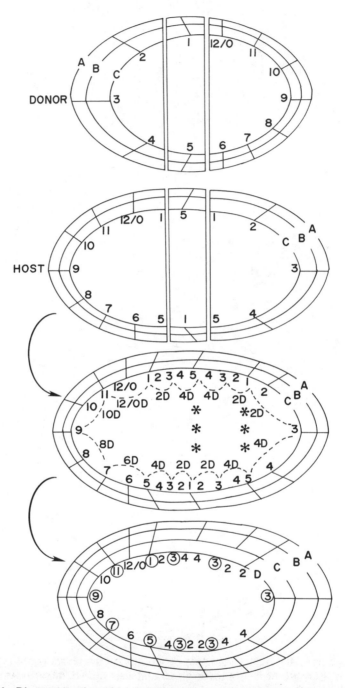

FIGURE 11. Diagrammatic polar coordinate description of limb outgrowth following the experimental grafting operation illustrated in Fig. 9. Diagrams illustrate the donor and host wing buds in end-on views and the positional disparity created in the host wing bud as the result of opposing donor and host dorsal and ventral tissue. At the time of the grafting operation, only proximal radial levels are present. Intercalation of cells with inter-

vening positional values (circled) occurs via the shorter circumferential route to eliminate the positional disparity. The circumferential intercalation which occurs results in the formation of symmetric sequences of positional values about the lines indicated by the asterisks. Cells at radial level C interact via short arcs (dashed lines) as described in Fig. 8 to produce the next distal radial level D. Because cells with identical positional values interact at the lines of symmetry, no progeny with these positional values are produced. While this "convergence" results in the loss of some of the circumferential positional values, it is postulated the final circumferential sequence has extra positional values, and the cells with these positional values give rise to the extra muscles observed in limbs resulting from this operation.

In conclusion, these two grafting operations illustrate how the rules of cellular behavior proposed by the polar coordinate model can be used to account for the final pattern of limb structures which arises as the result of positional disparity created by the manipulation of wing bud tissue.

Conclusions

Many questions remain to be answered before our understanding of chick limb pattern specification is complete. For example, the events that occur between the time an experimental manipulation is performed and the final pattern of differentiated structures is present are essentially unknown. How is positional information passed on to progeny cells during distal outgrowth? Are the developmental capabilities of these progeny cells restricted in any way? Further grafting operations like those described will help answer these questions, shedding light on this dimly perceived period of pattern specification.

The suggestion that local cell-cell interactions are intimately involved in pattern specification during chick limb development indicates positional information may reside on the cell surface or in the extracellular microenvironment. Experiments directed at the level of the cell surface may further elucidate the cellular and molecular nature of positional information. Recent advances in monoclonal antibody technology allow one to examine cell surface antigens. Monoclonal antibodies have been isolated which appear to reflect a cell surface expression of position in both the avian retina (37) and the *Drosophila* wing imaginal disc (38). This technology may be useful, allowing us to probe the molecular basis of pattern specification in the developing chick limb.

Acknowledgments

The author is currently supported by American Cancer Society Grant PF-2141. The investigations reported here were supported by a David Ross Fellowship awarded by Purdue University, West Lafayette, Indiana and in part by National Science Foundation Grants 77-08408-A1 and 81-10848 awarded to Laurie E. Iten.

General References

HINCHLIFFE, J.R., and JOHNSON, D.R. (eds.): *The Development of the Vertebrate Limb*, Oxford: Clarendon Press (1980).

FALLON, J.R., and CAPLAN, A.L. (eds.): *Limb Development and Regeneration, Part A*, New York: Alan R. Liss (1983).

References

1. Amprino, R.: Aspects of limb morphogenesis in the chicken. In: *Organogenesis*, (eds. DeHaan, R.L., Urspring, U.), pp. 255–281. New York: Holt, Rinehardt and Winston (1965).

2. Bryant, S.V., French, V., Bryant, P.J. Distal regeneration and symmetry. *Science* **212**: 993–1002 (1981).

3. Chevallier, A., Kieny, M., Mauger, A.: Limb-somite relationship: origin of the limb musculature. *J. Embryol. Exp. Morphol.* **41**: 245–258 (1977).

4. Chevallier, A., Kieny M., Mauger, A.: Limb-somite relationship: effect of removal of somite mesoderm on the wing musculature. *J. Embryol. Exp. Morphol.* **43**: 263–278 (1978).

5. Christ, B., Jacob, H.F., Jacob, M.: Experimental analysis of the origin of the wing musculature in avian embryos. *Anat. Embryol.* **150**: 171–186 (1977).

6. Cooke, J., Summerbell, D.: Control of growth related to pattern specification in chick wing-bud mesenchyme. *J. Embryol. Exp. Morphol.* **65** (suppl): 169–185 (1981).

7. Driesch, H.: *The Science and Philosophy of the organism*. London: Black (1908).

8. French, V., Bryant, P.J., Bryant S.V.: Pattern regulation in epimorphic fields. *Science* **193**: 969–981 (1976).

9. Hamburger, V., Hamilton, H.L.: A series of normal stages in the development of the chick embryo. *J. Morphol.* **88**: 49–92 (1951).

10. Iten, L.E.: Pattern specification and pattern regulation in embryonic chick limb bud. *Am. Zool.* **22**: 117–129 (1982).

11. Iten, L.E., Murphy, D.J.: Pattern regulation in the embryonic chick limb: supernumerary limb formation with anterior (non-ZPA) limb bud tissue. *Dev. Biol.* **75**: 373–385 (1980).

12. Iten, L.E., Murphy D.J.: Supernumerary limb structures with regenerated posterior chick wing bud tissue. *J. Exp. Zool.* **213**: 327–335.

13. Iten, L.E., Murphy, D.J., Javois, L.C.: Wing buds with three ZPAs. *J. Exp. Zool.* **215**: 103–106 (1981).

14. Iten, L.E., Murphy, D.J., Muneoka, K.: Do chick limb bud cells have positional information? In: *Limb Development and Regeneration, Part A*, (eds. Fallon, J.F., and Caplan, A.I.), pp. 77–88. New York: Alan R. Liss (1983).

15. Javois, L.C.: Pattern regulation and pattern specification in the embryonic chick wing. Ph.D. Thesis. Purdue University, West Lafayette, Indiana (1981).

16. Javois, L.C., Iten, L.E.: Position of origin of donor posterior chick wing bud tissue transplanted to an anterior host site determines the extra structures formed. *Dev. Biol.* **82**: 329–342 (1981).

17. Javois, L.C., Iten, L.E.: Supernumerary limb structures after juxtaposing dorsal and ventral chick wing bud cells. *Dev. Biol.* **90**: 127–143 (1982).

18. Javois, L.D., Iten, L.E., Murphy, D.J.: Formation of supernumerary structures by the embryonic chick wing depends on the position and orientation of the graft in a host limb bud. *Dev. Biol.* **82**: 343–349 (1981).

19. Kieny, M.: Les phases d'activeté morphogènes du mesoderme somatopleural pendant le développement précoce du membre chez l'embryon de poulet. *Annales d'Embryologie et de Morphogenèse* **4**: 281–298 (1971).

20. Kieny, M., Chevallier, A.: Autonomy of tendon development in the embryonic chick wing. *J. Embryol. Exp. Morphol.* **49**: 153–165 (1979).

21. MacCabe, A.B., Gasseling, M.T., Saunders, J.W., Jr.: Spatio-temporal distribution of mechanisms that control outgrowth and anterio-posterior polarization of the limb bud in the chick embryo. *Mech. Ageing Dev.* **2**: 1–12 (1973).

22. MacCabe, J.A., Knouse, E.R., Richardson, K.E.Y.: Verification of an *in vitro* bioassay for limb-bud polarizing activity. *J. Exp. Zool.* **216**: 103–106 (1981).

23. MacCabe, J.A., Saunders, J.W., Jr., Pickett, M.: The control of the anteroposterior and dorsoventral axes in embryonic chick limbs constructed of dissociated and reaggregated limb-bud mesoderm. *Dev. Biol.* **31**: 323–335 (1973).

24. Saunders, J.W., Jr.: The proximal-distal sequence of origin of the parts of the chick wing and the role of the ectoderm. *J. Exp. Zool.* **108**: 363–403 (1948).

25. Saunders, J.W., Jr.: The experimental analysis of chick limb bud development. In: *Vertebrate Limb and Somite Morphogenesis*, (eds. Ede, D.A., Hinchliffe, J.R., Balls, M.), pp. 1–24. Cambridge: Cambridge University Press.

26. Saunders, J.W., Jr., Gasseling, M.T.: Ectodermal-mesenchymal interactions in the origin of limb symmetry. In: *Epithelial-Mesenchymal Interactions*, (eds. Fleischmajer, R., Billingham, R.E.), pp. 78–97. Baltimore: Williams and Wilkins, Co. (1968).

27. Saunders, J.W., Jr., Gasseling, M.T.: New insights into the problem of pattern regulation in the limb bud of the chick embryo. In: *Limb Development and Regeneration, Part A*, (eds. Fallon, J.F., Caplan, A. I.), pp. 67–76, New York: Alan R. Liss (1983).

28. Saunders, J.W., Jr., Gasseling, M.T., Cairns, J.M.: The differentiation of prospective thigh mesoderm grafted beneath the apical ectodermal ridge of the wing bud in the chick embryo. *Dev. Biol.* **1**: 281–301 (1959).

29. Saunders, J.W., Jr., Gasseling, M.T., Gfeller, M.D.: Interactions of ectoderm and mesoderm in the origin of axial relationships in the wing of the fowl. *J. Exp. Zool.* **137**: 39–74.

30. Searls, R.L., and Janners, M.Y.: The initiation of limb bud outgrowth in the embryonic chick. *Dev. Biol.* **24**: 198–213 (1971).

31. Smith, J.C., Honig, L.S.: Growth and the origin of additional structure in reduplicated chick wings. In: *Limb Development and Regeneration, Part A*, (eds. Fallon, J.F., Caplan, A.I.), pp. 57–65, New York: Alan R. Liss (1983).

32. Summerbell, D.: A quantitative analysis of the effect of excision of the AER from the chick limb-bud. *J. Embryol. Exp. Morphol.* **32**: 651–660 (1974).

33. Summerbell, D.: Interaction between the proximo-distal and antero-posterior coordinates of positional value during the specification of positional information in the early development of the chick limb bud. *J. Embryol. Exp. Morphol.* **32**: 227–237 (1974).

34. Summberbell, D., Lewis, J.H.: Time, place and positional value in the chick limb bud. *J. Embryol. Exp. Morphol.* **33**: 621–243 (1975).

35. Summberbell, D., Lewis, J.H., Wolpert, L.: Positional information in chick limb morphogenesis. *Nature* (London) 244: 492–495 (1973).

36. Tickle, C., Summerbell, D., Wolpert, L.: Positional signalling and specification of digits in chick morphogenesis. *Nature* (London) 254: 199–202 (1975).

37. Trisler, G.D., Schneider, M.D., Nirenberg, M.: A topographic gradient of molecules in retina can be used to identify neuron position. *Proc. Natl. Acad. Sci. USA* 73: 2145–2149 (1981).

38. Wilcox, M., Brower, D.L., Smith, R.J.: A postion-specific cell surface antigen in the *Drosophila* wing imaginal disc. *Cell* 25: 159–164 (1981).

39. Wolpert, L.: Positional information and the spatial pattern of differentiation. *J. Theoret. Biol.* 25: 1–47 (1969).

40. Zwilling, E.: Development of fragmented and of dissociated limb bud mesoderm. *Dev. Biol.* 9: 20–37 (1964).

Questions for Discussion with the Editors

1. *Given the amount of experimental work which has been done on the developing chick limb, why do two such different views of limb development (the ZPA/progress zone model and the polar coordinate model) remain unresolved?*

Historically, the descriptions of limb development presented by the two models discussed here originated from two very different views of how cells recognize their position and local environment. On the one hand, it was proposed that cells passively receive information by "reading" the local concentration of a diffusible morphogen emanating from a special region of the limb bud. On the other hand, cells were viewed as playing a more active role. Local cell-cell interactions were thought to be involved in assessing the environment. Many of the experiments which have been done can be interpreted from either viewpoint. Investigators have found it very difficult to design a simple definitive experiment. The experiments that we have done to test the validity of the models have served to further our understanding of limb development, and present thinking has evolved to more common ground. For example, it was initially proposed that the long-range signaling mediated by the morphogen spanned 100 cell diameters (the width of the limb bud). This was in contrast to short-range cell-cell mediated interactions. It is currently proposed that the range of action of the morphogen is on the order of 20 to 30 cell diameters. It is also not unreasonable to propose that cells can contact each other via cell processes over a distance of 10 to 20 cell diameters. Given the similarities of these distances, it may be unrealistic to expect to distinguish between the two models on the basis of the distance over which interactions occur.

2. *The extensive series of conventional grafting experiments performed during the past two decades appears, in your opinion, to support the polar coordinate model as an explanation for chick limb pattern formation. Please offer suggestions for what types of analyses might provide the next generation of chick limb pattern specification experimentation.*

While we have formulated general principles that govern the development of pattern and models which account for observed experimental phenomena, our understanding of limb pattern specification is still at a very primitive stage. Certainly, continued work at the level of tissue interactions will provide more information about the events occurring between the time of the experimental manipulation and the fully differentiated state, a developmental period that has not been studied in depth. A better understanding of the cellular activities involved in the patterning process is also necessary. For example, we are just beginning to examine the role of growth in pattern regulation. Since presently we have no idea how positional information is recorded by a cell, interpreted, or passed on to progeny cells, work at the cellular and molecular levels is a necessity. If cell-cell interactions are an important part of the patterning process, monoclonal antibody technology, which allows us to probe the cell surface, may help us examine the cellular and molecular nature of positional information.

CHAPTER **25**

Specification of Feather and Scale Patterns

Danielle Dhouailly

THE INTEGUMENT of multicellular organisms is probably one of the most favorable organs for the study of how regular repetitive structures can arise during development, within an ensemble of apparently identical cells. It offers many different models of patterned ornamentation and is also directly accessible for observation and experimentation.

In amniotes, the skin is well equipped with cutaneous appendages that are distributed in an orderly fashion according to regional and specific patterns. In snakes and lizards, dorsal scales are small, and arranged in a hexagonal pattern, while ventral scales are large and rectangular, forming one longitudinal row in snakes and six longitudinal rows in lizards. Likewise, the hair follicles of mammals are not distributed nor are they shaped at random. However, the major part of our present knowledge of the establishment of cutaneous patterns in amniotes comes from study of the chick embryo, because of the ready availability of fertilized domestic bird eggs and the ease of operating on these embryos.

In birds, the integument can give rise to two types of appendages—scales and feathers. In the chick, while most of the body is covered with feathers, the feet bear scales and some areas, referred to as *apteria*, remain bare. Within the feathered and scaled areas, the appendages are grouped in more or less well-defined tracts—the feather tracts or *pterylae* and the scale tracts. It must be emphasized that while feathers may be found within scale tracts under various genetic conditions (ptilopodous breeds), scales never occur anywhere except on the foot. The arrangement of feathers within each tract is constant, and displays an open hexagonal pattern

581

(Fig. 1), each feather being surrounded by six more or less equidistant feathers. However, the outlines of each pteryla, as well as the number and shape of their constituent feathers, are region-dependent. Scales, unlike feathers, always display a close-packed pattern. Two major types of scales may be distinguished: large scutellate scales, broader than they are long (with respect to the proximo-distal axis of the leg), and small roundish tubercular scales. The scutellate scales are arranged in two (or three) alternate longitudinal rows on either the anterior face (Fig. 2), or the proximal part of the posterior face of the shank (tarsometatarsus), and in a single row on the upper face of each toe. The tubercular scales form a tight hexagonal pattern that covers the planter face of the foot. Neither the feathers nor the scales appear simultaneously but according to a strict temporal and spatial sequence within each tract.

In order to understand the formation of these different cutaneous patterns, it is necessary to find the answers to the following questions. Why do certain areas of the integument remain bare, while others become covered with feathers or with scales? How is the length, width and contour of each tract controlled? Within each tract, what are the mechanisms governing the subdivision of the integument surface? In other words, how are the size and shape of each appendage determined, and what mechanisms control their arrangement (although it could be presumed that the shape and arrangement may be linked in some way)? Is the initiation of appendages autonomous, or does a wave of morphogenetic influence exist within a prospective tract? In the latter case, what factors determine the formation of the primary appendages of each field?

In vertebrates, as opposed to invertebrates where the integument is composed of a single stratum of ectodermal cells, the situation is complicated by the fact that the skin is formed by two tissues, the epidermis and the underlying dermis. Indeed, it has long been known that scales and feathers, although they are entirely constituted of keratinized epidermal cells, arise from the morphogenetic collaboration of dermis and epidermis (26, 30). Consequently, the formation of cutaneous appendages in birds implies not only intratissue cell-to-cell epidermal interactions, but also intertissue dermal-epidermal, as well as intratissue dermal interactions. Furthermore, the possible influence of underlying organs must not be neglected.

Many data are available relating to the respective roles of the two skin components for each step of pattern specification in the chick embryo (3, 5–7, 15, 19, 21, 25, 28, 30, 31, 33), but little is known about the language used by the tissues to communicate. However, it is clear that the major components of the extracellular matrix—collagens and glycosaminoglycans—could play an important role in those events (4, 15, 16, 18, 22, 23, 40, 42). Recently, the use of retinoic acid (the acid form of vitamin A), which is known to interact with the mechanisms of cellular communication, offers new experimental opportunities for the clarification of regional specification in bird skin, which could lead to the understanding of the molecular basis of some of the involved interactions (2, 9, 10).

Before reporting and discussing the main experimental results in this field, it is useful to give a brief descriptive account of the morphogenesis of the two extensively studied patterns in the chick embryo: the spinal tract for feathered skin, and the anterior tarsometatarsal tract for scaled skin.

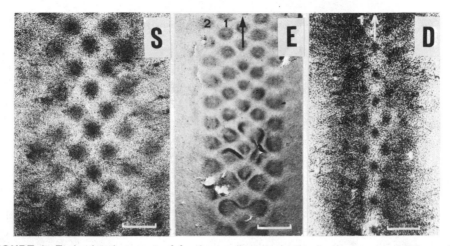

FIGURE 1. Early development of feather rudiments in the lumbar region of the spinal pteryla from 7.5-day chick embryos. Cleared mounted skin pieces. S: entire skin; E: epidermis; D: dermis. The samples of epidermis and dermis originated from the same skin fragment, after enzymatic treatment. Note that the second lateral row (2) of epidermal placodes is formed while only the first lateral row (1) of dermal condensations is constituted on each side of the middorsal initial row (arrow). Bar: 300 μ. Already observed on histological frontal sections by Sengel and Rusaouën: *Comp. Rend. Acad. Sci. Paris* **266**:795–797 (1968).

The Establishment of Feather and Scale Patterns in Chick Embryo

Morphogenesis of the Feather Pattern in the Spinal Pteryla

Between five and six days of incubation, the dermis differentiates in the dorsal region of the embryo. Its cellular density reaches a level of 2.60 nuclei/1000 μ^3, while the cellular mesenchymal density of the surrounding areas remain low (1.96 nuclei/1000 μ^3) (41).

The first feather rudiments appear in the midline of the lumbar region. The development of the field progresses toward the head from the caudal region and laterally from the midline. As more dermis becomes available laterally, new rows are added in alternate fashion. Finally, the feathers are arranged in a hexagonal pattern (Fig. 1), each feather surrounded by six others. However, this hexagon is not regular for two reasons. First, the different rows are not parallel, and form a fan-shaped tract wider in the thoracic than in the lumbar region. Secondly, following the formation of the first row of feathers, the pattern is stretched along the midline, as a result of growth by the embryo.

The feather rudiments comprise an epidermal circular placode (about 120 μ in diameter) which covers a dermal condensation, the cellular density of which is 5.52 nuclei/1000 μ^3 (41). Of the two feather rudiment components, the epidermal placode is the first to become recognizable (38). In the initial middorsal row, one to three placodes precede the appearance of dermal condensations. As the lateral rows successively differentiate, one row of epidermal placodes precedes each row of dermal condensations (Fig. 1).

Finally, by the time the last feather rudiments appear on the edges of the spinal pteryla, at about 8.5 days of incubation, the primary rudiments of the median row emerge above the skin surface and give rise to the feather buds which then elongate into feather filaments. From 12 days of incubation, the epidermal wall of these filaments becomes subdivided into longitudinal units, later giving rise to the barbs and barbules of the neoptile feather.

Morphogenesis of the Scale Pattern in the Anterior Tarsometatarsal Tract

The embryonic development of scales is quite different in several aspects from that of feathers. The scale is the simplest appendage in amniotes. The fully formed scutellate scale consists of an asymmetric elevation of the epidermis, filled with a mesodermal core. The scale epidermis comprises a thick upper epidermis, constituted by a large number of keratinizing cell layers and a thin inner epidermis that forms the junction with the next scale.

The scale primordia appear in a strict temporal sequence (12, 29). At about 8.5 days of incubation, one bifid large strip of dense dermis (3.31 nuclei/1000 μ^3) (12) forms on the anterior face of the shank. It corresponds to the two future rows of scales, which will be located on the external and internal regions of the shank. It should be noted that not infrequently the internal row may be subdivided into two rows of medium-sized scales. The first indication of scale development is the appearance at about 9 days of three oval placodes (up to 1000 μ in length), which belong to the internal row and are localized at the level of the metatarsal-phalangeal joint of the third toe (29). A little later, the first placodes of the external row appear contiguously and alternately to those of the internal row, at the base of toe 4. During day 10, additional placodes are formed proximally, and a few distally, to complete first the external row and then the internal row. At this stage, the dermal cell density is still uniform (Fig. 2). From 11 to 12 days of incubation, the placodes give rise to the scale ridges and then to the prominent humps (29). At this stage, a dermal condensation appears along and underneath the scale rims (Fig. 2), and the dermal cell density shows a gradient from the tip of the scale (5.19 nuclei/1000 μ^3) to the hinge region (2.73 nuclei/1000 μ^3) (12). This scale-tip dermal condensation is linked to the relatively limited growth of the scale. It is a transient structure, in contrast to the feather dermal condensation.

Effects of Retinoic Acid on the Patterning of the Foot Integument in Chick Embryo

The formation of feathers on chick foot scales can be obtained when a simple molecule, retinoic acid, is injected at appropriate stages into the amniotic cavity of chick embryos (11). The period of sensitivity of the foot integument to retinoic acid

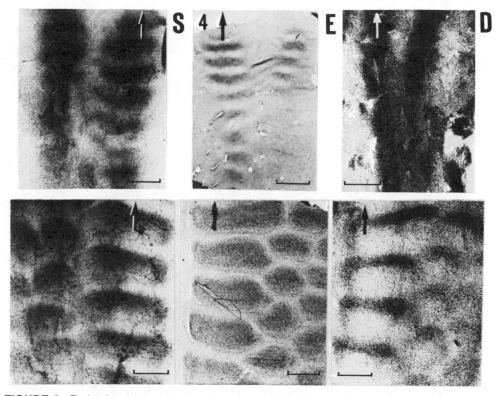

FIGURE 2. Early development of scale rudiments in the anterior tarsometatarsal tract from a 10- and an 11-day chick embryo. S: entire skin; E: epidermis; D: dermis. For each stage, the skin samples are originated from the same embryo. Note that the formation of the epidermal placodes precedes the organization of the dense dermis. The arrows indicate the tarsometatarsal row of scales in line with the fourth toe. Bar: 300 μ.

is strictly correlated to the early stages of scale morphogenesis: only those scales which become discernible as placodes at the time of injection or during the 24 subsequent hours are affected.

Morphological Observations

Twenty-four hours after a retinoic acid injection into the amniotic cavity of 10-day embryos, most of the scale placodes of the anterior face of the tarsometatarsus appear subdivided into several rudiments of more or less roundish shape (Fig. 3) (9, 10). Each of these abnormally shaped placodes tops a dermal cell condensation, and can therefore be interpreted as a feather placode. Thus the immediate effect of retinoic acid is to inhibit the formation of scale primordia by transforming them into feather primordia. However, this inhibition is temporary, so that, as soon as the drug is eliminated from the tissues, the scales resume their development. This

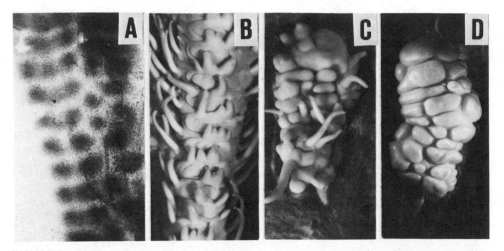

FIGURE 3. Effects of an injection of retinoic acid at 10 days of incubation on chick foot appendage pattern specification. A: Formation of abnormal roundish placodes 24 hr after the treatment. B: Formation of feathered scales 7 days after the treatment. C: Formation of feathered scales by a recombinant of treated epidermis and untreated dermis. D: Formation of scales only by a recombinant of untreated epidermis and treated dermis. Experiments are by Dhouailly: *in preparation*; Dhouailly *et al.: J. Embryol. Exp. Morphol.* **58**:63–78 (1980); Cadi *et al.: in preparation.*

resumption occurs proximally with respect to each row of feather placodes, so that the feather filaments are finally located, seven days later, at the distal tip of the scales (Fig. 3). They are thus not distributed according to the usual hexagonal feather pattern, but are aligned in short transverse and alternate rows, according to the alternate distribution of the rectangular scutellate scales. In a few extreme cases, the entire distal edge of the scale may be transformed into an abnormal brush-type feather resulting from the fusion of adjacent feather units. Most frequently, only one to three feathers are carried by each scale. These are not distributed at random, but in an ordered symmetrical fashion, in relation to the middle axis of the scale.

Determination of the Target Tissue of Retinoic Acid

The formation of feather rudiments instead of scale rudiments implies both epidermal and dermal changes. In order to find out whether retinoic acid acts directly on one or both skin components, heterotypic dermal-epidermal recombinants were prepared. For these recombinants, one of the two tissues was obtained from embryos treated with retinoic acid 24 hr or 48 hr before recombination, the other tissue being obtained from a normal, untreated 10- or 11-day embryo. The results are quite clear (2, 9): heterotypic recombinants involving treated epidermis and untreated dermis formed feathered scales (Fig. 3C), while the reverse combination; of untreated epidermis and treated dermis led exclusively to the formation of scales (Fig. 3D). Thus retinoic acid acts directly on the epidermis to cause the formation of feathers in the chick foot integument. The morphogenesis of heterotypic recombinants comprising a treated dermis demonstrates that the effects produced in the dermis, especially the formation of lens-shaped dermal condensations, are reversible, provided the dermis is associated with a normal, untreated epidermis.

Contribution to the Understanding of Pattern Specification in Tarsometatarsal Skin

The above results provided useful information. First, contrary to a previous hypothesis (26, 30), the tarsometatarsal dermal cells are invested with proper scale-forming properties before 12 days of incubation. Furthermore, they are able to recover those properties entirely even in cases where, after retinoic acid treatment, they have been involved in feather-type dermal condensations. Secondly, the retinoic acid effects on the dermis are mediated through the epidermis, which is thus shown to transmit a feather-forming message to the dermis. This possibility of a primary epidermal induction in normal feather development has already been foreseen by many authors (1, 38), on account of the appearance of epidermal placodes slightly preceding that of dermal condensations (see Fig. 1). Thirdly, identification of the mechanisms by which retinoic acid changes the morphogenetic properties of tarsometatarsal epidermal cells could lead to a better understanding of how the epidermal cell pattern is established.

The following hypothesis on the action of retinoic acid on tarsometatarsal skin could be advanced. In normal development, the dermal cells could cause the formation of scale fields in the epidermis by indirectly triggering protein synthesis involved in cell-to-cell contact properties. It must be remembered that it is only at this stage that retinoic acid is effective in producing feathered scales. It is well known that, among various other actions, retinoic acid or its derivatives could act as a carrier for the glycosylation of proteins. Retinoic acid treatment could thus cause specific alterations in the composition of cell surface glycoproteins. These alterations could lead to disruption of the tarsometatarsal, epidermal morphogenetic field. The epidermal scale placodes, which are in formation at the time of injection, would therefore lose the capacity to maintain their cell grouping, and then break up into smaller, circular feather fields. After elimination of retinoic acid from the epidermis, epidermal cells would recover their ability to communicate properly, and would then reconstruct a normal scale placode under the continuing influence of the scale dermis. Other examples of such abnormal partitions of morphogenetic fields (see Chapter 23), are also illustrative of this kind of action of retinoic acid.

It should be noted, however, that the partition of a scale field into smaller units should not be sufficient for the transformation of a scale into feathers: such a subdivision could only result in the formation of smaller scales. In view of the fact that feather-forming capacity is a primitive property of avian ectodermal cells, the argument could be put forward that normal differentiation of scales in birds requires the inhibition of feather-forming properties. Retinoic acid treatment could interfere with such an inhibition.

Finally, these results emphasize the role of the epidermis in specification of scale and feather differentiation. Thus, although the dermis plays an important role in the pattern distribution of the cutaneous appendages, the morphogenetic activities of the epidermal cells must not be ignored. With this in mind, in the following sections the steps occuring between the early stages in prepatterning of the future integument and the realization of the scale and feather patterns have been analysed.

The Different Steps of Cutaneous Appendage Pattern Specification in Chick Embryo

The Early Mesodermal Prepatterning Determinism

Primary experiments (27) have shown that the regional characteristics within different tracts and apteria are established very early during morphogenesis. The rotation of a block of superficial wing bud tissues (ectoderm plus mesoderm), corresponding to the future shoulder tract and to the region of the future elbow, results in shoulder tract defeciency and in a group of supernumerary feathers in the upper cubital apterium. By varying the stage of experimentation, it has been demonstrated that this tract specificity is progressively fixed between 3 and 4 days of incubation, for the limb buds.

Which of the two future skin components, ectoderm or mesoderm, is responsible for the alternative differentiation of an apterium or a pteryla? To answer this question, experiments have been performed consisting of heterotopic transplantation of blocks of ectoderm-free mesoderm. The results show that when a piece of somatic mesoderm from the prospective midventral apterium of a 2-day embryo is implanted in the dorsal region, a patch of glabrous skin develops inside the spinal pteryla (21), and conversely, the implantation of somatopleure of the presumptive leg bud in the midventral region leads to the development of an ectopic pteryla inside the midventral apterium (17). From these results, it is clear that the distribution of pterylae and apteria is established within the mesodermal cells at a very early stage of development. Furthermore, the specific characteristics of each pteryla are determined by the regional origin of the mesoderm. For example, when a strip of still unsegmented mesoderm from the thoraco-lumbar region is transplanted into the posterior cervical region, the feather pattern that it elicits in the cervical epidermis is of the thoraco-lumbar type (21). Likewise, when a fragment of presumptive thigh mesoderm of a 4-day embryo is grafted in an orthotopic position on the wing bud, it induces the formation of feathers, whose shape and arrangement are of the thigh type (3).

Let us now turn to the problem of scales versus feather determination. From the above results, it is not unreasonable to presume that the ability to form scales is restricted to the autopodial mesoderm of the leg bud. Surprisingly, toes covered with scales and feathered scales form when a piece of thigh mesoderm of a 3-day embryo is grafted subjacent to the apical ectodermal ridge of the wing bud (Fig. 4) (28). Thus under the influence of the wing apical ridge, thigh mesodermal cells are induced to develop as distal parts and to cause the formation of scales in the wing ectoderm. This results shows that at 3 days of incubation the thigh mesodermal

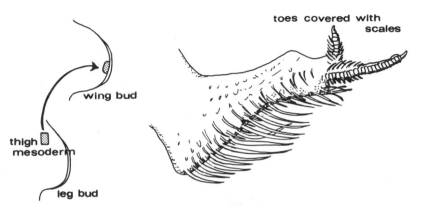

FIGURE 4. Determinism of early mesodermal prepatterning. Toes covered with scales and feathered scales, resulting from the implantation of prospective thigh mesoderm subjacent to the apical ectodermal ridge of the wing bud of a 3–day chick embryo. Drawn from experimental result of Saunders *et al.: Dev. Biol.* I:281–301 (1959).

cells are restricted to forming structures of a hind limb type; however, they are not irreversibly determined in the characters linked to their proximo-distal position in the bud (especially in the formation of feathers of the thigh type). Thus the development of scales depends both on the formation of distal parts of the limb and on the "leg quality" of the mesodermal cells. The differentiation of cutaneous appendages appears linked, in some way, to that of the underlying structures. Within this scheme, how could feather information in the leg resulting from heterotopic recombination of wing ectoderm and leg mesoderm be explained (37)?

The Early Ectodermal Regional Specification

Heterotopic recombinants of ectoderm and mesoderm of wing and leg buds of embryos of 4 days of incubation result in the development of a limb, the regional (wing or leg) quality of which is in accordance with the regional origin of the mesoderm. The leg ectoderm differentiates into normal feather wing tracts when it is recombined with a wing mesoderm. The reverse combination, of wing ectoderm and leg mesoderm, leads to the formation of feet covered either with feathers only, or with scales and feathered scales (37) (Fig. 5). However, these feathers are always arranged according to a typical scale pattern: the rows of large scales on the anterior face of the foot are replaced by rows containing an equal number of thick feathers, while thin feathers replace the smaller tubercular scales on the plantar face of the foot.

From these results, it is clear that the ectoderm of the wing bud and that of the leg bud are not equivalent with respect to their morphogenetic competence. Leg ectoderm is able to form either scales or feathers, depending on the origin of the mesoderm it is associated with. In contrast, wing ectoderm expresses feather-forming potentialities in spite of the leg-type general morphogenesis and of the scale-type distribution pattern imposed on it by the leg mesoderm: it is biased towards

FIGURE 5. Early ectodermal specification: Feathered feet resulting from the heterotopic recombination of wing bud ectoderm and leg bud mesoderm of a 4-day embryo. Redrawn from experimental result of Sengel and Pautou: *Nature* **222**:693–694 (1969).

feather formation, and responds to the mesodermal influence by forming feathers arranged in a scale fashion.

The hypothesis could be advanced that bird ectodermal cells, which possess both feather and scale information in their genome, could progressively lose their scale-forming competence through contact with an underlying feather-tract meso-dermis. The presumptive foot ectoderm, which is deprived of such contact, re-mains neutral, without bias towards either feather or scale morphogenesis. The results of heterogenetic leg bud recombinants between embryos of breeds with nor-mal or feathered feet [ptilopodous mutation, (14)], could be interpreted in the same way. The resulting composite legs all display the ptilopodous phenotype, whether it is the mesoderm or the ectoderm which originates from the mutant em-bryo. Therefore, it is conceivable that the gene substitution affects the properties of the mesodermal cells which can then give a feather bias to their overlying ectoder-mal cells.

Finally, scale and feather pattern specification appears to be determined long before the differentiation of the skin. Nevertheless, the emergence of feather and scale buds requires further reciprocal interactions between the two skin compo-nents, the epidermis and dermis.

The Formation of the Dermis

In the normal development of the spinal pteryla, it can be observed that subecto-dermal mesoderm cell density increases prior to the appearance of the successive rows of feather rudiments, laternally and progressively, from 5 to 7 days of incubation. Similarly, this cell density increases in the tarsometatarsal region at 9 days of incubation, just before the appearance of the scales while it remains low in the future apteria (41).

Is the formation of a dense dermis necessary for the subsequent feather or scale morphogenesis? There are a number of different procedures which prevent the normal densification of the dermis within the spinal pteryla. These include injection of hydrocortisone just at the time of the normal appearance of dense dermis in this region (43), or destruction of 3 to 6 pairs of somites of a 2-day embryo by x-irradiation (20), or the excision of a corresponding length of neural tube (35). All three result in the formation of abnormal transverse dorsal apteria. Likewise, the injection of hydrocortisone into 9-day embryos (24) causes the formation of feet covered with bare skin.

How do these different experimental procedures interact with the process of mesodermal cellular densification? In hydrocortisone-treated embryos, abnormally high synthesis of collagen may prevent the increase of cellular density (Demarchez, *unpublished result*): long before their normal appearance, numerous fibers of collagen are formed without any orientation. After x-irradiation, most of the dermatome cells are destroyed. In much the same way, after spinalectomy, a rapid disorganization of the somites occurs along the spinal lacuna. Thus, in the two latter cases, the populating of the subectodermal space is the result of an abnormal migration of cells, which may either lack the properties required for the construction of a dense dermis or may lose them.

It is possible to provoke artificially a densification of the subectodermal mesenchyme in the normally glabrous skin of the midventral apterium (36) and even in the somatopleure of the extraembryonic membranes (8), by implanting a fragment of living tissues or of inanimate materials into the corresponding areas of a 2-day embryo. The implant causes abnormal fusions between somatopleure and splanchnopleure, and may act as a mechanical obstacle, preventing the centrifugal extension of mesodermal cells and thus causing an abnormal densification of abutting cells. Such densifications result in the development of ectopic pterylae. These results lead to the concept that the mere densification of the subectodermal mesenchyme above a certain threshold entails the acquisition of feather tract properties.

What are the factors which initiate the densification of the dermis in normal development? As the formation of an appendage tract or an apterium is dependent on the regional origin of the mesoderm in the heterotopic transplantation experiments, as seen before, it seems reasonable to presume that it results from autonomous mesodermal cell capacities. However, a permissive ectodermal cell influence is necessary for the expression of dermal appendage-forming properties. This permissive role of the epidermis is well illustrated by analysis of the scaleless mutation, which causes the absence of scales and of most of the feathers. Heterogenetic limbs resulting from recombinations of the mesoderm and ectoderm from normal and mutant embryos (15) formed normal scales and feathers when the ectoderm originated from normal embryos, whereas the reciprocal combination did not. In the latter recombinants, a dense dermis formed, but did not acquire its appendage-inducing capacities. Thus scaleless mutant dermis is not directly affected by the mutation that results in a deficiency of the epidermis. In contrast, the normal epidermis exerts on the underlying dermis a morphogenetic action leading to the acquisition by the latter of the ability to participate in appendage morphogenesis.

How could this epidermal influence be exerted on the dermis? An intriguing coincidence has been revealed by means of immuno-histological studies of extracellular matrix components (22, 23). Prior to the formation of the dense dermis, a thin layer of collagen underlies the dermal-epidermal junction. Simultaneously with the densification of the dermis, this particular localization disappears, while collagen fibers, as well as fibronectin, spread throughout the depth of the dermis. This loosening of the collagen underlying the dermal-epidermal junction could be due to enzymatic activity of the epidermal cells.

It should be noted that this epidermal influence on the dermis is systemic in nature, since the site of formation of each appendage is entirely dermis-dependent, as shown by the results of heterotopic and heterospecific recombinations set out in the following discussion.

The Formation and Distribution of Epidermal Placodes

In bird skin development, there are two types of epidermal placodes, the feather and the scale placode. They differ not only in their general shape and dimensions, but also in the morphogenesis of the inner face of the epidermis. The feather placode possesses anchor filaments, which contain fibronectin (16), and do not occur in scale placodes.

Although the appearance of epidermal placodes precedes the patterning organization of dermal cells, the regional origin of the dermis controls their initiation, shape and distribution, as shown by the results of numerous heterotopic, homospecific and heterospecific skin recombinations (Fig. 6). The recombinants were prepared in two ways, by recombing prospective skin tissues from limb buds, or dermis and epidermis taken just before the appearance of appendage rudiments. The results are quite similar, so it is more interesting to distinguish the recombinants according to whether the two skin components originate, from the same region (feather-forming or scale-forming, or from a glabrous region and an appendage-forming region, or from two regions which differ in the type of appendages they carry.

Thus the anterior tarsometatarsal dermis induces scutate scale formation in foot pad epidermis (Fig. 6), while small tubercular scales are formed in the reverse combination (19). Likewise, in heterospecific recombinants between chick and duck, the size, contour, number, and arrangement of the scales are in conformity with the specific origin of the dermis (31). Similarly, the association of duck and chick feather-forming tissues leads to the differentiation of feathers whose shape, size, and distribution are determined by the dermis (5). The determining role of the dermis in feather pattern specification is also conclusively shown when dermal-epidermal recombinations are prepared by autoplastic reassociation, in which the epidermal antero-posterior axis is rotated by 90° with respect to the cephalo-caudal axis of the dermis (25). Under these conditions, the feather placodes always differentiate in rows parallel to the cephalo-caudal orientation of the dermis (Fig. 8).

Not only are the shape and arrangement of placodes controlled by the dermis,

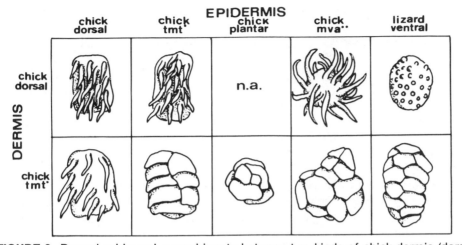

FIGURE 6. Dermal-epidermal recombinants between two kinds of chick dermis (dorsal and anterior tarsometatarsal), and different kinds of epidermis from chick and lizard embryos. Note that the distribution pattern of the appendages is always determined by the regional origin of the dermis. Likewise, feathers or scales are formed according to the regional origin of the dermis, except in two cases, where the epidermis does not possess the information to build feathers (lizard epidermis) or has lost the possibility to express the information to build scales (chick epidermis from featherforming regions). Experiments are by Sengel: *Ann. Sci. Nat. Zoo. II*:430–514 (1958); Sengel *et al.: Dev. Biol.* **19**:436–446 (1969); Linsenmayer: *Dev. Biol.* 27:244–271 (1972); Dhouailly: *Wilhelm Roux' Arch. Dev. Biol.* **177**:323–340 (1975); Cadi and Dhouailly: *unpublished.*

but also their quality, whether scale or feather, as shown by the results of heterotopic recombinations involving epidermis from the midventral apterium, a region which does not form appendages normally. This epidermis is, indeed, able to form either feather or scale placodes, arranged in their respective patterns, according to the regional origin (dorsal or tarsometatarsal) of the associated dermis (2, 39) (Fig. 6).

However, although it is clear that the dermis is responsible for the regional specificity of the cutaneous appendages, feather morphogenesis predominates over scale morphogenesis in heterotopic recombinants of dorsal epidermis and tarsometatarsal dermis (26, 30) (Fig. 6). This fact is due to the diminished competence of feather tract epidermis to build only feather placodes. This restriction, as seen before, is acquired very early during development, while the epidermis from the future scaly or glabrous regions remains bipotent.

Thus the response of epidermal cells to the morphogenetic cues originating from the dermis depends on its origin. This is further illustrated by the results of heterospecific interclass recombinations. For example, mouse epidermis forms hair buds arranged according to the scale pattern when it is associated with chick tarsometatarsal dermis (6), whereas lizard epidermis, under the same influence, build scales whose shape and arrangement are indistinguishable from those of the chick embryo (Fig. 6).

Furthermore, the competence of bird epidermis to form feather or scale placodes can be modified by chemical treatments or genetic mutations. Indeed, after retinoic acid administration (2, 9, 10) or in ptilopodous breeds (15), the tarsometatarsal epidermis loses its ability to build scale placodes. On the other hand, after hydrocortisone treatment (40), or in the scaleless mutation (1, 34), dorsal and tarsometatarsal epidermis lack the capacity to form either feather or scale placodes. It should be noted that the scaleless mutation affects not only the inductive properties of the epidermis, but also its capacity to build placodes.

Thus the formation of placodes and the number of their constituent cells which transform from undifferentiated cuboidal shape to columnar shape, depend both on the spatial distribution of the dermal signal and on its interpretation by epidermal cells. As this signal can be interpreted by foreign cells of another species, it belongs both to an instructive type (for pattern establishment) and to a permissive type (for placodal-type initiation). As no evidence of dermal heterogeneity has been found at this stage, either at cellular or at biochemical level, we do not know how these dermal messages are transmitted to the epidermis. Nevertheless, the cell density (12) and also the frequency of zones of close parallel apposition between the dermal cells and the basement membrane of the epidermis (4) differ between feather- and scale-forming dermis at stages prior to placodal morphogenesis. These differences could constitute part of the dermal signal. One other question arises: what mechanisms are triggered by this signal in the epidermal cells? From the study of retinoic acid effects, as seen before, it is presumed that they are related to changes in cell contact properties. This hypothesis is supported by the results of immuno-histological observations which have revealed that endogeneous lectin is concentrated in the placodal regions during feather morphogenesis (18).

However that may be, once the epidermal placodes are formed, as we shall see they in turn generate organization of the dermis.

The Organization of Dermal Cells and of the Dermal Extracellular Matrix

Following the appearance of epidermal placodes, the dermal cells acquire a region specific pattern. In feather-forming skin, they form lens-shaped condensations under each placode, while between them they assume an elongated bipolar shape, and take on a radiating orientation (Fig. 7) with respect to the condensations in formation (38, 42). In scale-forming skin, the cells localized in the proximal part of the rudiment form numerous cell elongations which are oriented in a transverse direction with respect to the proximo-distal axis of the shank (4), while those situated under the distal part of the placode form a slight condensation.

As dermal cells start to become organized, the distribution of fibronectin and of collagen, two major components of the dermal extracellular matrix, becomes heterogeneous, showing two distinct inverse patterns. In feather-forming skin, fibronectin is concentrated in the region of dermal condensation, whereas collagen deposits increase in the interplumar skin (18, 22, 23) (Fig. 7). The same link between the distribution of fibronectin and the condensation of dermal cells is observed in scale-forming skin, where the concentration of this compound is limited to the distal part of the scale rudiment (22). Furthermore, in feather-forming skin, the collagen fibers form a regular network (Fig. 7), the intersection of which coin-

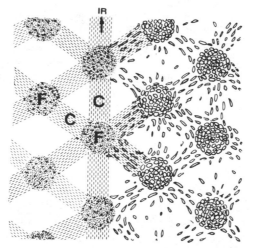

FIGURE 7. Diagrams illustrating the distribution of the two major components of the extracellular matrix of the dermis, on the left, and the arrangement of the dermal cells in and between condensations, on the right, in dorsal skin of a 7.5 -day chick embryo. IR: initial row of feather rudiments; C: collagen; F: fibronectin. Adapted from various papers, especially from Sengel and Rusaoüen: *Comp. Rend. Acad. Sci. Paris* **266**:795–797 (1968); Wessells and Evans: *Dev. Biol.* **18**:42–61 (1968); Ede *et al.: J. Embryol. Exp. Morphol.* **25**: 65–83. (1971); Stuart *et al.; J. Exp. Zoo.* **179**:97–118 (1972); Kitamura: *J. Embryol. Exp. Morphol.* **65**:41–56 (1981); Mauger *et al.: Comp. Rend. Acad. Sci. Paris* **294**:475–480 (1982); Mauger *et al.: Dev. Biol. in press.*

cide with the site of feather formation, and the lines of which correspond with the spatial orientation of the interplumar dermal cells (40). The lack of this collagen network in scaleless mutants (15), its abnormal configuration in talpid mutants (13), or its enzymatic experimental destruction (40) leads to the absence or malformation of appendage rudiments.

It is therefore likely that the increase of fibronectin, as well as the presence of anchor filaments in the case of feather-forming skin, are related to the aggregation of dermal cells. The distribution of collagen might exert a negative control over the outgrowth site of appendage rudiments (22, 23); it might also serve as a migration pathway for condensing cells (40). There are indications that dermal cell condensations arise from centripetal cell migration rather than from focal increase of cell proliferation, although cell counts have not indicated any decrease in cell density in interplumar skin (41). Indeed, colchicine treatment does not prevent dermal papilla formation (40); furthermore, macroscopic observations of cleared mounted skin pieces (Fig. 1) show that dermis is less dense between cell condensations than in undifferentiated unpatterned skin.

In addition to the fact that the formation of the placodes precedes that of the dermal condensations, there are several instances of their inductive role in dermal organization. For example, when dorsal dermis from a normal chick embryo is recombined with dorsal epidermis from a scaleless high-line embryo, the size of the dermal condensation which are formed reflects that of the abnormal placodes (1). Or again, when tarsometatarsal dermis is associated with feather-forming epidermis, from the dorsal region of a normal embryo (26, 30) or from the tarsometatarsal region of an embryo treated with retinoic acid (2, 9), it undergoes feather-type morphogenesis. Furthermore, the feather-type dermal condensations formed in the tarsometatarsal skin of treated embryos, disperse when the dermis is separated from its epidermis and recombined with an untreated tarsometatarsal or midventral epidermis (2, 9).

Thus the organization of dermal cells is under the control of the epidermis. As a causal relationship appears to exist between the increase of fibronectin, the decrease of collagen and the condensation of dermal cells, the deposition of these molecules might also be governed by the epidermal cells.

The feather-type or scale-type pattern of the dermis is in turn responsible for the outgrowth, architectural organization, and histogenesis of the feather and the scale, which constitute the last step of scale and feather morphogenesis (6). It should be noted that all the preceding steps can be passed in heterochronic recombinants where 6-day dorsal epidermis is recombined with 13- to 15-day tarsometatarsal dermis. In this case, scales are formed (26, 30), resulting from the moulding of the undifferentiated epidermis on top of a dermis which has assumed its final form and which has thus reached the stage of inducing the formation of the thick upper epidermis of the scale.

The Progressive Patterning and the Reorganizing Capacities of the Integument

The different cutaneous appendage patterns become gradually organized as the placodes of successive feather and scale rows are formed. This sequential differentiation raises new questions which, until now, have been ignored to simplify the discussion.

In the case of scale-forming integument, it is intriguing to note that a spatial relationship exists between the bone joints and the appearance of the scales (12, 29). Indeed the first scales form exactly at the location of the metatarsal-phalangeal joint. Then new scales form on the site of the phalangeal joints of the toes. Finally, scales are gradually added between the joint levels to complete the different rows. Does mechanical stretching of the skin over the joints (which move at this stage of development) have a role in the initiation of formation of the first scales? Is there a chain mechanism leading to the appearance of the intermediate scales?

Similar questions arise from the development of the feather pattern, but in this case they have led to experimental studies. In the spinal pteryla, the structure which underlies the first row of feathers to appear is the spinal cord. Plumage deficiencies produced by myelectomy (35) consist of transverse apteria. Furthermore, there is no case where lateral rows have developed in the absence of the median row. This result has led to the suggestion (32) that mechanisms linked to the presence of the axial structures are responsible for the initiation of the first row of appendages, which in turn could control the position of the feathers that will form the next row. When lateral halves of dorsal skin are cultured *in vitro* (25), the sequential appearance of successive rows is normal only in cases where the middorsal row is undamaged at the time of explantation. However, the presence of the pri-

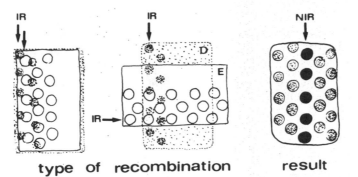

type of recombination   result

FIGURE 8. Reorganization and dermis-dependence of feather pattern in cultured dorsal skin explants. Note that the position of the new primary row (NIR) is dependent on orientation of the dermis (D) and on the stage of differentiation of the dermal cells at the time of recombination. IR: initial row of feather rudiments; E: epidermis. Experiments by Novel: *J. Embryol. Exp. Morphol.* **30**:605–633 (1973); redrawn with modifications from Sengel, in R. Porter and J. Rivers (eds.), *Cell Patterning,* Ciba Foundation Symposium 29 (new series), North-Holland, Amsterdam, pp. 51–70, (1975).

mary row is not necessary for the patterning of the integument: when it is eliminated by excision, in dorsal skin (25) or in femoral skin (19), or when it is destroyed by the splitting of dermis and epidermis (25) (which interrupts the normal sequence of dermal-epidermal interactions leading to the formation of a feather rudiment), a total disorganization, followed by a reorganization of the skin fragment pattern, occurs. Under these conditions, a new dominant row appears (Fig. 8), laterally in relation to the initial one and increasing in distance from it in proportion to the stage reached by the pattern morphogenesis at the time of explantation (25). This location of the new primary row can be easily explained (32): it corresponds to the site where the dermal cells are on the point of initiating the formation of epidermal placodes at the time of explantation whereas the other dermal cells have to retrogress, or have yet to reach this stage.

Thus, the patterning of skin appears in a wave-like fashion, the factors which initiate the chain of these events remaining hypothetic. Probably, the first row of dermal cells which become active corresponds to the cells which first condensed to form the dermis. In this way, the formation of the primary appendages in each tract may be linked to the sequential appearance of the dense dermis, which in turn appears to depend on the presence of the underlying structures. The latter may have a simple mechanical effect on the condensation of the mesodermal cells.

Conclusion

From the examination of events occurring from the early stages of embryonic development to the emergence of feathers and scales, it appears that the specification of integumental patterns results from a continuous dialogue between the two future skin components.

Let us briefly review the chronology of tissue interactions leading to the distribution and morphogenesis of feathers and scales:

1. The predetermination of feather pattern appears within 2 days of incubation, and depends on the regionalization of the mesoderm and on the subsequent restriction of the capacity of the ectodermal cells. The predetermination of scale pattern appears in the following two days, linked in some way to the specification of hind limb autopodial structures.
2. The differentiation of the dermis results from autonomous properties of the mesodermal cells, but nevertheless requires a permissive influence from the epidermis.
3. The dermis induces the formation of epidermal placodes and specifies their size and arrangement according to its regional and specific pattern.
4. The type and size of placode which is formed depends however on the regional origin of the epidermis. Epidermal cells from the feather tracts are only able to build feather placodes, while epidermal cells from the future scaled integument or from the future apteria are bipotent. The type of placode which is formed determine in turn the regional organization of the dermal cells and of the dermal extracellular matrix.
5. The organized dermis induces the transformation of the placodes into completed appendages. These later dermal messages determine the formation of the thick upper-scale epidermis or the outgrowth of the feather filament and the differentiation of barbs and barbules.

Many problems in the processing of these interactions are still to be resolved. First, the determination of feather or scale placode results from a restrictive induction of the mesodermal cells. Does the restriction of the epidermal competence occur as the result of a transcriptional or of a post-transcriptional control? Second, whereas valid hypotheses can be advanced to explain the mechanisms triggered in the effector tissue (changes in glycoprotein or collagen deposits), nothing is known about the nature of the triggering messages which control the formation of epidermal placodes or dermal organization.

General References

SAWYER, R.H.: The role of epithelial-mesenchymal interactions in regulating gene expression during avian scale morphogenesis. In: *Epithelial-Mesenchymal interactions in development*, (eds. Sawyer, R.H., Fallon, J.F.). New York: Praeger (1982).

SENGEL, P.: Morphogenesis of skin. In: *Development and cell biology series*, pp. 1–277, (eds. Abercrombie, M., Newth, D.R., Torrey, J.G.). Cambridge: Cambridge University Press (1976).

References

1. Brotman, H.F.: Epidermal-dermal tissue interactions between mutant and normal embryonic back skin: site of mutant gene activity determining abnormal feathering is in the epidermis. *J. Exp. Zool.*, **200**: 243–258 (1977).
2. Cadi, R., Dhouailly, D., Sengel, P.: *In preparation.*

3. Cairns, J.M., Saunders, J.W.: The influence of embryonic mesoderm on the regional specification of epidermal derivatives in the chick. *J. Exp. Zool.*, **127**: 221–248 (1954).

4. Demarchez, M., Mauger, A., Sengel, P.: The dermal-epidermal junction during the development of cutaneous appendages in the chick embryo. *Arch. Anat. Microsc. Morphol. Exp.*, **70**: 205–218 (1981).

5. Dhouailly, D.: Déterminisme de la différenciation spécifique des plumes néoptiles et téléoptiles chez le poulet et le canard. *J. Embryol. Exp. Morphol.*, **18**: 389–400 (1970).

6. Dhouailly, D.: Formation of cutaneous appendages in dermoepidermal recombinations between reptiles, birds and mammals. *Wilhelm, Roux Arch. Dev. Biol.*, **177**: 323–340 (1975).

7. Dhouailly, D.: Dermo-epidermal interactions during morphogenesis of cutaneous appendages in amniotes. *Front. Matr. Biol.*, **4**: 86–121 (1977).

8. Dhouailly, D.: Feather-forming capacities of the avian extra-embryonic somatopleure, *J. Embryol. Exp. Morphol.*, **43**: 279–287 (1978).

9. Dhouailly, D.: Effects of retinoic acid on developmental properties of the foot integument in avian embryo. In: *Embryonic development*, Part B: Cellular aspects, pp. 309–316, New York: Alan R. Liss (1982).

10. Dhouailly, D.: *In preparation.*

11. Dhouailly, D., Hardy, M.H., Sengel, P.: Formation of feathers on chick foot scales: a stage-dependent morphogenetic response to retinoic acid. *J. Embryol. Exp. Morphol.*, **58**: 63–78 (1980).

12. Dwyer, N.K.: Chick scale morphogenesis: early events in the formation of overall shank and individual scale shape. Master's thesis, University of Massachusetts, Amherst (1971).

13. Ede, D.A., Hinchliffe, J.R., Mees, H.C.: Feather morphogenesis and feather pattern in normal and talpid mutant chick embryos. *J. Embryol. Exp. Morphol.*, **25**: 65–83 (1971).

14. Goetinck, P.F.: Tissue interactions in the development of ptilopody and brachydactlyly in the chick embryo. *J. Exp. Zool.*, **165**: 293–300 (1967).

15. Goetinck, P.F., Sekellick, M.J.: Observations on collagen synthesis, lattice formation, and morphology of scaleless and normal embryonic skin. *Dev. Biol.*, **28**: 636–648 (1972).

16. Haake, A.R., Sawyer, R.H.: Avian feather morphogenesis: fibronectin-containing anchor filaments. *J. Exp. Zool.*, **221**: 119–123 (1982).

17. Kieny, M., Brugal, M.: Morphogenèse du membre chez l'embryon de poulet. Competence de l'ectoderme embryonnaire et extra-embryonnaire. *Arch. Anat. Microsc. Morphol. Exp.*, **66**: 235–252 (1977).

18. Kitamura, K.: Distribution of endogenous β-galactoside specific lectin, fibronectin and type I and III collagens during dermal condensation in chick embryos. *J. Embryol. Exp. Morphol.*, **65**: 41–56 (1981).

19. Linsenmayer, T.F.: Control of integumentary patterns in the chick. *Dev. Biol.*, **27**: 244–271 (1972).

20. Mauger, A.: Le développement du plumage dorsal de l'embryon de Poulet étudié à l'aide d'irradiations aux rayons X. *Dev. Biol.*, **22**: 412–432 (1970).

21. Mauger, A.: Rôle du mésoderme somitique dans le développement du plumage dorsal chez l'embryon de Poulet. II Régionalisation du mésoderme plumigène. *J. Embryol. Exp. Morphol.*, **28**: 343–366 (1972).

22. Mauger, A., Demarchez, M., Georges, D., Herbage, D., Grimaud, J.A., Druguet, M., Hartmann, D.J., Sengel, P.: Répartition du collagène, de la fibronectine et de la laminine au cours de la morphogenèse de la peau et des phanères chez l'embryon de poulet. *Comp. Rend. Acad. Sci. Paris*, **294**: 475–480 (1982).

23. Mauger, A., Demarchez, M., Herbage, D., Grimaud, J.A., Druguet, M., Hartmann, D.J., Sengel, P.: Immuno-fluorescent localization of types I and III collagen and fibronectin during feather morphogenesis in the chick embryo. *Dev. Biol.* (1982, *in press*).

24. Moscona, M.H., Karnofsky, D.A.: Cortisone induced modifications in the development of the chick embryo. *Endocr.*, **66**: 533–549 (1960).

25. Novel, G.: Feather pattern stability and reorganization in cultured skin. *J. Embryol. Exp. Morphol.*, **30**: 605–633 (1973).

26. Rawles, M.E.: Tissue interactions in scale and feather development as studied in dermal-epidermal recombinations. *J. Embryol. Exp. Morphol.*, **11**: 765–789 (1963).

27. Saunders, J.W., Gasseling, M.T.: The origin of pattern and feather germ tract specificity. *J. Exp. Zool.*, **135**: 503–528 (1957).

28. Saunders, J.W., Gasseling, M.T., Cairns, J.M.: The differentiation of prospective thigh mesoderm grafted beneath the apical ectodermal ridge of the wing bud in the chick embryo. *Dev. Biol.*, **1**: 281–301 (1959).

29. Sawyer, R.H.: Avian scale development. I. Histogenesis and morphogenesis of the epidermis and dermis during formation of the scale ridge. *J. Exp. Zool.*, **181**: 365–384 (1972).

30. Sengel, P.: Recherches expérimentales sur la différenciation des germes plumaires et du pigment de la peau de l'embryon de Poulet an culture *in vitro*. *Ann. Sc. Nat. (Zool.)*, **11**: 430–514 (1958).

31. Sengel, P.: The organogenesis and arrangement of cutaneous appendages in birds. *Adv. Morphog.* **9**: 181–230 (1971).

32. Sengel, P.: Feather pattern development. In: *Cell patterning*, Ciba Foundation, Symposium 29, pp. 51–70, Amsterdam: ASP, (1975).

33. Sengel, P.: Morphogenesis of skin. In: *Developmental and cell biology series*, pp. 1–277. (eds. Abercrombie, M., Newth, D.R., Torrey, J.G.). Cambridge: Cambridge University Press (1976).

34. Sengel, P., Abbott, U.K.: *In vitro* studies with the scaleless mutant: Interaction during feather and scale differentiation. *J. Hered.*, **54**: 254–262 (1960).

35. Sengel, P., Kieny, M.: Sur le rôle des organes axiaux dans la diffeérenciation de la pteryle spinale de l'embryon de Poulet. *Comp. Rend. Acad. Sci. Paris*, **256**: 774–777 (1963).

36. Sengel, P., Kieny, M.: Production d'une pteryle supplémentaire chez l'embryon de Poulet. I. Etude morphologique. *Arch. Anat. microsc. Morphol. Exp.*, **56**: 11–30 (1967).

37. Sengel, P., Pautou, M.P.: Experimental conditions in which feather morphogenesis predominates over scale morphogenesis. *Nature, Lond.*, **222**: 693–694 (1969).

38. Sengel, P., Rusaouën, M.: Aspects histologiques de la différenciation précoce des ébauches plumaires chez le Poulet. *Comp. Rend. Acad. Sci. Paris*, **266**: 795–797 (1968).

39. Sengel, P., Dhouailly, D., Kieny, M.: Aptitude des constituants cutanés de l'aptérie médio-ventrale du Poulet à former des plumes. *Dev. Biol.*, **19**: 436–446 (1969).

40. Stuart, E.S., Garber, B., Moscona, A.A.: An analysis of feather germ formation in the

embryo and *in vitro*, in normal development and in skin treated with hydrocortisone. *J. Exp. Zool.*, **179**: 97–118 (1972).

41. Wessells, N.K.: Morphology and proliferation during early feather development. *Dev. Biol.*, **12**: 131–153 (1965).

42. Wessells, N.K., Evans, J.: The ultrastructure of oriented cells and extracellular materials between developing feathers. *Dev. Biol.*, **18**: 42–61 (1968).

43. Züst, B.: Le développement du plumage, d'après l'analyse des malformations cutanées produites par l'administration d'hydrocortisone à l'embryon de Poulet. *Ann. Embryol. Morphol.*, **41**: 155–174 (1971).

Questions for Discussion with the Editors

1. *Cell-cell contact and/or extracellular matrix interactions apparently play major roles in feather and scale specification. Is there any reason for postulating that gradients of diffusable morphogens might also be involved at one or another stage of pattern specification?*

There is no experimental evidence that diffusable morphogens may interact either in intertissular dermal-epidermal interactions or in intratissular interactions.

2. *Is it certain that the progressive patterning of the integument represents a wave mechanism which requires physical contact? Or, like amphibian somite morphogenesis (e.g., Cooke and Elsdale, J. Embrol. Exp. Morph. 58: 107), might a kinematic wavefront of cell patterning which does not require the propagation of a stimulus from cell to cell be involved?*

Observation shows that in the spinal tract, feather rudiments are progressively laid down from the primary median row towards the edges of the integumentary field. This propagation of skin patterning reflects that of formation of the dermis at 5 days of incubation, dermal cells being determined to initiate placode formation as soon as they have reached a certain level of densification. However, it is only an impression that skin patterning depends on a wave mechanism. Indeed, a metallic screen inserted between medium and lateral dorsal regions at 4 days of incubation has no effect at all on lateral feather development (21), a result which demonstrates that the propagation of a stimulus from cell to cell is not required for dermal cell densification. In fact, this mesodermal cell ability appears determined as early as 2 or 3 days of incubation, as has been shown by the results of several transplantation experiments.

CHAPTER **26**

Pattern Regulation in Shark Dentitions

Wolf-Ernst Reif

SHARK DENTITIONS are complex organs which undergo developmental processes throughout ontogeny. They replace their teeth every few days or weeks (7). In any given jaw, a large number of replacement teeth is present in addition to the functioning teeth. Ontogenetic changes of the patterns can thus be directly seen. Because of technical difficulties, regeneration experiments have never been carried out with shark dentitions. Morphogenesis and regeneration capacities can, however, be analyzed by comparing large numbers of jaws which show normal development or which exhibit anomalies induced by accident. The pattern of the dentition can be reduced to a few parameters: tooth size, number of teeth, tooth shape, and polarization of tooth shape. Polarization of tooth shape is defined by the direction into which the main cusp (or the main cusps) of the tooth points. (The cusp points either toward the distal end of the jaw or towards the symphysis. In the symphysis, left and right quadrants of a jaw are fused.) All other aspects of tooth morphology, number of cusps, size and shape of cusps, shape and vascularization of the root etc. are included in the parameter "tooth shape." Factors which control tooth shape are unknown. Some sharks have a completely homodont detition. That is, not only does the dentition of any given growth stage consist of teeth with identical shape, but the shape also does not change during ontogeny. Other species have a highly heterodont detition with a strong tooth shape gradient at any given growth stage and display very strong ontogenetic changes.

Shark dentitions are always organized into tooth families, i.e., a functional tooth and its successors. Number of tooth families, i.e., number of functional teeth at a given growth stage, and size of tooth families, i.e., size of functional teeth, and are closely interrelated factors. The available space on a jaw can either be filled with a few large teeth or with many small teeth. Increase in size of the jaws can either be compensated for by the addition of new tooth families or by an increase in size of the existing tooth families.

Tooth families are founded during embryogenesis long before the animal hatches from the egg or is born. This first developmental phase of the dentition is followed in many genera by a phase of counting of the tooth families. Supernumerary tooth families are exterminated by a suppression mechanism, and from that phase on the number of tooth families remains constant during ontogeny in these genera. For the functioning of the dentition, it is essential that they form a densely paved pattern. If gaps between teeth occur during normal development, they are spontaneously filled by singular teeth, which are here called stop-gaps. These stop-gaps often have an anomalous shape and they have no successor. It is not so essential that they can bite, but it is important that they fill the pattern and protect the underlying soft tissue. Tooth polarity is regulated independently of tooth shape. This is shown in embryonic dentitions but also by the fact that injuries in the dentition can result in a reversal of tooth polarity. The teeth with reversed polarity are mirror images of the original teeth. In other words, except for polarity they have the normal morphological properties. A regulation model for tooth polarity reversal is indicated in Chapter 3.

Despite the fact that it is still impossible to perform surgical experiments with shark dentitions, they have proved to be a very interesting developmental system. The pattern of the dentitions is easy to analyze; it consists of simple geometric properties like tooth shape, tooth size, and relative positions of teeth. The dentition undergoes development throughout ontogeny. Because each functioning tooth is followed by several replacement teeth in any given jaw, developmental processes can be directly analyzed by "reading" the dentition from the functional tooth to the last formed replacement tooth. Developmental systems which consist of repetitive units, *sensu lato*, are a common phenomenon in the organic kingdom. Each segment often undergoes rather a different development from its neighbor. Segmentation itself, however, is a process which involves the whole organism or the whole organ. Pattern regulation in repetitive systems is discussed in Chapters 1 and 11. Tooth families in shark dentition can be regarded as analogs to body segments. It is interesting to see how "segmentation" is brought about in the embryonic dentition and how each individual "segment" develops during ontogeny.

Dental Lamina and Morphogenetic Pattern of the Dentition

Like dentitions of all other jaw-bearing vertebrates (*Gnathostomata*), dentitions of sharks are formed by a *dental lamina*. This is an ectodermal fold which develops during embryogenesis (Fig. 1). In sharks it is situated behind the jaws. Teeth are formed at the anterior interface between the ectodermal fold and the surrounding mesoderm. In the same way as the fold deepens during embryogenesis, new teeth are added at the basal end of the dental lamina (Fig. 1a–c). Near the end of embryogenesis, the deepening of the fold ends and the teeth are transported into functional position by a conveyor-belt system situated between the jaw cartilage and the dental lamina [Fig. 1d; (3, 5)]. Tooth germs are continuously transported up-

Development of Dental Lamina in Elasmobranchs

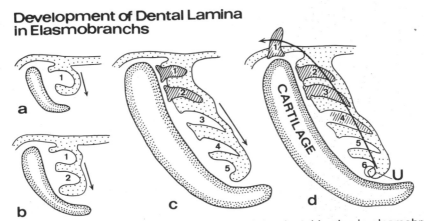

FIGURE 1. General scheme for embryogenesis of the dental lamina in elasmobranchs. Dotted areas: ectoderm (surrounded by mesenchyme, white). Tooth germs are formed in the course of the infolding of the ectoderm (a-c). At the end of embryogenesis, the tooth exchange mechanism begins to work (d), moving tooth generation 1 into a functional position. At the same time, the germ of generation 6 is formed at the lower end of the dental lamina. The arrows in a, b, and c indicate the direction of infolding of the dental lamina. The arrow in d indicates the tooth transport. U denotes primordial tissue. [From Reif (3), by permission.]

wards in the fold throughout ontogeny. Hence one must assume that the cell clusters which differentiate into tooth germs are derived from the basal part of the dental lamina, which comprises ectoderm and mesoderm and which is called "primordial tissue" ["U" in Fig. 1d; (3)].

Shark dentitions can be treated as two-dimensional patterns. One axis (the horizontal axis = x-axis) is parallel to the usually curved functional margin of the jaw. The second axis (vertical axis = t-axis) is perpendicular to the first axis and is parallel to the direction of tooth replacement. Shark dentitions are always organized into *tooth families*.

The t-axis is parallel to the tooth families, whereas the x-axis crosses all tooth families. The morphogenetic stability of the tooth families leads to the assumption that the primordial tissue is compartmentalized and consists of compartments competent for tooth formation ("A" in Fig. 2), separated by compartments incompetent for tooth formation ("B" in Fig. 2). "A" is called a protogerm (3). "A" plus "B" can be regarded as an analog to a segment. As sharks replace their teeth every few days or weeks (7), a protogerm produces about 200 teeth in a lifetime of 10 years. If one plots all teeth in their relative position produced by a dental lamina during ontogeny, a two dimensional pattern will result. In this case, the x-axis runs across all tooth families and the t-axis can be interpreted as a time arrow which points from the small teeth formed in young animals to the large teeth formed in adult specimens. Such a x-t-plot will be called the "morphogenetic pattern of the dentition." Because of great technical difficulties the morphogenetic pattern of the dentition is not completely known in any shark species. In fact only in a single genus has the development of the dentition been traced with a large number of specimens from

Overall Dentitional Pattern and Insertion of New Elements

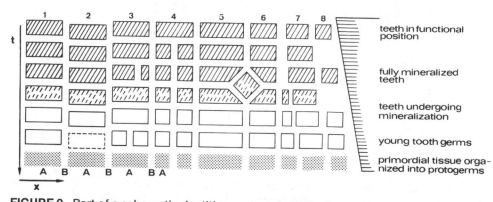

FIGURE 2. Part of a schematic dentition as seen from the lingual side (interior side of the jaw). Upper horizontal demarcation: functional edge of the jaw. The arrow t (= time arrow) points from older to younger teeth. x is the direction along the jaw. The dental lamina is in the plane of drawing. The oblique line indicates the distal end of the dentition. The organization of the primordial tissue into protogerms is hypothetical and cannot be seen in anatomical or histological preparations. Tooth families 1 and 2 undergo normal development. Their teeth are staggered with respect to one another. Teeth of family 3 split spontaneously which is the case in most dentitions, but this splitting of individual teeth does not lead to a new tooth family. Family 4 has split into two autonomous parts. Because of their small sizes, teeth of families 5 and 6 create an open gap in the dentition which is filled by a stop-gap. Later 5 and 6 fill the gap themselves. Gaps between 6 and 7 and between 7 and the distal end of the dentition lead to the insertion of two new tooth families (the distal one is called 8). The diagram is based on morphogenetic studies of more than 10 shark and ray genera. [From Reif (3), by permission.]

the embryonic stage to old age (3). In 10 species not only adult dentitions are available, but also those of embryos (Reif, *unpublished*). Since 5 to 10 (sometimes up to 30) replacement teeth for each tooth family are attached to the jaws, a significant part of the whole morphogenetic pattern is exhibited in any given shark jaw.

The numbers of tooth families vary from 1 per jaw (upper and lower) in the duck-billed rays to more than 300 in the whale shark. In most sharks the tooth family number is between 20 and 30. (Skates and rays are regarded as a specialized group *within* the modern sharks. They will not be treated specifically here.)

Because of surgical difficulties no experiments have so far attempted to elucidate the morphogenetic and regulatory mechanisms which control the activity of the primordial tissue. The accessability of the primordial tissue deeply below the inside of the jaws is very limited. Sharks are very difficult to hold in captivity for a long time span, and they are very sensitive to injury in captivity. However, anesthesization of animals poses no major problems and hence regeneration of surgically removed scales was successfully studied (4).

The morphogenetic pattern of the dentition has four different parameters each of which will be treated separately: (1) shapes of the teeth in relation to their x-t-coordinates; (2) number of tooth families (this is equivalent to the segmentation of the primordial tissue); (3) size of tooth families; (4) polarity of the teeth (in many shark genera the cusps are deflected in a lateral-posterior direction). So far the morphogenetic behavior of the primordial tissue can only be deduced from comparing a large number of jaws with normal development and from anomalies occurring in animals caught in the wild. Most of the anomalies were probably caused by injuries.

Control of Pattern

Tooth Shape

A change in tooth shape along the axes of the x-t-plot will be called tooth-shape gradient. In some dentitions the gradient along the x-axis at any given growth stage is very small (e.g., spiny dog-fish, Fig. 3). This is called *monognathic homodonty*. If the gradient along the t-axis is small, the dentition has by definition an *ontogenetic homodonty*. In the spiny dog-fish, for example, the ontogenetic gradient is zero. Tooth shape (and also tooth size relative to jaw size) remain constant. Hence any given protogerm produces always the same tooth shape though the absolute size increases considerably. The other extreme is the bullhead-shark which has both strong monognathic and ontogenetic gradients, i.e., monognathic and ontogenetic *heterodonty* (Fig. 4). Any given protogerm produces a wide spectrum of different tooth shapes. Tooth shape varies with the x-t-coordinates. So far no anomalies with respect to tooth shape have been found. It would, however, be very interesting to graft primordial tissue from one position to another in the same animal to see how tooth shape is controlled.

Squalus acanthias

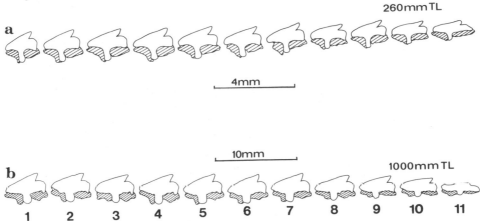

FIGURE 3. The spiny dog-fish has a homodont dentition that does not change during ontogeny. In this diagram, the dentition of a mature embryo (a) is compared to that of an adult (b). (Note the different scales.) TL: total body length.

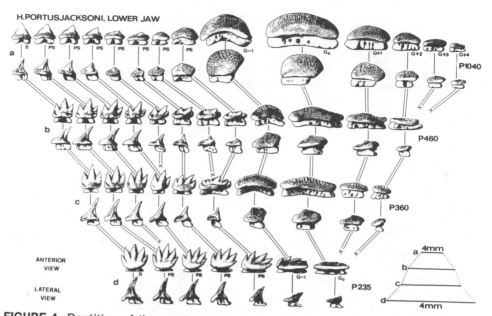

FIGURE 4. Dentition of the southeast Australian bullhead shark *Heterodontus portus-jacksoni*. P 235 (= body length 235 mm) is a newly hatched individual, P 1040 (= body length 1040 mm) is an adult. For each of the different growth stages (note the different scales.), the dentition of the lower left ramus is shown. The teeth are seen from the labial side (upper row of each growth stage) and from the distal side (lower row). S, symphyseal tooth. New tooth families are inserted during ontogeny between S and the crushing teeth (G) and distal to the crushing teeth. To guarantee big crushing teeth, these families are never "allowed" to split. [From Reif (3), by permission.]

Tooth Size and Number of Tooth Families

It will be shown that these two parameters are closely related in their regulation because available space can either be filled by the addition of new tooth families or by an increase in tooth size. As far as is known, shark dentitions fall into three categories with respect to numbers of tooth families; (a) number of tooth families is species-specific and remains constant during ontogeny (spiny dog-fish; tiger shark). (b) Number of tooth families varies within small limits from specimen to specimen within a species, but probably does not increase during ontogeny (many grey sharks, *Carcharhinidae*). (c) Number of tooth families increases during ontogeny (Bullhead shark, *Heterodontus*). The exact tooth family number for a given growth stage is not definitively predictable, but depends on local regulations and on overall trends (3).

Embryogenetic Development of Tooth Families

In those reptiles and bony fishes which have been studied so far [see Reif, (5) for references] the dentition in each jaw quadrant is built up from a single initiator tooth which is the founder of one tooth family and also induces more tooth families until the available space is filled up. The standard method to study such a development process is to cut series of histological sections of successive growth stages and to reconstruct three-dimensional diagrams. This method, however, has failed in the sharks (5). One reason for the failure was the complex geometry of the dental lamina. The other reason was that the catshark which was studied has very small embryos. As an alternative method, anatomical preparations of whole jaws of almost full-term embryos were made. The relative position of the teeth with respect to the x- and t-axes is used as major reference of the time sequence of the developmental processes, keeping in mind that the tooth transport by the conveyor belt starts only very late in embryogenesis.

In the species studied so far (several genera of the family *Carcharhinidae*) the initiator teeth of families S (= symphyseal = parallel to the jaw symphysis), 2, 4, 6 etc. are formed first (Fig. 5, 6). Then the initiator teeth of the odd-numbered families are formed. The relative position of the teeth of the dental lamina can be explained by an inhibitor-field model: Every tooth forms an inhibitor field around itself, which prevents the induction of more teeth in the close vicinity. This mechanism leads automatically to the diagonal tooth rows seen in Figs. 5 and 6. The diagonal rows themselves lead to an alternating tooth replacement pattern. After the initiator teeth of the odd-numbered tooth families have been formed, the first generation of the replacement teeth in the even numbered tooth families are induced, and so forth. There is one additional complexity found in the lower jaw. Here tooth families, S, 3, 5, 7 etc. start first, then families 2, 4, 6, 8 etc. and then the initiator tooth of family 1 is inserted at the same time when the replacement teeth of families S, 3, 5, 7 etc. are formed.

Prionace glauca, embryo, 205mmTL

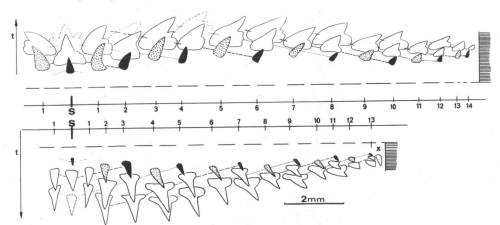

FIGURE 5. Right half of the embryonic dentition of a blue shark as seen on the inside of the jaw. The teeth have not yet started to move into a functional position. Horizontal interrupted lines: functional edges of the jaws. The teeth are arranged in diagonal rows (dotted lines and interrupted lines). S: symphyseal tooth family. Black: teeth formed during developmental stage 1 of the dentition; stippled: teeth formed on stage 2. Note that family 1 of the lower jaw is added later to the dentition; t: time arrow.

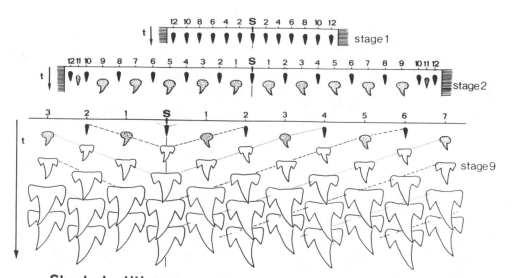

Shark dentitions: Embryogenetic development

FIGURE 6. General scheme of embryonic dentition development based on several genera of the grey sharks, *Carcharhinidae*. For explanations see Fig. 5.

COUNTING OF TOOTH FAMILIES

At developmental stage 3 (Fig. 5), all tooth families are established and their number should remain constant in all dentitions of type (a) and type (b). However, during the first stages, supernumerary teeth are formed, which are then eliminated in such a way that they are not "allowed" to form successor teeth. In some cases elimination takes place only after the supernumerary teeth have formed one or two successors. Hence a process of counting of the tooth family number must take place. The supernumerary teeth are found either in the symphyseal region (Fig. 7c), or in the posterior quarter of the quadrant (Fig. 8a), or posterior to the last regular tooth family (Fig. 7a, 8a, 9). The shapes of the supernumerary teeth are quite normal. Supernumerary teeth occur in most preparations of embryonic dentitions studied so far.

POSTNATAL REGULATION OF TOOTH SIZE

In dentitions of type (a) and type (b), increase in jaw size is compensated solely by the increase in tooth family size. This is shown in Fig. 3 where the size of the tooth family relative to the whole dentition (and relative to the body size) remains constant. In dentitions of type (c), growth of the tooth families is regulated in such a way that the relative size of the families decreases. This leads to open gaps in the dentition. They are filled (according to the inhibitory field model by an automatic induction) in such a way that new tooth families are established either by insertion or by splitting of a family (Fig. 2, 4). At first the newly inserted tooth family is very small, but it soon grows in such a way that it fits into the pattern. From the very beginning, teeth of newly inserted tooth families have quite normal shapes.

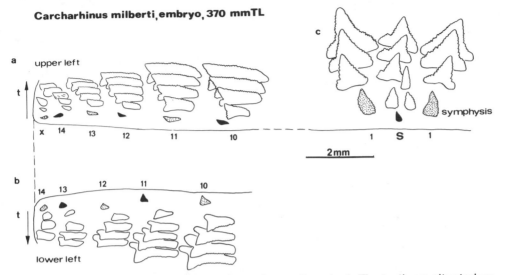

Carcharhinus milberti, embryo, 370 mmTL

FIGURE 7. Parts of the embryonic dentition of a sandbar shark. The teeth are situated on the inside of the jaw. Horizontal lines: Functional edges of the jaws. For information see Fig. 5. Note the supernumerary teeth at the distal end of the upper left quadrant (marked with an x) and in the symphyseal region.

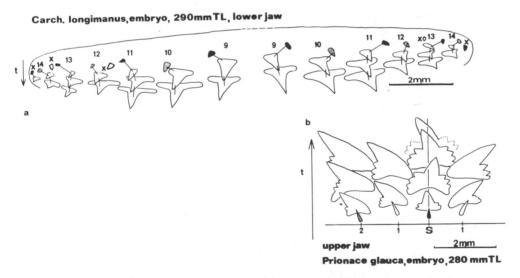

Carch. longimanus, embryo, 290 mm TL, lower jaw

Prionace glauca, embryo, 280 mmTL

FIGURE 8. (a) Part of lower-jaw dentition of an oceanic whitetip shark. For further explanations see Fig. 5. Note the supernumerary teeth marked with an x. Because of a counting process, supernumerary embryonic teeth are not "allowed" to establish a tooth family, rather they are exterminated immediately or after they have formed one to two successors. (b) Symphyseal region of a blue shark embryo. For further explanations see Fig. 5. The polarity vector of the symphyseal family points first to the right and then to the left. This means that polarization can be spontaneously reversed during embryogenesis.

Galeocerdo cuvier, embryo, 550mm TL

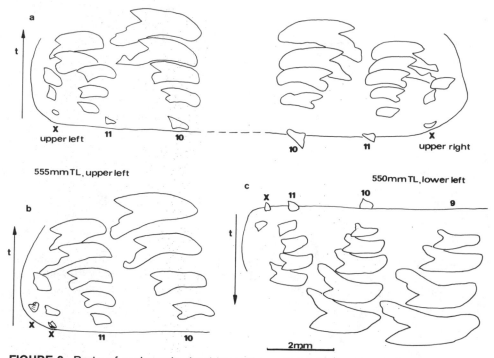

FIGURE 9. Parts of embryonic dentitions of two tiger sharks (a: specimen 1; b and c: specimen 2). For further explanations see Fig. 5. Horizontal demarcations: functional edges of the jaws. The tooth transporting mechanism has just started to work. The first generation of the tooth families are in a functional position (a and b), or are already shed (family 9, Fig. c). Note that the families 9, 10, and 11 expand drastically in a distal position in order to suppress the supernumerary teeth.

In those dentitions where the number of tooth families is kept constant (types a and b) gaps are often filled by "stop-gaps" (individual teeth which often have anomalous shapes), which have no successors. After the stop-gap, the space is filled by an increase in size of the adjacent tooth families (Fig. 10a, b). Stop-gaps often have anomalous shapes and show no polarity.

An anomalous situation with respect to the control of tooth family number is shown in Fig. 10c–f [see also (6)]. The dentition is of a tiger shark of 460 cm total length. This is the largest tiger shark whose dentition is available. Tiger shark dentitions belong to type 1. All tiger sharks I have studied have a tooth formula of 11-S-11/11-S-11. The formula says that there is one symphyseal tooth family in the upper and one in the lower jaw. Each quadrant has 11 tooth families. Bigelow and Schroeder (1), however, report tooth family numbers between 9 and 12 per quadrant. In this large tiger shark, the jaw cartilage had grown disproportionately compared to the growth of the dentition and also compared to the body length. During most of its life the animal must have had the standard tooth formula with 11 teeth per quadrant. The original tooth families 9, 10, and 11 can still be clearly identified. Only in the last part of ontogeny when the jaws started to grow dispropor-

Insertion of new teeth and new tooth families

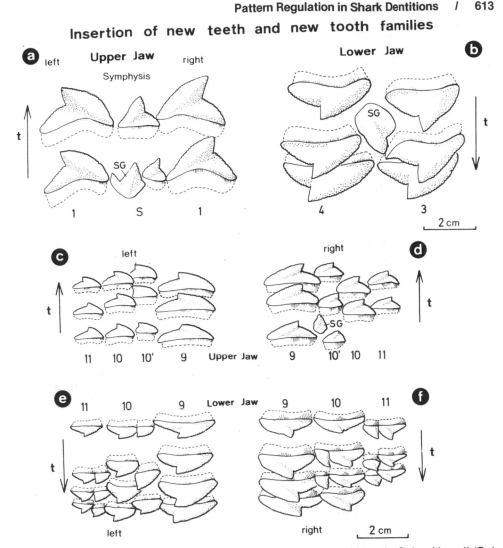

FIGURE 10. (a–f) Tiger shark, *Galeocerdo cuvier*, 4.60 m total length, Oahu, Hawaii (Collection of the Department of Zoology, University of Hawaii at Manoa, Honolulu, no catalogue number). Views of the inside of the jaws with replacement teeth and functional teeth. The functional teeth were projected on the same optical plane as the replacement teeth. Arrows indicate the time arrows (see Figs. 2 and 5). The teeth at the bases of the arrows are the functional teeth. Continuous lines around the teeth indicate the crowns; discontinuous lines indicate the roots. Because of disproportionate growth of the jaw cartilage in relation to the tooth set, gaps appear in the dentition, which are either filled with stop-gaps (SG; a, b, d) or filled by the splitting of tooth families (c–f; families 10′ in c and d). [From Reif (6), by permission.]

tionately did gaps result leading to the insertion of additional teeth and the splitting of tooth families.

In the lower left quadrant (Fig. 10e), family 10 appears to have split. At first (top) the splitting of the protogerm is not complete; hence the two parts of the resulting tooth are still fused at their bases. In the next generation, the division of the protogerm is complete. In accordance with the inhibitory field model, the two, now independent, germs are not induced directly adjacent to each other; instead, the right (anterior) germ is formed later than the left one, which automatically leads to a staggering of the teeth [cf. (3), Figs. 33c–e]. Shortly after family 10, family 11 starts to split in the same way.

The differences between the left and the right quadrant indicate that the splitting of the tooth families is not preprogrammed in every detail. It appears to be a reaction to local conditions. In the lower right quadrant (Fig. 10f), family 11 splits first. The details of the splitting are exactly as in the lower left quadrant and soon afterwards family 10 starts to split. The time difference between two tooth generations of the same tooth family varies from a few days to several weeks, depending on the species and on the age of the individual (7). After the second generation, the splitting of family 10 is "revoked," so that the next tooth generation of family 10 is not split, perhaps because the third tooth of family 9 extends in a posterior direction, thus filling available space at the expense of family 10.

All cases of splitting in this specimen have a normal polarity of the fragments and the crown of the anterior fragment always lacks a posterior shoulder during the first two (or three) generations. Whereas the lower jaw displays the process of splitting directly, the development of the upper jaw is ahead of the lower jaw (probably by several months). Thus one can only speculate that in the upper left quadrant the family 10' originated by *splitting* of the original family 10, rather than by *insertion*. When the animal died, 10' was probably only two tooth-generations old because the posterior shoulders were only weakly developed.

The upper right quadrant (Fig. 10d) is even more difficult to interpret because the functional teeth of 10 and 11 have been shed. One can assume that 10' has just split off from 10; it has normal polarity, but has not yet developed a posterior shoulder. Apparently 10' has not completely filled the available space, hence a stop-gap (SG, showing no polarity) developed between 9 and 10'. In the next generation, both 9 and 10' have grown to fill the space (hence the stop gap has no successor). Because of disproportionate growth of the jaw cartilage, gaps also opened in other regions of the dentitions (Fig. 10a, b).

To summarize, growth of tooth families is preprogrammed to a certain degree; so is number of tooth families in dentition types (a) and (b). If by disproportionate growth of the dentition open gaps occur, they are filled with new tooth families in dentitions of type (c), but usually by stop-gaps in dentitions of types (a) and (b). After the stop-gap, the adjacent tooth families increase in size to fill the gap. In dentitions of type (c), addition of new tooth families is possible only in certain regions. In other regions it is prohibited (Fig. 4). As was shown, the prohibition to split tooth families is released in tiger-sharks of old age despite the fact that they have dentitions of type (a).

Polarity

In many shark genera, tooth shapes are polarized. Their cusps are deflected in a lateral-posterior direction parallel to the jaw margin. Developmental studies suggest that polarity and tooth shape are regulated independently.

Embryogenetic Development of Polarity

In the species studied so far (most of which belong to the grey sharks, *Carcharhinidae*), the teeth first formed in any dentition are simple slender hooks which are curved inwards (i.e., in a lingual direction). They are unpolarized and have no specific characters (Fig. 5). Only gradually the typical tooth shape and polarity develop in all the tooth families. Most species with polarized shapes have a symphyseal tooth family which perhaps plays a highly special role. In some species, the symphyseal teeth remain unpolarized throughout ontogeny. Their cusp points in a vertical direction. These symphyseal teeth are usually much smaller than their neighbors.

In several species, however, the symphyseal tooth family becomes polarized. Whether the symphyseal tooth cusp points to the right or to the left seems to be a completely statistical phenomenon. In blue-sharks (*Prionace glauca*), the polarization of the symphyseal family is determined later than in the postsymphyseal families. In Fig. 8b, the polarization is first to the right and then to the left, where it seems to be finally established. The same phenomenon can also be observed in tiger-shark embryos. Polarization in the upper jaw is completely independent from that in the lower jaw. In a litter of 15 tiger-shark embryos, the following combinations were observed (first the upper jaw and then the lower jaw will be given): right/right: 3 specimens; left/left: 4 specimens; left/right: 4 specimens; right/left: 2 specimens; right/undetermined: 2 specimens. It is quite likely that the polarization of the symphyseal tooth family is under the competing influences of both neighbors. This can be shown as follows: One jaw of a tiger-shark (University of California, Museum of Paleontology, No. 22867) has two symphyseal families in the lower jaw, with the right families polarized to the right and the left family polarized to the left. Very rare cases occur where the polarizing influence of the postsymphyseal neighbors is balanced. This leads to anomalous extra tooth cusps: A symphyseal family in one dog-fish specimen has two tooth cusps (instead of one); one points to the right and one to the left.

Polarization in Post-Embryonic Stages

The polarization pattern seems to be very stable throughout ontogeny. It is, however, an important question whether polarization is retained when a protogerm is split in order to produce an extra tooth-family. The empirical data to answer this question are very scarce because, as it happens, species with polarized tooth shapes have a conservative tooth family number. The only exception known to me is the old tiger-shark described above which splits its posterior tooth families, but retains its polarity.

Injury induced polarity reversals confirm the assumption made above that polarization and tooth shape are regulated independently. All jaws discussed below are from specimens caught in the wild. The reasons for the injuries of the primordial tissue are unknown, but in some cases stings from sting-ray or fish hooks may be responsible.

The now classical example of tooth pattern reversal, which was partly described by Compagno (2), is a specimen of *Galeorhinus zyopterus* (Fig. 11a, b). Family 11 is almost a mirror image of family 10, though it has a slightly more up-

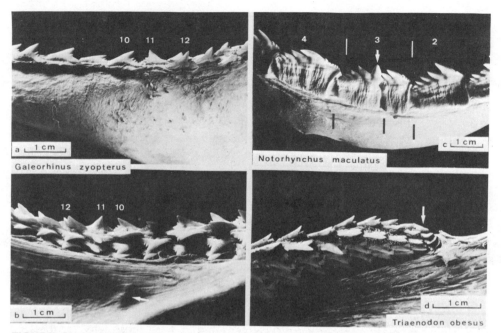

FIGURE 11. (a, b) *Galeorhinus zyopterus*, oil shark, adult male; 1.4–1.6 m total length; Morro Bay, California (L. J. V. Compagno private collection, No. 0114); lower left quadrant; (a) jaw seen from the outside, (b) jaw seen from inside. The polarity of tooth family 11 is reversed, probably the result of an injury. The injury is inferred from a thickening (a callus) on the inside of the jaw (arrow). (c) Seven-gilled shark, *Notorhynchus maculatus*, length and locality unknown (Natural History Museum of Los Angeles County, no catalogue number). Lower jaw, right quadrant, seen from outside. Tooth family 3 is split; the black lines indicate the limits of both fragments. The arrow points to the anterior cusp of the posterior fragment; this cusp shows a tendency to change polarity. (d) White-tipped reef shark, *Triaenodon obesus*, adult, ca. 1.1–1.4 m; Tagus Cove, Isabella (Albemarle) Island, Galapagos Islands (Stanford University, SU-12588); lower jaw right quadrant, seen from inside. The arrow points at the most distal tooth family, showing a polarity reversal. [From Reif (6), by permission.]

right cusp. It is not immediately obvious that 11 must have resulted from a splitting of 10. However, one indication of a former injury is provided by a slight thickening of the cartilage, probably a callus, below 10 (see arrow).

In *Carcharhinus amblyrhynchus* (specimen B, Fig. 12c, d), the splitting of family 9 is clearly injury induced and not caused by a gap between 9 and 10. The posterior shoulder of the anterior fragment of 9 is missing. Probably, the posterior fragment of 9 has formed by the splitting off of shoulder material. The injury held responsible for this is clearly documented by a callus on the inside of the jaw cartilage (arrow). The cusp of the posterior fragment of 9 is deflected in an anterior direction. In comparison with other examples, one can assume that this bending of the cusp would have been retained if both fragments of 9 had regenerated and reached normal size within a few tooth generations.

Direct evidence for an injury is provided by another specimen of *C. amblyrhynchus* (specimen A, Fig. 12a, b): a whole section of the dental lamina is miss-

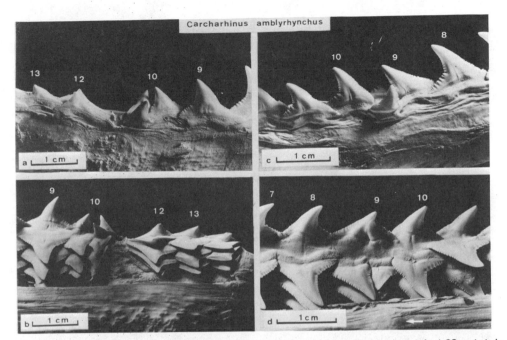

FIGURE 12. (a, b) *Carcharhinus amblyrhynchus*, gray reef shark, specimen A, 1.69 m total length, male, Ducic Atoll, Pacific (Bernice P. Bishop Museum, Honolulu, no catalogue number); upper jaw, left quandrant; (a) from outside; (b) from inside. The dental lamina was heavily injured and this probably led to the extinction of family 11 and to the splitting of family 10. (c, d) *C. amblyrhynchus*, specimen B, length and locality unknown (Waikiki Aquarium, Honolulu, Hawaii, no catalogue number); upper jaw, left quadrant. An injury seems to have caused the splitting of family 9. This injury can be inferred from a thick callus on the inside of the jaw (arrow). [From Reif (6) by permission.]

ing. One has to assume that there was a tooth family in this open section. The tooth family posterior to this gap consequently was given the number 12. The injury has led not only to the extinction of family 11 but also probably to the splitting of family 10. As shown in Fig. 12a, b, the posterior fragment of 10 is bent in an anterior direction while the anterior fragment shows normal polarity. Both fragments have undergone regeneration and gradual expansion into the gap. Also family 12 has extended into the gap from the posterior side.

Clear indications of former injuries are displayed by specimens A and B of *Triakis semifasciata* (Fig. 13). In both jaws the lower left quadrant is badly damaged (by a fish hook?). In specimen A (Fig. 13a–c), there is one family with a tooth pattern reversal (right arrow in Fig. 13c); behind this is a large gap, then a very broad tooth which appears enlarged in order to fill the gap (middle arrow). Behind this is a family with small anomalous teeth (left arrow) which show polarity reversal, and another large gap.

In specimen B, the posterior part of the dentition in the lower left quadrant has been completely exterminated, probably by a fish hook. In the upper left quadrant (fig. 13e, f), directly opposite the scar in the lower jaw, there is a single tooth pattern reversal. Its cause is not obvious, but it might be attributed to the same fish hook. (Both specimens of *Tr. semifasciata* are from Moss Landing, California,

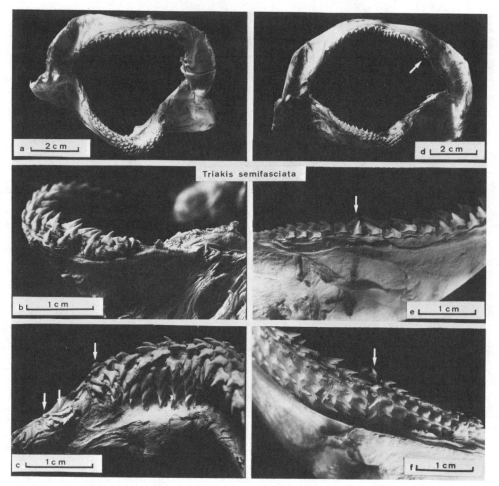

FIGURE 13. (a-c) *Tr. semifasciata*, specimen A; 1.32 m total length, adult male, Moss Landing, Calif. (Stanford University, Shark Derby number SD-27). (a) heavily injured lower left quadrant; (b) lower jaw from the left side; (c) lower left quadrant from above. The injury caused malformations: polarity reversal (right arrow), a gap (extinction of several tooth families), a family with anomalous tooth shapes (*middle arrow*), another polarity reversal (indicated by *left arrow*), and another big gap. (d–f) *Tr. semifasciata*, specimen B, 1.30 m total length, adult male, Moss Landing (Stanford University, Shark Derby number SD-102). The posterior half of the dentition was completely destroyed by an injury (d). The same injury appears to have caused the polarity reversal in the upper left quadrant (*arrows* in d, e, f). [From Reif (6) by permission.]

where this species is regularly caught with hook and line during shark fishing tournaments.)

No cause is evident for the pattern reversal of the last tooth family in the lower right quadrant of *Triaenodon obesus* (Fig. 11d). It could be due to an injury of the dental lamina, which caused the most distal tooth to split, but then healed and left no trace on the cartilage.

Notorhynchus maculatus (Fig. 11c) is the only case known so far in which the protogerm of a multicuspid tooth is split. In Fig. 11c anterior and posterior ends of both teeth are marked with black lines on the jaw. Normal tooth crowns of *N. maculatus* consist of an anterior shoulder with serration denticles, and of 5 to 7 cusps which are bent in a posterior direction. The anterior fragment in Fig. 11c consists of an anterior shoulder, and of 3 cusps which are deflected posteriorly. The posterior fragment has no anterior shoulder (as yet). It has 6 cusps, all deflected in a posterior direction except for the most anterior one which stands upright. Hence, as in the preceding examples, the posterior fragment shows a tendency to change polarity.

A MODEL FOR THE POLARITY SWITCH

A model to explain the tooth pattern reversal has to account for several different observations.

First, tooth polarity does not become visible earlier than the third to fifth tooth generation in embryonic dentitions. In the first tooth generations, the cusps are upright.

Second, in the very early tooth generations, the polarity of the symphyseal family can change spontaneously before it becomes permanently established. Polarization of the symphyseal family seems to be induced by a competing effect of the left and the right postsymphyseal neighbor. In a large sample of jaws, the arrow of polarization shows a random distribution.

Third, polarity reversal seems to occur only when a protogerm is split by an injury; it does not occur when a protogerm is split as part of normal morphogeny of the dentition.

Fourth, it is very likely that protogerms can also be split by an injury without a polarity reversal.

Fifth, it is always the posterior fragment of the tooth family that shows the polarity reversal.

In an earlier paper (6), I showed that a double-gradient system can be designed that adequately describes the observations made above in an abstract way. I assumed that each competent compartment, in other words each protogerm (A in Fig. 2) has its own double gradient. The anterior border of the compartment is the source of substance A and the sink of substance P. The posterior border of the compartment is the sink of substance A and the source of substance P. Asymmetrical gradients of A and P lead to a polarization of the tooth germ. When the compartment is split into two compartments by an injury, the newly formed boundaries of the compartments take over the role of sources and sinks, thus establishing double-gradient systems in both new compartments. It can be shown geometrically that under certain conditions the posterior member of the new compartments has a reversed polarity. However, the drawback of such a double gradient system is that its biological reality is very difficult to prove. In fact, a double gradient system of a size of a protogerm, up to 4 cm long, has never been shown to exist. Additionally, the behavior of a double-gradient system after injuries is very difficult to predict. Hence reference to a new model for the polarity switch based on other assumptions is provided in Chapter 3.

General References

GRADY, J.E.: Tooth development in sharks. *Arch. Oral Biol.* 15: 613–619 (1970).

PEYER, B.: *Comparative odontology*, 347 pp. Chicago: University of Chicago Press (1968).

References

1. Bigelow, H.B., Schroeder, W.C.: Sharks. In: *Fishes of the Western North Atlantic*, Mem. Sears Foundation for Marine Research, No. 1, pp. 59–576 (1948).

2. Compagno, L.J.V.: Tooth pattern reversal in three species of sharks, *Copeia*: 242–244 (1967).

3. Reif, W.-E.: Morphogenesis, pattern formation and function of the dentition of *Heterodontus* (Selachii), *Zoomorphologie* 83: 1–47 (1976).

4. Reif, W.-E.: Wound healing in sharks. Form and arrangement of repair scales, *Zoomorphologie* 90: 101–111 (1978).

5. Reif, W.-E.: Development of dentition and dermal skeleton in embryonic *Scyliorhinus canicula. J. Morphol.* 166: 275–288 (1980a).

6. Reif, W.-E.: A mechanism for tooth pattern reversal in sharks: The polarity switch model, *Wilhelm Roux' Arch.* 188: 115–122 (1980b).

7. Reif, W.-E. McGill, D., and Motta, P.: Tooth replacement rates of the sharks *Triakis semifasciata* and *Ginglymostoma cirratum. Zool. J. Anat.* 99: 151–156 (1978).

Questions for Discussion with the Editors

1. *Would you expand on your idea that shark tooth families are analogous to body segments in insect embryos? Could a similar analogy be made to amphibian chick somite segments?*

Yes, a similar analogy could be made to somite segments. The basis of the analogy is the repetitive pattern consisting of a predictable number of units (segments, teeth etc.). Two different mechanisms can, theoretically, lead to such a repetitive pattern. (a) From a founder cell (or cell cluster) successively new cells (or cell clusters) are budded off until the final number is established. Each of the cells (or cell clusters) develop into one unit. (b) A synchronous partitioning of an embryonic tissue takes place, which directly leads to the final number of units. Details of the mechanisms are highly theoretical so far, but according to the literature, mechanism (a) is found in dentitions of teleost fishes and reptiles and in the segmentation of the leeches and mechanism (b) is found in the shark dentition and in the segmentation of insects and vertebrate somites. Nothing whatsoever is known of a possible counting mechanism of the unit numbers.

2. *What mechanism might account for the production and subsequent elimination of supernumerary teeth?*

If segmentation of the dentition arises from a spontaneous, synchronous partitioning of the primordial tissue in the young embryo into protogerms long before the first teeth are formed, it is not surprising that the number of teeth of the first generation is not quite identical in all specimens. The question that we cannot solve so far is how the counting of the teeth is done. The suppression of the replacement teeth of the supernumerary teeth is probably done by means of the inhibitory field mechanism. This is a plausible assumption in the light of the observation that "regular" tooth families physically expand in order to suppress the "supernumerary" ones. It is unknown how the animal "decides" exactly which families are "regular" ones and which are "supernumerary" ones.

Index

This outstanding volume is the first text/reference to cover comprehensively all important topics in pattern formation. In 26 carefully assembled chapters, an international group of leading experts offer more than an up-to-date introduction to this rapidly evolving discipline; they provide a systematic description of major research areas in pattern formation and a much-needed assessment of the concepts and models guiding current research. Treating a wide range of organisms, from ciliate protozoa and *Volvox* to higher plants and animals, the contributors approach pattern formation from several perspectives, including modeling and theory and various types of experimental systems.

Pattern Formation also contains special sections that present the reader with an insider's view of the discipline. Beginning the volume are revealing "off-the-record" discussions that editor, George M. Malacinski, had with contributors. These give rare insight into the thought processes of many of the most distinguished men and women working in this area. And at the end of each chapter, Malacinski poses two specific questions to which the authors respond in detail.

With the broad scope and conceptual treatment of a textbook and the detailed information of a reference, this profusely

(continued on back flap)